EQUIVALENCE
Elizabeth L. Scott at Berkeley

EQUIVALENCE
Elizabeth L. Scott at Berkeley

Amanda L. Golbeck

University of Arkansas for Medical Sciences

USA

CRC Press
Taylor & Francis Group
Boca Raton   London   New York

CRC Press is an imprint of the
Taylor & Francis Group, an **informa** business

A CHAPMAN & HALL BOOK

CRC Press
Taylor & Francis Group
6000 Broken Sound Parkway NW, Suite 300
Boca Raton, FL 33487-2742

© 2017 by Taylor & Francis Group, LLC
CRC Press is an imprint of Taylor & Francis Group, an Informa business

No claim to original U.S. Government works

Printed on acid-free paper

International Standard Book Number-13: 978-1-4822-4944-6 (Paperback)
International Standard Book Number-13: 978-1-138-08669-2 (Hardback)

Visit the Taylor & Francis Web site at
http://www.taylorandfrancis.com

and the CRC Press Web site at
http://www.crcpress.com

Printed and bound in the United States of America by
Edwards Brothers Malloy on sustainably sourced paper

To my stars: Craig and Dan.

In memory of Betty: An exemplar and inspiration still.

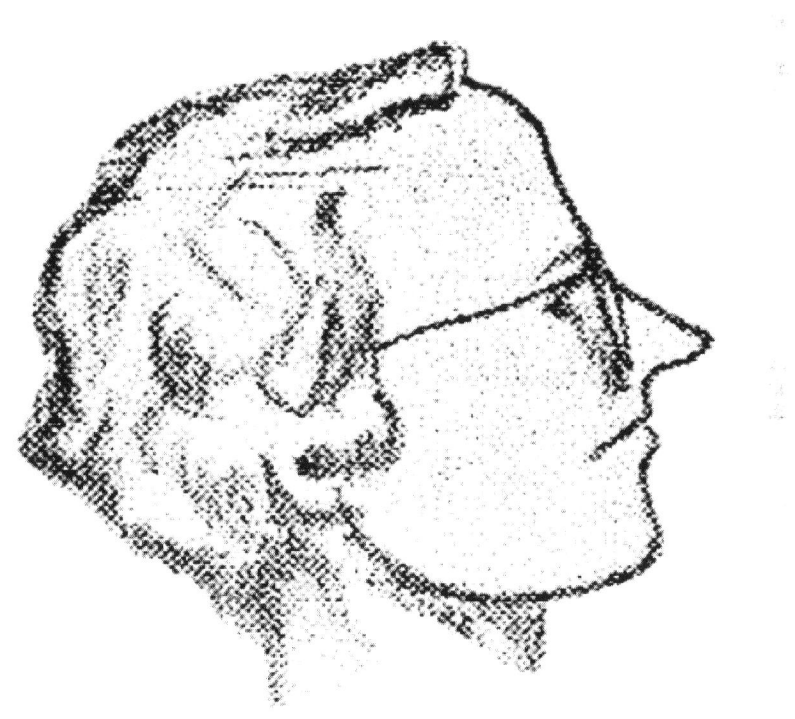

Black and white caricature of Elizabeth L. Scott (Betty) from her high school yearbook.
Senior Souvenir (Oakland: University High School, 1935)

Shall we go,
you and I
while we can?

Through
the transitive nightfall
of diamonds

Lyrics from *Dark Star*.
(Robert Hunter, Ice Nine Publishing, 1967)

Contents

Preface

This book is a story of intellect, love, hard work, service, tenacity, societal engagement, principle, perseverance, and more. It is a combined biography and microhistory that centers on the late (1917-1988) Elizabeth Leonard Scott's use of statistical reasoning to promote the status of women in academe and science. Scott was a founding professor in, and first woman chair of, the department of statistics at the University of California at Berkeley. She is known for her scientific research in mathematical statistics, astronomy, and biostatistics. She is especially remembered for her work toward a vision of equivalence for women and men in academe. I knew Scott personally in the late 1970s.

The Committee of Presidents of Statistical Societies (COPSS) recognized Scott's persistent efforts to advance the careers of academic women in the form of the Elizabeth L. Scott Award. COPSS is comprised of the past presidents, presidents, and president-elects of the major statistical professional organizations in North America: The American Statistical Association, Eastern and Western North American Regions of the International Biometrics Society, Institute of Mathematical Statistics, and Statistical Society of Canada. The operating principles for the award committee state the following:

> This award shall recognize an individual who exemplifies the contributions of Elizabeth L. Scott's lifelong efforts to further the careers of women in academia. An astronomer by training, she began to work with Jerzy Neyman in the Statistical Laboratory at Berkeley during World War II and had a long, distinguished, career as a professor at Berkeley. She worked in a variety of areas besides astronomy including experimental design, distribution theory, and medical statistics. Later in her career, Dr. Scott became involved with salary inequities between men and women in academia and published several papers on this topic. In addition to her numerous honors and awards, she was president of the Institute of Mathematical Statistics (IMS) and the Bernoulli Society, vice-president of the American Statistical Association (ASA) and International Statistical Institute (ISI), and elected an honorary fellow of the Royal Statistical Society (RSS) and fellow of the American Association for the Advancement of Science (AAAS). In recognition of her lifelong efforts in the furtherance of the careers of women, this award is granted to an individual who has helped fos-

ter opportunities in statistics for women. [Accessed on 3/14/2016 at http://community.amstat.org/copss/awards/scott]

Past recipients of the Scott Award include eminent statisticians F.N. David, Donna Brogan, Grace Wahba, Ingram Olkin, Nancy Flournoy, Janet Norwood, Gladys Reynolds, Louise Ryan, Lynn Billard, Mary E. Thompson, Mary Gray, and Kathryn Chaloner.

As I pondered this preface, I was notified that I had won the 2016 COPSS Elizabeth L. Scott Award. Here are some excerpts from my acceptance speech (August 3, 2016):

I'd like to take a few minutes to tell you about some things less known ... some things that I discovered while writing a book about Scott that is being published by Chapman and Hall. Here are a dozen things that might surprise you:

1. *Scott wasn't the only person in her family who had a PhD in astronomy. Her aunt also had earned a PhD in astronomy from Berkeley in 1913. Now there is an observatory named after her aunt at the Smithsonian Institution in Washington DC.*

2. *All of Scott's degrees were in astronomy, but she didn't aspire to observe through the telescopes. Rather, she wanted to work with the measurement data, and became an expert in statistical aspects of astronomy.*

3. *Scott published early, and a lot. She published 12 papers in seven years about the orbits of comets. This was already before she finished her PhD coursework. As a statistician, she went on to publish in a broad range of areas.*

4. *Scott was involved as a statistician in World War II in the strategic bombing campaign against Germany. She was so involved in the war effort that it took her 10 years to finish her PhD.*

5. *Then the department of defense tried to recruit Scott to a permanent position. In 1947, Vassar College also tried hard to recruit her to the faculty as an astrophysicist, promising her the department chairmanship and observatory directorship. But Scott decided to stay at Berkeley and become a statistician.*

6. *Scott was very kind and generous to many. In 1963, she did fundraising for Martin Luther King, Jr. Every Saturday, she brought bakery into the office for statistics students and faculty. She was even known to give students her own money when they ran out of theirs.*

7. *In 1965, Scott was instrumental during the Berkeley Free Speech Movement. She kept many students out of jail. She knew activist-leader Michael Rossman personally and asked the judge that he be released upon her recognizance.*

8. *At Berkeley, only men used to be allowed into The Faculty Club. Women had to climb through windows to attend faculty meetings there. In 1969, Scott wrote a letter of complaint to the president of the Club. A result of her letter was that the club opened its membership to women.*

9. *In 1970, Scott's research uncovered that only 2% of full professors at Berkeley were women. For 22 years, she was the only woman in a tenure track appointment on the Berkeley statistics faculty.*

10. *Some thought Scott was scary; some thought she was a holy terror. But as you were talking to her, you often thought she was falling asleep. She would then surprise you by opening her eyes and interjecting a lucid critique of the research problem at hand. She had been listening intently.*

11. *Scott helped to start the "equal pay for equal work" movement in this country. In 1974, her national research uncovered that, among those with regular faculty appointments, women who had equal qualifications to men were being paid much less. This difference was not explained by interruptions in women's careers.*

12. *In 1988, Scott was the keynote speaker at a pioneering conference, called Pathways to the Future, which produced many women statistics leaders. Conference leaders Nancy Flournoy and Lynne Billard still remember that Scott blew everyone's socks off at that meeting with her data analysis.*

Ten years before she died, Scott already knew that work on the status of academic women would not be completed in her lifetime. Which brings us to today.

Previous Scott Award winners know that women in statistics are still being underutilized and under-recognized. They have contributed to elevating the status of women in our profession in so many different and important ways. I am in awe of these statisticians and deeply honored to be in their company.

I remain grateful to the late Elizabeth Scott, who was my advisor and mentor when I was a statistics and biostatistics graduate student at Berkeley. Her work has awakened and inspired me.

The Scott Award is important. It carries the promise that Scott's work will continue.

In spite of the existence of this high level and visible award, there is little published biographical information on Scott. With few exceptions, the published articles are brief and absent details of the struggles that Scott and others experienced in the workplace and their efforts to promote gender equivalence. Convinced that her life and accomplishments needed to be well documented and made public for the scientific and lay communities, some ten years ago I began to conduct research to chronicle the career and examine the legacy of Scott.

Equivalence: Elizabeth L. Scott at Berkeley is the result of these efforts. The audience for the book is individuals with interests in statistics, biostatistics, astronomy, higher education, the history of science, women's and equity studies, and others. The book fills an important gap in the historical record while aiming to provide inspiration for statisticians who will use their craft for social advocacy. "Part I: The Betty Book" sets the stage: 1968 was the year that academic women on the Berkeley campus began to organize, and it was the year Betty became involved in equity issues for academic women. For Scott, this included articulating a research question: "Why are there so few women on the faculty?" "Part II: Shaping the Life" and "Part III: Clusters of Impact" are biography. They provide the context and background to understand Scott's vantage point. These parts of the book explore Scott's family history, educational background, personal life, work life, and scholarly activities. They fill out the picture of Scott as a whole person and answer important questions about her inspirations, motivations, and the raising of her consciousness toward using statistical reasoning to promote the status of women in academe. "Part IV: The Status of Academic Women at Berkeley," "Part V: Getting on the Agenda," "Part VI: Affirmative Action," "Part VII: Salary Equity Studies," and "Part VIII: AAUP Higher Education Salary Evaluation Kit" are microhistory. They place the reader inside ten years of the academic women's movement, told from the vantage point of Scott. They examine how Scott became a leader within the movement on the Berkeley campus, for the Carnegie Commission on Higher Education, on national science committees, for the American Association of University Professors (AAUP), and within the higher education community. They specifically examine the data gathering and statistical approaches that Scott adopted for her work. "Part IX: Conclusion" reviews Scott's work on the status of academic women over the last ten years of her life.

Throughout this book, I refer to Elizabeth L. Scott by her first name, "Betty." This is intended to engender a sense of familiarity to the reader as she or he gets to know Scott through this book. With the exception of Betty's relatives, I refer to all other women and men in the book, including Jerzy Neyman, by their last names. Reserving the familiar form of reference for the person who is the subject of this book is meant to honor her and bring you closer into her involvement in the struggles.

Equivalence: Elizabeth L. Scott at Berkeley draws heavily from information in archival records. Most of these records were located at the UCB Bancroft Library. The university archivist made available to me the very large Elizabeth Scott Collection, consisting of 185 linear feet (148 boxes) that contain all of Scott's professional files. I reviewed every piece of paper in her collection. At Bancroft, I also accessed the Jerzy Neyman Collection, Chancellor Collection, and Oral Histories. In addition, I accessed the Status of Women and Ethnic

Minorities Committee and other relevant materials at the Berkeley Division of the UC Academic Senate, the Local History Collection at the Oakland Public Library, the Phoebe Waterman Collection at the McHenry Library (UC-Santa Cruz), the Caroline E. Furness Collection at the Vassar College Library, the AAUP Collection at the Gelman Library (George Washington University), and the Military Personnel Records at the National Personnel Records Center. My research also used original face-to-face, telephone, or e-mail interviews, visits to the observatories where Scott did astronomical measuring, and articles published in professional journals or on the Internet.

My approach conforms to the viewpoint of social history, where the experiences of more ordinary women academics (as opposed to the most highly celebrated academics such as, say, Nobel Prize winners) are used to study changing gender relations in the past. In research, we find it useful to differentiate between process variables and outcome variables. While the account in this book presents both process and outcomes, there is emphasis on the microhistory of process when recounting Betty's activities on the status of academic women. I want the reader to feel the intensity of struggles that Betty and others engaged in on a daily basis to try to lessen discrimination of women in academic employment. I hope the reader will thereby be catalyzed to do more to try to advance equality in employment. Betty fortunately was a prolific correspondent, and she apparently saved every piece of paper and dated it, making it possible to understand and write her story with the detail that I include in this book. Nevertheless, sometimes the outcomes of the struggles are unknown, and for this I apologize and hope that future researchers will fill in some of these gaps.

Scott fought for many highly qualified women. This book is in memory of all women who were:

- discouraged from applying to graduate school because they were expected to become homemakers;

- overlooked for fellowships and other financial support because they were married;

- relegated to lesser positions because they were trained in the same field as their husbands;

- denied tenure ladder positions because it was presumed they would marry and leave the academy;

- discouraged from applying for academic promotions because they were women;

- paid lesser academic salaries because they were presumed to be geographically immobile;

- needing but had no access to child care on campus;

- given seats on the instruction committee instead of the budget committee;

- expected to have their babies in the summer;

- paying higher insurance premiums because they were women; and

- physically ejected from membership clubs that only admitted men.

It is in memory of all women researchers who were not allowed to be principal investigators and had to work under a nominal male who reaped their rewards; and all women astronomers who were prohibited from observing at the biggest telescopes because they were women. Scott fought for all of these women and more. She fought the fight for 20 years, persistently and selflessly, until the day she died.

Because of Scott and her advocacy collaborators, today women are generally encouraged to apply to graduate school. They fare better in competitions for graduate support. Today we have conflict of interest rules rather than nepotism rules. Women have better chances of fair consideration for tenure-track positions. They have better chances of being promoted. They have salaries that are more equitable. There are more women on major academic senate committees. Individuals have better access to maternity leave. There is better access to campus childcare. Men and women now pay the same insurance premiums. Men's clubs have been opened to women. Women on the research track are now allowed to be principal investigators. Qualified women astronomers are able to use the most advanced equipment.

But there is still much that we must do. In our day and age, academic women fare better; but generally they don't fare equally with men. Women have better chances for success, recognition, and rewards; but generally they don't have equivalent chances with men. Policies are written, but generally they could be more family-friendly. Women are still being underutilized in the workplace. Women still don't have enough role models and champions. Few women are full professors in doctorate-granting departments of statistics and biostatistics. Fewer women are academic administrators. Few women win national awards for research and scholarship. There is still a glass academic ceiling. Together we have conquered many explicit biases, but many implicit biases remain. Most of us realize this as we mature in the profession.

Now that this book is written, I pause to reflect on the love I have for my alma mater and believe this to be a Berkeley book in every way. Berkeley's motto is: "Let there be light!" We were taught to shed light on the truth, even if it creates discomfort, because only then will we be able to learn and improve the human condition.

About the Author

Author Amanda L. Golbeck met Elizabeth L. Scott in 1976 while a graduate student in anthropology, statistics, and biostatistics at UC-Berkeley. Scott was Golbeck's graduate advisor in statistics from 1976 through 1983, and Golbeck completed her Master's degree in statistics under Scott's direction in 1977 before earning her PhD in biostatistics at Berkeley in 1983. Currently Golbeck is Professor of Biostatistics and Associate Dean for Academic Affairs in the Fay W. Boozman College of Public Health at the University of Arkansas for Medical Sciences in Little Rock. The author has been a tenured faculty member at a number of universities, and has had a notable academic administration career that includes positions as vice president and academic dean for a liberal arts college, and vice president for academic affairs for a statewide system of post-secondary education. She is an elected fellow of the American Statistical Association, an elected member of the International Statistical Institute, a past president of the Caucus for Women in Statistics, and a country representative to the International Statistical Institute Committee on Women. She has had a long-term interest in gender equity issues in academe that stems from her early association with Scott. Golbeck was the lead editor of Leadership and Women in Statistics (Chapman & Hall/CRC Press), and she has had a number of published articles on gender issues in the statistics profession. In 2016, the Committee on Presidents of Statistical Societies (COPSS) selected Golbeck to receive the Elizabeth L. Scott Award.

Acknowledgments

There are many people to thank for their help with this book. The Phoebe W. Haas Charitable Trust and the Waterman Fund of the Philadelphia Foundation, via David Haas, provided funding for the research. Craig Molgaard encouraged me to write this book and reviewed many drafts. Dan Molgaard contributed astronomy and LaTex expertise to bring this book to its final submission. Susan Ervin-Tripp provided important recollections and material in the early stages. Ingram Olkin reviewed a partial first draft and championed the book. Deborah Bennett and Brian Yandell reviewed the final draft. Juliet Shaffer provided an interview and reviewed the first half of the book. Kjell Doksum, David Draper, Bill Kahn, and Dennis Pearl provided interviews and reviewed chapters containing their comments. Mary Gray, Joseph Gastwirth, Sanford Kadish, Herma Kay, and Lucy Sells provided interviews. Peter Bickel, Lynne Billard, John Curtis, Nancy Flournoy, Albert Lo, and Sylvia Paull provided e-mail contributions. Many provided access to archival materials: David Farrell, Kathi Neil, and the UC-Berkeley Bancroft Library Staff; Christine Bunting at UC-Santa Cruz Special Collections and Archives; Dean Rogers at Vassar College Libraries; Jennifer King at George Washington University Special Collections Research Center; and Kathleen Digiovanni at the Oakland Public Library. Evelina Badery translated to English Betty's French Letters to Jerry. Elinor Gates provided information about the telescopes at Lick Observatory. Danielle Porter provided assistance with permissions; Alan Crawford, Rachel Edelman, Elise Gustovson, Michael Jordan, Jostein Lillestl, Lillian Lin, Terry Speed, and Philip Stark provided assistance with pictures. Emma Bukacek provided web assistance. Ruth Licitra, Kerry Ryan, and Julie Stevens provided student assistance. Yulia Gel and Kelly Zou reviewed the first half of the book. James Raczynski provided resources to ready the book for production. The cover contains pictures from the IMS (Betty) and NASA and STScI (Hubble Finds Dark Matter Ring in Galaxy Cluster). My sincere thanks to all, including any I might have inadvertently forgotten to name, for their generosity. Finally, I would like to thank John Kimmel of Chapman & Hall for being a pleasure to work with on this project.

Acronyms

AAAS, American Association for the Advancement of Science
AAS, American Astronomical Society
AAU, The Association of American Universities
AAUP, American Association of University Professors
AAUW, American Association of University Women
AAVSO, American Association of Variable Star Observers
ACE, American Council on Education
ASA, American Statistical Association
ASUC, Associated Students of the University of California
AWIS, Association for Women in Science
AWM, Association of Women Mathematicians
CalTech, California Institute of Technology
CCHE, Carnegie Commission on Higher Education
CCW, Coalition of Campus Women
CEEB, College Entrance Examination Board
CEEWISE, Committee on the Education and Employment of Women in
    Science and Engineering
Chancellor's CSAW, UC-Berkeley Chancellor's Committee on the Status of
    Academic Women
Chancellor's CSNAW, UC-Berkeley Chancellor's Committee on the Status of
    Non-Academic Women
COPSS, Committee of Presidents of Statistical Societies
ERA, Equal Rights Amendment
ETS, Educational Testing Service
FTE, Full-time equivalent
HAC, Harvard Announcement Cards
HEW, Department of Health, Education, and Welfare
IMS, Institute of Mathematical Statistics
ISI, International Statistical Institute
LAW, League of Academic Women
NAS, National Academy of Sciences
NASA, National Aeronautics and Space Administration
NASULGC, National Association of State Universities and Land Grant
    Colleges
NCES, National Center for Education Statistics
NIH, National Institutes of Health
NOW, National Organization for Women

NRC, National Research Council
NSF, National Science Foundation
OCR, Office of Civil Rights
ONR, Office of Naval Research
Policy CSAW, UC-Berkeley Policy Subcommittee on the Status of Academic
    Women
SAC, UC Statistical Advisory Committee
SEC, UC-Berkeley Salary Equity Committee
Senate CSAW, UC-Berkeley Committee on the Status of Academic Women
Stat Lab, UC-Berkeley Statistical Laboratory
STEM, Science, Technology, Engineering, and Mathematics
SWEM, UC-Berkeley Committee on the Status of Women and Ethnic
    Minorities
System CSW, UC President's Committee on the Status of Women
UC, University of California
UC-AFT, UC branch of the American Federation of Teachers union
VC, UC-Berkeley Vice Chancellor
VCA, UC-Berkeley Vice Chancellor for Administration
VCAA, UC-Berkeley Vice Chancellor for Academic Affairs
VCR, UC-Berkeley Vice Chancellor for Research
VPA, UC Vice President for Administration
VPAA, UC Vice President for Academic Affairs
WEAL, Women's Equity Action League
WFC, UC-Berkeley Women's Faculty Club
WFCSC, UC-Berkeley Women's Faculty Club Study Committee
WFG, UC-Berkeley Women's Faculty Group

Part I

The Betty Book

1

Caught in the Thick of It (1968)

Why are there so few women on the faculty? Why are there so few working toward and obtaining their PhDs?
-Elizabeth L. Scott

She is still present at the University of California at Berkeley. You will find her in the UC-Berkeley Library and Bancroft Library in over 100 places, as an author under the names Elizabeth L. Scott or E. L. Scott. Although her publication was a frequent event, in the mid 20th century she herself was what a statistician would call a rare event. She was a woman in science who had a PhD and held the title of Professor. If she were alive today, you would call her Dr. Scott or Professor Scott. In her time, you would more often have politely called her Miss Scott. But for those of us who knew her and worked with her, she was and remains simply "Betty."

This is "The Betty Book."

Betty Scott (1917–1988) was a renowned Berkeley mathematical statistician who is known for her influential work in diverse areas of science. She followed in the footsteps of her maternal Aunt Phoebe and pursued postsecondary training in astronomy at UC-Berkeley; but unlike Phoebe who aspired to be an observational astronomer, Betty gravitated toward theoretical astronomy. When World War II happened as Betty was working on her PhD in astronomy, she was tapped by celebrated mathematical statistician Jerzy Neyman (1894–1981) to do war work. She threw herself into this work with a deep sense of patriotism learned from growing up in a family of military officers. This war work and other work at The Statistical Laboratory immersed Betty in mathematical and applied statistics ("ten thousand hours of practice") and significantly influenced the trajectory of her career. After the war, Betty completed a dissertation on statistical problems in astronomy and joined the mathematics and then the statistics faculty at UC-Berkeley. Starting out as one of only a handful of women on the faculty in the sciences, Betty had a long and prosperous career at Berkeley, rising to the position of professor, managing for a time the top-ranked Berkeley statistics department as its chairperson, and serving in high level positions in statistics professional associations, including being the first woman elected president of the Institute of Mathematical Statistics. In the course of her career, Betty did remarkable research on general statistical methods, as well as on statistical problems in astronomy, bioscience and health, weather modification, and gender equity. She did not shy away from controversial societal problems; in fact, she was drawn to them. She was exemplary at using her expertise in statistics to advocate for positive social change, especially for women in academe. She became the national expert at evaluating academic salaries, and many women and minorities used her methods to successfully argue for adjustment of their salaries. In all, Betty spent twenty years of her life using statistics to advocate for equivalent status for women and men in academe.

This book tells the story.

1.1   At Work

The year 1968 was when the academic women's movement began. What was Betty doing in 1968? She was at work as usual with Neyman, doing mathematical statistics. Betty and Neyman worked on astronomy research whenever they got the chance. They were highly effective at establishing and maintaining professional relationships across the globe. By 1968, one of the many people with whom they were globally connected was prominent Polish astronomer Wlodzimierz Zonn. Neyman's home country had been Poland. Zonn was director of the Astronomical Observatory at the University of Warsaw and director of the Polish Astronomical Society.

Neyman and Betty had a special interest and "vivid activity" in trying to improve the situation of Zonn and other astronomers and scientists in Poland under the then communist regime. In fact, it has been said that Neyman produced something like a "big beat" in Polish astronomy. He had written to the Polish Ambassador Michalowski suggesting that Zonn be sent to the US and other countries as an emissary of Polish astronomy. This advocacy failed to result in Zonn's appointment as emissary. But the Polish government did commit to donating a 2-meter telescope with all the necessary equipment and buildings to the University of Warsaw. Zonn credited Neyman as being an important catalyst for the acquisition.

While he valued Neyman as a scientist and advocate, Zonn especially valued Betty's specific expertise in astronomy, where she excelled Neyman. Betty had conducted doctoral work in the distinguished Department of Astronomy at UC-Berkeley. She had done measurement work at the great observatories on Mount Hamilton and Mount Wilson in California. She had been elected three years earlier to the International Astronomical Union. She was well published in astronomy. Among her distinguished astronomy collaborators was C. Donald Shane, whose name is synonymous with the Lick Observatory, historically the first permanently occupied mountaintop astronomical observatory, on Mount Hamilton, southeast of San Jose.

In February of 1968, Zonn sent a draft of a professional paper to Neyman. The subject was not mathematical sciences. It was astronomy and in particular it was double galaxies. Zonn intended to publish it in the Polish peer-reviewed earth and planetary sciences journal, *Acta Astronomica.* Zonn specifically asked Neyman to show the paper to Betty.[1] Zonn also sent a postcard directly to Betty to ask her to please read it and send him her remarks.[2] Betty read the paper carefully and critically. She thought the paper had merit but was somewhat skeptical of the conclusions. Betty told Zonn so, even though she knew he would be disappointed by her assessment.

Zonn's reaction to Betty's honesty, sincerity, and directness was characteristic in her many relationships. Zonn responded positively to Betty's professionalism, he welcomed her communication as constructive, and he valued her appraisal. He digested her comments, and then he heartily returned to her his best wishes.[3] Their professional relationship would continue. Many others from many fields would also be exposed to Betty's honest, sincere, and direct approach. For her, it was as if the stars did not blink, and neither did she.

In 1968, in the midst of yet another California drought, Betty's attention was focused on her statistical evaluation of the efficacy of cloud seeding, carried out in collaboration with Neyman and others. She was being very productive. One of her articles on cloud seeding had just appeared in the *Proceedings of the National Academy of Sciences.* She was working hard on additional cloud seeding experiment studies. This was work that she would publish the next year in three separate articles in *Science* and an additional article in the *Proceedings of the National Academy of Sciences.* The work was highly controversial at the time (see Chapter 8). Betty couldn't afford to blink.

1.2   UC and the Urban Crisis

Beyond the seemingly safe boundaries of Betty's world of academe, there was a genuine urban crisis happening in the US in 1968. Cities were ablaze. This was the year of the assassination of Dr. Martin Luther King, Jr., whom Betty had supported both in spirit and financially. There were race riots, and violence seemed to be everywhere. The problems of society were increasingly being identified with problems of inequality. Blacks had historically been treated as lesser citizens in the US, segregated and restricted to the worst neighborhoods and schools and subjected to oppressive policing. But then in the 1960s the

---

[1] 1/29/1968 memo from University of Warsaw Astronomical Observatory Director Wlodzimierz Zonn to Jerzy Neyman re a number of matters.

[2] 2/10/1968 postcard from University of Warsaw Astronomical Observatory Director Wlodzimierz Zonn to Betty Scott.

[3] 1/29/1968 memo from University of Warsaw Astronomical Observatory Director Wlodzimierz Zonn to Jerzy Neyman re a number of matters.

urban black economy went into crisis mode as manufacturing jobs that paid decent wages disappeared, causing unemployment to skyrocket and the vicious cycle of poverty to intensify. The social fabric was strained, the streets became more dangerous, and there was looting and burning of stores. At the same time, there was affluence among whites who were fleeing to the suburbs.

President Lyndon B. Johnson asked, "What happened?"

President Johnson's Kerner Commission in March 1968 gave him a brutally honest answer: "Our nation is moving toward two societies, one black, one white – separate and unequal. Discrimination and segregation have long permeated much of American life; they now threaten the future of every American."[4]

In 1968 UC was a statewide system that had campuses in Berkeley (the oldest and flagship campus), San Francisco, Davis, Riverside, Los Angeles, Santa Barbara, San Diego, Irvine, and Santa Cruz. Immediately after the Kerner Commission report came out, the head of the UC system, President Charles J. Hitch, declared that the university should be "large in spirit." The times demanded it. The university should take on the urban crisis.

In May of 1968, Hitch delivered a special report to the UC regents, who governed the university. The report was titled: "What We Must Do; The University and the Urban Crisis." He declared the crisis to be not only a moral, economic, and racial one, but also an educational one. He reflected that what was happening would require work over a long term and would not be resolved until (as he put it in language that was not at all gender neutral) "every man is allowed his full measure of human dignity." Hitch challenged the university to act immediately and proposed that the three core missions of the university – research, public service, and education – be directed toward the urban crisis. He proposed that a focus within the university's research mission should be on determining the causes and resolutions of the urban crisis. Decades before community-based participatory research became popular, Hitch proposed that a focus within the university's public service mission should be on community partnerships, especially involving contact with school children that would positively impact the surrounding communities. And a focus within the university's education mission should be on educational partnerships that increased access to higher education. Hitch then proposed specific and immediate first actions. These included additional funding for educational opportunity programs, a review of participation of individuals from minority groups in graduate training, a call for proposals to improve K-12 education, and establishment of a position of fair employment coordinator to further increase the number of minorities employed by the university.

President Hitch's report concluded with an invitation to members of the university community to offer creative and constructive ideas to help solve the

---

[4]Kevin Boyle. "The Fire Last Time," *The Washington Post*, July 29, 2007, accessed September 27, 2011, http://www.washingtonpost.com/wp-dyn/content/article/2007 /07/27/AR2007072701672.html.

urban crisis.[5] Doras Briggs, who was assistant to the UC system vice president for physical planning and construction, decided to accept this invitation.

1.3 Berkeley Women and the Urban Crisis

Briggs' idea was to involve women faculty in the search for a solution to the urban crisis. As Briggs wrote, "In thinking about his message, it seemed to me there is no more vital group to tackle this challenge than the women of the faculty. You are leaders in your chosen fields and are proven creative thinkers. You also have special qualities of compassion and human warmth which are so much needed today."

And so in June of 1968, Briggs sent a letter to Betty and other women on the Berkeley campus who she perceived to be leaders. One went to May Diaz, who was an associate professor of anatomy and vice chairman of the Department of Anthropology. Another went to Ann Heiss, who was a lecturer in Education and assistant research educator in the Center for Research and Development in Higher Education. Barbara Kirk was a lecturer in the Counseling Center. Herma Kay was a professor of Law. Anne Low-Beer was a biochemist in the Division of Biology and Medicine. Mary Lou Norrie was an associate professor of Physical Education. Roberta J. Park was head of women's physical activities in Physical Education. Margaret Thal-Larsen was a specialist in Industrial Research at the Institute of Industrial Relations and president of the Women's Faculty Club (WFC). And there were six others.

Betty found herself among an equally impressive group of women. Norrie, for example, was the first department chair of Physical Education at UC-Berkeley. Kirk was to become dean of Counseling. And Kay was to become the first woman dean of the Law School.

In her letter, Briggs brought President Hitch's special report on the urban crisis to the attention of Betty and the select group of women. At this time, the UC system office was located in University Hall. This was across from the West Gate of the Berkeley campus. Briggs was in the backyard of Betty and these women. She either knew them personally and had confidence in them, or else the campus administration had recommended them highly to her.

Briggs asked Betty and the select group of women to read the president's report, think about it, and then come to a luncheon in the boardroom at the WFC. The WFC, established in 1919, is located in the heart of the Berkeley campus and was a convenient meeting place for women. Briggs asked Betty and the 14 women to bring their ideas and suggestions to the meeting. The

---

[5]Charles J. Hitch, "What We Must Do: The University and the Urban Crisis," special report, University of California, 1968.

idea was to frame these into a statement that UC-Berkeley Chancellor Roger W. Heyns could then forward to UC system President Hitch.

Briggs' vision was that this would not be a time-consuming effort that would culminate in a lengthy report. She thought it would take only three or four lunch hours worth of time. Little did she know then that these efforts would ultimately result two years later in a lengthy and legendary report to the UC-Berkeley academic senate, the faculty governing body. Betty would emerge as a leader in the production of the report.

Briggs called the group she assembled in 1968 the *Women's Faculty Club Study Committee on the University and the Urban Crisis*[6] (WFCSC). The WFCSC would plant the seed for the legendary report.

The first meeting of the WFCSC took place on July 19, 1968 and was attended by nine of the women, including Betty and Briggs. Ideas flowed. The problems identified were many. The women thought the possibilities for making an impact were broad. Educate each other. Institute a counseling degree program. Expand university job training programs. Adapt university systems to people's needs. Examine their attitudes. Learn from minority women leaders. Investigate teaching techniques.

The ideas kept coming. Make inventory of present programs. Design programs for women. Keep careful records of success. Do early preparatory work. Study academic careers for women. Study continuing education of women. Change attitudes toward vocations. Undertake parent education. Set up community advising. And, last but not least: Remain hopeful, and do not despair!

It is interesting to note that even Betty's contributions to the discussion were delivered in the male-oriented language of the time. Betty commented: "In the field of mathematics (and probably in other fields as well), unless a student has gone to a good school, he is so far behind it is almost impossible for him to catch up."

It was Thal-Larson who, at that meeting, first suggested the WFCSC concentrate on what they could do for women. Betty, using all of her orientations to scientific inquiry, followed almost immediately with an articulation of primary research questions, a style so common today in the science, technology, engineering, and mathematics (STEM) fields, but an approach pioneered by Betty and other statisticians of her generation. These were,

"Why are there so few women on the faculty?" and

"Why are so few working toward and obtaining their PhDs?"

Betty knew that good research questions needed to be formulated at the start of any study, and that they would drive the study. She followed by calling for more research on "the emotional damage done to women beginning as early as age three when they are steered away from intellectual pursuits – they have

---

[6]6/21/1968 letter from UC Assistant to the Vice President Mrs. Doras Briggs to Betty Scott and 14 others.

to play with dolls or engage in other 'feminine' activity. We are losing ground in our efforts to interest women in academic careers," she observed.[7]

In the interim between the first and second meeting of the WFCSC, Briggs distributed another report. This was President Hitch's most recent report to the regents about his initiative on the university and the urban crisis. Hitch reported widespread and warm interest and enthusiasm for the initiative both within and outside the university. He reported on the first steps that had been taken. Leadership had been appointed for the initiative. Planning had commenced for a conference on improving access to higher education. Data were being collected on current and planned research and service activities related to the initiative. Outreach activities of the Agricultural Extension Service were being expanded into additional disadvantaged areas of the state of California.[8]

But there was no mention of women's issues in Hitch's report. There was no evidence in his report that the WFCSC was on the radar screen.

Betty was appointed to the position of chairman of the Department of Statistics effective July 1, 1968, succeeding Henry Scheffé. Betty was the fifth chair of the department, the first woman to hold the position. At that time, chairs in academe were called chairman reflecting the preponderance of men. And so Betty signed her letters and referred to herself as "Chairman." Her position gave her some clout on the campus, especially relative to the women.

The UC-Berkeley Statistics Laboratory, or Stat Lab, had been organized in 1939 as part of the Department of Mathematics and was the precursor to the Department of Statistics; it continued as a research unit after the statistics department was established in 1955. Betty began her affiliation with the Stat Lab in the year it was founded, and she retained her affiliation while she was department chair. In 1968 the Stat Lab was working on important problems, such as carcinogenesis and weather modification; however, as Neyman noted, the contribution of the lab to the solution of any of the pressing social problems that fell under the urban crisis initiative amounted to "exactly zero."[9,10]

1.4 Berkeley Women Begin to Organize

The second meeting of the WFCSC took place at the end of July, two weeks after the first meeting, in the library of the WFC. Members pressed for clarification of the purpose and goals of the group. Should it be a "thought"

---

[7]7/19/1968 minutes of the WFCSC.

[8]7/24/1969 letter from UC Assistant to the Vice President Doras Briggs to the WFCSC.

[9]Note: Jerzy Neyman had been brought to UC-Berkeley in 1938 not just to develop a statistics program, but also to assist faculty across the campus with the statistical aspects of their research. It was in 1939, only a year after his arrival to Berkeley, that the consulting activity was given official status under the name The Statistical Laboratory.

[10]6/20/1968 letter from Jerzy Neyman to UC-Berkeley Vice Chancellor L.L. Sammet.

or "action" group? Should it concentrate exclusively on women or not? And should it focus exclusively on the Berkeley campus or not? Briggs reminded the group that the original call was for the committee to be a "thought" group, but that it could later morph into an "action" group. Kay said she would like to see the group do something about interesting women in higher education.

Betty chimed in, saying: "We should recognize there is great pressure against women going into education. The same kind of pressure as is exerted on Negroes." The group then ruminated on the roadblocks to education for women, that the roadblocks were part of the American belief system, and the committee should work to eradicate them. It was acknowledged that many departments refused to accept fully qualified women for graduate school. The meeting concluded with a motion to confine the activities of the WFCSC to issues within the university.

There was also a call to members of the committee to submit written suggestions for the focus of the committee.[11] A noteworthy submission came from Park who wrote, among other things: "The suggestion that this committee explore ways of improving the role and status of women faculty in higher education (and especially this campus) has merit – especially as it may be related to encouraging competent women form [sic] all spheres of society to pursue professions in institutions of higher learning." Other suggestions in the submissions appeared to be as broad as those discussed in the first meeting of the committee. Park even suggested that the committee engage in the stereotypical women's work of keeping the campus attractive and tidy.[12]

A subsequent meeting of the WFCSC took place toward the end of August.[13] Briggs and Kirk had drafted a plan for a series of lectures to help the group learn what was already going on within the university related to the urban crisis. They laid out six months of lectures, starting in October and continuing through March. Proposed topics included the Upward-Bound Program, development of race relations, thinking and planning for minorities in graduate education, plans for the urban crisis initiative, minority apprenticeship programs, equal opportunity programs, and other possibilities.

Notably, these topics focused on minorities, not on women as Thal-Larson had advocated. Nor on women in higher education as Betty and Kay had advocated. Nor on the status of women especially on the Berkeley campus as Park had advocated.

In spite of the agenda, once the women got to the meeting and began discussions, the focus took a decided shift from minorities to women in higher education. Members of the committee liked the proposed lecture series. But, significantly, Kay suggested the WFCSC should tailor the series to what the committee wanted as its focus. The committee had been roused by President

---

[11]7/30/1968 minutes of the WFCSC.

[12]8/7/1968 memo from UC-Berkeley Head of Women's Physical Activities Roberta J. Park to UC Assistant to the Vice President Mrs. Doras Briggs re WFCSC.

[13]Note: This subsequent meeting of the WFCSC, held on August 21 in the WFC board room, was described as a fourth meeting. Records for a third meeting were not found.

Hitch's initiative, but they still hadn't decided whether they were going to be a "prestige group" or a "pressure group." The committee was cognizant it had no official status. It needed to decide what role it was going to play.

Kay then put the "women" focus back on the table. She suggested that if the WFCSC were to focus on the educational problems of women, then they needed to be a pressure group. And they needed to define the problem. There were few women on the Berkeley faculty, and Kay was one of the few women members of committees of the Berkeley Division of the Academic Senate. She had been appointed to the Senate Policy Committee earlier in the year. This was a relatively new committee, created to advise the chair of the senate and recommend issues for senate study. Kay offered the key idea that she could suggest formation of a senate committee to work on the educational problems of women.

Betty chimed in that she was in full accord.

Low-Beer suggested they begin their work by undertaking a study to determine the present status of women. Heiss came to the meeting armed with some statistics. Women held only 18% of faculty positions nationally, and only 7% within universities. But women held only 2% of the faculty positions at UC-Berkeley! She agreed to pull together additional statistics for the next meeting of the committee. Norrie agreed to compile a list of all academic women appointees by rank within departments. Heiss suggested the group tackle the nepotism rule, which she felt ought to be broken, because it was preventing wives who could teach from doing so because their husbands were teaching. Norrie suggested the proposed lecture series include a session on the facts and figures that Heiss had assembled.[14] It seemed that the WFCSC had found its focus, and this focus was already evidence-based.

During the five weeks between meetings, Briggs circulated to the committee a report that crossed her desk. The report mentioned that 58% of the budgeted faculty positions at Berkeley in the previous year (1967–68) were tenured, and that this percentage had been rising. Briggs worried that this could be an unfortunate trend for women. She worried that it would make it increasingly difficult for women to gain tenure.[15]

The fifth meeting of the WFCSC took place over lunch at the beginning of October. There was more discussion about the speaker series. Some thought they should be planning a focused series for the closed group committee, rather than piggyback onto the WFC's open, regular Lunch and Learn series. But it turned out that the committee still hadn't officially settled on a focus. Thal-Larsen pushed them. The speaker series should be directed at the focus, whatever that was to be. Thal-Larsen pushed them again. Heiss responded by advocating that the WFCSC steer away from racial minorities and instead work for all women, from all racial groups. Kirk said the committee should

---

[14]8/21/1968 minutes of the WFCSC.

[15]9/16/1968 memo from UC Assistant to the Vice President Doras Briggs to Members of the WFCSC re report on the percentage of budgeted faculty positions on the Berkeley campus that were tenured.

decide whether to limit itself to university women. Heiss suggested they do something for the wives who lived in University Village married student housing.[16] Heiss offered that many of these women "are effectively cut off from intellectual life and many feel it keenly."

The focus on university women was solidifying.

Then the discussion turned again to questions about data. What are we collecting now? How many women are in graduate school? What are the statistics? What obstacles do women face? As she had done in an earlier meeting, Heiss offered to gather more statistics. Kay suggested they conduct anonymous surveys of women students. Heiss stressed the importance of statistics and the insights that can be gained from looking at them.

The focus on data, especially quantitative data, was solidifying.

Heiss remarked that bright women were pretty invisible and needed to be made visible. At the previous meeting, Low-Beer raised the question about getting the WFCSC some kind of official status. It was raised again at this meeting. At the fourth meeting, Kay had suggested formation of a senate committee to work on the educational problems of women. The fifth meeting ended with Kay declaring she would indeed propose to the academic senate that they establish an official committee to investigate the status of university women.[17]

By virtue of her participation on the select WFCSC, Betty was getting caught up in the national women's movement. The National Organization for Women (NOW) had formed two years earlier, in 1966, as the "NAACP (National Association for the Advancement of Colored People) for Women."[18] As the letterhead indicated, NOW was working for "full equality for women in truly equal partnership with men." Betty wanted to join in that work. Toward the beginning of October, she received a letter from the San Francisco office of NOW welcoming her as a new member and acknowledging receipt of $7.50 ($46.47 in 2010 dollars) to cover her membership for the year 1968. NOW indicated they were glad Betty was going to be working with the organization "to achieve our goals speedily."[19] Betty became one of the 1200 members of this then-small national organization. The organization had a commitment to welcoming male members, which suited Betty's philosophies.[20]

---

[16] Note: University Village was former World War II temporary military housing that was taken over by the university in 1956. The university turned it into a married student housing project that was completed six years later with 500 new units.

[17] 10/2/1968 minutes of the WFCSC. Note: The meeting took place in the WFC board room.

[18] Gail Collins, *When Everything Changed: The Amazing Journey of American Women from 1960 to the Present* (New York: Little, Brown and Company, 2009), 85.

[19] 10/8/1968 letter from National Organization for Women Secretary-Treasurer Inka O'Hanrahan to Betty Scott.

[20] Gail Collins, *When Everything Changed: The Amazing Journey of American Women from 1960 to the Present* (New York: Little, Brown and Company, 2009), 191.

1.5 A Complicated Set of Problems

A subsequent meeting of the WFCSC was scheduled for mid November. Each woman was to bring her ideas about what the committee's "concerns" should be. The meeting, which would be the sixth, was actually held two weeks early, on October 22. Two memoranda had come out from President Hitch's office.

The first was to members of the Committee on Finance for their October 17 meeting. It was a plan and budget, for 1969–70 through 1973–74, for its research and public service program relating to the university and the initiative. The plan and budget, which among other things included a request for $3 million ($18.5 million in 2010 dollars) in new funds and a small grant program administered in the Office of the President, were to be submitted to the UC regents for approval in November, and then to the executive and legislative branches of the State of California government. The 1969 budget for the urban crisis initiative was $24 million (including the $3 million request for new funding), which adjusting for inflation would be over $141 million in 2010 dollars.[21] It was no small amount, and the women took notice.

The other memorandum was President Hitch's progress report to the regents on the urban crisis initiative. He reported the enrollment of disadvantaged students at the university had increased, dates had been set to mid December for the conference of California educators to consider how to improve access, and minority employment was expanding but needed to go further. He reported the "urban crisis is such a complicated set of problems that it requires the fullest study, analysis, and evaluation... and that the great range of problems that make up the urban crisis is and will be a time-consuming, challenging, and frequently frustrating experience."[22]

This complicated set of problems included the problems women faced in the university, even if this had not yet reached the consciousness of President Hitch and the broad university community. And Betty and the women who were beginning to work on the problems of university women would soon begin to feel the time-consumption, the challenge, and the frustrations.

The sixth meeting of the WFCSC still did not produce a definitive focus for the group. President Hitch's conference on access to higher education was discussed and interest was expressed in having members attend. Briggs approached the system vice president for educational relations, who expressed pleasure that the women of the WFCSC were interested in the conference and said he would be glad to invite anyone the committee wanted to name.

---

[21] 10/11/1968 memo from UC President Charles J. Hitch to Members of the Committee on Finance re Plan and Budget for the Program of Research and Public Service Relating to the Urban Crisis, 1969–70 through 1973–74.

[22] 10/18/1968 memo from UC President Charles J. Hitch to the Regents of the UC re Progress Report on the Urban Crisis Program. Note: The dates for the conference were set to December 13–15.

The committee continued to discuss the speaker series and various statistics about women on the Berkeley campus. They inquired when the 1968 Statistical Summary would come out, and it was subsequently sent to all members. Briggs ended the meeting by suggesting the members of the committee come to official agreement on their focus at their next meeting, scheduled for mid November.[23]

On the same day as the sixth meeting, Thal-Larsen sent a handwritten note to Briggs that summed up what their goals ("rather than concerns") should be: "1. equal educational opportunities for women on this campus, and 2. equal employment opportunities for women on this campus, as the above signifies equal opportunities for all women, the provision of special facilities, procedures or the like to ensure real equality of opportunity... e.g., child care facilities, special concessions as to readmissions, guarantees against racial discrimination, special counseling, etc."[24]

The last official communication of the year from Briggs to the committee contained various reports of interest to the members. These included notes from Lunch and Learn sessions, copies of reports having to do with the urban crisis initiative, and various statistical summaries of Berkeley women students, staff, and faculty members.[25] The WFCSC would catalyze the formation and mobilization of a larger and more comprehensive group of academic women in 1969 (see Chapter 13).

1.6   Thick Politics and Early Exhaustion

So much was happening in California in 1968. Ronald Reagan had been elected to the governorship of California two years prior, at least to some extent on an anti-Berkeley platform in which he denounced campus protests. And the next year, with the assistance of the FBI, Reagan removed Clark Kerr as president of the UC.[26] Kerr, who has been called "one of the most revered leaders in public higher education in the 20th century" and "the dean of the higher education community not only in California, but in America," went on to lead the Carnegie Commission on Higher Education.[27] In October of 1968

---

[23] 10/22/1968 minutes of the WSCSC. Note: The meeting was originally scheduled for November 12 but took place on December 3 due to a flu outbreak.

[24] 10/22/1968 handwritten letter from UC-Berkeley Institute of Industrial Relation, Industrial Research Specialist, Margaret Thal-Larsen to UC Assistant to the Vice President Doras Briggs.

[25] 11/22/1968 memo from UC Assistant to the Vice President Mrs. Doras Briggs to WFCSC re information items.

[26] "Clark Kerr," *Wikipedia, The Free Encyclopedia*, http://en.wikipedia.org/wiki/Clark_Kerr (accessed September 26, 2011).

[27] "Former UC President Clark Kerr, a national leader in higher education, dies at 92," *UC-Berkeley News*, December 2, 2003, accessed January 1, 2012, http://www.berkeley.edu/news/media/releases/2003/12/02_kerr.shtml.

Reagan made a move that violated faculty authority and would have gutted the principles of shared governance. He attempted to stop an experimental course on race issues featuring a series of lectures by Black Panther Eldridge Cleaver.[28] Some 200 students were arrested around this issue.[29]

In November of 1968, Betty was again involved in international professional association work. This time she was working on nominations for the International Association of Statistics in the Physical Sciences.[30] From parenthetic remarks in her correspondence, we get an early glimpse of the fatigue she felt from her gender equity work, even though unbeknown to her, this work was just beginning. She wrote to the executive secretary of the association at The Hague, Netherlands: "These have been very hectic times in Berkeley. Our Governor Reagan has made a lot of trouble for the University and I have been caught in the thick of it. There does not seem to be any end, unfortunately."[31]

---

[28] John Kifner, "Eldridge Cleaver, Black Panther Who Became G.O.P. Conservative, Is Dead at 62," *New York Times*, May 2, 1998.

[29] "Where Were You in '68?" *UC-Berkeley News*, June 4, 2008, accessed September 26, 2011, http://berkeley.edu/news/berkeleyan/2008/06/04_wherewereyou.shtml.

[30] *The Committee on Probability and Statistics in the Physical Sciences*, accessed July 8, 2015, http://www.aueb.gr/bs-cpsps/index.php?cid=1. Note: The IASPS is regarded as the predecessor of the Bernoulli Society, an association of the International Statistical Institute.

[31] 11/11/1968 letter from Betty Scott to International Association for Statistics in Physical Sciences Executive Secretary E. Lunenberg.

Part II

Shaping the Life

2

Boots and Saddles (Before 1932)

Artillery adds dignity to what would otherwise be a vulgar brawl.
-Frederick the Great

Although it was for her steps alone, Betty's path was shaped in part by the unique paths of members of her family. Betty summed up her major family influences by saying, simply and with humor, that she was an "army brat."[1]

Betty was largely of English, Scottish, and English-French ancestry. She came from a line of Watermans, Herricks, and Leonards on her mother's side and Scotts and Duvals on her father's side. Much is known about Betty's mother's side, but little about her father's. It is known there were military leaders on both sides with high aptitudes for, and appreciations of, mathematics.[2] With these strong family influences, and also many for which there is no surviving record, Betty was shaped and poised to lead the examination of evidence in answer to the question she articulated in 1968, namely: Why are there so few women on the UC-Berkeley faculty?

2.1 Grandfather

Charles Waterman was Betty's maternal great grandfather. He immigrated to America from Farnham, England, and established one of the original farms in the Alpine Township just north of Grand Rapids, Michigan. In 1870, Charles was 48 years old and already a US citizen, with family wealth considerably higher than most of his neighbors.[3] A family wealth advantage would eventually support Betty's decision to work in the university: In the 1970s,

---

[1] Florence N. David, "Obituary: Elizabeth Scott, 1917–88," *Journal of the Royal Statistical Society A (Statistics in Society)* 153, no. 1 (1990), 100.

[2] A.L. Golbeck, "Four Principles of Leadership for Statisticians: A Note on Elizabeth L. Scott," in *Leadership and Women in Statistics*, ed. A.L. Golbeck, I. Olkin, and Y.R. Gel (London: Chapman & Hall/CRC Press, 2015).

[3] US Census Bureau, "1870 US Federal Census, JC Waterman," www.ancestry.com. Note: Charles declared the value of his family's real estate to be $15,000 (about $250,000 in 2010 dollars) and the value of their personal estate to be $2,800.

women who advanced into positions in universities were usually from high socio-economic status families.[4]

Phoebe Herrick Waterman was Betty's maternal great grandmother. She was born in Schenectady, New York. In 1870, she was 38 years old and married to Charles, had two sons, and was keeping house. Fredric K.E. Waterman was the older son. Fred remained in Alpine and made farming his career, as was common among eldest sons within traditional English farming families.

John Charles "JC" Waterman (1857–1939) was the younger son and Betty's grandfather. JC was four years younger than his brother Fred. JC first attended the country school associated with the Alpine Township, then high school in Grand Rapids, followed by a year of college in 1876–77 at the University of Michigan. After that JC embarked on a military career, following a tradition that was common among the younger sons.[5]

The Waterman family was sufficiently well connected within Michigan that JC won an appointment from his state to attend the US Military Academy at West Point. JC entered the academy in 1877 and graduated four years later.[6,7]

Around the time of his graduation, JC married a girl two years his junior named Clara Leonard who was also born in Michigan and was his high school sweetheart.[8,9] JC and Clara had five children.[10] The oldest was Betty's maternal aunt Emma "Phoebe" Waterman (1882–1967), who was born at Fort Totten in the Dakota Territories. Phoebe is an important part of Betty's story (see Chapter 6). The youngest was Betty's mother Elizabeth Waterman (1892–1992), who was born at Fort Yates, also on the frontier, and was 11 years younger than Phoebe.

Upon graduation from West Point, JC received his first posting which was to the rebuilt 7th Cavalry. This was the same 7th Cavalry that General Custer led to the disastrous Battle of the Little Bighorn in 1876 in Montana, a battle thought by many to be the worst disaster the US military ever had.[11] JC was assigned to the 7th Cavalry only five years after this infamous battle. The unit, which would have still been recovering from the battle due to heavy casualties, was on what was considered to be dangerous frontier duty. Railroads were

---

[4]Susan Ervin-Tripp, interview with author, June 13, 2010.

[5]US Census Bureau, "1870 US Federal Census, JC Waterman;" "1880 US Federal Census, Fred K.E. Waterman;" "1890 US Federal Census, Frederick Waterman."

[6]US Military Academy, "List of Cadets Admitted into the USMA, West Point (1902)," www.ancestry.com.

[7]Note: JC graduated 38th out of the 53 Cadets. He went on to have a notably long, diversified, and distinguished military career. Eventually he attained the rank of colonel, becoming one of the nation's highest-ranking field officers. See "The West Point Class," *New York Times* (West Point, NY), June 10, 1881. See also "Register of Graduates and Former Cadets" (West Point, NY, 2010).

[8]Note: The 1910 US Federal Census states that Clara Leonard Waterman's father was born in Vermont and her mother was born in Ohio.

[9]"Register of Graduates and Former Cadets" (West Point, NY, 2010).

[10]US Census Bureau, "1910 US Federal Census."

[11]Note: This battle had cost the 7th Cavalry a total of five of its companies (268 men), as well as its commanding General.

being extended farther into the western territories, and more and more lands were being opened to homesteading as American Indians were being collected onto reservations.[12] JC was ordered to the Dakota Territory to help suppress the American Indians and enable the new settlement.

JC's service was mainly at Fort Totten in northeast North Dakota, although he also spent periods at Camp Belcourt and Fort Meade or in the field. In his time in the 7th Cavalry, JC mainly performed duty with his troop (policing the reservation and guarding transportation routes), although he also served for periods of time as Indian supplies inspector, troop commander, and post topographical officer. In 1888, as relations between the American Indians and the settlers in the Dakotas began to settle down, JC was relieved from frontier duty and sent to Fort Riley in Kansas. In all, JC served ten years in the 7th Cavalry, until 1891.

In the year before his service to the 7th Cavalry ended, JC was called up again to the Dakota Territory, together with the whole 7th Cavalry, and a third of the US Army (some 9000 soldiers). In addition to their rifles, the Cavalry brought with them two light 42-mm rapid-firing rifled guns made by the Hotchkiss Company in France.[13] The mobilization was an orchestrated move to "overawe" the American Indians and quickly slam the door on any American Indian uprising. JC participated in three consecutive days of military engagements with the American Indians at Pine Ridge and White Clay Creek. Two weeks after the engagements, the remaining rebellious American Indians surrendered. JC was promoted to first lieutenant and attached to the 8th Cavalry at Fort Meade and Fort Yates, and then at Fort Keogh, Montana. This was more frontier duty.

During the Pine Ridge engagements, JC reported as a second lieutenant through Captain Henry J. Nowlan to First Squadron Major Samuel M. Whitside, who reported to Colonel James W. Forsyth. Whitside did not have a history with either Custer or Little Bighorn, whereas Forsyth had served alongside Custer in the Civil War. In the first day of engagement, JC helped to bring Chief Big Foot and his band to the army camp: This happened under Whitside who, having high cultural intelligence and having had experience with American Indians, provided care to the ailing Big Foot and allowed the band to keep their weapons as they came to the camp. In the second day of engagement, JC helped to oversee the band at Wounded Knee Creek: Now at camp, Forsyth who had no experience with American Indians was in command, and his order to disarm the band is what began the trouble and created a tragedy. In the third day of engagement, JC participated at White Clay Creek. This turned out to be another botched mission by Forsyth. In just three days, JC

---

[12] "North Dakota Legends: Old West Timeline of North Dakota," *Legends of America*, accessed August 28, 2012, http://www.legendsofamerica.com/nd-timeline.html.

[13] "Hotchkiss Gun," *The Hotchkiss Mountain Gun*, accessed October 3, 2012, http://www.hotchkissmountaingun.com. Note: These were sometimes also known as Hotchkiss cannons. They were breech-loaded and fired one-piece high-explosive ammunition over long ranges. They were light enough to be transported over long distances.

saw first hand in Whitside an example of a compassionate, culturally sensitive leader, and in Forsyth an example of how quickly situations can unravel under poor leadership, where grudges and inexperience of leaders can lead to unfavorable actions. JC saw many decisions made (and events happen) on the basis of rumor and politics instead of hard facts and evidence. He also saw first hand the effects of poverty, disadvantage, and inequality. So did Betty's aunt Phoebe, who was nine years old at the time of the engagements.[14,15]

In 1893 at age 36, JC was detailed to 15 months of continuing education at the School of Application for Infantry and Cavalry at Fort Leavenworth, Kansas. The school, which was essentially a military graduate school, was at the time exclusively for West Point graduates like JC. The idea was for junior officers to retire from the field to focus on the practical application of theories and methods they had learned at West Point. Instruction was given in areas of military art, engineering, law, infantry, cavalry, and military hygiene. The school had space for only a relatively few young officers, and it was looking for those who, like JC, were qualified in algebra, geometry, and trigonometry.[16,17]

Later, in 1897, JC was among the 100 or so military officers (from all three arms of the service, although mostly infantry) each year who were detailed for duty as college professors. Their mission was to educate young people for possible future military duty.[18] These faculty posts were considered plum military assignments. JC was detailed to a faculty position in the Department of Military Science and Tactics at Cornell College in Mount Vernon, Iowa.[19] Women were eligible along with men to take classes from Betty's grandfather.[20] JC's receipt of such a post indicates he was good in mathematics. It was sometimes possible to get extra compensation (salary or a residence) within such a post for teaching non-military subjects, especially mathematics; one officer is even known to have taught astronomy.[21,22] So Betty wasn't the first college professor in her family.

---

[14]Note: JC was relieved from frontier duty in July 1888, but then was recalled in November 1890 to the Pine Ridge Reservation. In this recall, JC was part of the largest army mobilization since the Civil War. He was a second lieutenant in Troop I of the First Squadron. The three days of engagements were December 28, 29, and 30, 1890. JC's superiors had reason to be biased toward the Sioux, as six of the current officers had fought under General Custer. A reporter at the scene, Charley Allen, reflected that the only thing on the minds of the Seventy Cavalry was vengeance for what happened at Big Horn.

[15]Heather C. Richardson, *Wounded Knee: Party Politics and the Road to an American Massacre* (New York: Basic Books, 2010).

[16]E.F. Townsend, "Annual Report (Including Reports of Instructors)" (annual report, US Infantry and Cavalry School, Fort Leavenworth, KS, 1895).

[17]H.S. Hawkins, "Annual Report (Including Reports of Instructors)" (annual report, US Infantry and Cavalry School, Fort Leavenworth, KS, 1895).

[18]"Army Instruction at Colleges," *New York Times* (Washington, DC), December 5, 1894.

[19]"Register of Graduates and Former Cadets" (West Point, NY, 2010).

[20]"To Teach Military Science," *New York Times* (Washington, DC), April 23, 1892.

[21]"Army Officers at Colleges," *New York Times* (Washington, DC), February 6, 1893.

[22]"Timeline of Cornell College History," www.cornellcollege.edu/150/timeline.shtml. See also "Army Officers at Colleges," *New York Times* (Washington, DC), February 6, 1893.

The normal term for military officer college instruction duty was three years. JC's college service was only a little over a year, cut short by the proclamation of the Spanish-American War in 1898. One week after war was declared, JC was detailed to Des Moines, Iowa, his responsibility being to distribute supplies and provisions to the volunteers mustered into service for the war. Ten months after that, JC was promoted to the rank of captain and his mustering duties expanded to include both Indiana and Iowa troops. These duties were also short lived, as the formal Spanish-American War lasted less than four months; The Treaty of Paris gave Cuba back to the Cubans, and it gave Guam, Puerto Rico, and the Philippines to the US.[23],[24]

Next JC was reassigned to the 7th Cavalry to carry out a series of international assignments. First he went to Cuba in 1899 for two years where he commanded almost 100 men on a mission to scout and maintain security after Cuban independence.[25] Next JC went to the Philipines in 1905 for two years on an assignment to oversee an addition to a two-squadron post that would make it suitable as a regimental post. Then JC went back to the Philippines as a major in 1910 for about a year. There he was detailed as paymaster at the headquarters division in Manila, a post that recognized his aptitude for numbers and mathematics.[26],[27] Betty's mother Elizabeth, who was in her formative teenage years, was with the family on these international assignments, and Betty's aunt Phoebe was with the family for part of the time.

Then in 1911 JC was ordered to San Francisco, California, where he was detailed as paymaster at the Presidio. He later served with the Texas Border Patrol during the Mexican Expedition of 1916–1917, commanded the post at Fort Riley, and then retired as a colonel in 1919 at age 62 after some 40 years of military service. He retired to a house on the Severn River, about eight miles north of Annapolis, making frequent trips to see his children and grandchildren, including Elizabeth and Betty in Berkeley, which brought him right across the bay from his old post at the Presidio in San Francisco.[28]

---

[23] "Mustering Officers Chosen: The Detail for New York and Other States Announced by the War Department," *New York Times* (Washington, DC), May 3, 1898.

[24] James C. Waterman, "Annual Report of the Association of Graduates (1939)" (annual report, West Point, NY, 1939), 147.

[25] Note: In Cuba, 1899–1900, JC commanded at Pinar del Rio and Guanajay, and Columbia Barracks. Then he commanded at Chickamauga Park, GA, a place being used to marshal troops and ready them for service in Cuba. Then JC was detailed to recruiting duty in his hometown of Grand Rapids, MI, for two whole years while his troop remained in Chickamauga Park.

[26] Bill Taylor, e-mail message to author, February 16, 2008. Note: The Treaty of Paris had given the Philippines to the US, but the people wanted to be independent. Fighting between the US and the Philippines began in 1899 and continued through 1913.

[27] 6/19/1906 handwritten letter from Phoebe Waterman to Vassar College Instructor Caroline E. Furness re Phoebe's activities in the Philippines (Vassar College Archives Furness Folder 1.13). Note: This letter describes the building in some detail.

[28] James C. Waterman, "Annual Report of the Association of Graduates (1939)" (annual report, West Point, NY, 1939), 147. Note: JC retired at his own request.

2.2 Uncle and Father

John Julius "JJ" (1887–1962) was JC's son and Betty's uncle. JJ graduated from the US Military Academy at West Point in 1910, 29 years after his father. JJ served in World War I. He was a commanding officer of a field artillery battalion that was part of the American Expeditionary Force. His service with the 76th Field Artillery at the Battle of the Marne from July 5 to August 2 earned him an AEF Meritorious Service Citation Certificate and a silver star for gallantry in action in France. He was gassed during this period to the point of blindness, but returned to serve with his brigade on the final push into Germany. His health never recovered, but just like his father, JJ attained the rank of colonel before he retired. Again just like his father, JJ took an assignment as a college professor of Military Science and Technology. Although disabled when he retired, JJ was recalled to active duty from 1940–1942 at the Tuskegee Institute in Alabama, and from 1942–1946 at the Alabama Polytechnic Institute (API), which is now known as Auburn University.[29],[30]

Betty's paternal grandfather was born in Virginia; her paternal grandmother was born in Alabama and had the maiden name of Duval. Richard Christian Scott (1887–1955), also born in Virginia, was their son and Betty's father.[31],[32]

Richard graduated from the US Naval Academy in Annapolis in 1911. However, because he had poor eyesight, he was disqualified from sea duty two years later. It was then that he transferred to the US Army and was commissioned as a second lieutenant.[33] Richard was an officer of field artillery.[34] It is known that he served in the Mexican Expedition. In 1917 he was assigned as first lieutenant to the 18th Field Artillery at El Paso, Texas.[35] In 1918 he was appointed to the rank of captain.[36] It is also known that Betty's father participated in World War I with the rank of major.[37]

---

[29] Official Military Personnel File, "John Julius Waterman" (personnel file, Military Personnel Records Center, St. Louis, MO).

[30] "John Julius Waterman," *Military Times: Hall of Valor*, accessed December 5, 2014, http://projects.militarytimes.com/citations-medals-awards/recipient.php?recipientid=768 62. Note: API was participating in the war effort by offering continuing education courses to train people for sorely needed technical professions.

[31] Beverly A. Boyko, e-mail to author.

[32] US Passport Applications, Elizabeth Leonard Scott, June 28, 1978.

[33] Robert W. Scott, "Loxley R. Scott 1945," accessed February 2, 2014, http://apps.westpointaog.org/Memorials/Article/15170/.

[34] Terry McClure, letter to author, March 21, 2008. Note: Not much is known about Richard's military life. The Official Military Personnel File for Richard C. Scott was likely destroyed by the 1973 fire at the Military Personnel Records Center in St. Louis, MO.

[35] "Obituary," *New York Times* (Oakland, CA), June 10, 1917.

[36] *New York Times*, June 16, 1918.

[37] US Census Bureau, "1930 US Federal Census, Richard C. Scott."

Richard Scott married Elizabeth Waterman in 1916 when he was 28 and she was 23 years old.[38] Both JC Waterman and Richard Scott were serving in Texas the year Betty was born, and so it is highly likely that Betty's mother and father met through the military connections of family or friends. In 1920 Richard was a 32-year-old officer at Fort Sill, Oklahoma, which is within a day's ride from the Mexican border.[39] Betty's parents were renting a house in Fort Sill and her mother was a homemaker.

And so Betty had two members of her family who were officers of field artillery, her father Richard Scott and her maternal uncle JJ Waterman. Field artillery was a mathematical exercise that included all of the following: arithmetic, algebra, geometry, trigonometry, approximation, aids to calculation, graphing, and probability. Student officers of field artillery were required to master numerous mathematical ideas including, among others, alignment chart, azimuth, center of impact, deflection, diagram, dispersion, grad, mil, nomogram, and probable error. Students were required to solve problems including, among others, approximating a square root by the algorithm, determining visibility on a map with the contour lines given, or interpolating with second differences. Precision was an imperative.

Specific examples of how these mathematical ideas were being used in field artillery in the early 20th century are given below.[40]

1. *Near the muzzle of the three-inch gun the rifling makes one turn per 25 calibers. If the projectile leaves the muzzle with a velocity of 1,700 feet per second, how many rotations about its axis does it make per second? Ans. 272.*

2. *The total length of the French 155 mm. cannon is 2.332 m. Express this in calibers. Ans. 15.05.*

3. *In rising above the earth's surface the temperature falls about 6° C. for each kilometer of altitude up to eleven kilometers and thereafter remains practically stationary. What increase in elevation (to the nearest ten feet) will lower the temperature 1° F.? Ans. 300 ft.*

4. *For a range of 3,000 yards the 3-inch gun is elevated at an angle of 5° 5.3' above the line joining gun and target. When fired the projectile actually leaves the gun at an angle of 92.3 mils about this line. How much is the former angle increased by the "jump" of the gun as it is fired? Ans. 1.8 mils.*

5. *A scout measures the angle between a line to an object C and the straight road along which he is passing. He proceeds along the road until, 1,300 yards farther, a line to C makes with the road an angle twice as great. How far is he then from C? Ans. 1,300 yds. Note.*

[38]Obituary: Elizabeth Waterman Scott. Oakland Public Library.

[39]*The New York Times*, 11/30/1919.

[40]Albert A. Bennett, "Reviewed work: Elementary mathematics for field artillery, by Lester R. Ford," *The American Mathematical Monthly* 26, no. 8 (1919), 353.

> *This is the method used by sailors in "doubling the angle on the bow."*

6. *A reconnaissance officer has an accurate map of the surrounding region but does not know his exact position on the map. He sees two objects A and B, which are shown on the map. He measures the angle between the two points and draws lines through A' and B', the map positions of the two points, to meet at the angle found at some point C' on the map. Show that the true position is on the circle drawn through A', B' and C'. If he recognizes a third point whose map position is known he can get a second circle on which his position lies. His true position is at the intersection of these circles. Note. This is the method of "Italian Resection" used to locate a position with a plane table.*

7. *In maps based on English units 1 in. to the mile, 3 in. to the mile, 6 in. to the mile, and 12 in. to the mile, are commonly used scales. Find the representative fractions of these maps? Ans. (a) 1/163360, (b) 1/21120, (c) 1/10560, (d) 1/5280.*

8. *Find the area of a shrapnel pattern 100 yds. long and 35 yds. wide. Ans. 2,749 sq. yds.*

9. *Find the danger space for a horse 15 hands high when the trajectory makes an angle of 252 mils with the ground. Ans. 20 ft.*

As artillery people, JJ Waterman and Richard Scott had to be more than good horsemen, they also needed to be good mathematicians.

2.3   Childhood

The brown-haired, brown-eyed Betty[41] was born at Fort Sill Military Reservation in Oklahoma on November 23, 1917. Her mother Elizabeth's maiden name was Leonard, a name she gave to Betty as a middle name. Since Betty and her mother had the same first and last names, Betty used the middle initial, L, to distinguish herself from her mother. In 1920, Betty was two years old and her family was still living at Fort Sill. The family had a 25-year-old live-in servant – a divorcée from Georgia – whose three-year-old daughter was undoubtedly one of Betty's playmates.[42]

By 1930, Betty's father had retired from the military and become a lawyer in private practice in Oklahoma City. It is not known whether he formally studied law at the university or whether he became a lawyer under the apprentice system that operated at the time. Betty was 12, and the family was

---

[41] 6/28/1978 departure data, Passport Application, Elizabeth Leonard Scott.
[42] 1920 US Federal Census, Richard C. Scott.

living in a rented home. Betty now had three younger brothers. Richard C. Scott, Jr. was 11, John was 10, and Loxley was 8.[43] All three served in the military.

So by the time Betty joined the Berkeley faculty, she already had considerable experience holding her own in a male environment.

Richard C. Scott, Jr. (1919–unkn), Betty's oldest brother, served in the military during World War II. It is known that he sailed at age 23 on the famous aircraft carrier USS Wasp (CV-7). He sailed on the same ship a month later. His military branch is not identified. Perhaps he was in the navy, or else was one of the marines being transported by the navy. It is widely reported on the Internet that the Wasp was lost on September 15, 1942. This is inconsistent with the dates of sailing for R.C. Scott, Jr. that are given in the muster rolls.[44] In any case, it is possible that R.C. Scott, Jr. was aboard the Wasp in the battle that took the aircraft carrier down.

John W. Scott (1920–2006), Betty's middle brother, served during World War II and the Korean War. He enlisted in the army as a private in 1943 at Fort Devens, Massachusetts for the duration of World War II. At the time of enlistment, JW was an unmarried college student, majoring in business or public administration. JW eventually became an officer, serving as a lieutenant colonel in the US Army Air Force.[45,46]

Loxley Radford Scott (1921–1997), Betty's youngest brother, also served during World War II and the Korean War. He went to West Point Preparatory School at The Presidio in San Francisco. He then went to West Point, just like his uncle and grandfather, entering from California in 1941 and graduating four years later. At the time of enlistment, Loxley's civil occupation was as a motion picture actor or director, or entertainer. He was assigned to the Coast Artillery Corps branch. His first assignment was to Japan as part of the postwar occupation force, followed by assignments at The Presidio in San Francisco, then Korea where he earned a Silver Star as a battalion operations officer, then various assignments around the US. Loxley achieved the rank of captain in the US Army. He retired from the military in 1961, but then earned a master's degree in education from San Jose State University, and used it to teach math and science to junior high school students in Salinas, California

---

[43]1930 US Federal Census, Richard C. Scott; 1930 US Federal Census, Elizabeth L. Scott. Note: The address was 517 Northwest 30th Street, Oklahoma City, OK. Betty's three younger brothers were each a year apart in age.

[44]WWII U.S. Navy Aircraft Carrier Muster Rolls, 1939–1949, roll MIUSA2006_082862. Note: Brother Richard sailed on the USS Wasp on September 30 and November 1, 1942.

[45]U.S. Veterans Gravesites, ca. 1775–2006. John W. Scott.

[46]US World War II Army Enlistment Records, 1938–1946. John W. Scott. Note: At the time, the air corps was part of the army. The US Air Force separated in 1947.

for over 20 years. In his retirement from teaching, one of the things that he studied was astronomy.[47,48,49]

Betty's family moved to Berkeley sometime between 1930 and 1932, during the Great Depression. One of the reasons for the move was that her family expected all four of their children to go to college, and they knew that UC-Berkeley was a good university with low cost.[50] Betty's father continued to practice law after the move. Richard was admitted to the State Bar of California in November 1932.[51] At some point Richard was in a car accident where he lost one eye and the other was injured. Following the accident, Betty's mother, who had been a homemaker, sometimes worked (especially during World War II), and she assisted Betty's father in his practice. Betty explained it like this: "So she would read to him. She actually participated..." in his practice.[52,53]

Betty walked with a limp, a kind of shuffle.[54] Some conjecture she had a mild form of polio or some other disease as a child, but this is not substantiated.[55] In any case, Betty grew up to be a large, 5' 7"woman.[56]

2.4   Family Influences

Women in Betty's time were not eligible to study at West Point or become field artillery officers. Had they been, one can easily imagine her in such a military career. Betty was a natural leader. She was fearless, mathematical, and comfortable with men.

Betty's maternal grandfather Colonel JC Waterman lived with Betty and the Scott family in Berkeley in the last years of his life. He was around 80 years old and retired at the time, while Betty was around 20 years old and an undergraduate at UC-Berkeley. JC's saddle was kept on a sawhorse in the living room of Betty's house during those years. Betty had opportunities from

---

[47] "Loxley Radford Scott – Salinas, CA," *LocateGrave*, acessed on October 12, 2012, http://www.locategrave.org/l/963725/Loxley-Radford-Scott-CA.

[48] Robert W. Scott, "Loxley R. Scott 1945," accessed February 2, 2014, http://apps.westpointaog.org/Memorials/Article/15170/.

[49] US World War II Army Enlistmeent Records, 1938–1946. Loxley R. Scott.

[50] Scott, Elizabeth L. Influences, Challenges, and Problems. Speech to unidentified audience, 3 double spaced typewritten pages. Around 5/1976.

[51] MEMREC, e-mail to author, March 3, 2008.

[52] 6/10/1978 letter from Ms. Mina Edelston, attendee at the UC-Berkeley Conference for Women in Engineering and Computer Science on 5/13/1978, to Betty Scott with attached questionnaire completed by Scott.

[53] Scott, Elizabeth L. "An Interview with Elizabeth Scott," an oral history conducted by Suzanne B. Riess, in "The Women's Faculty Club of the University of California, Berkeley, 1919-1982," Oral History Center, The Bancroft Library, University of California, Berkeley, 1983.

[54] Brian S. Yandell, telephone interview with author, June 24, 2014.

[55] Dennis Pearl, telephone interview, February 17, 2014.

[56] 6/28/1978 departure date, Passport Application, Elizabeth Leonard Scott.

her grandfather and mother to learn international perspectives, human rights lessons, and the value of higher education. Both were in the Philippines and had witnessed an insurrection being suppressed, seeing firsthand that people's human rights are sometimes denied them, even by Americans.

Betty never moved out of her parents' home. Her gender equity work was surely informed by her father's and mother's professional engagement with the law. The law is malleable and interpretive, a living document. It requires precise documentation. Betty dated every piece of paper in her files, greatly facilitating the construction of the microhistories that comprise this book.

3

Aunt Phoebe's Telescope (1882–1967)

Please remember me most kindly to the Mountain.
-E. Phoebe Waterman

Emma Phoebe Waterman – later E. Phoebe Waterman Haas (1882–1967) and known as "Phoebe" – was Betty's oldest maternal aunt. She lived at a time when female US citizens were not allowed to vote. Phoebe spent her childhood on the frontier. It was a place absent of light pollution, where she would have been able to see the stars in a way that most people today could only imagine. Phoebe was schooled on the frontier until she was ready to go to high school. Then her parents, who had a high value for education, sent Phoebe to live with her grandparents back in Grand Rapids, Michigan so she could attend a regular high school.[1] Betty and Phoebe had many things in common relating to their fierce intellectual interests in astronomy, but they had very different life trajectories that reflected their differing times.[2]

3.1 Astronomy Education

Phoebe attended Vassar College, graduating with a bachelor's degree in astronomy in 1904. She was smart and was elected to Phi Beta Kappa.[3] She immediately went on to graduate school at Vassar with a fellowship, earning her master's degree in astronomy in 1906.[4] Vassar at the time was a great place for women to study astronomy. It had its own observatory, a strong scholarly reputation in astronomy evidenced by an output of five or six publications in

---

[1] Thomas R. Williams, "Phoebe Haas – An AAVSO Volunteer," *American Association of Variable Star Observers* 20, (1991), 18.

[2] Note: Unless otherwise indicated, correspondence to Vassar College in this chapter is taken from Vassar College Archives Furness Folder 1.13.

[3] 12/6/1910 handwritten letter from Phoebe Waterman to Vassar College Astronomy Instructor Caroline Furness.

[4] UC-Berkeley official transcript, E. Phoebe Waterman, two handwritten pages; 1901 Vassarian Yearbook; Fellowships, Scholarships and Prizes section of the Vassar Miscellany, Vol LX, No. 9, 1 June 1911, p. 743.

astronomy each year,[5] and more astronomy graduates than any other college or university in the US.[6] The faculty was strong, led by astronomer Mary Watson Whitney (1847–1921), who like Betty was interested in promoting both science and gender equity.[7,8] Astronomer Caroline E. Furness (1869–1936), at the time a young member of the faculty,[9] quickly became Phoebe's primary mentor. In 1908, Phoebe published her master's thesis with Furness on the orbit of a comet.[10]

Phoebe's experience at the women's college was reportedly very different from what Betty's experience would be at a public university. The astronomy students at Vassar tended to see themselves as special vis à vis the other students, with a sense that they were comrades working together to unlock the secrets of the universe. They tended to have a much more glamorous or romantic view of astronomy. In the words of historian John Lankford, "It is improbable that undergraduates, working at the Student's Observatory at Berkeley, under the supervision of Armin Otto Leuschner (1868–1953) and his graduate students, ever viewed their experiences in quite this [romantic] way." With such an atmosphere, it is no wonder that Phoebe remained at Vassar to earn her master's degree rather than enrolling in a public university. Lankford continues: "Working closely with female mentors, these young women formed their basic scientific tastes and styles. In comparison to male astronomers, many of these women found the undergraduate experience far richer and more rewarding. Undergraduate mentors would remain more important than male dissertation supervisors and the undergraduate years would take a central place in the memories of these women."[11] Such would be the case with Phoebe, who remained close to Furness until her mentor's death.

Before finishing her master's degree, Phoebe followed her parents to the Philippines where her father JC Waterman was stationed in the military. Phoebe wrote her master's thesis there while she home-schooled Betty's mother Elizabeth Waterman. Elizabeth was in her freshman year of high school and loved horses, which was not surprising because their father was a

---

[5]Furness, Caroline E. The Vassar Observatory. Vassar Miscellany, Vol. XLI, No. 6, 1 April 1912.

[6]John Lankford, *American Astronomy: Community, Careers, and Power, 1859–1940* (Chicago: University of Chicago Press, 1997), 312.

[7]Mary Watson Whitney, Vassar Encyclopedia. Accessed on 12/18/2013 at http://vcencyclopedia.vassar.edu/alumni/mary-watson-whitney.html.

[8]Furness, Caroline E. The Vassar Observatory. *Vassar Miscellany*, Vol. XLI, No. 6, 1 April 1912.

[9]Caroline E. Furness, Vassar Encyclopedia. Accessed on 12/18/2013 at http://vcencyclopedia.vassar.edu/alumni/caroline-e-furness.html.

[10]Furness, Caroline E. and Waterman, Emma Phoebe. Definitive Orbit of Comet 1886 III. *Kiel Verlag der Astronomischen Nachrichten*, No. 14, pp. 27-35, 1908.

[11]John Lankford, *American Astronomy: Community, Careers, and Power, 1859–1940* (Chicago: University of Chicago Press, 1997), 308, 313.

cavalry officer.[12,13] Phoebe wrote many detailed letters to Furness and Whitney back at Vassar, reporting on her life and the turmoil in the Philippines. Most interesting was her message: "There are exactly four Americans, there; the Mrs. Taylor whom we visited being the one American woman in all that Catubig valley. It is a week's march to another woman, over or across pathless, jungle covered hills, and swamps, or a long ride down the river and a longer wait in the town at its mouth for the erratic supply transports and native boats. We were the first girls she had seen since the end of August!" The Philippines was another education for Phoebe, where she saw a good amount of poverty, ignorance, lack of ambition, and superstition. Phoebe would not have realized it at the time, but her experiences of rustic life on the frontier and in the Philippines would prepare her well for the rustic life of an astronomical observer in remote locations. At the Philippines, she did some observing and recording without a telescope, but she was discouraged at the lack of astronomy research partners, missed the astronomy work she had done at Vassar, and longed to be back to help Furness with more comet research.[14]

Phoebe assessed her employment options as she planned to go back to the US. She could try to get a position at a YWCA or else at a research observatory. But just then, a temporary faculty position at Vassar became available to fill in for one of the regular instructors who went on leave in the second semester of the 1907–08 academic year. Phoebe was highly regarded among the faculty and was offered the position. She became an acting instructor in observational astronomy and assistant to her mentor Furness at the observatory.[15,16] While on the faculty at Vassar – together with Furness and a German mathematician and optics specialist – Phoebe published a determination of the orbit of another comet.[17] When the position ended, Phoebe again followed her family to where her father was stationed, this time at Fort Riley, Kansas. She was unemployed and occupied her time reading and recreating. She wasn't doing much astronomy, which she regretted.[18]

---

[12]12/6/1910 handwritten letter from Phoebe Waterman to Vassar College Astronomy Instructor Caroline Furness.

[13]6/19/1906 handwritten letter from Phoebe Waterman to Vassar College Astronomy Instructor Caroline Furness.

[14]3/21/1906 handwritten letter from Phoebe Waterman to Vassar College Astronomy Instructor Caroline Furness.

[15]Furness, Caroline E. The Vassar Observatory. Vassar Miscellany, Vol. XLI, No. 6, 1 April 1912.

[16]12/6/1910 handwritten letter from Phoebe Waterman to Vassar College Astronomy Instructor Caroline Furness.

[17]Hans Boegehold, Caroline Ellen Furness, and Emma Phoebe Waterman, "Bestimmung der Bahn des Kometen 1825 I," *Astronomische Nachrichten Abhandlungen* 2, no. 14 (1908), 35.

[18]10/21/1908 handwritten letter from Phoebe Waterman to Vassar College Astronomy Instructor Caroline Furness.

3.2   Computer Work

At Phoebe's time, Vassar was highly successful at placing their astronomy graduates into positions at observatories doing practical measurement work in astrophysics. These were contract positions for women with the title of "Computer." The annual starting pay was low ($750, which is about $18,000 in 2010 dollars, plus a month's vacation – about half of the starting pay of men at the observatory), and there was no expectation of or plan for upward mobility. The hours were long (six days a week). The women computers used tools to measure data from glass photographic plates (where they often had to stand and lean forward in physically straining positions for long periods of time), and then they classified, reduced, and cataloged the data for the men to interpret and write up for publication. The work was concentrated, tedious, and routine, involving hundreds of thousands of stars, and the women were not supposed to see or care about the big picture of the research. The men rationalized that women were well suited to this kind of detail work, citing similarities to stitching or embroidery, thereby keeping women from observing at the telescopes. There was an expectation that women computers would find a spouse among the men observers.[19,20,21] Harvard College Observatory Director Harlow Shapley explained: There was "a tremendous amount of measuring. I invented the term 'girl-hour' for the time spent by the assistants. Some jobs even took several kilo-girl-hours. Luckily Harvard College was swarming with cheap assistants; that was how we got things done."[22]

Phoebe took notice of the computer positions and thought that, "for a girl" such a position was "as well as any one could hope to do." She hoped Furness could place her in California, either at the Carnegie Mount Wilson Solar Observatory offices in Pasadena, or at Lick Observatory on Mount Hamilton. In January 1909, at around age 26, Phoebe joined the computing division at Mount Wilson.[23,24] The five workers in the computing division there – all women – formed a close knit group, and some of them even lived together. Phoebe was very overqualified for the job as she already had a master's de-

---

[19]Websites on Women in Astronomy. Accessed on 10/29/2013 at http://astro. berkeley.edu/gmarcy/women/history.html.

[20]John Lankford, *American Astronomy: Community, Careers, and Power, 1859–1940* (Chicago: University of Chicago Press, 1997), 290; 318; 322; 328; 340.

[21]Furness, Caroline E. The Vassar Observatory. Vassar Miscellany, Vol. XLI, No. 6, 1 April 1912.

[22]Harlow Shapley, *Through Rugged Ways to the Stars: Reminiscences of an Astronomer* (New York: Charles Scribner and Sons, 1969), 94.

[23]6/19/1906 handwritten letter from Phoebe Waterman to Vassar College Astronomy Instructor Caroline Furness.

[24]Hale, George Ellery. Directors Report, Carnegie Institution of Washington Year Book, No. 9, p. 162, 1909.

gree in astronomy and most computers had only high school degrees.[25,26,27] Phoebe raised questions about the work and wanted to do big picture work, but was prevented from doing so within her job title or at this employer.[28]

Phoebe worked as a computer for several years but she had gotten restless after only nine months on the job doing support work,[29] and after two and a half years she wanted to do her own astronomy work. While at Mount Wilson, Phoebe worked for some big name astronomers: Walter Sydney Adams (1876–1956), Harold Delos Babcock (1882–1968), George Ellery Hale (1868–1938), and Jacobus Cornelius Kapteyn (1851–1922). She supported some really exciting astronomy work, in her own words: "My work here has been to assist in the determination of solar rotation from a comparison of the spectra of the east and west limb, and to classify, measure and reduce stellar spectra, for radial velocity." But coming from Vassar, she had been trained to do observing, not just computing, and at Mount Wilson she was only allowed to do the computing. She aspired to observe with the biggest telescopes, photograph the astronomical objects, and participate in the most important professional conferences in astronomy. Phoebe knew she needed to pursue a PhD degree in order to achieve these aspirations.[30,31] She told Furness: "...Ooh, I do want so much a position as astronomer, part of my work with the instruments and part with the reduction of my plates, as the men here have. I never did want the teaching. But it is very bold and presuming of a woman to think of such a position I suppose!! [sic] They think so here."[32] But Phoebe came from a family that had high regard for education, and her parents fully supported her decision to go back to school for the PhD and pursue "head work."[33,34]

---

[25] 9/16/1909 handwritten letter from Phoebe Waterman to Vassar College Astronomy Instructor Caroline E. Furness.

[26] 1910 United States Federal Census. E Phoebe Waterman, at www.ancestry.com.

[27] John Lankford, *American Astronomy: Community, Careers, and Power, 1859–1940* (Chicago: University of Chicago Press, 1997), 340, 343.

[28] 9/16/1909 handwritten letter from Phoebe Waterman to Vassar College Astronomy Instructor Caroline E. Furness.

[29] 9/16/1909 handwritten letter from Phoebe Waterman to Vassar College Astronomy Instructor Caroline E. Furness.

[30] John Lankford, *American Astronomy: Community, Careers, and Power, 1859–1940* (Chicago: University of Chicago Press, 1997), 322.

[31] 12/6/1910 handwritten letter from Phoebe Waterman to Vassar College Astronomy Instructor Caroline Furness.

[32] 5/14/[1911] letter from Phoebe Waterman to Vassar College Astronomy Instructor Caroline Furness.

[33] 2/19/1911 handwritten letter from Phoebe Waterman to Vassar College Astronomy Instructor Caroline Furness.

[34] 1/28/1912 handwritten letter from Phoebe Waterman to Vassar College Astronomy Instructor Caroline Furness.

3.3   Astronomy Doctoral Studies

Phoebe chose UC-Berkeley over the University of Chicago because Berkeley at the time had free tuition. Chicago had the bigger reputation, but Phoebe reasoned: "As long as I can get the courses I want at Berkeley, I suppose it would be foolish to go clear to Chicago, just for the sake of the name."[35] Also, Phoebe's father was stationed at The Presidio, and if she went to UC-Berkeley, she could take the ferry to San Francisco and easily see her family on weekends.[36] And so Phoebe wrote an application to UC-Berkeley Astronomy Department Chair Armin O. Leuschner to pursue her PhD, indicating an interest in astronomical spectroscopy.[37] Leuschner shared Phoebe's application with Lick Observatory Director William Wallace Campbell (1862–1938) who judged Phoebe to be a "very attractive candidate"– with solid background under Furness and Whitney at Vassar and under Hale and Adams at Mount Wilson – and pledged to provide opportunities at Lick for her work in spectroscopy. Phoebe was "anxious to begin."[38,39] She had learned the importance of physics in modern astronomy while at Mount Wilson, and so she made plans to study more physics at Berkeley.[40]

In Phoebe's time, a student aiming to study astronomy at UC-Berkeley could apply and compete for one of ten fellowships that were not restricted as to field of study, and three fellowships that were restricted to astronomy. In addition, Vassar had a few alumni fellowships that were not restricted as to place or field of study. Phoebe was so determined to pursue the PhD that she declared to Furness she was going to do it with or without fellowship support.[41,42,43] Phoebe wasn't successful at obtaining a Berkeley fellowship. Many of the male faculty at the time reportedly thought women weren't a good investment for fellowships or tenure ladder positions.[44] But Phoebe was

---

[35]n.d. [prior to enrollment at UC-Berkeley] handwritten letter from Phoebe Waterman to Vassar College Astronomy Instructor Caroline Furness.

[36]1/28/1912 handwritten letter from Phoebe Waterman to Vassar College Astronomy Instructor Caroline Furness.

[37]6/8/1911 letter from Lick Observatory Director William Wallace Campbell to Phoebe Waterman.

[38]n.d. [prior to enrollment at UC-Berkeley] handwritten letter from Phoebe Waterman to Vassar College Astronomy Instructor Caroline Furness.

[39]1/8/1911 handwritten letter from Phoebe Waterman to Vassar College Astronomy Instructor Caroline Furness.

[40]5/14/1911 letter from Phoebe Waterman to Vassar College Astronomy Instructor Caroline Furness.

[41]W.W. Campbell, "Fellowships in the Lick Observatory," *Popular Astronomy* 22, (1914), 194.

[42]n.d. [prior to enrollment at UC-Berkeley] handwritten letter from Phoebe Waterman to Vassar College Astronomy Instructor Caroline Furness.

[43]1/8/1911 handwritten letter from Phoebe Waterman to Vassar College Astronomy Instructor Caroline Furness.

[44]John Lankford, *American Astronomy: Community, Careers, and Power, 1859–1940* (Chicago: University of Chicago Press, 1997), 294; 331.

successful at obtaining fellowships from Vassar for both years of her PhD work.

Phoebe began her doctoral work in astronomy at UC-Berkeley in fall 1911.[45] Once in the program, Berkeley astronomy faculty were known to treat women fairly. Phoebe found the department to be "energetic" and "wide awake." She found it to be "well off," having excellent equipment. She also found the facilities for the study of physics to be excellent; among other things, they included a spectroscopic lab that took up a whole floor. She looked forward to getting to Lick Observatory to further her work in stellar spectroscopy.[46,47] She did some more work on orbits of comets as part of her coursework. By the end of the first year, Phoebe completed her required courses in astronomy and mathematics. She did very well: Leuschner told Campbell that Phoebe was "one of the most unusually well equipped women we have ever had at Berkeley. She is brilliant, quick, and accurate and disposes of her work with promptness and accuracy."[48]

3.4   Lick Observatory Work

Phoebe proposed to start her work at Lick in the summer of 1912, at the completion of her first year in the doctoral program.[49] She acknowledged some trepidation coming from a women's college: "They give a woman just the same work there, as the men, & judge them on the same basis. I am getting used to the different standard a little – for it surely is a different one, & quite a different thing from measuring up against women."[50] Phoebe's fellowship was only a partial one, and she looked to Campbell for some additional financial support. She thought she could spend some time in the computing division to earn some extra money; she would feel at home there given her experience at Mount Wilson.[51] Campbell told Phoebe he would try to provide her with some small amount of financial support for the summer out of the observatory budget, offering that she could analyze his Polaris radial veloc-

---

[45]10/8/1911 handwritten letter from Phoebe Waterman to Vassar College Astronomy Instructor Caroline Furness.

[46]n.d. [presumably Phoebe's first semester at UC-Berkeley] handwritten letter from Phoebe Waterman to Vassar College Astronomy Instructor Caroline Furness.

[47]10/8/1911 handwritten letter from Phoebe Waterman to Vassar College Astronomy Instructor Caroline Furness.

[48]Thomas R. Williams, "Phoebe Haas – An AAVSO Volunteer," *American Association of Variable Star Observers* 20, (1991), 18.

[49]1/23/1912 card from Phoebe Waterman to Lick Observatory Director William Wallace Campbell.

[50]n.d. [presumably Phoebe's first semester at UC-Berkeley] handwritten letter from Phoebe Waterman to Vassar College Astronomy Instructor Caroline Furness.

[51]11/22/1911 handwritten letter from Phoebe Waterman to Lick Observatory Director William Wallace Campbell.

ity plates.[52] But Phoebe's dissertation topic remained unclear.[53] She at first proposed using objective prisms to measure radial velocities, but Campbell reported that the topic, while important, was already taken.[54,55] In the end, she did a dissertation in the area of classification research. It was titled "The Visual Region Spectrum in Brighter Class A Stars." Basically, she obtained spectra, analyzed the presence of lines in the visual region, and then compared them to the lines in the photographic region, as a way of testing whether the visual region had value for the purpose of classification. It was good work, and Campbell agreed to publish it in the *Lick Observatory Bulletin*.[56,57,58] Phoebe's dissertation was reportedly used "very frequently in the assembly [gathering] room" at Lick Observatory.[59]

As Phoebe got started on her dissertation research in the summer of 1912, there was a problem with housing at Lick. Six months earlier there was a magnitude 6.5 earthquake that destroyed the residence hall where graduate students normally stayed. Campbell suggested Phoebe might have to stay in a tent, "but at the best this would demand patience, from the occupants, with primitive conditions."[60] This didn't phase Phoebe, as she was born and raised on the frontier. As she said: "I have lived in a tent long enough to know how to make myself quite comfortable in one."[61] Campbell offered Phoebe either a single or a double tent,[62] but there were so few women in astronomy that she didn't know of another woman with whom to share. Phoebe asked for and was given clarification on her living arrangements. She could bring her small dog, a neutered English-Irish setter, but she would need to keep it away from the observatory. She should bring warm clothing for night observing.[63,64] In

---

[52] 12/13/1911 letter from Lick Observatory Director William Wallace Campbell to Phoebe Waterman.

[53] 1/28/1912 handwritten letter from Phoebe Waterman to Vassar College Astronomy Instructor Caroline Furness.

[54] 5/4/1912 handwritten card from Phoebe Waterman to Lick Observatory Director William Wallace Campbell.

[55] 5/9/1912 letter from Secretary for Lick Observatory Director William Wallace Campbell to Phoebe Waterman.

[56] UC-Berkeley official transcript, E. Phoebe Waterman, two handwritten pages.

[57] William W. Campbell, "Dr. Emma Phoebe Waterman," *Astronomical Society of the Pacific* 25, (1913), 168.

[58] E. Phoebe Waterman, "The Visual Region of the Spectrum of Brighter Class A Stars," *Lick Observatory Bulletin* 243, (1913), 17.

[59] 3/26/1914 letter from Lick Observatory Secretary to Director William Wallace Campbell.

[60] 12/13/1911 letter from Lick Observatory Director William Wallace Campbell to Phoebe Waterman.

[61] 6/2/1912 handwritten card from Phoebe Waterman to Miss Standen, Lick Observatory Secretary to Director William Wallace Campbell.

[62] 2/3/1912 handwritten letter from Phoebe Waterman to Lick Observatory Director William Wallace Campbell.

[63] 5/4/1912 handwritten card from Phoebe Waterman to Lick Observatory Director William Wallace Campbell.

[64] 5/9/1912 letter from Secretary for Lick Observatory Director William Wallace Campbell to Phoebe Waterman.

the end, she didn't have to live in a tent after all, but instead was given a room with a family.[65,66] Phoebe took a commercial "stage" – or automobile service – to get from San Jose to Lick Observatory on Mount Hamilton. It was and still is an effort to make the 30-mile trip up the mountain. The trip is not for the faint of heart. But Phoebe didn't see the roads as treacherous: She saw the trip as stunningly beautiful, fun, and pleasurable.[67]

Phoebe had a five-member dissertation committee. It included Campbell, Leuschner, physicist Frederick Slate (1852–1930), spectroscopist E.P. Lewis (1863–1926), and mathematician M.W. Haskell (1863–1948).[68] Phoebe was also helped by members of the Lick Observatory staff. She wrote in her dissertation: "...I wish to make acknowledgment of Dr. Campbell's generosity in the use he has permitted me of the equipment of the Observatory; of the kindness of the members of the Observatory staff, especially of Mr. Wright who has helped me often, and of Mr. Merrill, many of whose plates I have used; and of my obligation to the Alumnae of Vassar College through whose fellowship the year's work has been made possible."[69]

Together Phoebe and Estelle Glancy were the first women to earn PhDs in astronomy at UC-Berkeley. Glancy was a graduate of Wellesley College and had been at UC-Berkeley for five years working as a computer before Phoebe arrived. The two became close friends in the PhD program. They spent time together on the mountain at Lick Observatory.[70,71,72] They were roommates together in town, and they recreated together. They graduated together on May 14, 1913, along with four men.[73] Glancy was technically the first woman graduate of the UC-Berkeley Astronomy PhD program, but only because her last name came before Phoebe's in alphabetical order, and they graduated within minutes of each other. Lick Observatory Archives Curator Dorothy Schaumberg sums it up this way: "...I do not think it detracts from Phoebe

---

[65] 5/9/1912 letter from Secretary for Lick Observatory Director William Wallace Campbell to Phoebe Waterman.

[66] 6/2/1912 handwritten card from Phoebe Waterman to Miss Standen, Lick Observatory Secretary to Director William Wallace Campbell.

[67] 5/15/1913 handwritten card from Phoebe Waterman to Lick Observatory Director William Wallace Campbell.

[68] 1913 Dissertation signature page.

[69] 2/15/1913 Summary by E. Phoebe Waterman.

[70] 1/28/1912 handwritten letter from Phoebe Waterman to Vassar College Astronomy Instructor Caroline Furness.

[71] Dr. Estelle Glancy, September 1952. PDF of brief handwritten autobiography for the years 1918–1951. Accessed on 11/9/2013 at http://www.dickwhitney.net/DrEstelleGlancy HandWrittenTextSept1952.pdf.

[72] 6/2/1912 handwritten card from Phoebe Waterman to Miss Standen, Lick Observatory Secretary to Director William Wallace Campbell re Phoebe's move to the top of the mountain.

[73] 5/16/1913 letter from unknown [apparently Lick Observatory Director William Wallace Campbell] to Phoebe Waterman.

that she was one of two. They were both 'first.' (It was a struggle for women in those days.) And they were both products of the Lick Observatory!"[74]

As she graduated, Phoebe was working on a big picture problem on stellar evolution that she would publish a few months later in the *Publications of the Astronomical Society of the Pacific*. Phoebe described stellar evolution as "the order of development of worlds so huge and so remote that the mind fails utterly to realize their dimensions, the laws of a progress so slow as to be beyond detection in any individual case. . . "[75] As she graduated, Phoebe made a plan for her future that involved Glancy. Both women wanted careers as astronomers rather than as teachers, but the job market at the time was tough, with supply exceeding the demand.[76] Glancy recalled: "Phoebe said: 'Let's get jobs at the same observatory.' I answered: 'Who would want two women?' 'Maybe Dr. Perrine would take us,' she said. Dr. Perrine was the American Director of the Argentine National Observatory at Cordoba. In a cooperative mood, I said: 'Well, you can ask him.' Lo and behold, he answered: 'Come.'" In September, the two women got on a boat headed for Argentina with five-year appointments at Cordoba Observatory.[77,78,79]

3.5   Life Career Balance

Between 1859 and 1940, there were 426 women employed in astronomy, and half had professional careers lasting five years or less that were usually truncated by marriage.[80] This would be the case with Phoebe. While Glancy ended up spending her entire five-year appointment at Cordoba, Phoebe ended up spending only three months there.[81] While on the boat to Argentina, Phoebe fell in love with fellow passenger Otto Haas, an emerging industrial tycoon on his way to South America to promote a new chemical product that trans-

---

[74]6/12/2002 e-mail from Curator Dorothy Schaumberg of the Lick Observatory Mary Lea Shane Archives to Hollister Knowlton of the William Penn Foundation.

[75]E. Phoebe Waterman, "The Present Status of the Problem of Stellar Evolution," *Publications of the Astronomical Society of the Pacific* 25, no. 149 (1913), 189.

[76]John Lankford, *American Astronomy: Community, Careers, and Power, 1859–1940* (Chicago: University of Chicago Press, 1997), 10.

[77]5/15/1913 handwritten card from Phoebe Waterman to Lick Observatory Director William Wallace Campbell.

[78]Dr. Estelle Gancy, September 1952. Personal account of the years from 1918 to 1951. Accessed on 11/21/2013 at http://www.dickwhitney.net/DrEstelleGlancyHandWrittenText Sept1952.pdf.

[79]William W. Campbell, "Dr. Emma Phoebe Waterman," *Astronomical Society of the Pacific* 25, (1913), 168.

[80]John Lankford, *American Astronomy: Community, Careers, and Power, 1859–1940* (Chicago: University of Chicago Press, 1997), 287; 293.

[81]Dr. Estelle Gancy, September 1952. (Handwritten) Personal account of the years from 1918 to 1951. American Optical History. Accessed on 11/21/2013 at hhttp://www.dickwhitney.net/AOHistoryLensDesignersGlancyWellsleyInfo.html.

formed hides into leather.[82] Phoebe soon resigned her position at Cordoba and came back to the states. Phoebe married Haas in 1914, becoming Phoebe Waterman Haas and a home-oriented mother of two boys.[83]

Phoebe retained her great love for astronomy after starting her family, which is epitomized in her statement to Campbell: "Please remember me most kindly to the Mountain."[84] Over the years, she kept up relationships with the staff at Lick Observatory and other prominent astronomers.[85] Then, six and a half years after she finished her PhD and became established in her home life, Phoebe tried to get back into astronomy research. When she pitched new research ideas to Campbell and was stalled by lack of access to necessary equipment,[86] she tried to work on a smaller problem related to her dissertation. New data were also needed for this smaller problem, and Campbell asked one of his graduate students, who happened to be Shane, to collect the needed data. But then Campbell published the result without including either Phoebe or Shane as authors. This must have been disappointing to Phoebe given that six months earlier when she raised the question to Campbell she envisioned publishing the result. Phoebe found it hard to get her foot back in the door to astronomy research.[87,88,89] A scientist at heart, she nevertheless continued to raise more research questions and share them with Campbell.[90]

In 1928 at age 47, burning to return to "head work" after starting her family, Phoebe consulted again with Furness, who had established herself as a foremost expert in variable stars.[91,92] Furness suggested that Phoebe could do research in variable stars from her home. All Phoebe would need is a telescope. The observing could be done from Phoebe's backyard. Phoebe should contact the Harvard Observatory Recorder for the American Association of Variable Star Observers (AAVSO). A result was that she acquired a 4" Clark

---

[82]Sheldon Hochheiser, *Rohm and Haas: History of a Chemical Company* (Pennsylvania: University of Pennsylvania Press, 1986), 3; 12.

[83]Thomas R. Williams, "Phoebe Haas – An AAVSO Volunteer," *American Association of Variable Star Observers* 20, (1991), 18.

[84]6/4/1920 handwritten card from Phoebe Waterman Haas to Lick Observatory Director William Wallace Campbell.

[85]4/14/[presumably 1922] handwritten card from Phoebe Waterman Haas to Lick Observatory Director William Wallace Campbell.

[86]1/25/1920 handwritten card from Phoebe Waterman Haas to Lick Observatory Director William Wallace Campbell.

[87]7/20/1920 letter from Lick Observatory Director William Wallace Campbell to Phoebe Waterman Haas.

[88]Lick Observatory Director William Wallace Campbell [author listed, apparently mistakenly, as W.C.C.]. Note on Lines in the Visual Spectrum of α Cygni. Lick Observatory Bulletin, No. 332; Berkeley: University of California Press, p. 108, 1921.

[89]6/10/1920 letter from Lick Observatory Director William Wallace Campbell to Phoebe Waterman Haas.

[90]7/28/1920 handwritten card from Phoebe Waterman Haas to Lick Observatory Director William Wallace Campbell.

[91]Caroline Ellen Furness, *An Introduction to the Study of Variable Stars* (New York: Houghton Mifflin Company, 1915).

[92]Caroline E. Furnesss. Vassar Encyclopedia. Accessed on 10/29/2013 at http://vcencyclopedia.vassar.edu/alumni/caroline-e-furness.html.

refractor.[93] This was an excellent telescope: At the time, Alvan Clark & Sons telescopes had the best quality lenses and were telescopes of choice in all of the major observatories in this country. With the telescope, Phoebe became a volunteer astronomical observer for Harvard Observatory and AAVSO. Phoebe reported 338 variable star observations over a five-year period through 1933.[94]

William Hammond Wright (1871–1959) became and would remain a champion of Phoebe's. Wright had previously headed the D.O. Mills Expedition which established an observing station on Cerro San Cristobal, Santiago Chile. This was in 1903. Wright returned to Lick Observatory in 1906 as an astronomer. Wright undoubtedly was an important resource to Phoebe when she decided upon graduation to accept a position at Cordoba. Phoebe kept in touch with Wright over the years. In 1922, Wright arranged for Lick Observatory to put Phoebe on the mailing list to receive copies of the *Lick Observatory Bulletin* even though she was no longer employed as a professional astronomer. Phoebe was thrilled at the arrangement, which is known to have continued for at least 12 years. When Leuschner put together a list of all UC-Berkeley astronomy PhD graduates, indicating their career paths, he indicated that Phoebe got married and left astronomy. Wright seriously objected to this characterization of Phoebe, because she had kept up with the literature, maintained relationships with many in the astronomy community, and prolifically observed and reported on variable stars for the AAVSO. In 1936, Wright told Leuschner: "I know of no one who has kept alive a keener interest in the science... [Phoebe] has carried her enthusiasms for, and knowledge of astronomy as a powerful cultural influence into her very charming family, and to her friends. In doing so it seems to me that she has fulfilled and is fulfilling the purposes of scientific education quite as effectively as though she were actively engaged in professional work."[95,96,97,98,99,100]

The Haas family business prospered from, among other things, introducing Plexiglas to the American market, and in 1944 it had profits of over $43 million (this was over $527 million in 2010 dollars).[101] Phoebe had become

[93]Note: It is not certain whether AAVSO loaned the telescope, or whether Phoebe purchased it at an estate sale.

[94]Thomas R. Williams, "Phoebe Haas – An AAVSO Volunteer," *American Association of Variable Star Observers* 20, (1991), 18.

[95]3/2/1922 letter from Lick Observatory Secretary [unnamed but most likely Miss A.G. Marshall] to Phoebe Waterman Haas.

[96]3/15/1922 handwritten card from Phoebe Waterman Haas to Lick Observatory Secretary Miss A.G. Marshall.

[97]3/23/1922 letter from Lick Observatory Secretary [unnamed but most likely Miss A.G. Marshall] to Phoebe Waterman Haas.

[98]4/18/1934 letter from Lick Observatory Secretary Leslie G. Potwin to Phoebe Waterman Haas.

[99]5/11/1934 letter from [presumably Lick Observatory Director Robert Grant Aitken] to Phoebe Waterman Haas.

[100]Thomas R. Williams, "Phoebe Haas – An AAVSO Volunteer," *American Association of Variable Star Observers* 20, (1991), 18.

[101]Ernie Weiler, "The Haas Story," *Neuer Pennsylvanicher Staatsbote*, 5 (2010).

very wealthy. She gave gifts annually to AAVSO starting in 1941,[102] formed the William Penn Foundation with her husband in 1945,[103] inspired her family to high levels of philanthropy,[104] and over an 11-year period beginning in 1953 contributed her expertise in astronomy to help stabilize AAVSO when its existence became threatened.[105] In 2013, the Smithsonian Institution named a public observatory the Phoebe Waterman Haas Public Observatory at the National Air and Space Museum in Washington, DC, exactly 100 years after Phoebe earned her PhD in astronomy at UC-Berkeley.[106]

---

[102]Thomas R. Williams, "Phoebe Haas – An AAVSO Volunteer," *American Association of Variable Star Observers* 20, (1991), 18.

[103]The William Penn Foundation. Accessed on 12/29/2013 at www.williampennfoundati on.org.

[104]Waldemar A. Nielsen, *Inside American Philanthropy: The Dramas of Donorship* (University of Oklahoma Press, Norman, 1996).

[105]Thomas R. Williams, "Phoebe Haas–An AAVSO Volunteer," *American Association of Variable Star Observers* 20, (1991), 18.

[106]Introducing the Phoebe Waterman Haas Public Observatory. Accessed on 12/29/ 2013 at http://blog.nasm.si.edu/astronomy/introducing-the-phoebe-waterman-haas-public-observatory/.

4

Becoming an Outlier (1932–1939)

The best of my last year's students, an astronomer. I do not know, however, whether she will stay. Her attachment to astronomy seems to be considerable.
-Jerzy Neyman

Author Malcolm Gladwell studied outliers and determinants of success. He pointed out that the usual stories about successful people, which focus on intelligence and ambition, do not tell the whole story. The determinants of success, he argued, are much more complex and depend on the social context in which the person existed, as well as on various serendipities. It is therefore important toward understanding someone's success to uncover their salient contextual advantages and embedded serendipities.[1]

Betty was an outlier. She was a successful woman in science at a top research university during a time when there were almost no such women in science. Betty was certainly intelligent and ambitious, but she also had some hidden advantages and extraordinary timely opportunities that contributed greatly to her success. One of her advantages was her family. As previously discussed, Betty came from a family that was mathematical and fearless, one that had significant international experience and orientations, and one that had high values and expectations for education and leadership.

Betty had other advantages. Her family had the resources and practical intelligence to successfully move to California during the Great Depression. They enrolled Betty in the most scholastically oriented high school in the East Bay, one that was joined with UC at the hip and was highly progressive for the times by, among other things, promoting girls' interests and achievements in science. As an undergraduate at UC-Berkeley in astronomy, Betty had access to an instructional observatory right on the campus that was so well equipped she was able to conduct professional research there. Betty had the opportunity to develop her interests in astronomy within a world class department and using state-of-the-art equipment at the great Lick Observatory.

---

[1]Malcolm Gladwell, *Outliers: The Story of Success* (New York: Little, Brown and Company, 2008).

Serendipitously, one of Betty's academic advisors in astronomy played an important role in recruiting the renowned statistician Jerzy Neyman to UC-Berkeley and championing him to young Betty. Betty showed up in Neyman's class in his first semester of teaching, and Neyman was immediately so impressed with Betty's abilities and performance that he hired her as his research assistant. This is how Betty began her statistics practice. Betty had four publications before she began graduate school. These publications focused squarely on measurement, qualifying Betty as an expert in measurement and making her a lifelong advocate for data that are well measured.

4.1   Move to California

As previously mentioned, the Scott family moved from Oklahoma to California between 1930[2] and 1932.[3] Betty was a teenager, around age 14 or 15. It was the "Dust Bowl." It was at a time when the Oklahoma governor called out the National Guard to halt oil production in an effort to stop falling oil prices. Many Oklahomans moved out of state in the 1930s due to the condition of the state with its mass unemployment, mortgage foreclosures, huge deficits, and bank failures. Many like the Scott family moved to California.

Betty lived in two different residences after moving to California and while in high school. These were used for only a few years and served to transition the family from Oklahoma to California, until they could secure a more permanent residence in Berkeley. Both were located within one and a half miles of the UC-Berkeley campus.[4] Betty would eventually settle down in a home near Claremont. She would spend the rest of her life in Berkeley. Betty's association with the UC began as soon as she moved to Berkeley and her family elected to send her to University High School in Oakland.

4.2   University High School Advantage

University High was a renowned public college-prep school that had an extremely strong academic reputation. It was a progressive alternative school that didn't have its own district. Rather, students from many districts in virtually every city in the East Bay not just elected to go there, but even

---

[2]1930, United States Census of Population and Housing. Note: The Census placed the family in Oklahoma in 1930.

[3]1932 California Voter Registration lists. Note: The lists placed the family in California in 1932.

[4]1932, 1934, and 1936 California Voter Registration lists.

pulled strings because they were scholastically oriented. Betty's classmates were accordingly very bright and motivated. The classes at University High repeatedly tested ahead of classes at the other schools in the state on college entrance examinations. Most of the students went on to college at the UC. The school was called a "beacon of excellence" and enjoyed a national reputation.

Betty's high school was the practice site for the UC School of Education. It was financed and operated under a joint agreement between the UC and Oakland Board of Education; the director of practice teaching at UC was also the principal of University High. UC provided partial support for the school's staff and helped to select them, and the city board elected and paid for the teachers based on recommendations by the UC president. By design, the school was an experimental laboratory for many theories and philosophies of education. Students like Betty learned to "nurture their gifts" while teacher-education students learned to teach under the careful supervision of UC education faculty. Betty and the other students benefited from being schooled in this "living laboratory of education," which was unique in the geographic area in its time.

Betty was extremely fortunate to attend such a progressive school. She had small classes (classrooms permitted only 20 to 25 students), allowing her and other students to have close relationships and stimulating interactions with their teachers. There was a sense of family among the students, student teachers, and instructors. The instructors set a "brisk pace academically," and the students and teachers "were able to establish and maintain an academic momentum seldom experienced in high schools of that era."

It didn't hurt that Betty and the other students went to school every day in a building that was described as a "modern architectural masterpiece." Designed by the same person who designed the landmark Claremont Hotel, the building opened its doors in 1923. It was in Spanish colonial revival style, "tempered by small details that are distinctly American": "What could be more suitable against the background of our rolling hills, dark oaks and eucalyptus than a low-lying, picturesque building with walls of golden, pinkish cream and a roof of red brown tile?" The mansion covered an entire block – 8.9 acres – and had "virtually every amenity one could hope to find in a showcase educational center."[5,6,7] The building, still in existence, is a masterpiece of functional design.

---

[5]Newspaper clipping file, "University High School," Oakland Public Library. Note: University High School was founded in August 1914. It was closed about ten years after Betty's graduation, in 1946 after World War II due to declining enrollment (attributed to changing demographics and competition among districts), bringing the "significant educational experiment" to an end. The building is located at 5714 Martin Luther King Jr. Way in Oakland, CA.

[6]George Arthur Rice, Education: Berkeley, 1882–1962. University of California: In Memoriam, April 1964. Accessed on 12/4/2013 at http://texts.cdlib.org/view?docId=hb6g500784&doc.view=frames&chunk.id=div00016&toc.depth=1&toc.id=.

[7]Jay, Robert. *The Architecture of Charles W. Dickey: Hawaii and California*. University of Hawaii Press.

Before Betty arrived at the school, the method of "teaching through problem situations" had already been introduced. This was group experience, where teachers and students alike would be expected to participate using procedures such as these:

An introductory discussion, ending in the statement of problems.
Planning how to proceed.
Adapting experience and gaining additional information.
Generalizing and organizing a group experience.
Drilling, testing, applying ideas in new fields – a procedure adapted to both individuals and groups.

It is easy to recognize these procedures, which were designed to enable students to find joy in learning, as fundamental to the practice of research. The methods certainly contributed to Betty's development as a lifelong learner and researcher. Betty would adopt this style of problem-based teaching when she became a professor.

4.3   Math Advantage

It is no wonder that Betty liked math and was good at it. All one has to do is look at the school handbook's description of mathematics instruction:

When you visit the mathematics classrooms, we trust you will find that our children "like arithmetic," "like algebra," "like geometry," and that the upper classes like the more advanced elective study that has lured them on ... there is evidence of dexterity with fundamental mathematical skills and evidence of independent ability in problem-solving... we endeavor to "generate brain-power."

Student teachers were asked to look within the mathematics classrooms for how students were challenged on problems that interested them and how they gained satisfaction in learning how to meet those challenges:

Note the process of developing and maintaining fundamental skills: Are they using scientifically organized drill material? Does the drill appear to have a definite objective goal? Is it timed wisely? Is it individual – each pupil keeping a record of his progress? Note the attention given to problem solving. Are the problems real to the student? Is the informational material of vital importance in the world outside of school and introduced either by the students or to the students in a manner to challenge their interest? Can you detect the conscious use of any technique in problem solving?

Between her field artillery family and this mathematics program, it is no wonder Betty was interested in mathematics applied to real world problems, and that she developed real world problems into her teaching.

The school also had two more long-term goals for mathematics learning. One was to develop the "ability to think through a problem clearly." The other was the "ability to use symbols effectively in expressing thought." The handbook went on to justify these goals as follows:

> *The technique of the scientific research worker often resolves into the organizing of data in mathematical form in order to find a symbolic solution of the problem in an equation, formula, or graph. Perhaps in these classes some efficiency engineer is getting his early training; can you trace in class procedure an effort to build the foundations for later technical practice?*

The lessons were of five types: inductive development lessons which involved "the discovery of a general truth as a result of having observed its verification in particular cases"; drills or pure-practice lessons; review lessons; deductive development lessons, which involved leading "the student to the solution of problem situations by referring to principles rules, definitions, axioms, and generalizations already studied"; and appreciation lessons.

Although the school didn't require any mathematics for graduation, Betty took one unit each of elementary algebra and plane geometry as these were required for college entrance. She also took one unit of intermediate algebra, one half unit of trigonometry, and one half unit of solid geometry. Over 40 years later, Betty recalled that when she was in high school, girls were forbidden to take mechanical drawing, and they were also "almost forbidden" to take trigonometry. She recalled being the only girl in her trigonometry class. It was in these classes that Betty got her early training and was able to build a solid foundation for later research practice in astronomy and statistics.

4.4   Science Advantage

Betty was able to get the very best science education because the school emphasized science departments. The infrastructure for the study of science was impressive. There were seven science laboratory-classrooms, three for general science and biology, two for chemistry, one for physics, and one for physiology. These were spread across two stories of the building and situated around a courtyard that contained a large walled outdoor science garden, which served as a lab for the study of open-air science. The science rooms, which allowed for discussion, lab exercises, etc. in the same room, were a departure from tradition where lab rooms and lecture rooms were separate. The laboratory-classrooms had tables with running water, gas, and electricity. They had lockers. It was reported that the equipment in the science laboratory-classrooms would "rival in many ways that of many colleges." With this kind of science infrastructure, and a democratic school atmosphere, Betty was able to make huge strides in male-dominated subjects like mathematics and science.

Betty took one unit of physics. There was a change in the physics classes in spring 1934 when Betty was in her junior year at the school:

> *Girls who elect physics not infrequently find it difficult and uninteresting, because they are very much in the minority, scattered through the day in the several classes, and instruction is designed largely to meet the needs and fit in with the more extensive background of the boys who take this course. This term, after a careful analysis of the schedules of the girls in the physics classes, it was found possible to program all of them in single sections of Physics I and II, in each of which they made up about half the class. There by dint of their numbers their interests are less likely to escape attention and the units of the course are more likely to be planned with a view to their background and preparation than when they were submerged minorities.*

Also, the school organized a Vocation Day that included discussion groups specifically for girls led by physicians, lawyers, architects, and other professionals.[8,9,10,11,12] These changes indicate Betty was in an environment where girls were not being ruled out in science or traditionally high-valued vocations. She emerged from an environment where she could have high expectations for personal achievement at a time when other schools were limiting girls' visions for their futures.

4.5   High School to College

When asked some 45 years later when it was that she got "turned-on" to math/science/engineering, Betty said she always liked these subjects. She

---

[8]University High School: A Handbook. Oakland, California, 1929. Reprinted from the *University High School Journal*, 9(2), 1929. Note: It is assumed that this handbook, which predates Betty's attendance by three years, is generally consistent with the ones for the years Betty attended the school.

[9]6/24/1978 letter from Betty Scott to UC-Berkeley College of Engineering Dean A.M. Hopkin.

[10]Hoge, James W. The Department of Mathematics: A Handbook. *University High School Journal* 13(3):143-183, June, 1934.

[11]Conrad, Clinton C. University High School – 1914–1934: A Study in Growth and Development. *University High School Journal* 13(3):105-125, June, 1934.

[12]Elizabeth L. Scott undergraduate transcript, UC-Berkeley, 1939.

would say she had general abilities, i.e., she had no special capabilities as a child or adolescent that reinforced her interest in these subjects.[13]

Betty was an Honor Student and on the Honor Board. In addition to the already-mentioned four units of mathematics and one unit of physics, Betty took three units of history, four units of English, three units of Latin, and three units of electives. She was very active in extracurricular activities: She was a Member of the Girls' League Council and Class Councils, Commissioner of Publicity for the H 12 Class, Secretary of the Latin Club, Member of the Rainbow Club, G.A.A., and Debating Society, and worked on the Sales Committee, Dramatic Night, and Bulletin Board.[14] Upon graduation in 1935, Betty became one of thousands of University High alumni who emerged as civic leaders and distinguished citizens.

Immediately after graduation, Betty entered UC-Berkeley. It was a logical choice to stay at home to attend college. The Great Depression was still on and tuition was low. Also importantly, Betty was already oriented to the UC and sold on its excellence given its relationship to University High.

Serendipitously, as Betty would choose an astronomy major, UC-Berkeley was the third largest producer of baccalaureate degrees for women in astronomy in the US between 1900 and 1940, exceeded only by Vassar College and Radcliffe College. Berkeley was also the third largest producer of master's degrees, and the largest producer of PhDs, for women in astronomy.[15] The Berkeley astronomy community was generally welcoming to women.

In 1935, Betty was already a member of two scholarly societies. One was Sigma Xi, The Scientific Research Society. Founded in 1886, Sigma Xi's mission is "to honor excellence in scientific investigation and encourage a sense of companionship and cooperation among researchers in all fields of science and engineering."[16] The other was Pi Mu Epsilon, Honor Society in Mathematics. Founded in 1914, Pi Mu Epsilon's mission is "to promote scholarly activity in mathematics among the students in academic institutions."[17]

---

[13] 6/10/1978 letter from Ms. Mina Edelston, attendee at the UC-Berkeley Conference for Women in Engineering and Computer Science on 5/13/1978, to Betty Scott with attached questionnaire completed by Scott.

[14] "Senior Histories: Elizabeth Scott," *Senior Souvenir* (University High School), 1935, 21.

[15] John Lankford, *American Astronomy: Community, Careers, and Power, 1859–1940* (Chicago: University of Chicago Press, 1997), 312, 321.

[16] Sigma Xi. Accessed on 2/8/2013 at http://www.sigmaxi.org/about/overview/index.shtml.

[17] Pi Mu Epsilon. Accessed on 2/8/2013 at http://www.pme-math.org/.

4.6   Tunnel Road House

Between 1936 and 1938, when Betty was in college and around 20 years old, the Scott family purchased a large house at 34 Tunnel Road in Berkeley.[18] It was a two-story stucco house with Tudor accents, built in 1921 with four bedrooms, four bathrooms, and 2,824 square feet of living space. Betty lived in this house for the remaining 50 years of her life. She hosted many gatherings there. Betty's mother would sometimes come down and sit at these gatherings. It is hard to remember Betty without remembering her presence at the well-attended gatherings that occurred at the Tunnel Road house.

Betty's Tunnel Road house was in a prime location, right across the street from the famous historic, romantic, castle-like Claremont Hotel, which has...

> ... *a picturesque quality reminiscent of the Age of Romanticism. In the 1930s, the entire second floor was flanked by a large porch where visitors would sit, take walks and admire the surroundings and spectacular sunsets. Some will remember The Claremont's Garden Room "high atop the hill" where such famous bands as Count Basie, Louis Armstrong and Tommy Dorsey performed. During this period, lawn sports such as tennis, badminton and croquet were gaining popularity and the resort's first tennis courts and pool were built.*[19]

Betty could look across the park-like grounds and have a direct frontal view of the grand hotel.

One might wonder why the family chose a residence on busy Tunnel Road, now a major commuter route. In the late 1930s this stately house was in a quiet neighborhood in a prime location in terms of transportation. The Key System was in place, and Claremont was one of the main arteries of this transit system which was a predecessor to the current BART system. Using the Key System, Betty and her family could easily travel on commuter train from a terminal between the tennis courts at the Claremont Hotel to other communities in the East Bay and via a ferry connection to San Francisco.

Many important dinners took place at Betty's house over the years, such as the "40-yr Hoopla" celebration of Neyman's 40th anniversary at UC-Berkeley.[20] According to Juliet Shaffer, "Betty had big parties at her house – there was lots of food – she was very generous that way... She had a fairly big living room and kitchen. There was always someone helping in the kitchen.

---

[18]1936 and 1938 California Voter Registration lists. Note: The lists first placed the family at the Tunnel Road house in 1938.

[19]The Claremont Resort and Spa. History: Born of Romanticism. Accessed on 12/6/2013 at http://www.claremontresort.com/history-romanticism.shtml. Note: The Claremont Hotel is at 41 Tunnel Road.

[20]8/11/78, Jerzy Neyman's Calendar, Academic Year 1977–78.

She was always too busy at the parties: I didn't have too much interaction with her at the parties because she was the hostess and busy being hostess." [21]

Graduate students felt like it was a huge honor to be asked to Betty's home. Women graduate students felt like it was very special to be asked to a social event hosted by a woman professor, as women professors were scarce. Attendees were given a copy of a hand-drawn map to the home to make sure they didn't get lost. Parking was difficult, and so attendees were instructed to park across the street in an overflow hotel parking lot. Betty was a highly observant, yet reserved, hostess. She was quietly in charge.

4.7 Neyman Serendipity

One of Betty's undergraduate academic advisors was Astronomy Professor C. Donald Shane. Shane had substantial interests in mathematics in addition to astronomy. In fact, his first faculty appointment was as an instructor in mathematics at UC-Berkeley. Four years later he moved to the astronomy department, where he spent the rest of his career and also for a time served as director of the Lick Observatory.

In 1937, Shane took the initial step to bring Jerzy Neyman to Berkeley. Shane had been serving on the university budget committee (described by Betty as "a high-powered committee") and called its attention to the fact that there were about 14, or maybe 17, lower-level statistics courses in about as many different departments. They ranged from statistical astronomy (which, according to Betty, "is not really statistics as such") to zoology. The campus needed a professor who could teach statistics at a higher level.

Shane convinced the budget committee to recommend to the UC president that an assistant professor of mathematics be hired to teach higher-level statistics courses, improve the quality of teaching of lower-level statistics courses, and reduce proliferation of statistics courses. Shane was appointed to a search committee with Physics Professor Raymond T. Birge as chair. Birge "quickly came up with Neyman." [22]

Neyman was recruited out of a faculty position at University College London where he had a renowned collaborative relationship with statistician Egon Pearson. Now Neyman had to decide whether to separate geographically from this collaboration and other work. He had to decide whether to come to America and to Berkeley in particular. It helped that he had just the year before given short courses at ten universities across the US. He had tasted America, and he liked it. Neyman in the winter of 1937 received "satisfactory" offers from both the University of Michigan-Ann Arbor and UC-Berkeley. He ended

---

[21] Juliet Shaffer, interview with author, August 3, 2014.
[22] 6/9/1978 commencement address by Betty Scott.

up going to Berkeley. Neyman arrived at Berkeley to begin teaching in the fall of 1938. Shane saw Neyman's arrival as a real opportunity for Betty. As few astronomers had any real knowledge of statistics, Shane advised Betty to connect with Neyman. Betty took Shane's advice, and this is how her own, in Shane's words, "very successful career" began.[23]

Betty met Neyman just after he arrived to his professorship in the mathematics department at UC-Berkeley. It was his first semester of teaching, in fall 1938. Neyman was a 44-year-old statistician with a large international reputation. Betty was a 20-year-old first semester undergraduate senior at Berkeley. She was a honors student studying astronomy.

Betty's undergraduate coursework included mostly astronomy, physics, and mathematics, with some languages including French and German. She had taken the first semester of Math 120, The Theory of Probability, which was an upper division course, in fall of 1937. The course, in Betty's words, emphasized "elementary problems about who will win a tennis match under this condition or that condition, from the worst instructor I ever had; he did all the things we tell our Teaching Assistants not to do – he mumbled, he faced the blackboard and wrote in front of his chest, quickly erasing before there was any hope of seeing anything." Betty got a "B" in the class. Because of this instructor, she did not sign up for the second semester of the course in the spring. She took another astronomy course instead.

When Neyman arrived, he took over the teaching of Math 120. Since Betty had taken Math 120A the previous year, she was not eligible to enroll in it again for credit, "in spite of its many vicissitudes." But she could audit the course. Since she already knew some of the material, she waited a few months to attend the class. Some of her astronomy mates were enrolled in the class, and they told Betty that some new material was coming up.

> *I went to class – Professor Neyman came in and asked a question, addressing it to a student near the front. The student answered 'I don't know,' the next student said 'I don't know' and so they continued until the finger pointed at me. I said 'I don't know.' [Neyman:] Why don't you know? [Betty:] This is the first time I've come to class. [Neyman:] Why haven't you come to class – you should attend classes. [Scott:] I'm auditing because I've already... [Neyman:] There is no point in auditing unless you're a serious student...*

This was the first interaction between Betty and Neyman.[24]

After that, Betty attended the class regularly and even prepared for the course examination. Neyman's instructions were to "(a) Better attempt fewer questions, but answer them properly. (b) Write clearly. (c) The passing grade can be obtained for satisfactory answers to four questions. (d) The grade of A can be obtained for good answers to six questions." Then came nine problems,

---

[23]8/8/1981 handwritten letter from former Lick Observatory Director Donald [Shane] to Betty Scott.

[24]6/9/1978 commencement address by Betty Scott.

including some definitions, conditions, formulas, descriptions, proofs, and calculations.[25] Betty saved the exam and would later emulate the Neyman style in her own courses when she became a professor.

Betty was intellectually taken. It was a no-brainer that she would enroll in the second semester of the course from Neyman in the spring semester of 1939. This was the last semester of Betty's senior year. She satisfied the requirements of Neyman's midterm examination by writing clearly and providing satisfactory answers to at least three of the five questions provided by Neyman, one of which had to do with proving a Markov theorem.[26] On her final examination, she again wrote clearly and provided satisfactory answers to at least two of the five questions provided by Neyman.[27] Actually, her answers were more than satisfactory: They were excellent. With a more than competent instructor, Betty earned an "A" in the class.

Betty published early, in 1939, while still an undergraduate. It happened while she was taking Professor R.T. Crawford's theoretical astronomy course. The students in the course spent a good deal of time at Student's Observatory. Betty found that the observatory was being used for both instruction and professional research, especially the determination of orbits of newly discovered comets. This is exactly the kind of research that led to Betty's first publication. Together at Student's Observatory, Betty and two fellow students calculated orbital elements and ephemerides of the Cosik-Peltier Comet. This was the first comet of the year in 1939.[28] The three students published the measurements in the Harvard College Observatory Announcement (HAC) series.[29] Betty was only 22 years old.

Discoveries in astronomy fall into one of two categories that depend on how fast they change over time. The first category includes objects that change little over time: most asteroids, dwarf novae, eclipsing binary systems, and pulsars. The second category includes objects that change unpredictably, quickly, and often suddenly over time: comets, near-Earth asteroids, new stars, planetary cloud and surface features, supernovae, and "similar matters." The HACs were set up for the quick and economical reporting of transient characteristics such as those that fall into the second category.[30] Astronomers could send their recordings directly to the Harvard Bureau of Astronomical Telegrams, where they would be archived and communicated to a broad group of interested professionals in the Western Hemisphere.[31]

---

[25]December 1938 Course Examination Mathematics 120A – The Theory of Probability, Instructor in charge, Mr. Neyman.

[26]n.d. [Spring 1939]. Mathematics 120B. Mid-term examination. [Mr. Neyman]

[27][Spring] 1939. Final Examination. Course 120B. [Mr. Neyman]

[28]R.T. Crawford, "Note on the Cosik-Peltier Comet (a 1939)," *Publications of the Astronomical Society of the Pacific* 51, no. 300, 120.

[29]Bartlett T.J., Panofsky H.A.A., and Scott E.L. Comet Cosik-Peltier, elements and ephemeris. Harvard College Observatory Announcements, No. 471, 1939.

[30]Shapley H. Harvard College Observatory Announcement Card 1, March 12, 1926.

[31]Note: In 1965, Harvard Observatory became the hub for astronomical information reporting worldwide.

The first HAC was issued in March 1926, 13 years before Betty published in the series. From 1926 through 1964, the HACs were an indispensable communication tool for western astronomers. Astronomer Dorrit Hoffleit (1907–2007) captured the importance of the HACs in the following excerpt from her poem, "Minkowski's 'Nova'":

> *Minkowski found a "nova"*
> *On a plate of long ago,*
> *So at once he wrote to Harvard*
> *To have us check that this was so.*
> *Sixth magnitude he'd found the star*
> *Near Orion in Gemini.*
> *A one so bright in such a place,*
> *Where all good amateurs do look,*
> *Could hardly have escaped their watch.*
> *Yet no Harvard Announcement Card*
> *Ever told of its discovery.*
> *So something drastic must be wrong![32]*

A total of 1,674 cards were published over the 39 years. Every card contained the date of publication and name of the editor. The typical card contained the names of the people who computed the ephemerides and where they worked, dates on which the observations were made, and professor or other professional who oversaw the observations. Typical measurements included common ephemeris (position values) such as Right Ascension and Declination. These gave astronomers a general location of the object within the Celestial Sphere, a spherical coordinate system with the Earth at its center. The communications were through the US Postal Service on standard 3-1/2 x 5" postal cards that had the postage (1¢ in Betty's time) pre-printed on them. Cards were only created and circulated when events needed to be reported.[33]

Together with others and also under the general supervision of Crawford, Betty published measurements on two more comets in the HAC series before she entered graduate school in the fall of 1939. Calculations included orbital elements and ephemeris of Hassel's Comet and of Comet c 1939 (Vaisala).[34] Betty and her student collaborators also published a note on the Vaisala Comet in the *Journal of the Astronomical Society of the Pacific*. Other astronomers had previously calculated the orbit of Vaisala under the assumption of a ten-year period. Betty and her collaborators had accurate observations

[32] Hoffleit, Dorrit. *Misfortunes as Blessings in Disguise: The Story of My Life*. Cambridge: The American Association of Variable Star Observers, 2002.

[33] Amanda L. Golbeck and Dan G. Molgaard, "A historical note on the Harvard College Observatory Announcement Cards: Elizabeth L. Scott at the intersection of statistics and astronomy," *Proceedings of the 59th World Statistics Congress*, (2013), 5297.

[34] Kaster K.P., Bartlett T.J., Scott E.L., and White R. Comet Hassel, elements and ephemeris. Harvard College Observatory Announcements, No. 480 (and Union Astrom. Intern. Circ., No. 762), 1939. Bartlett T.J., Panofsky H.A.A., and Scott E.L. Comet Vaisala, elements and ephemeris. Harvard College Observatory Announcements, No. 482, 1939.

over a fairly long arc. They were able to use these to compute an orbit without making any assumption about length of period. The note contained the orbit and concluded the period was slightly less than 11 years.[35]

4.8 Klumpke Prize and Graduation

At the end of her senior year, Betty was awarded the Dorothea Klumpke Roberts Prize in Astronomy for 1938–39. Klumpke, a pioneer woman astronomer, established the prize in memory of her mother, father, and husband. It was given "For outstanding scholarly achievement by either an upper-level undergraduate student, or an outstanding graduate student early in his/her research career."[36] This cash prize of $80.00 (about $1,225 in 2010 dollars)[37] was a good sum of money for an undergraduate to receive in 1938.

With Klumpke prize in hand, Betty graduated from UC-Berkeley in astronomy on May 20, 1939. In all, she took 33 units of math, 29 units of astronomy, 17 units of physics, 12 units of French, ten units of chemistry, eight units of German, six units of political sciences, three units of English, and two units of psychology. Her grades were mostly A's and B's, and she graduated with honors. Betty would stay on at UC-Berkeley, entering graduate status in astronomy in August of 1939, right after completing her bachelor's degree.[38]

In July of 1939, the summer immediately before Betty began graduate school, Neyman found money in his new budget to hire Betty. He described Betty to Egon Pearson as "...the best of my last year's students, an astronomer." Neyman was already thinking about Betty over the long term: "I do not know, however, whether she will stay. Her attachment to astronomy seems to be considerable."[39] Betty would over the next few years, in fact, devote herself with equal attachment to both astronomy and Neyman's laboratory work. Their intellectual bond would be strong.[40]

---

[35]Bartlett T.J., Scott E.L., and Panofsky H.A.A. Note on periodic comet Vaisala. *Publications of the Astronomical Society of the Pacific* 51:173, 1939.

[36]Student Awards and Prizes, University of California Berkeley Astronomy Department. Accessed 3/22/12 at http://astro.berkeley.edu/academics/awards.html.

[37]5/5/1939 letter from UC-Berkeley Registrar Thomas B. Steel to Betty Scott. Attachment: University of California Prizes 1938–39, Berkeley, CA, September 1938.

[38]Elizabeth Leonard Scott. Undergraduate transcript, graduation A.B. May 20, 1939. UC-Berkeley.

[39]Reid C. Neyman: From Life. Springer-Verlag, 1982, p. 166. Note: On her 1939 half-page typed biography, Betty lists her occupation as Research Assistant, Experimental Biology.

[40]1/17/1947 letter from Betty Scott to Smith College Astronomy Professor Marjorie Williams re a possible teaching position at Smith College.

5

Ten Thousand Hours of Practice (1939–1946)

...ten thousand hours of practice is required to achieve the level of mastery associated with being a world-class expert – in anything.
-Daniel Levitin

A basic formula for success is: Success equals a measure of innate talent plus many hours of preparation. For example, the Beatles were innately talented musicians, but the way they became successful was by perfecting their music, playing live all night seven days a week in Hamburg. Likewise, Betty was an innately talented scientist, but she perfected her science by working long hours on projects at the Student's Observatory, Lick Observatory, and Stat Lab. She put in the required 10,000 hours of practice necessary to achieve the status of world class expert in astronomy and statistics.

Betty was a hard worker. In graduate school, she often held two jobs simultaneously. She did this by working every day, evening, and weekend. She had a long list of positions: research assistant in mathematical statistics from 1939–1941; volunteer assistant at Lick Observatory in summer 1939; volunteer assistant at Mount Wilson Observatory in summer 1941; and research statistician under the National Defense Research Committee in spring 1942 through fall 1945. While she was working full time for the war effort, she also worked as a teaching assistant in astronomy. She took this position in 1941, and it continued to 1946, but she did it as a volunteer from 1942–1944 while she had graduate fellowships. She was a teaching assistant in mathematics for a semester in 1944 and in 1946. Beginning in 1946, she was both an associate in astronomy and an instructor in the Extension Division.[1]

Back in summer 1939 before beginning graduate school, Betty was an assistant at Lick Observatory in addition to working for Neyman. The UC-Berkeley astronomy department assigned students to assist various professional astronomers. Betty was assigned to assist Joseph Moore[2] until July

---

[1]1/17/1947 letter from Betty Scott to Smith College Astronomy Professor Marjorie Williams.

[2]Note: Joseph Haines Moore (1878–1949) was an astronomer whose specialty was the measurement of stellar radial velocities (rate of approach or recession of an object) by using

29, Edward Fath[3] from August 1–25, and Ferdinand Neubauer[4] at other times.[5] Moore was assistant director at Lick Observatory and an expert in measurement of stellar radial velocities using a photographic spectroscope. While assisting Moore, Betty measured spectrographic plates of the star α Geminorum.[6] Although it was volunteer work, Lick did pay some of Betty's expenses. For July, she was reimbursed for ten postcards (25¢), a map (15¢), ice cream ($1.70), electricity (27¢), and stage freight (27¢). For August, she was reimbursed for postage (35¢), electricity (11¢), and a stage ticket ($1.50).[7]

That summer, Betty and a fellow student computed and published a set of elements and ephemeris measurements on Comet Rigollet.[8]

5.1   Year One – Getting up the Mountain

In fall 1939, Betty began taking graduate courses at UC-Berkeley. She took a two-semester course, Advanced Topics in Probability and Statistics, co-taught by Neyman and George Pólya.[9] In the first semester, she learned about testing statistical hypotheses. Neyman had already developed confidence interval methods and introduced them to the class in the middle of the second semester. Betty also took a two-semester course, Statistical Astronomy, taught by Robert Trumpler.[10] Trumpler, who became known for his work on star clusters, was making long-term plans to develop a book on statistical astronomy based on his course.[11]

---

a photographic spectroscope (spectrograph). He went to Lick Observatory in 1903, was promoted to Director in 1942, and transferred to a faculty position at Berkeley in 1945.

[3]Note: Edward Arthur Fath (1880–1950) was a German-born astronomer who held faculty positions at a number of American colleges and universities. He graduated with a PhD in astronomy from UC-Berkeley and over his career did much of his observing at Lick Observatory. He made important discoveries about spiral nebulae and light spectra.

[4]Note: Ferdinand Johannes Neubauer (1886–1952) was an Austrian-born astronomer who earned a PhD at the UC. His specialty was stellar radial velocities. He went to Lick Observatory in 1922, was in charge of the Lick Observatiory station in Santiago, Chile from 1923 to 1929, and then returned to Lick to become an associate astronomer and the observatory librarian.

[5][summer] 1939 list from the UC-Berkeley Department of Astronomy.

[6]Note: Alpha Geminorum is also known as Alpha Gem or Castor. It is one of the brightest nighttime stars and the second brightest in the constellation Gemini.

[7]8/31/1939 receipt from Lick Observatory to Betty Scott.

[8]Panofsky H.A.A. and Scott E.L. Comet Rigollet, elements and ephemeris.*Harvard College Observatory Announcements*, No. 500, 1939.

[9]Note: George Pólya (1887–1985) was a Hungarian mathematician with expertise in combinatorics, number theory, numerical analysis, and probability theory. He was a professor at ETH Zurich and then Stanford University. Pólya cared deeply about teaching and understanding mathematics.

[10]Note: Robert Julius Trumpler (1886–1956) was formerly a Lick Observatory astronomer. He took a faculty position at Berkeley the year before in order to accommodate the schooling of his five children, who could only be schooled on Mount Hamilton through the eighth grade.

[11]Scott, Elizabeth Leonard. University of California Graduate transcript, 1949. Note: The course numbers were Math 260AB and Astro 218AB.

That year, Betty and fellow graduate students calculated and published elements and an ephemeris of the faint, diffuse Comet Friend.[12,13]

As Christmas break approached, Betty looked forward to time away from formal coursework. She wanted to use her time wisely and do a working vacation. Betty wrote to Moore. Did he have a chance to get some spectrographic plates of α Geminorum? If so, could she come to Lick during her break to measure these plates? Could she come on weekends so her work wouldn't interfere with the computing work she was doing for Neyman at the Stat Lab? Was Moore able to get several plates on the same night, as he suggested last summer, so she could determine if there was any short period variation?[14]

Moore replied that what he had were for some plates of γ Geminorum. Betty was welcome to come up to Lick whenever was best for her.[15] Over the next two weeks, Moore managed to pull together three sets of plates: three plates from November 2 covering one hour and 10 minutes; three plates from December 29 covering one hour and 16 minutes; and four plates from January 2 covering two hours. Moore had hoped for more, and that he could have had sets covering longer periods of, say, six to eight hours. But there had been bad weather and he wasn't having any luck. He pledged to keep trying.[16] Two weeks later, Moore did manage to get a better span of plates. He now had 16 plates from January 18 covering seven hours, and they were of higher quality. He would make these available to Betty at her convenience.[17]

Twenty-three-year-old Betty lived with her mother, who consequently had a say about Betty's plans to drive up to Lick Observatory to measure the plates. It is unclear what Betty's mother thought about the nature of the road, with its 365 curves and many sharp turns (four hairpins) in the last seven miles of the 4,209-foot ascent. Having spent a childhood on the frontier, perhaps she didn't think much of it. But her mother didn't think Betty had enough driving experience to make such a long trip by herself, and she didn't think Betty would know what to do if something went wrong.

In order to appease her mother, Betty asked Phoebe Neubauer, the eldest daughter of one of the Lick Observatory associate astronomers, to make the trip with her. Neubauer agreed. The weekend of February 10, 1940, was the first that Neubauer had free, and so that's when Betty proposed to travel.[18] Betty had started the spring semester with two classes on Saturday morning and others on Monday. So she proposed to drive up the mountain on Saturday

---

[12]Crawford R.T. Note on Comet Friend, 1939n. *Publications of the Astronomical Society of the Pacific* 51(304):360, 1939.

[13]Fell P.E., Scott E.L., White R., Irwin J.H.B., and Panofsky H.A.A. Comet Friend, elements and ephemeris. *Harvard College Observatory Announcements*, No. 512 (and Union Astronom. Intern. Circ., No. 799), 1939.

[14]12/1939 letter from Betty Scott to Lick Observatory Assistant Director Dr. J.H. Moore.

[15]12/16/1939 letter from Lick Observatory Assistant Director J.H. Moore to Betty Scott.

[16]1/3/1940 letter from Lick Observatory Assistant Director JHM [JH Moore] to Betty Scott.

[17]1/24/1940 letter from Lick Observatory Assistant Director JHM [J.H. Moore] to Betty Scott.

[18]1/30/1940 letter from Betty Scott to Lick Observatory Assistant Director J.H. Moore.

afternoon, drive back down on Sunday afternoon or evening, and eat at the Lick boarding house. Betty told Moore she and Neubauer were "not going to do any pushing on the Mount Hamilton Road – I hope, anyway."[19]

Betty apologized to Moore for not greeting his wife the last time on the mountain. It had been hectic. Betty usually was careful, comparing her most recent set of measurements with those she had done previously, to see how well they agreed. That time she was in such a hurry to finish measuring her last plate that she didn't have time to make the comparison.[20]

Now back on the mountain, Betty worked steadily over the weekend. Even so, she didn't finish measuring all of the γ Geminorum plates. She had seven left to do, so she proposed to return in the middle of March, on a Sunday. She thought she could finish them in part of a day. Betty had wanted to go up the mountain sooner. But her mother was being "most uncooperative" and wouldn't let Betty go by herself. So in March Betty persuaded one of her little brothers to accompany her. If Moore could just leave the plates and the star's card on a table in his office, then Betty could get them without troubling anyone. Also, she wouldn't be troubling anyone about meals this time.[21]

Two months later, in the middle of May, Betty again asked Moore if he had been able to obtain any more plates or information on the star. She was starting a two-week "vacation" and could come up to Lick again to continue, and perhaps finish, her work on the star.[22] But the weather had been bad, and Moore had not been able to obtain any more plates of the star that spring. He offered that he would try again in the fall.

Betty hadn't been able to find a period that fit the velocities she had measured on γ Geminorum. Moore suggested she look at those published by others, and try to find one that would fit all of them: "This would be well worthwhile if you have the time for it."[23]

Also in mid May, Betty had a statistical idea she wanted to test:

> *In connection with the testing distributions that I am doing here I should also like to see if the last digits of my screw readings are randomly distributed in a rectangular distribution (that is, if equal numbers of 1's, 2's, 3's... occur along, in series, etc.). I think they should be with a large number of measurements. I having [sic] been doing this here with ordinary tables of random numbers, but it would be more interesting to do it for something real for if I find that I use too many 6's and not enough 7's it would mean that my measurements are much less accurate than I realize.*[24]

[19] 2/5/1940 letter from Betty Scott to Lick Observatory Staff Member Leslie [Potwin].

[20] 1/30/1940 letter from Betty Scott to Lick Observatory Assistant Director J.H. Moore.

[21] 3/13/1940 letter from Betty Scott to Lick Observatory Assistant Director J.H. Moore.

[22] 5/17/1940 letter from Betty Scott to Lick Observatory Assistant Director J.H. Moore.

[23] 5/20/1940 letter from Lick Observatory Assistant Director [J.H. Moore] to Betty Scott.

[24] 5/17/1940 letter from Betty Scott to Lick Observatory Assistant Director J.H. Moore.

Betty was thinking like a statistician. She was worrying about bias, or systematic error, in her astronomical measurements.

5.2 Year Two – Summer at Mount Wilson

The only course Betty took in the 1940–1941 academic year was Shane's two-semester graduate astrophysics course.[25] The course attracted some of the finest young talent in physics.

Betty calculated and published the orbit and ephemeris of Comet Okabayasi.[26] In another publication, she reported the computation of a preliminary orbit of Comet Okabayasi. This comet was moving through the Leo and into the Great Bear constellations. Betty hypothesized it was identical to Comet 1926, Comet Blathwayt. She planned to use observations over a longer arc in order to examine the hypothesis.[27] Then Betty and a fellow graduate student computed and published the improved elements and ephemeris. By then the comet was called Comet Okabayasi-Honda.[28]

After spring semester 1941, Betty again turned her attention to γ Geminorum. She preferred a Sunday, but almost any day would work for her to go back up Mount Hamilton to measure more plates. This time Betty would be spared the drive up Mount Hamilton because, as fortune would have it, the Lick staff was taking the Hartmann Spectrocomparator down to their office in the flatlands of San Jose, and the instrument would be there for two or three weeks. Betty could use it there to measure and compare the plates.[29] Moore was also happy to bring down for Betty anything having to do with γ Geminorum: spectrograms, cards, books, etc. In fact, Moore had four more spectrograms for her to examine, from two nights in April. He suggested Betty could use them to examine the probable period of the star.[30]

Next Betty headed to Southern California, to Mount Wilson Observatory. This must have been very exciting for her because the largest telescope in the world at the time, the Hooker 100", was at Mount Wilson. Many major discoveries in astronomy had already been made using this telescope. Only the year before, Edwin Hubble's work there had resulted in his being awarded the

---

[25]Scott, Elizabeth Leonard. University of California Graduate transcript, 1949.

[26]Scott E.L. Comet Okabayasi, elements and ephemeris. *Harvard College Observatory Announcements*, No. 540 (and *Union Astronom. Intern. Circ.*, No. 826), 1940.

[27]Scott E.L. Note on Comet Okabayasi (1940 e). *Publications of the Astronomical Society of the Pacific* 53:34, 1941.

[28]Scott E.L. and Panofsky H.A.A. Comet Okabayasi-Honda, improved elements and ephemeris. *Harvard College Observatory Announcements*, No. 571 (and *Union Astronom. Intern. Circ.*, No. 847), 1941.

[29]5/17/1941 letter from Betty Scott to Lick Observatory Assistant Director J.H. Moore.

[30]5/22/1941 letter from Lick Observatory Assistant Director [Dr. J.H. Moore] to Betty Scott.

Gold Medal of the Royal Astronomical Society.[31] Betty spent two months at Mount Wilson that summer. According to Director Walter S. Adams,[32] Betty assisted "in several investigations in stellar statistics."[33]

5.3   Year Three – Beginning War Work

Betty took a heavier graduate course load in the 1941–42 academic year. She took a two-semester course, Celestial Mechanics, from Shane. In fall 1941, she additionally took Differential Equations, Special Topics in Mathematics, and Spectroscopy. In the spring of 1942, she additionally took Orbits.[34]

Betty published two more sets of calculations. One was the elements and ephemerides of Comet du Toit-Neujmin computed with a fellow graduate student. The other corrected two elements presented in an earlier publication.[35]

By the beginning of 1942, World War II was in full swing. A major controversy was the proper approach to aerial bombardment of industrialized nations, specifically Germany and Japan. Data were required to evaluate the effectiveness of air power on aerial targets, and statisticians were needed as analysts. Some 400 operations research analysts or statisticians across the country provided this expertise to the war effort.

Princeton University had a National Defense Research Committee contract to conduct bombing research for the Air Force, and Neyman was offered a subcontract.[36] The Stat Lab would prepare probability tables upon which bombing policy would be based. The US Army leadership considered the work to be of "very great importance" and "extremely urgent."

Right after securing his subcontract in March 1942, Neyman hired 25-year-old Betty to work with him on the Army ballistics research in the Stat Lab. Betty joined the war effort without hesitation.[37] This was the plan: She would

---

[31] Note: Edwin Powell Hubble (1889–1953) was one of the most important observational cosmologists of the 20th century. He was influential in establishing the field of extragalactic astronomy, and his work helped convince most astronomers that the universe is comprised of many galaxies. The Hubble telescope was named after him.

[32] Note: Walter Sydney Adams (1876–1956) was a Syria-born astronomer and expert in stellar spectra. He was appointed Director of Mount Wilson Observatories in 1923.

[33] Adams, Walter S. Director's Report. Year Book No. 41, Carnegie Institution of Washington, 1941–42, p. 6.

[34] Scott, Elizabeth Leonard. University of California Graduate transcript, 1949.

[35] Scott E.L. and Stahr M.E. Comet du Toit-Neujmin, elements and ephemeris. *Harvard College Observatory Announcements*, No. 597 and 603, 1941.

[36] Note: "The air arm of the United States Army changed its name several times. From 1926 until 1941 it was known as the Army Air Corps; from 1941 until 1947... it was called the United States Army Air Forces (USAAF). It gained its independence from the army in 1947, and became the United States Air Force (USAF)." William W. Ralph, "Improvised destruction: Arnold, LeMay, and the Firebombing of Japan," *War in History* 13, no. 4 (2006), 496.

[37] Reid C. *Neyman: From Life*. Springer-Verlag, 1982, pp. 180–181, 183–186.

continue her teaching assistantship in astronomy, and add the war responsibilities as a research assistant; she would work part time on the war effort with a monthly salary of $45;[38] and this arrangement would continue until June 30, after which Neyman would hire her full time with a salary of $142 a month (almost $2,000 in 2010 dollars).[39] Betty co-supervised the four full-time equivalent "Computers," those who did computations using "ordinary" electric calculators. When she agreed to do work to support the war efforts, this slowed down Betty's PhD work substantially. But with a West Point and field artillery family, Betty had in her blood an above-average understanding of the value of national service toward the greater good.

The war years set a baseline for the Stat Lab in terms of congeniality. At one point in her career there was a banquet for Betty. She wrote: "I was completely taken aback by your wonderful kindness and beautiful gifts at the banquet and want to thank you all most heartily. And I regret that I did not have enough presence of mind to tell you that it is my colleagues who are really accomplishing what gets done." Betty wrote this poem to express her gratitude:

> *An overwhelmed lady named Scott*
> *Owes thanks to the colleagues she's got,*
> *To LeCam and Genelly,*
> *Haynes, Blom and Jeannie,*
> *To all of the Lab, the whole lot!*
>
> *She thanks les savants eminents*
> *Venus au Labo comme le vent,*
> *The warmest we owe it*
> *To the inspiring poet –*
> *Retournez au Labo dans cinq ans!*

Thus Betty thanked the eminent scholars, who came to the lab as the wind![40]

5.4   Year Four – Lick Fellowship

In April 1942, just as her war work was starting, Betty was selected to be a Lick Observatory Fellow in Astronomy for the 1942–43 academic year. Her talent and dedication to astronomy now affirmed, she was ready to enter the dissertation stage of her graduate work. The astronomy department gave only

---

[38]Note: Another source says Betty received $50 a month. A page is missing in the 3/18/1942 letter from Neyman to Sproul. Bancroft Box JN7, NDRC Budget.

[39]3/18/1942 letter from Jerzy Neyman to UC President Dr. Robert Gordon Sproul.

[40]n.d. poem written by Betty Scott.

two of these fellowships each year, with a stipend of $650 each.[41] Betty was pleased, honored, and grateful to receive one of them. But given her war work, she would need to postpone the start of her fellowship.[42]

Betty talked with Neyman. Neyman in turn talked with the dean. The conclusion was reached that the war work was so important that Betty could delay the start of the Lick Fellowship until August. Astronomy Department Chair Shane suggested Betty write to Lick Observatory Director W.H. Wright to work out her schedule at Lick during the fellowship period.

Betty wasn't sure about her dissertation topic, which made it difficult to contemplate her schedule. She did know, however, that there were more courses she wanted to take, and she needed to start studying for her preliminary exam. She proposed to Wright that she stay at Berkeley "for the present if this is possible and convenient to you."[43] Wright congratulated Betty on receiving the fellowship, wished her "a happy and profitable tenure," and immediately endorsed the plan to delay the start of the fellowship. Wright actually expected Betty to stay on the Berkeley campus for at least the beginning of the fellowship. But, once she decided on her thesis topic, she should consult with Shane for advice about whether and when to go up to Lick.

August 1942 came. Betty still wasn't ready to start the Lick Fellowship. She was working long hours on the defense subcontract. Neyman needed her to work for another month on pressing "exploratory" conceptual problems before the burden of the work could shift to the computations stage. After that, Betty thought she would no longer be needed on the defense project and would be ready to start the fellowship. She asked new Lick Observatory Director J.H. Moore if she could please start the fellowship in September instead.[44] Moore understood the importance of Betty's war work and immediately approved her request for the additional one-month delay. In fact, he suggested she delay the start of the fellowship until she could be spared from the defense project work.[45] He also immediately wrote to the UC-Berkeley graduate division dean asking his permission for the delay.[46]

Betty finally entered into the Lick Fellowship on October 1. At the same time, she continued to dedicate a good amount of time (25%) to the war effort. She also taught navigation one night a week to Navy Air Navigation School instructors in Livermore, California.

---

[41]The UC Graduate Division, Northern Section, 1941-1942, Berkeley, CA, November 1940, Fellowships and Graduate Scholarships 1942–1942. Accessed on 12/30/2012 at http://www.archive.org/stream/fellowshipsgra1940caliuoft/fellowshipsgra1940caliuoft_djvu.txt.

[42]9/7/1942 letter from Lick Observatory Director J.H. Moore to UC-Berkeley Graduate Division Dean C.B. Lipman.

[43]4/11/1942 letter from Betty Scott to W.H. Wright.

[44]9/4/1942 letter from Betty Scott to Lick Observatory Director J.H. Moore with cc to UC-Berkeley Graduate Division Dean Lipman.

[45]9/7/1942 letter from Lick Observatory Director J.H. Moore to Betty Scott.

[46]9/7/1942 letter from Lick Observatory Director J.H. Moore to UC-Berkeley Graduate Division Dean C.B. Lipman.

In mid December 1942, Moore contacted Betty and asked for clarification about whether she had started the Lick Fellowship. He asked whether she wanted to continue to make use of it, and whether she wanted to continue with the war work. Moore then presented another option for Betty to consider: She might consider taking a position at Lick Observatory as an assistant for the spring semester. "We should very much like to have you," he wrote, "but of course do not wish to interfere in any way with the war effort."[47]

Betty responded awkwardly to Moore that she didn't know how to answer. She thought her Army Air Forces research problem would be completed, and the navigation course would definitely end in about a month. But lately it looked like the Stat Lab might be "greatly expanded." Now Betty had no idea when her war work would end. She wrote to Moore, declining his kind offer: "I rather feel that I should do whatever I can towards the war effort."[48]

5.5   Year Five – University Fellowship

Immediately following the Lick Fellowship, Betty was awarded a prestigious University Fellowship in Astronomy for the 1943–1944 academic year.[49] Betty signed up for special topics or individualized instruction courses, as she would over the next six years until she completed her dissertation. By the end of 1943, Mathematics Chair Griffith Evans thought that Neyman, Betty, and others in the Stat Lab were contributing most of the new statistical ideas to the war effort.

Immediately after the University Fellowship year, Betty took a formal two-year leave of absence from her graduate program. She continued her war work and dissertation work.[50]

5.6   Year Six – Qualifying Exam

During the leave-of-absence, Betty attended to a few important graduate school matters. By the end of fall semester in 1944, she needed to establish a PhD qualifying examination committee. This examination would test her eligibility for admission to candidacy for the PhD degree. The committee would decide whether the examination would be written or oral or both, and whether it would be given in one or several parts. The committee members and subjects would be chosen based on Betty's interests.

For Betty, the subjects included practical astronomy, celestial mechanics,

---

[47]12/18/1942 letter from Lick Observatory Director J.H. Moore to Betty Scott.

[48]12/29/1942 letter from Betty Scott to Lick Observatory Director J.H. Moore.

[49]Scott, Elizabeth Leonard. University of California Graduate transcript, 1949.

[50]Reid C. *Neyman: From Life*. Springer-Verlag, 1982, pp. 192–193. Note: The requested leave of absence was from fall semester (September) 1944 through fall semester 1946.

astrophysics, introduction to atomic structure, probability, and differential equations. The committee included an academic astronomer, an astronomer from Lick Observatory, a mathematical statistician, a physicist with expertise in statistics, and a mathematician with expertise in astronomy. Professors Crawford, Moore, and Neyman were on Betty's committee, along with Physics Professor Raymond Birge, and Mathematics Professor Thomas Buck.[51,52,53] Over the next year, Betty's relationship with Trumpler grew and he was added to the committee,[54] eventually becoming the official chair.

5.7   Year Seven – Ending War Work

World War II ended in 1945, and Betty's war work was finally over. By the time they were done, the Stat Lab under Neyman's direction and with Betty's participation had worked on an impressive list of problems. Most of the problems had been "messy." By the time the war ended, Betty and others working at the Stat Lab had become experts in practical statistics.

With the end of the war, Neyman was able to turn his attention again to the development of the statistics program in the mathematics department at UC-Berkeley. He had been the only faculty member in statistics, but that was about to change. The 1945–46 UC-Berkeley course catalog indicates that six statisticians were hired to teach six statistics courses in the mathematics department. It was not yet a highly credentialed group, but these individuals were highly talented. Betty was not yet part of the group.[55,56]

---

[51] Note: Raymond Thayer Birge (1887–1990) was a physics professor who was instrumental in development of the field of physical chemistry and who had a strong interest in statistics. He served as chair of the Berkeley physics department.

[52] Note: Thomas Buck (1881–1969) was a mathematics professor. His dissertation was on mathematical astronomy and at Berkeley he taught the mathematics courses of interest to students of astronomy.

[53] 12/21/1944 memorandum from UC-Berkeley Acting Dean of the Graduate Division James P. McBaine to Betty Scott.

[54] 1/5/1945 memorandum from UC-Berkeley Acting Dean of the Graduate Division James P. McBaine to Betty Scott.

[55] Note: Associates in mathematics included Edward William Barankin (1920–1985) who would become a process theorist and stochastic process expert, Erich Leo Lehmann (1917–2009) who would become a statistical inference expert, and Frank Jones Massey Jr. (1919–1995) who would become a biostatistics expert. All had master's degrees and were just beginning their careers. Lecturers in mathematics included Mark Wrede Eudey (unkn.) with an A.B., Evelyn Fix (1904–1965) with an M.A., and Pao-Lu Hsu (1909–1970) with a PhD who would become an expert in multivariate statistics. Lehmann, Massey Jr., Eudey, and Fix were all working on PhD degrees under Neyman.

[56] Mathematics. UC-Berkeley Course Catalog, 1945–46, pp. 302–309. Note: This catalog account differs from the account given in Speed, Terry, Pitman, Jim, and Rice, John. A Brief History of the Statistics Department at the University of California at Berkeley, 26 pages, p. 2 (accessed on 2/23/2014 at http://arxiv.org/ftp/arxiv/papers/1201/1201.6450.pdf); and Reid, Constance. *Neyman: From Life.* Springer, p. 199–200, 1982. The Reid account also includes Barankin, Eudey, Fix, and Hsu; however, Reid does not mention Massey Jr. and also includes John Gurland and Betty.

In 1946, Betty and a fellow student published three more HACs on the Comet Pajdusakova-Rotbart, one including elements and ephemeris, a second including improved elements based on the mean of two positions, and a third including improved ephemeris. The first card included the comment that this comet might possibly be identical to an earlier-reported comet. The second card dismissed the comment because the data did not support it.[57]

These announcements would be Betty's final contributions to the HAC series. The cards had been a fruitful vehicle for Betty to communicate scholarship activity over an eight-year period. In all, Betty published 12 announcements in the series between 1939 and 1946, during her senior year as an undergraduate and while she was taking courses toward her PhD degree and working for Neyman in the Stat Lab. These publications documented her expertise in measurement and prepared her for the next stage of her research career.

5.8   Year Eight – Teaching and Research

In December 1946, Betty was ready to go back up to Lick Observatory to do more work on γ Geminorum. She had noticed that some measurements of the star's radial velocities were "extremely discordant" and wanted to check them. When she showed the measurements to UC-Berkeley astronomer Harold Weaver, he also advised her to check them.

Betty wrote to Shane asking if she could go to Lick over the Christmas break. She would once again need to use the Hartmann.[58] Shane asked Betty if she could postpone her visit until at least January 7, because the person in charge of the dining hall was away and the assistant was taxed with trying to provide services to almost a dozen people. Or perhaps Betty could come up some weekend after classes resumed in January.[59] But Betty was teaching at that time, and she was actively presenting her research at professional meetings, so she had to coordinate her visit to Lick with these activities: She taught on Tuesdays, Wednesdays, Thursdays, and Saturdays; classes would resume after the Christmas break on Thursday, January 9; the week-long semester recess would begin on February 7; and the fourth week of January she had to go to a meeting back east. Betty asked if she could go to Lick on Saturday, January 11, after her class, and work through Monday afternoon. Or alternatively, could she come at the semester recess?

---

[57] Cunningham L.E. and Scott E.L. Comet Pajdusakova-Rotbart, elements and ephemeris. *Harvard College Observatory Announcements*, No. 749 (and *Union Astronom. Intern. Circ.*, No. 1042), 1946. Cunningham L.E. and Scott E.L. Comet Pajdusakova-Rotbart, improved elements and ephemeris. *Harvard College Observatory Announcements*, No. 751 and 752 (and *Union Astronom. Intern. Circ.*, No. 1049), 1946.

[58] 12/26/1946 letter from Betty Scott to Lick Observatory Director C.D. Shane.

[59] 12/30/1946 letter from Lick Observatory Director C.D. Shane to Betty Scott.

Betty decided she would actually prefer to go up in February, either during the recess or on a weekend after registration for spring semester classes. But she was willing to come up before that if there was a more immediate need for the measurements. She was happy to come when Shane thought it would be best.[60] Shane suggested Betty come some time during the recess and told her they would be very glad to see her on the mountain.[61]

5.9   A Symmetric Intellectual Relationship

During graduate school, Betty may have benefited greatly from Neyman's teaching of statistics, but Neyman also benefited greatly from Betty's teaching of astronomy. Even this early on, Betty and Neyman had a symmetric intellectual relationship. In June 1946 Neyman wrote to Professor Otto Struve at the University of Chicago's Yerkes Observatory:

> *I am very grateful to Dr. Chandrasekhar[62] for his reprints, which interest me very much. As frequently occurs, I have overlooked his work for some time, and only recently happened to see one of his papers. This was due to a nice lady, Miss Elizabeth Scott, who combines studies in astronomy with a prolonged and active interest in statistics. Also, during the war we worked together on problems of bombing. Now she approaches the stage of getting a Ph.D. degree meant to be on a statistical thesis in Astronomy. In this connection I regret that your father[63] was as complacent as he was and did not force me to learn more than I did. As it is, I read Smart, Eddington, and scattered papers in Journals which Miss Scott feeds to me. Incidentally, I was delighted to find in your own journal some signs that, on occasion, the Astronomers have temperaments comparable to those of some statisticians (Littleton vs Luyten).*
>
> *Anyway, I read Chandrasekhar. In time I hope to learn from Miss Scott the rudiments of modern astronomy and, if I do, she will surely deserve her degree! (Or do they give degrees in Astronomy for success in teaching this subject to staticians? [sic]) The trouble is that what we call modern statistics was developed under strong pressure on the part of biologists. As a result, there is practically nothing done by us which is directly applicable to problems of astronomy. Also it is quite a job for*

---

[60]1/3/1947 letter from Betty Scott to Lick Observatory Director C.D. Shane.

[61]1/7/1947 letter from Lick Observatory Director C.D. Shane to Betty Scott.

[62]Note: Otto Struve recruited astrophysicist Subrahmanyan Chandrasekhar to the University of Chicago. Chandrasekhar won the Nobel Prize in physics in 1983.

[63]Note: Otto Struve came from a family of Russian astronomers. His father was Ludwig Struve, a professor of astronomy at the University of Kharkov, the same university where Jerzy Neyman was educated as a mathematician. In 1950, Otto Struve took a position at UC-Berkeley as chair of the astronomy department.

me to learn what the problems of real interest to Astronomers are. On reading Smart, I got the impression that the text of the two drift versus allipsoidel hypothesis, using "modern" methods might be of interest. On the other hand Miss Scott protests that, in spite of the recent papers, such as that of Luyten, the two drift hypothesis is about as out of date as the theory of phlogiston and insists on teaching me galactic dynamics and such.[64]

At the same time, Neyman was Betty's coach, mentor, and champion, and created professional opportunities for Betty even before she completed her PhD. For example, Betty was young, and her earliest professional correspondence accordingly lacked the finesse that comes with experience. Neyman prompted Betty to be clear, and to go through proper channels.[65] For example in 1946 Neyman was invited to give an address at an Institute of Mathematical Statistics (IMS) meeting in Atlantic City, New Jersey. Neyman arranged for the chair of the program committee to also invite Betty to participate. Betty would give a 15-minute statement that would supplement Neyman's address. The UC-Berkeley astronomy department saw this as an opportunity for Betty to form a bridge between the astronomy and statistics departments on the campus. And so the astronomy department petitioned the university's committee on research to support Betty's travel to the meeting: "Through her thorough training in both Astronomy and Statistics she is particularly qualified for this."[66]

Not yet done with her PhD, Betty was already acknowledged at UC-Berkeley to be the perfect bridge between astronomy and statistics.

---

[64]6/25/1946 letter from Jerzy Neyman to Professor Otto Struve.

[65]n.d. [11/1946?] handwritten note from Jerzy Neyman to Betty Scott.

[66]11/25/1946 letter from UC-Berkeley Department of Astronomy Chairman S. Einarsson to UC-Berkeley Committee on Research Chairman Professor R.T. Birge. Note: The meeting was scheduled to take place on January 24–26, 1947.

6

A Rising Star
(1947–1954)

You are not likely to get a chance... in a place with the undoubted advantages of Vassar College, and with the possibility of becoming director of the Observatory within a few years... and [becoming] chairman in the course of time.
-Maud Makemson

Betty began doctoral work in 1939 and completed it in 1949. It took ten years due to interference of war. Betty prepared a dissertation with two parts. The first part, *Contribution to the Problem of Selective Identifiability of Spectroscopic Binaries*, was more astronomy. The second part, *Note on Consistent Estimates of the Linear Structural Relation Between Two Variables*, was more mathematical statistics. Trumpler was listed as her official advisor, but it is clear that Neyman acted as co-advisor. Betty actually earned her PhD in astronomy, but the UC-Berkeley statistics department claims Betty as one of its own, listing her among its PhD alumni.[1]

Graduate students in astronomy are required to take certain mathematics courses. At Betty's time, these included differential equations "and some computing... or method of least squares... or probability." Betty's opinion was that it was good to have more, especially if pursuing statistical astronomy, in which case the student should take at least two years of undergraduate level probability and statistics.[2] Betty's own mix of graduate courses was actually heavier in mathematics (41 units) than astronomy (27 units). This was evidence of her growing interest in mathematics even while she was working on a PhD in astronomy, and evidence of support she was given in astronomy to pursue these interests.[3] Upon completing her PhD, Betty's primary academic interest solidified to statistical problems in astronomy.

---

[1] Note: The first 11 PhDs awarded in statistics at Berkeley were between 1945–1949, with Neyman listed as the official advisor on all but Scott's. Accessed on 5/5/2015 at http://statistics.berkeley.edu/people/alumni/phd.

[2] 3/25/1949 letter from Betty Scott to Carleton College Mathematics Professor Kenneth May.

[3] Scott, Elizabeth Leonard. University of California Graduate Transcript, 1949.

6.1 Prospects at Vassar

At the beginning of 1947, two years before Betty completed her dissertation, she began to contemplate various career employment opportunities. She wrote to Professor Marjorie Williams, who was at Smith College Observatory and the newly elected president of the American Association of Variable Star Observers, asking about a permanent teaching position.[4]

Betty also wrote to Vassar College Observatory Director and Astronomy Professor Maud W. Makemson (1891–1977). Makemson was an expert in celestial mechanics and a pioneering researcher in the area of archeoastronomy. She earned her own PhD in astronomy at UC-Berkeley 17 years earlier. Furthermore, Makemson was nurturing good students like Vera C. Rubin, who would go on to formulate the concept of dark matter. Betty told Makemson she would be in Atlantic City and offered to come to Vassar for an interview.[5] A Western Union telegram was delivered to Betty's door as soon as she arrived in Atlantic City, offering her the interview. Her expenses would be paid, including an overnight stay.[6] Betty visited Vassar, and the interview went well.

As soon as Betty returned to Berkeley, she attended to getting a copy of her transcript and biography to Makemson. Betty asked UC-Berkeley Astronomy Chair Sturla Einarsson (1879–1974) and Trumpler for letters of recommendation. But Betty had some questions and concerns. She hoped to receive a travel allowance as part of her offer. She also hoped to receive a higher salary than was discussed at the interview, because she had "quite a few years of experience in teaching and research." Betty concluded the letter of negotiation by saying: "I want to thank you for being so nice to me on Monday. I do hope that I can come to Vassar."[7]

Vassar at the time was experiencing budget uncertainties. Makemson didn't think the astronomy department would be greatly affected, or their faculty reduced to just one person. Their lower division enrollments were "quite satisfactory," even if enrollments in the advanced courses were somewhat small. But a decision on Betty's appointment had to be postponed until the budget problems were ironed out. That would take at least a month.

On top of the budget problems, there was another potential "difficulty." Makemson had been given permission from the administration to recruit an astrophysicist. Makemson thought Betty had a very fine academic record and

---

[4] 1/17/1947 letter from Betty Scott to Smith College Astronomy Professor Marjorie Williams.

[5] 1/18/1947 letter from Betty Scott to Vassar College Observatory Director and Astronomy Professor Maud Makemson.

[6] 1/22/1947 telegram from Vassar College Observatory Director and Astronomy Professor Maud Makemson to Betty Scott.

[7] 2/1/1947 letter from Betty Scott to Vassar College Observatory Director and Astronomy Professor Maud W. Makemson.

had received a very fine recommendation from Trumpler (Einarsson's letter was still in transit). But at the same time Makemson thought Betty was "unfortunately" not an astrophysicist. Makemson wrote to Betty:

> *It is not that I distrust your ability to teach courses in astrophysics, but it is rather a question of where the interest is focused. The astrophysics side of our curriculum is pretty much neglected, only one student is taking The Sun this semester, and there were only three in the first term. The president believes that we should put forth a strong effort to build up astrophysics, and this means establishing relations with the physics department. Can you imagine a physics department which does not at least recommend astronomy?... Obviously it should not be so, and the trend now is to bring departments together in related studies, and this offers us a very real and good opportunity.*

Makemson described this situation to Betty, first as a problem, but then as a potential challenge.

Did Betty want to take up the challenge? Was her interest in astrophysics "keen enough?" Or would she prefer a position in statistics, "which would undoubtedly pay a great deal more?" What would be the lowest salary Betty would accept? The highest that Vassar would likely give a new PhD was $2,500 (about $24,000 in 2010 dollars). Makemson was willing to try to get that for Betty.

Makemson generously considered Betty's interests in addition to the interests of Vassar. She wrote to Betty: "... if you should get a definite and more attractive offer, my personal advice would be not to be influenced by the possibility of a job at Vassar. I should enjoy having you in my department, and I amm [sic] sure that the students would like you, and that you would be an ornament to the department. But if you can do better for yourself, you should by all means do so."[8]

Betty remained interested in the position at Vassar and continued to pursue it. She told Makemson she enjoyed the trip to Vassar and reassured Makemson about the astrophysics issue. Betty indicated she had eight courses in astrophysics, including Astronomy 103AB (using the book: Russell, Dugan and Stewart, *Astronomy*, Ginn and Company, two volumes, 1926 and 1927), Astronomy 117AB (astrophysics), Astronomy 217AB (astrophysics), Physics 111A (atomic structure), and Physics 211A (spectroscopy).

Betty mentioned she was working on the radial velocity of γ Geminorum, commenting that it seemed to have irregular variations. Betty also reported a discussion she had with Einarsson about the astrophysics issue. He had noted that, in Betty's case, statistics was a side issue, and with her preparation Betty could easily have chosen to write a dissertation in astrophysics.[9] In the

---

[8] 2/12/1947 letter from Vassar College Observatory Director and Astronomy Professor Maud Makemson to Betty Scott.

[9] 2/23/1947 letter from Betty Scott to Vassar College Observatory Director and Astronomy Professor Maud Makemson.

process of trying to secure a position for Betty, what Einarsson and Betty did not disclose was that statistics was an *important* side issue, one that would eventually become, with astronomy, a solid dual focus in Betty's career.

In spring 1947, Betty was assigned to teach a course, Introduction to Astronomy, for the UC continuing education program. Her final examination had two parts. Part 1 required the student to answer any four out of six questions. For example: "Name and describe the parts of a comet. Discuss three methods of determining the sun's temperature. Why don't the three methods give precisely the same answer?" Part II required the student to answer any six out of eight. For example: "Describe the history of our calendar, the present calendar, and proposed calendars. Why do we change the calendar? State three proofs that the Earth is a ball."[10]

6.2   Competing Offers

In March 1947, Neyman was offered a position at Columbia University. He used the offer to leverage greater independence for his Stat Lab from the mathematics department and decided to remain at Berkeley. The Office of Naval Research (ONR) had provided the Stat Lab with post-war funding, and there weren't any strings attached. Neyman and his Stat Lab collaborators, including Betty, could work on whatever they wanted. Berkeley statistics under Neyman's leadership had emerged as one of the foremost academic centers for statistics in the US.[11] In the Stat Lab, Betty not only found herself planted in an idyllic intellectual work environment, she also found herself in one of the American centers of the statistics universe.

The letter from Vassar finally came. Betty was offered an appointment as Instructor in Astronomy for the academic year 1947–1948. It came with a $2,500 salary, the highest possible for a new PhD.[12] Betty was honored.

While waiting for Vassar, Betty took Makemson's advice and continued to look elsewhere for an attractive offer. One began to materialize right in her own backyard, at UC-Berkeley. Neyman didn't want Betty to go to Vassar or anywhere else. He was working on a position for her, a research position "closely connected" with the work on her dissertation, one that would allow Betty to finish her dissertation in a more timely fashion: At Vassar she would have a heavy teaching load, especially of elementary level courses, that would likely retard completion of her dissertation.

Betty had already developed a strong research orientation. She was really tempted by the possibility with Neyman at UC-Berkeley. But the Neyman offer was still tentative, and Makemson had been so gracious to Betty. Betty

---

[10] 6/14/1947 Final Examination by Miss Scott.

[11] Constance Reid, *Neyman: From Life* (Berlin: Springer-Verlag, 1982), 216.

[12] 4/4/1947 letter from Vassar College President Sarah Gibson Blanding to Betty Scott.

wanted to return the favor to Makemson and explained, in some detail, her complex situation and her thinking about it. Betty further explained how important it was to her to complete her PhD.

Betty asked Makemson for a one-month delay. She knew Makemson would need to look for other candidates in the meantime, and that the position might go to someone else. Betty wrote: "Should my hopes of the position here fail to materialize by the middle of May and should you then be in the position to repeat your offer to me, I shall be very glad to accept and to do my best."[13] Betty frequently pledged to colleagues that she would *do her best*. It was an expression that defined her orientation to her professional activities.

At this point Makemson became even more interested in Betty. At Makemson's suggestion, Vassar offered Betty a residentship in a dormitory (student residence hall) that would pay at least $500 in addition to Betty's regular teaching salary. It "is quite a privilege":

> *Don't turn it down, or write that you have not made up your mind, but make up your mind on the merits of the case, whether you get the job in California or not. I don't like to think that you will come to us as a last resort.*
>
> *You are not likely to get a chance at a salary equivalent to $3000 a year, in a place with the undoubted advantages of Vassar College, and with the possibility of becoming director of the Observatory within a few years. What we want, as I explained, is someone who will stay here and really put her heart into the work of building up the department on the astrophysics side, and will become chairman in the course of time. If this is not what you want, the sooner you decide the better. I don't want you to take the job just because you can't find anything else. Why don't you talk it over with Dr [sic] Trumpler? He would probably advise as I am doing, but in any case, his advice would be very sound.*[14]

Betty's head must have been spinning.

But then Neyman managed to solidify the position for Betty in the Stat Lab. It involved mostly research but had a small amount of teaching. Her research duties would include work on her dissertation. Now Betty had a difficult choice to make: Vassar or UC-Berkeley. Both offers were certainly attractive. Betty consulted members of the Berkeley astronomy department: "They were kind enough to consider my problem carefully although each was careful not to make any decision for me."

---

[13] 4/15/1947 letter from Betty Scott to Vassar College Observatory Director and Astronomy Professor Maud Makemson.

[14] 4/25/1947 letter from Vassar College Observatory Director and Astronomy Professor Maud Makemson to Betty Scott.

6.3   The UC-Berkeley Decision

At Vassar, Betty reasoned, she might be so busy with teaching and social duties that she might not be able to complete her dissertation and PhD at all, and then the vision of an observatory directorship or department chair would disappear. If this happened, it would have a profound effect on her entire future.[15] At UC-Berkeley she would be able to complete her dissertation and her PhD, and in a timely fashion. At Berkeley she might even be able to pursue a research career, rather than be tethered to an isolated women's college position that would greatly restrict or negate her research ambitions.

The call of Neyman and Berkeley turned out to be stronger than the call of Makemson and Vassar. It also turned out to be stronger than any of the other calls that Betty received, even one as attractive as Air Intelligence Specialist (Mathematician) with the Strategic Vulnerability Branch of the US Air Force. This Pentagon position would have paid a salary of $7,102 ($68,562 in 2010 dollars),[16] more than twice what she was offered at Vassar. But Betty accepted the Berkeley position. Betty was sorry that Makemson and Vassar had waited for her to make a decision, and then that her decision was to remain in California.[17]

Makemson would later read the first paper that Betty and Neyman did together: "Consistent estimates based on partially consistent observations."[18] This paper arose out of Betty and Neyman's presentations at the IMS meetings in Atlantic City. It had to do with some issues that can arise when estimating a straight line with two stochastic variables. Makemson was very impressed with the paper, describing it as "very learned and important." She admitted that Betty had made the best choice for herself and acknowledged Vassar's loss. Makemson promised to contact Betty the next time she was in Berkeley.[19]

Later, in 1949, Betty was offered a faculty position in the University of Washington Department of Mathematics' Laboratory of Statistical Research. She was "pleased and honored" by the offer, but turned it down.[20] And even later, in 1953, the US military was still trying to recruit Betty, this time to work at the Office of Naval Research as Head of the Statistics Branch, with a starting yearly salary of $9,600 ($77,351 in 2010 dollars). But Betty would stay at Berkeley. She didn't want to detach herself from her cooperative

---

[15] 5/15/1947 letter from Betty Scott to Vassar College President Sarah Gibson Blanding.

[16] 12/5/1947 letter from US Air Force Administrator John S. Patton to Betty Scott.

[17] 5/15/1947 letter from Betty Scott to Vassar College Observatory Director and Astronomy Professor Makemson.

[18] Jerzy Neyman, Elizabeth L. Scott, "Consistent Estimates Based on Partially Consistent Observations," *Econometrics* 16, no. 1 (1948), 1.

[19] 11/30/1948 handwritten postcard from Vassar College Observatory Director and Astronomy Professor Maud Makemson to Betty Scott.

[20] 1/29/1949 letter from Betty Scott to University of Washington Mathematics Laboratory of Statistical Research Professor Z.W. Birnbaum.

research activities at Berkeley and then return to them later. In her words: "...it would be very difficult to return and collect all the threads." Betty had a strong sense of self. It was her conviction that she "could not fill the [ONR] position satisfactorily." She told them so: "If I thought that I could do a good job then I would not mind sacrificing some of my present activities."[21]

The bottom line was that Betty knew she could do better pursuing her research interests at UC-Berkeley than anywhere else.

6.4   Lecturer in Mathematics

In 1948–49, Betty received her first appointment as a lecturer. She was one of seven people recruited by Neyman to teach in his budding statistics program within the mathematics department.[22] Because Betty was still working to complete her dissertation, she was assigned to teach only one course, Descriptive Statistics. The course description read:

> *Collective and individual characters. Mathematical statistics as theory of collective characters. Means. Measures of dispersion. Frequency curves. Moments. Sheppard's corrections. Pearson curves. Curves of Charlier. Methods of fitting. Stochastic explanation of various distributions. Multivariate distributions. Static regressions and correlations. Applications.*

The three-unit course had both lecture and lab components.[23]

In 1949–50, while still a lecturer, Betty taught both semesters of an upper division sequence on Statistical Inference that had an associated laboratory course. It was intended for students in natural and social sciences and engineering, covering the basic concepts and tools of probability and statistics without complicated mathematical proofs.[24]

In August 1949, Betty completed a form for the graduate division that included the title of her dissertation. She envisioned a dissertation that had two parts. One would be on the problem of binaries that are disconnected on the surface. She had finished this paper. She was still working on the other one, which was on selective identifiability of spectroscopic binaries, a problem related to work done by Trumpler, her dissertation chair. Betty got in touch with Trumpler, asking whether he thought she was ready to have her dissertation defense in September. She knew she was late in getting her draft dissertation to him. Would he have time to read it? In reality, Neyman

---

[21]7/23/1953 letter from Betty Scott to Office of Naval Research Department of the Navy Mathematical Sciences Division Director Dr. Mina Rees.

[22]Constance Reid, *Neyman: From Life* (Berlin: Springer-Verlag, 1982), 199. Note: Betty had not yet completed her PhD degree, but neither had recruits Edward Barankin, Mary Euday, Evelyn Fix, John Gurland, or Erich Lehmann.

[23]Mathematics. UC-Berkeley Course Catalog, 1948–49, pp. 365-376.

[24]Mathematics. UC-Berkeley Course Catalog, 1949-50, pp. 347-357.

knew more about what Betty was doing than Trumpler: "I have spoken to Mr. Neyman and he is agreeable, but of course he knows what I am doing," she told Trumpler.[25] Betty would complete her degree on September 10, 1949.

6.5   Remarkable Research

Back at the end of July 1948, University of Chicago Yerkes Observatory Director Otto Struve read one of Betty's unpublished papers, one that had arisen from some questions he had raised and for which he had supplied the data. Struve was impressed. He told Betty he had "read with utmost interest" her "remarkable analysis," proclaimed her paper to be "exceedingly valuable," and encouraged her to publish it, even suggesting where it might be submitted for publication and particular strategies for submission.[26] Struve thought Betty's paper would shed new light on an interesting question in pure astronomy, but it would also introduce astronomers to procedures they would want to know more about and draw "attention to the power of statistical analysis." Struve himself was eager to learn more from Betty about statistical hypothesis testing and approached her with a number of probing questions. He also encouraged Betty to continue this line of work and suggested he might call on her again for help at some point in the future.[27]

Betty wrote and rewrote the manuscript that described her approach and results, incorporating suggestions by staff at Lick Observatory. She published the work in two papers on the distribution of the longitude of periastron of spectroscopic binaries. Both were published in the *Astrophysics Journal*, before she completed her PhD, and with Betty as sole author. Shane wrote: "Please accept my congratulations on an excellent piece of work."[28,29,30]

Some of Betty's own research also arose from consulting problems that came through Neyman in the Stat Lab. For example, Shane had assembled counts of extra galactic nebulae. In April 1948, he approached Neyman, wanting to know what Neyman could make of the counts.[31] Neyman shared them

---

[25]8/5/1949 letter from Betty Scott to UC-Berkeley Astronomy Professor Robert J. Trumpler.

[26]7/26/1948 letter from University of Chicago Yerkes Observatory Director Otto Struve to Betty Scott.

[27]9/20/1948 letter from University of Chicago Yerkes Observatory Director Otto Struve to Betty Scott.

[28]10/19/1948 letter from Lick Observatory Director C.D. Shane to Betty Scott.

[29]Elizabeth L. Scott, "Distribution of the Longitude of Periastron of Spectroscopic Binaries," *Astrophysical Journal* 109, (1949), 194.

[30]Elizabeth L. Scott, "Further Note on the Distribution of the Longitude of Periastron," *Astrophysical Journal* 109, (1949), 446.

[31]4/6/1948 handwritten letter from Lick Observatory Director C.D. Shane to Jerzy Neyman.

with Betty, who in turn found a problem in them of her own.[32] Even though Shane brought the original problem to Neyman, Betty had no problem asserting to Shane her own interest in the counts, including the background to her interest, what specific research question she was investigating, and how she planned to treat the problem. She had no problem asking Shane for better data, which in this case meant counts relating to a larger rectangle. She decided to develop a new hypothesis test for these types of data. Betty told Shane she was "burning to work on it."[33] Betty conducted this research in part under the Office of Naval Research contract to the Stat Lab.[34] She prepared to present the research at the IMS meeting in Seattle in late November.

In the middle of March 1950, Betty had another problem in stellar statistics. This would be a continuation of work she did at Lick the summer before on radial velocity of a star. This time the work would require use of a microphotometer to measure the widths of lines on a spectrogram. Lick had this kind of instrument, but theirs was in "deplorable condition." So Betty wanted to go to Mount Wilson, where she had done some summer work nine years earlier, to make the measurements and tracings. "We" (presumably the Stat Lab) would pay all expenses.[35] Astronomer Paul Merrill at Mount Wilson was happy to have Betty use their microphotometer. It was a complicated and delicate instrument, and staff would give Betty instruction on how to use it. The only possible expense would be photographic paper, but Betty wouldn't need to pay for the paper unless she used a large amount of it.[36] A plan was formulated for Betty to use the microphotometer for several days beginning June 30.[37]

Now that Betty had her PhD, she began efforts to obtain her own extramural funding. She sent a proposal to the ONR to work on statistical methods in astronomy and astrophysics. Her work was interdisciplinary. The ONR directed Betty's proposal to their Advisory Committee on Astronomy, which "recommended" it "as a worth while project." They "were unanimous in agreeing on the high merit of the proposed research," but didn't like the fact that it wasn't pure astronomy, and so they didn't give it a high priority score. There were too many "urgent" proposals for pure astronomy that there simply weren't enough funds to grant Betty a contract, they maintained.[38] Her

---

[32] 11/4/1948 letter from Betty Scott to Lick Observatory Director and Professor C.D. Shane.

[33] 11/4/1948 letter from Betty Scott to Lick Observatory Director and Professor C.D. Shane.

[34] 9/26/1948 letter from Betty Scott to Lick Observatory Director and Professor C.D. Shane.

[35] 3/18/1950 letter from Betty Scott to Mt. Wilson Observatory Astronomer Paul W. Merrill.

[36] 3/22/1950 letter from Mt. Wilson Observatory Astronomer Paul W. Merrill to Betty Scott.

[37] 6/19/1950 letter from Mt. Wilson Observatory Astronomer Paul W. Merrill to Betty Scott.

[38] 4/12/1950 letter from Office of Naval Research Advisory Committee on Astronomy Vice Chairman Leo Goldberg to Betty Scott.

interdisciplinary research was a round pin and the ONR astronomy committee decided to be a square hole.

6.6   Instructor in Mathematics

In 1950, Betty was promoted to the rank of instructor in mathematics. In the 1950–51 academic year, she taught her first graduate course, a sequence in Advanced Statistical Inference with an associated laboratory course. It was a continuation of the upper division Statistical Inference course. The subject matter was again offered without complicated mathematical proofs.[39]

In spring 1951, Betty had a bone to pick with Frank Massey who she knew quite well from the Stat Lab war work. Massey had just circulated a draft revision of his and Dixon's book, *Introduction to Statistical Analysis*.[40] This book sequenced the topic of statistics before the topic of probability, which was a reversal of the usual sequence. It would become highly influential.[41] Betty thought the book was excellent in many respects. But she was very unhappy with how he handled Neyman in the references in the draft second edition. She scolded Massey: "We all think that your treatment of Neyman is shocking! What did he ever do to deserve this?...when we saw what you had done I 'hit the ceiling' and I am not alone. Neyman was deeply hurt." She provided a long and detailed example of what can happen when references are misleading, saying "misleading references may be worse than none." Betty apologized for possibly sounding somewhat harsh. "But I feel very strongly that you are the sort of person whom we can expect to behave nicely," she wrote.[42] Betty was Neyman's champion, just as he was hers.

6.7   Assistant Professor of Mathematics

In 1951, Betty was promoted to the rank of assistant professor of mathematics. She was also appointed to be the mathematics department advisor for statistics. Now Betty was teaching a full load for the first time, and she was assigned to teach courses at all levels. Together with Barankin and Fix, Betty co-taught the statistics program's only lower division course, Elements

---

[39]Mathematics. UC-Berkeley Course Catalog, 1950-51, pp. 340-351.

[40]Wilfred J. Dixon and Frank J. Massey, Jr., *Introduction to Statistical Analysis* (New York: McGraw-Hill, 1949).

[41]Ingram Olkin, e-mail to author, September 9, 2013.

[42]4/2/1951 letter from Betty Scott to University of Oregon Statistics Professor Frank J. Massey, Jr.

of Probability and Statistics. Students across the campus could choose to take an introductory statistics course from applied subject matter experts in other departments, or else this course from statistics experts in the mathematics department. Neyman may have been hired with the idea that he would teach upper level statistics courses, but he occasionally co-taught this introductory course. The course was for students who wanted to specialize in statistics, as well as those who wished to learn basic concepts for general education. It covered relative frequency, discrete probability, and hypothesis testing. Illustrations were provided from genetics, bacteriology, industrial sampling, and public health.

At the upper division level, Betty taught Second Course in Probability and Statistics, a continuation of the lower division course. Topics included expectation, variance, correlation, regression, the law of large numbers, least square estimation, and confidence intervals. At the graduate level, Betty taught Statistical Problems in Experimentation, which had the upper division course in statistical inference as a prerequisite. Topics included mathematical models of experimental problems, random and systematic designs, complex experiments, randomized blocks, Latin and Graeco-Latin squares, biological assay, and recent developments in experimental design. The course incorporated both lectures and laboratory work.

In 1951, when Betty was still an assistant professor, there were about as many courses at the upper division level.[43] The department offered a major in mathematical statistics in addition to the major in mathematics. It was at a time when statistics enrollments were growing and the statistics faculty were reconsidering the mathematical statistics major.[44] By the time Betty was recruited to the UC-Berkeley faculty, Griffith Evans had been chair for 14 years (since 1934) and had developed the mathematics department into a world class academic unit.

The department was quite large. Twenty-six were in tenure-ladder appointments, i.e., as professors, associate professors, and assistant professors; and 20 were in other kinds of appointments such as instructor, lecturer, or visitor. The department was almost all male.[45]

---

[43]Note: Others teaching statistics in 1951–52, in addition to Betty and Neyman, were Fix, Loève, Gopinath Kallianpur (1925–2015) who was at UC-Berkeley in a one-year Lecturer position and would become a Distinguished Professor at the University of North Carolina-Chapel Hill, UC-Berkeley Associate Professor of Agricultural Economics George Michael Kuznets (1909–1986) who would become a leading expert in empirical studies of demand for California produce, and three of Neyman's current PhD students including Terry Allen Jeeves (n.d.), Harry Meachum Hughes (n.d.), and Lucien Marie LeCam (1924–2000). LeCam would go on to become an expert in the asymptotic theory of statistics.

[44]Mathematics. UC-Berkeley Course Catalog, 1951-52, pp. 364-375.

[45]Note: In addition to Neyman and Evans the other male full professors in the department were mathematical astronomer Thomas Buck (1881–1969), modern algebraist Alfred L. Foster (1904–1994), number theorist Derrick Henry Lehmer (1905–1991), probabilist and mathematical statistician Michel Loève (1907–1979), calculus of variations and partial differential equations expert Charles Bradfield Morrey Jr. (1907–1984), analysis and foundations of mathematics expert Anthony P. Morse (1911–1984), generalist Raphael Mitchel

Only three of the 23 tenure-track appointments were women:[46] one full professor, Sophia Levy McDonald (1888–1963); and two assistant professors, Betty and Evelyn Fix. McDonald was the first woman ever to achieve the most senior rank in the mathematics department (in fall 1949). Although she was a potential role model for Betty, especially as both had backgrounds in astronomy, and Betty as a student had McDonald as a substitute instructor, there is no evidence the two tried to collaborate on their research.[47,48] In previous years Pauline Sperry (1885–1967) was among the associate professors, and Betty also had her as a substitute teacher; but when Betty became a faculty member in the mathematics department, Sperry was no longer on the faculty, having been fired because she refused to take a loyalty oath (see Chapter 12).[49]

Some were blatant about treating Betty differently in professional relations because she was a woman. Among them was Mayo Clinic Physician and Statistician Joseph Berkson. Here is one of his letters to Betty: "I am writing this to you as more [sic] reliable for this sort of thing than any male I would write to... Will you please let me know whether a communication to Professor M.G. Kendall addressed to Berkeley will get him [sic]? I owe him a letter which I should like to have him receive before the Minneapolis meeting. Best regards. Be seeing you soon." Betty certainly was reliable, but it is arguable that this characteristic was due to her gender.[50]

In 1952–53, Betty again taught introductory statistics, this time with Robert Fleming Tate who was working on his dissertation under Neyman. Betty also taught upper division Theory of Probability and Statistics, and again graduate-level Advanced Statistical Inference, this time with Fix.[51] Since Betty was now teaching a full load and at all levels, she was well positioned in terms of teaching for promotion to associate professor.

Robinson (1911–1995), and logician and philosopher Alfred Tarski (1901–1983). Also among the 13 male senior faculty were three Emeritus professors, including logician Benjamin Abram Bernstein (1881–1964), generalist John Hector McDonald (1874–1953), and differential equations expert Charles Albert Noble (1867–1962).

[46]Mathematics. UC-Berkeley Course Catalog, 1951-52, pp. 364-375.

[47]Lenzen V.F., Einarsson S., and Evans G.C. Sophia Levy McDonald, Mathematics: Berkeley. In Memorium. Accessed on 1/26/2014 at http://texts.cdlib.org/view?docId=hb338nb1j4&doc.view=frames&chunk.id=div00013&toc.depth=1&toc.id=. Note: Levy McDonalds academic specialty was theoretical astronomy and numerical analysis. She, like Betty, had a PhD in astronomy from UC-Berkeley.

[48]Armin Otto Leuschner and Anna Estelle Glancy, "Tables of Minor Planets Discovered by James C. Watson," *Memoirs of the National Academy of Sciences* 14, no. 3 (1920). Note: Levy McDonald published with Leuschner and others.

[49]Pauline Sperry. Accessed on 2/19/2014 at http://www-history.mcs.st-and.ac.uk/Biographies/Sperry.html. Note: Sperry's specialty was in projective differential geometry. She was the first woman to achieve the rank of assistant professor and then associate professor in the UC-Berkeley mathematics department.

[50]8/16/1951 letter from Mayo Clinic Division of Biometry and Medical Statistics Physician Joesph Berkson to Betty Scott.

[51]Mathematics. UC-Berkeley Course Catalog, 1952-53, pp. 207-220.

In 1954–55, Betty took a leave of absence[52] to obtain additional training in mathematics. She went to Paris to study at the Institut Henri Poincaré (see Chapter 11).

6.8   Trumpler's Book

As previously mentioned, Betty began publishing as an undergraduate. Up through 1948, the year before she completed her PhD, she had 13 publications, the last of which was her first publication with Neyman. From the year 1949 through 1954, a six-year period, Betty produced another 15 publications about half of which were co-authored with Jerzy Neyman, and this would be the pattern. Twelve of these were in the *Annals of Mathematical Statistics, Astrophysics Journal, Astronomy Journal, Proceedings of the National Academies of Science, Proceedings of the International Congress of Mathematicians,* and *Proceedings of the Berkeley Symposium on Mathematical Statistics and Probability.* Others were book chapters or white papers. One of the book chapters was in a book authored by Trumpler, Betty's dissertation advisor.

Trumpler finally completed his book on statistical astronomy in 1951, the year he retired, and it was published in 1953. The book had parts on elements of statistical theory, statistical description of the galactic system, stellar motions in the vicinity of the sun, luminosity – spectral type distribution, space distribution of stars, and galactic rotation.

The synergy of Betty's involvement as a graduate student in both Advanced Topics in Probability and Statistics and Statistical Astronomy in the 1939–40 academic year was of immeasurable value to Trumpler in relationship to his nascent book concept on statistical astronomy. The book needed a section on elements of statistical theory. Within that section, the book needed a chapter on testing hypotheses.

Betty was up to date, having become an expert in both astronomy and statistics, and wrote the entire chapter on testing hypotheses. It appeared in the book as Chapter 1.8. It was 26 pages of the final published book. In the chapter, Betty segmented her discussion into a general statement of the problem, distinction between phenomenal hypotheses and statistical hypotheses, background of testing statistical hypotheses, hypothesis of constant radial velocity, the likelihood ratio principle and Student's t-test, and Chi-square tests. Betty knew how to speak statistics directly to the astronomer, and she did so with clarity and completeness. She included concepts, context, history, theory (derivations, relationships), application, examples, tables, and graphs, as well as extensive notes and bibliography.

---

[52]Mathematics. UC-Berkeley Course Catalog, 1954-55, pp. 224-238.

Betty also meticulously scrutinized and edited the entire book draft. She provided 24 single-spaced pages of typed comments to the book's authors, Trumpler and Harold F. Weaver.

What did Betty get for all of her fine intellectual contributions? Instead of repaying her with a co-authorship, Trumpler and Weaver gave Betty a small acknowledgment in the last paragraph of the book's preface and an easy-to-miss footnote at the bottom of the first page of the chapter she had written. Immediately prior to the acknowledgment of Betty, the authors referred to the student of statistical astronomy as "him," viz, "introduce him" and "give him knowledge." They then acknowledged Betty by saying this: "We gladly acknowledge our indebtedness to Dr. Elizabeth L. Scott... She has contributed the entire chapter 1.8, has read the whole manuscript, and made many helpful suggestions." Immediately following the acknowledgment of Betty, and in the same paragraph, the authors thanked their wives for typing the manuscript. The footnote at the bottom of the first page of Betty's chapter acknowledged that she "kindly contributed" the chapter.

New York University Professor W. Edwards Deming (1900–1993), renowned for his work on statistical quality control and now perhaps the world's best known statistician, was one of the reviewers of the book. He wrote this to Betty: "I send this with fear and trembling, as it is only another horrid incompetent book-review. Your chapter is certainly magnificent: in fact, I like the whole book."[53] A few months later he added this: "It is a work of great care and thoroughness, but it is your contributions in the book... that give it the modern touch of statistical theory as a basis for learning about nature more economically and with greater reliability."[54]

Betty's star had risen, even if it was not always allowed to shine to its full brightness.

6.9   A Retrospective: Betty and Phoebe

When Betty's Aunt Phoebe passed away in 1967 at age 85, Betty had had her Aunt Phoebe for 50 years. Phoebe is an important but quiet part of Betty's story.

Betty and Phoebe had many things in common. They were highly intelligent, were used to being in the company of men, studied astronomy as undergraduates, and had aspirations to do the same kind of creative "head work" men were allowed to do. Both had parents who supported their ambitions to earn PhDs, had fellowships to support their doctoral work, earned PhDs in

---

[53] 1/6/1955 handwritten letter from New York University Professor W. Edwards Deming to Betty Scott.

[54] 4/22/1955 letter from New York University Professor W. Edwards Deming to Betty Scott.

astronomy from UC-Berkeley, and chose Berkeley because of its high reputation, low cost, and location close to family. Neither had female role models in astronomy at Berkeley. Both did dissertation work at Lick Observatory, did well in their graduate studies, and used the Student's Observatory at Berkeley as their professional address while at Berkeley. They were members of a faculty, were more interested in research than teaching, had early immersions in other cultures, published orbits of comets, and were highly generous with their time and money, as their means would allow. Both knew astronomers Leuschner, Merrill, Wright, and Shane. And at least some of these knew Betty and Phoebe were related. For example, when Wright wrote to Betty in 1942, he remembered Phoebe and told Betty he was going to Philadelphia and might see Phoebe. "If I do your ears will probably tingle," he teased Betty.[55]

The two women also had differences. Phoebe's undergraduate experience at a small leading women's college gave her more of a romantic view of astronomy than Betty experienced at a large public university. Phoebe's early international experience was on a military base in the third world, whereas Betty's was at a university in a world class city. Phoebe's faculty position was temporary, whereas Betty's was permanent. Phoebe worked in a gender ghettoized position, while Betty worked more as an equal alongside men scientists. Phoebe decided to pursue a PhD degree because of this employment experience, whereas Betty decided to pursue a PhD out of her undergraduate experience. Phoebe competed successfully for university fellowships in competition with women, whereas Betty competed successfully with men. Phoebe was interested in doing observational astronomy, whereas there is no evidence that Betty had interests in astronomical observing. Phoebe married and raised a family, an experience Betty never had. Phoebe did most of her research as a volunteer, whereas Betty did her research as a regular faculty member.

Many of these differences are a reflection of the different times in which the two women lived. With 35 years between them, Betty and Phoebe lived at different times for women in science. The stars of success were arranged in different constellations. Overall, Phoebe was a variable star in astronomy, a star whose brightness in science varied over time with her personal circumstances. She was a star who fought to remain active in research after she built her family, pursued science as a volunteer, and worked within the constraints of the professional jungle gym. Betty on the other hand was a stable star, a star whose brightness in science was relatively unchanging over time in some part due to her decision not to build a family. She was a star who pursued science as a full time career and collaborated as an equal with astronomy giants.

---

[55]4/15/1942 letter from [W.H. Wright] to Betty Scott.

6.10   A Retrospective: Phoebe's Influence

Phoebe had both direct and indirect influence on Betty's interests in astronomy. It is known the two women were in contact. Even though they lived across the country from each other – Phoebe in Philadelphia and Betty in Berkeley – the families got together when they could, often annually. Betty didn't just get her PhD from the same place and in the same field as her aunt. Phoebe had broken a huge gender barrier in science by being a member of the first graduating class at Berkeley where women earned doctoral degrees in astronomy. Yet Betty was reluctant to acknowledge Phoebe's influence. Betty would not allow herself to be in Phoebe's astronomical shadow, even though Phoebe was an academic pioneer for women and astronomers.

Betty gave many speeches about her career to young women. But it is difficult to find speeches in which she mentioned Phoebe. In one speech written around 1976, Betty described the influences on her choice of career, her greatest challenges, and her greatest problems:

> *I'm an old-timer. I went to college at the end of the depression. Employment prospects were dim, no matter what. I studied what I was interested in – astronomy and art. My parents expected us to go to college. Indeed, one reason that they had moved to Berkeley was so that all four of us could go to a good university at low cost. My mother's oldest sister had a doctorate in astronomy but I knew her only as a friendly aunt whose children we played with. She was no longer active in astronomy, we never talked about astronomy before I was in college but I knew that astronomy existed and that there were women astronomers.*[56]

Thus Betty acknowledged knowing Phoebe had a PhD in astronomy, and so Betty knew women could earn doctoral degrees in astronomy. In another draft, Betty described her choice of undergraduate major in relationship to Phoebe:

> *My horizon then was vague; I knew that art and astronomy existed but I had no real contact with them even though my mother's oldest sister had a doctorate in astronomy and I had heard stories of her activities and problems. But that was long ago and she was far away, no longer active in astronomy. I soon learned that women were not allowed to use the large telescopes at the Mt. Wilson and Palomar Observatories (and no woman has yet used the 200-inch telescope), not because special size or strength are needed to handle large telescopes but just because of prejudice. Many persons told me of the problems and even advised me not to obtain a doctorate in astronomy, always with the best of intentions. The faculty at*

---

[56]Scott, Elizabeth L. Influence choice of career; greatest challenges; greatest problems. First draft of Speech: Influences, Challenges, and Problems. One handwritten page. Around 1976.

Berkeley were always most helpful and I liked astronomy more and more as I turned to theoretical astronomy and then to statistical astronomy.[57]

Thus Betty acknowledged hearing stories of Phoebe's "activities and problems," but was otherwise somewhat dismissive of her aunt's influence. Betty maintained she had no special female role models.[58]

Around this time, Mary Lea Shane, a PhD astronomer whose career in astronomy was aborted upon her marriage to C.D. Shane, was collecting information about Lick Observatory. Her efforts eventually resulted in the Mary Lea Shane Lick Observatory Archives, containing 195 linear feet of historical documents about the observatory.[59] In 1980, Mary Shane as part of her archival efforts wrote to Betty asking specifically about her Aunt Phoebe. Here is Betty's response:

Yes, she was Phoebe Waterman and received her PhD at Berkeley in 1912 or 1913, I have forgotten exactly when. Professor Einarsson knew her and used to tell me stories about her performance on the qualifying exam. I remember the year [W.W.] Campbell died; Aunt Phoebe was visiting us because our Grandfather was then staying with us. She planned to see Campbell but was just too late.

Aunt Phoebe was the oldest and my mother the youngest in a family of five children, 10 years between Aunt Phoebe and my mother. It is very hard for me to say how much she influenced my going into Astronomy. Certainly indirectly but not directly. Whenever we were together, she would entertain my brothers and me with stories about the good old days, especially stories about my mother. We knew that she had a PhD in Astronomy, and we knew that she met Uncle Otto on a ship traveling to Argentina to take a position as astronomer there (in the National Observatory I think) and that she did not find a position as an astronomer in the United States. We regarded the shipboard meeting as very romantic and somehow astronomy did not enter the picture. Otto Haas was a manufacturing chemist, deeply engaged in building Rohm & Haas. At first they were manufacturing chemicals to tan leather, etc., but about this time Dr. Rohm invented Plexiglas. As time went on the company expanded tremendously. Uncle Otto was something of a German authoritarian and Aunt Phoebe devoted a lot of time and effort to the problems of Rohm & Haas. Their two sons, one a year or so older and one a year younger than I am, were trained from the start to enter the business, one as a PhD chemist and the other as a chemical engineer.

---

[57]Scott, Elizabeth L. Influences, Challenges, and Problems. Speech to unidentified audience, three double-spaced typewritten pages. Around 5/1976.

[58]6/10/1978 letter from Ms. Mina Edelston, attendee at the UC-Berkeley Conference for Women in Engineering and Computer Science on 5/13/1978, to Betty Scott with attached questionnaire completed by Scott.

[59]Lick Observatory Archives at the University of California Santa Cruz. Accessed on 12/2/2013 at http://library.ucsc.edu/speccoll/lick-observatory-archives.

When I had already started to study Astronomy Aunt Phoebe sometimes talked to me about her experiences, especially those at Mt. Wilson before she came to Berkeley. She continued her interest in Astronomy and was engaged in observing variable stars, although not with the level and intensity that we would expect from a full-time astronomer. She died several years ago but again I do not remember the exact date. If you would like, I or you can ask for more details from my cousins whom I still see sometimes. The younger one still lives in the big old house...

Going back to your first question, I am sure that Aunt Phoebe influenced me in many ways. For ten years, from age four to fourteen, we visited my mother's parents at their summer place on the Severn River near Annapolis. My cousins were often there too but Aunt Phoebe usually went to Germany with Uncle Otto on business. So, I knew her quite well and I knew that she had a degree in Astronomy but I do not remember talking to her about Astronomy until I was already studying it at Berkeley. I am often asked how I came to study Astronomy. I really do not know. I certainly did not make telescopes as a child. At the last moment I was influenced by the Trumplers, especially Cecile [the daughter of Trumpler] who then became a classmate. Just at that time we moved to Piedmont Avenue, only a few doors from the Trumplers. I remember that I was already planning to take Astronomy and that Cecile recommended that I take Astronomy 103 even though I had never had any Astronomy before. She also recommended that I take Astronomy 104 from Leuschner, the last year before his retirement.[60]

Thus Betty acknowledged that Phoebe was unable to find a position in astronomy in the US, worked as a computer at Mount Wilson, and was a variable star observer. Betty acknowledged Phoebe told stories about these experiences, but Betty was nonetheless dismissive of any direct influence.

Having acknowledged hearing these stories, Betty surely knew of UC-Berkeley and its excellence when she was a young girl, in part because Phoebe had already attended Berkeley as a graduate student. Betty surely knew that women could do astronomical observing and research at Lick Observatory, because Phoebe had already done so. Betty surely knew that women could have a lifelong love of astronomy and science, just like Phoebe. Betty's cousin John Haas reported that: "Betty knew my mother, her aunt Phoebe, and admired her. Betty, like my mother, had a great aptitude for math and statistics. Betty wanted to follow her footsteps in statistics and astronomy."[61]

But Betty nevertheless stopped short of giving Phoebe any direct credit for her interests in astronomy. This seems incongruous with the fact that, between 1900 and 1940 only 39 women in the United States earned doctoral degrees in astronomy, only 14 of these were earned at UC-Berkeley, and Phoebe was

---

[60] 2/23/1980 letter from Betty Scott to Charles Donald Shane.
[61] John Haas, e-mail to author, May 12, 2008.

among the first of these.[62] In Betty's time, there were very few women full professors at UC-Berkeley, and the number of women professors in science was negligible, both at Berkeley and nationally.[63] Also, it seems that Betty's closest graduate students were not aware of Betty's Aunt Phoebe,[64] nor were her colleagues.[65] In the words of former UC-Berkeley graduate student and UC-Santa Cruz Professor of Applied Mathematics and Statistics David Draper, Betty was facing "substantial headwind" at the time in even being on the Berkeley statistics faculty.[66] Betty may have felt that giving any direct credit for her interests in astronomy to a relative would compromise her own hard-won stature. She didn't want to be discounted as unimportant or anomalous as were most women in science of her time. In either her conscious or subconscious, it may have been a risk Betty was not willing to take.

But memories do play tricks. Betty wrote in the above speeches that, as an undergraduate at UC-Berkeley, she studied astronomy and art. A review of her undergraduate transcript reveals that Betty never took a single course in art while a student at Berkeley.[67]

Both women, Phoebe and Betty, were products of their times, and the evolution of educational circumstance. Both were brilliant, and both were dedicated to the life of the mind. Circumstances drove them to different end points, but each contributed seriously and consistently to improving the worlds they observed. It was the legacy they shared.

---

[62] John Lankford, *American Astronomy: Community, Careers, and Power, 1859–1940* (Chicago: University of Chicago Press, 1997), 321.

[63] Margaret W. Rossiter, *Women Scientists in America Volume 2: Before Affirmative Action 1940–1972* (The Johns Hopkins University Press, 1995). ch. 6.

[64] Dennis Pearl, telephone interview with author, February 17, 2014. Note: Pearl indicated that Betty never talked with him about her Aunt Phoebe.

[65] Juliet Shaffer, e-mail to author, August 5, 2014. Note: Shaffer indicated that Betty never mentioned her Aunt Phoebe.

[66] 10/31/2014 personal communication from UC-Santa Cruz Professor of Applied Mathematics and Statistics David Draper.

[67] Elizabeth Leonard Scott. Undergraduate transcript, graduation A.B. May 20, 1939. UC-Berkeley.

Part III

Clusters of Impact

7

Championing Science (1939–1988)

I tell young women not to publish extensively with an older man, because it is often assumed that he did the important work. It is a real problem.
-Juliet Popper Shaffer

Research is what Betty liked to do best. Her scholarly publications, compiled into a list at the end of this book, are the official record of how Betty communicated her research to her peers (see Billard and Ferber).[1] Betty also left behind many archival materials – letters, abstracts, newspaper clippings, transcripts of radio shows, and the like – in which she provided thoughts about, and descriptions of, her research that make her science accessible to a general audience. These descriptions fill in gaps in our understandings of Betty as a science researcher and communicator and provide the basis for this chapter and the next.

Betty thought of her research as being "quite varied." It could be "in almost any direction."[2,3] In 1977 Betty's provost commended her for working on topics "so diverse in range";[4] and her dean was "most impressed with the wide range" of Betty's statistical applications.[5] Basically, Betty worked on things she liked, and she liked big projects. She thought that: "[The] interesting part is what comes out," and there are "Rewards!" She saw one of the rewards as being able to work with "interesting people."[6] She did not see the rewards as monetary, saying she "never did anything for financial gain."[7]

---

[1]L. Billard and M.A. Ferber, "Elizabeth Scott: Scholar, Teacher, Administrator," *Statistical Science* 6, no. 2 (1991), 206.

[2][?1977] "Brief Sketch of Activities" by Betty Scott.

[3]3/5/1977 Elizabeth L. Scott. In: New Directions for Women: Portraits of 21 Women in Science. Participants in a conference at Yale University.

[4]8/24/1977 letter from UC-Berkeley Provost Roderic B. Park to Betty Scott.

[5]2/17/1977 letter from UC-Berkeley Dean of Physical Sciences Leonard V. Kuhi to Betty Scott.

[6]6/10/1978 letter from Mina Edelston, attendee at the UC-Berkeley Conference for Women in Engineering and Computer Science on 5/13/1978, to Betty Scott with attached questionnaire completed by Betty Scott.

[7]6/10/1978 letter from Mina Edelston, attendee at the UC-Berkeley Conference for Women in Engineering and Computer Science on 5/13/1978, to Betty Scott with attached questionnaire completed by Betty Scott.

Betty's research was supported through the Stat Lab. Partial financial support came from the ONR, National Science Foundation (NSF), US Army Research Office, National Institutes of Health (NIH), Department of Health, Education and Welfare (HEW), US Public Health Service, Environmental Protection Agency, Carnegie Commission on Higher Education, American Association of University Professors (AAUP), and at least one private foundation. The most consistent source was the ONR, which provided funding to the Stat Lab in a long series of one-year contracts.[8]

The alignment of the Stat Lab with Betty's intellectual interests is evidenced by the fact that the course catalog description of the lab included mention of a relationship with Lick Observatory until 1968. In 1980, Neyman gave this description of the history of the Lab:

> *Our Stat. Lab. has been established in the 1930s and, since that time continued its existence, primarily as a research unit funded by a variety of federal agencies, but also, partly, engaged in teaching. In 1955 the Stat. Lab. generated the Statistics Department... However, we continue our semi-independent activities, with the scientific personnel composed of three individuals, Professor L.M. Le Cam, Professor E.L. Scott and myself. The official ties among us consist in being Co-Principal Investigators on our projects. Our common research interests begin with probability and mathematical statistics and extend to two domains of what may be called "societal problems," problems of public health and weather modification.*[9]

Betty started out in the Stat Lab as a research assistant and ended up being a major player, a co-principal investigator.

7.1   Themes and Controversies

There were five themes within Betty's broad range of research. One was astronomy, especially "distribution of galaxies in space and expansion of the universe." One was general statistical methods. A third was weather modification, especially design of experiments to determine whether cloud seeding increases rainfall. A fourth was bioscience and health, especially how malignant tumors start and grow.[10,11] A fifth was women's studies, especially how to estimate underpayments of women relative to men in academe.

---

[8]7/28/1980 letter from Jerzy Neyman to UC-Berkeley Provost Professor Roderic B. Park. Note: It is unclear if the Office of Naval Research funded any of Betty's work in the area of women's studies.

[9]7/28/1980 letter from Jerzy Neyman to UC-Berkeley Provost Professor Roderic B. Park.
[10][?1977] "Brief Sketch of Activities" by Betty Scott.
[11]3/5/1977 Elizabeth L. Scott. In: New Directions for Women: Portraits of 21 Women in Science. Participants in a conference at Yale University.

Three of Betty's research themes – weather modification, bioscience and health, and women's studies – involved applications of statistics to societal problems. Betty was genuinely and passionately interested in research that had societal relevance.[12] Especially when it came to the gender equity research, Betty felt she sometimes tried too hard. But the effort was well worth it, because her equity studies were known to be, as one colleague put it, "extremely painstaking in their accuracy and seminal in their effect."[13]

By at least 1980, Betty realized that almost everything she was working on was "very controversial."[14] This has been well documented in the case of her work on weather modification (see next chapter), but it was also the case for her work on women's studies (see the next parts to this book). Betty said this in 1976 at a women in science conference:

> *Statistics has the wonderful advantage, or perhaps distracting disadvantage, that it overlaps many fields, so that there are many different problems to work on. Perhaps the long tradition of women to study problems that are interesting and important from their own point of view without worrying about the personal gain attached (since there has rarely been any such gain for women) leads women not to be afraid of working on problems that are controversial. Certainly, I seem to have more than my share: carcinogenesis and other cancer problems, weather modification, and affirmative action, to name a few. I try to find the facts and have lots of fun in the process![15]*

The more the work focused on public policy issues, the more controversy it generated, or so it seemed.

Betty began her academic career as an astronomy researcher in the late 1930s. In the 1940s she put most of her effort into astronomy studies. In the 1950s and 1960s she divided her time between astronomy, general statistical methods, weather modification, and bioscience and health. She began her women's studies research in the late 1960s. By 1980, Betty was still working on weather modification, but she was putting most of her research effort into the skin cancer studies, about four hours a day, and she was also still putting considerable effort into gender equity studies.[16]

Jerzy Neyman had a rule: Authors of a study should be listed in alphabetical order by last name. Neyman came out of the mathematics tradition, and that is how they did it. There are other scholarly traditions, such as listing

---

[12]6/3/1978 letter from Betty Scott to Sonoma State College Mathematics Professor Rick Luttmann.

[13][n.d.] script introducing Betty Scott as a speaker on the topic of "Historical Trends in the Appointment and Promotion of Women Faculty."

[14]10/30/1980 memorandum from Betty Scott to UC-Berkeley Statistics Professor David Brillinger.

[15]Elizabeth L. Scott, "Panel Talk," *Proceedings of the Conference on Educating Women for Science: A Continuous Spectrum* (Mills College, Oakland, CA, 1976), 32.

[16]10/30/1980 memorandum from Betty Scott to UC-Berkeley Statistics Professor and Chair David Brillinger.

authors in order of decreasing contribution. Or where the most senior author, as the mentoring author, is listed last. At least one female statistician, who has had long-term collaborations with a man, selects at random the first author of each paper that results from the collaboration.[17] But this isn't how Neyman did it. Although as a statistician he should have been comfortable with such random ordering of authors, his rule was that every study associated with the Stat Lab would have authors listed in alphabetical order.

Betty's primary research collaborator was Neyman. Across her career, half of her publications were with him. The effect was that Neyman was always listed as an author before Betty. People who come out of other traditions may look at the ordering and assume Betty's contributions to the research were always secondary to Neyman's. This is a historical disadvantage for Betty. However, it should be noted that 24 of the 62 publications she did with Neyman were in the area of statistical astronomy, which was Betty's specific training (Neyman wasn't trained in astronomy), and 27 of the others were in weather modification which generally required extensive data analysis, the oversight of which was one of Betty's fortes.

Juliet Shaffer said this about the situation: "The trouble with Betty as a researcher is she did so much with Jerzy Neyman. I tell young women not to publish extensively with an older man, because it is often assumed that he did the important work. It is a real problem. I have a feeling men didn't think."[18]

7.2   Modern Statisticians, Old Equipment

Understanding Betty as a researcher requires understanding her conception of the field of statistics. Betty described statistics as "one of the new sciences." Her elevator speech was: "The main concern of studies on statistics is 'how to find out,' what should be done to obtain a better understanding of what is going on and why." She also subscribed to the idea that statistics is the theory of decision-making when faced with uncertainty.[19,20,21] Statistics was a growth field at the beginning of 1977. Statistics opportunities were expanding, "perhaps more than in any other field" because statistics was "being used to a

---

[17] J.D. Singer, "Fostering the advancement of women in academic statistics," in *Leadership and Women in Statistics*, ed. Amanda L. Golbeck, Ingram Olkin, and Yulia R. Gel (London: Chapman & Hall/CRC Press, 2015), 413.

[18] Juliet Shaffer, interview with author, August 5, 2014.

[19] [?1977] Brief Sketch of Activities by Betty Scott.

[20] 3/5/1977 Elizabeth L. Scott. In: New Directions for Women: Portraits of 21 Women in Science. Participants in a conference at Yale University.

[21] Ingram Olkin, e-mail to author, March 30, 2014. Note: This was a description that had been around since at least 1947 in relationship to the work in decision theory by Abraham Wald (1902–1950) and Jacob Wolfowitz (1910–1981) at Columbia University and Erich Lehmann at UC-Berkeley.

greater extent in more fields all the time." She saw the greatest opportunities in the applied sciences, especially in health sciences.

Over time, Betty developed great respect for data-handling statisticians, who she called modern statisticians, and who she contrasted with "design" men. By 1976, she had "extensive" experience with "mountains of data," and formed definite opinions about quality and timeliness in the handling of data.[22] She wanted more data for everything she studied, and was always worrying about getting rid of study biases and data errors.[23] When she collaborated with researchers from other institutions, the data sets were often almost impossible and a headache to transport from one computer platform to another given the state of machine computing at the time. Here is an example:

> *Their [Temple University's Skin and Cancer Hospital in Philadelphia] Wang [computer] has no possibilities of making cards or tapes. They have a Wang 2200 with a Wang disk 2270-2 which makes hard sector diskettes. He offered to send us a diskette but I have not yet found a way to read it. It will not interface with CDC in any easy or hard way although is possible to go around the barn somehow. I have talked to San Francisco Wang office... Finally I was referred to San Mateo headquarters... He was helpful and knew what he was talking about. Trouble was I did not have all needed information. There are several possibilities. One is to use IBM floppy equipment at UC (if we have ???). Another is to investigate the protocol for IBM at UCB or UCSF. If it is bisynchronous [sic] rather than asynchronous (spelling?) one can set up an emulator and telecommunications (a modem plus telephone). There is some jargon about having to initialize on IBM system (Wang cannot initialize but can communicate with IBM) and then create some 3740 data files (a 3740 diskette). Fisher et al regard punch cards as items of the past, beneath their dignity. Anyway, these IBM 3740 data files could be used almost anywhere. It seems that many hospitals have this Wang system (apparently UCSF Parnassus does). My request to you is to try to talk to some consultants or engineers in Computer Center (Shryl may know whom) to find out where UCB is and what they recommend. Then call... again if you wish after [you] have some information.*

Also, the data often came to Betty with errors that needed correcting.[24]

"New science" statisticians needed strong computer skills. Betty had a deep understanding of the need, and was an early advocate for this kind of training. She gave many speeches to young people who were thinking about pursuing a statistics career. She advised them to get a strong background in both probability and statistics. But that was not enough. They should also

---

[22]6/15/1976 letter from Betty Scott to University of Washington School of Medicine Pathologist Earl Benditt.

[23]8/28/1976 handwritten letter from Betty Scott to UC-Berkeley Stat Lab Assistants Marcella and Shyrl.

[24]4/17/1979 letter from Betty Scott to UC-Berkeley Stat Lab Assistant Penny.

get a strong background in "one or more fields of application that interest them." Since many of these applications involved data analysis, she advised them to "build strong skills in machine computing."[25,26]

In Betty's time, many other processes involved in research were also difficult. For example, in 1980 letters were still dictated, and a secretary was needed to write what was dictated into a notebook using shorthand, and then type the letters from the shorthand. Betty was allocated only two hours of dictation time per week. The secretary had no trouble typing Betty's short letters, which typically had to do with student recommendations. These usually came out perfect. But many of Betty's letters were long – over three pages long. These typically had to do with Betty's research or requests for money. The problem was that by the time the secretary got to these longer letters she could no longer read her own shorthand. Betty pleaded with her department chair for more secretarial help.[27]

7.3   Statistical Astronomy

Betty's first work in astronomy had to do with determining orbits of comets (see Part II). Next Betty worked for a long time in the area of statistical astronomy. She used "statistical tools to answer important questions in that field"... "using astronomy as the motivation for and application of statistical ideas." Betty distinguished the use of statistics in astronomy from bone fide statistical astronomy. In her view, the latter involved "statistical studies of the motion and the distribution in space of stars and galaxies," where information from galaxies was used to determine the relationship between apparent or absolute luminosity and redshift, and then this relationship was used to determine the distribution (positions and distances) of stars and galaxies. Distances so determined have been called Hubble distances. "It is just this rather narrow domain of statistics in astronomy that astronomers call *statistical astronomy*," she explained.[28]

Betty's research did not come without struggles. Many times mathematical derivations can seem impossibly difficult. For example, in one of their handwritten draft manuscripts about magnitude-redshift relation for cluster galaxies, Betty and Neyman worked on an equation for expectation, an aver-

[25][?1977] Brief Sketch of Activities by Betty Scott.

[26]3/5/1977 Elizabeth L. Scott. In: New Directions for Women: Portraits of 21 Women in Science. Participants in a conference at Yale University.

[27]10/30/1980 memorandum from Betty Scott to UC-Berkeley Statistics Professor David Brillinger.

[28]E.L. Scott, "Statistics in Astronomy in the United States," in *On the History of Statistics and Probability*, ed. D.B. Brown (New York: Dekker, 1976), 319.

age computed using probabilities rather than sample data. In the middle of a page of derivations, they wrote: "Now, Lord help us!"[29]

Also, mathematical statisticians often provide data illustrations to demonstrate the usefulness of their new methods. Betty and Neyman consistently developed relationships with applied researchers and asked to use their data for such illustrations. For example, they "derived a statistical test of the hypothesis that a cluster of objects (stars or galaxies) is expanding." Aware that someone was about to publish data on the Coma Cluster that would be suitable, they inquired: Would he be willing to let Betty and Neyman use his data even before they were published?[30]

Betty's work in statistical astronomy focused on the distribution of galaxies in space and "centered on the premise that the universe and its elements were products of random processes." Together with Neyman and C. Donald Shane, she was the first to model stochastic clustering and study the various properties of the spacial distribution of galaxies. This work got started when Shane brought a problem to Betty and Neyman at the Stat Lab in 1950 and asked for their statistical advice. Here is Betty's story:

> We devised a stochastic model of kth-order clustering [of galaxies], a four-dimensional point process projected onto the two-dimensional photographic plate. We estimated the parameters in the model from the observed quasi-correlations and other moments, taking into consideration the uncertainties in the counts due to the difficulties in distinguishing a faint galaxy from a star on the astrographic plates... Our first-order clustering model fitted the observations from the 10' x 10' counts very well. We also fitted the counts from cells combined into a 6° x 6° square very well, but unfortunately the numerical estimates of the parameters of the clustering were quite different in the two cases. As the expected number of galaxies per cluster cannot depend on the size of the cell in which the galaxies are counted, we must conclude that the first-order clustering model needs refinement, for example, by including the evidence for subclustering and superclustering which point to higher order clustering. But, by now [1976] astronomers are convinced that galaxies tend to occur in clusters.

In Betty's mind, there were two primary follow-up questions. One was: How well determined is the relationship between apparent luminosity and redshift, and how precise is its scale? The second had to do with selection bias: How can the fact that astronomical observers select the objects they observe be factored into the results?

---

[29]n.d. [?1953] handwritten manuscript titled Magnitude-Redshift Relation for Cluster Galaxies in the Presence of Instability and Absorption, by Jerzy Neyman and Elizabeth L. Scott.

[30]1/15/1953 letter from Betty Scott to Mount Wilson and Palomar Observatories Astronomer Edison Pettit. Note: Edison Pettit (1889–1962) was an astronomer with specialties in solar astronomy, planetary details, and instrument development, who made his career at Mount Wilson Observatories.

Betty and Neyman removed assumptions of normality, and they made the assumption that "...the galaxies in nearby and in distant clusters had the same relative abundances and the same space distribution of absolute luminosity, and also the same selection probability function for each type as had been estimated from the field galaxies and galaxies in small groups..." This allowed them to carry out estimation procedures:

> *They found remarkable agreement between their predictions and the actual catalog distribution for almost every type and kind of cluster, indicating that the space abundances, the space distribution of luminosity, and the selection probability functions vary markedly from one morphological type to another, but that they are the same for each type for field galaxies and for cluster galaxies, at least for those clusters within the Humason-Mayall-Sandage catalog. The striking empirical differences in the catalog distributions, well known to astronomers, are all easily explained by the differing selection probabilities operating at increasing distances.*

Betty herself also found that:

> *...the expected Hubble relation is not a straight line; the amount of deviation [from a straight line] will depend on the dispersion in the space distribution of luminosity. Also, a cluster will be more likely to have a galaxy bright enough to be observed if the cluster has an unusually large number of members. Both of these effects – the tendency for a distant galaxy to be unusually luminous and the tendency to belong to a very rich cluster – will curve the Hubble relation and thereby confuse the cosmological tests. Astronomers have tried to obviate the effect by searching out objects with small dispersion in luminosity. As astronomers are working here at the limit of their instruments, this task is not easy and it is probably more convenient and more accurate to calculate the statistical refinement in the Hubble relation.*[31]

Billard and Ferber explained Betty's contribution this way:

> *Clusters of galaxies are important for measuring the geometry of the universe. When more distant clusters are observed, they tend to contain more galaxies. This observational bias means that galaxies of any rank will be intrinsically brighter in these more distant systems. This observational bias or selection effect was first pointed out by Elizabeth Scott and is commonly known as the "Scott effect."*[32]

Betty's work on spacial distribution of galaxies resulted in over 20 papers. Sometimes these papers were with other astronomers, sometimes these were with Neyman, and sometimes she was the sole author.

---

[31] E.L. Scott, "Statistics in Astronomy in the United States," in *On the History of Statistics and Probability*, ed. D.B. Brown (New York: Dekker, 1976), 319.

[32] L. Billard and M.A. Ferber, "Elizabeth Scott: Scholar, Teacher, Administrator," *Statistical Science* 6, no. 2 (1991), 206.

In Betty's words: "We stopped when there were no additional observations."[33] As she turned her attention to problems in cancer and the status of academic women while continuing her work on cloud seeding, Betty remained excited about the many unsolved problems in astronomy and that many of these were statistical problems.[34]

Betty's last paper in astronomy was in 1976. It was a review of statistics in astronomy in the US. Within this broad topic, Betty focused on determinations of distance and distributions in space: "The distances to the objects that astronomers study today are determined through a complex chain where most of the links are statistics," she wrote. Whereas trigonometry could be used to determine distances to close objects, statistical techniques needed to be used for more distant objects. Betty's paper described these statistical techniques, many of which were the products of her own work.

In spring 1981, Betty was back at work on the distribution of galaxies. She wrote this in her 1984 sabbatical report:

When I was a student, it was thought that galaxies are uniformly distributed in space, with perhaps a few clusters to add 'poetic interest.' Neyman, Shane and I published a series of papers showing that the distribution of galaxies in space is consistent with the assumption that all galaxies are in clusters (and even the clusters may themselves be clustered). Recently, there has been discussion of the voids between the clusters. This is not just a matter of semantics, and there is some emphasis on thinking of clusters as sheets. Such studies require more careful analysis than they have yet received, and this is what I have started working on. The problems require heavy computations on large data sets. I am still in the organizing stage.[35]

But there would not be time in Betty's life for more astronomy papers.

UC-Berkeley Statistics Professor David R. Brillinger (1937–) described Betty and Neyman's work on spatial point processes as "path breaking," especially their work in astronomy, and acknowledged their influence on his own work. He also acknowledged Betty for inviting visiting scholars when she was department chair; these visitors became important collaborators.[36]

Betty was a serious scientist. She wasn't carried away by fads, nor influenced by magic or superstition. In 1975, she and 185 other scientists disavowed astrology and its practices by signing a statement in *The Humanist*

---

[33]4/28/1988 UC-Berkeley Faculty Research Grant Application for 1988-89 by Betty Scott regarding the project Spacial Distribution of Galaxies: Clusters versus Voids versus Sheets versus Bubbles versus Wormholes.

[34]E.L. Scott, "Statistics in Astronomy in the United States," in *On the History of Statistics and Probability*, ed. D.B. Brown (New York: Dekker, 1976), 319.

[35]5/21/1984 report from Betty Scott, "Report on Sabbatical Leave, January 1 to June 30, 1981."

[36]Victor M. Panaretos, "A conversation with David R. Brillinger," *Statistical Science* 26, no. 3 (2011), 440.

stating that there was no verified scientific basis to support such beliefs.[37,38] Betty urged the publication of studies on astrological phenomenon that "use the available data properly," and that involved using appropriate statistical tests.[39] She was occasionally asked to review articles involving astrology experiments, one of which had to do with the birth dates of sports champions and the position of Mars. She approached these reviews as she would any other scientific review, raising issues about sample sizes and power, discussing sensitivity of tests, questioning sampling designs ("Science cannot proceed with hidden aspects"), demanding discussion of the effects of specifying hypotheses after looking at the data, realizing the losses of power in subset analyses ("Philosophically, the statistician is not allowed to tamper with the data when testing a hypothesis"), etc.[40,41,42]

7.4   General Statistical Methods

Betty's first publication with Neyman was in 1948. It was a mathematical statistics study of "consistent estimates based on partially consistent observations," appearing in the journal *Econometrics*. The examples given were from astronomy, with but a nod to the applicability of the methods to the field of economics ("...likely to be applicable to some economic problems...similar set-ups may be applicable in studies of economic phenomena"). This publication set the stage for publication of the math stat part of Betty's dissertation a few years later in *The Annals of Mathematical Statistics*: "Note on consistent estimates of the linear structural relation between two variables." Betty's second publication with Neyman came the next year, in 1951. This was "on certain methods of estimating linear structural relations," also published in *The Annals*.

Even Betty's work within the theme of general statistical methods was broad. The chapter on hypothesis testing that she wrote for the Trumpler and Weaver textbook, *Statistical Astronomy* (see Chapter 6), falls within this theme. Across the span of her career she took on a variety of topics, including bias correction, asymptotically optimal tests, clustering, sub-clustering, tests of composite hypotheses, outliers, correlation and causation, and analysis of irregular effects. Much of this work was inspired by Betty and Neyman's many years of work on weather modification.

[37] 8/11/1976 letter from Sune V. Malmgren of Redondo Beach, CA to Betty Scott.

[38] "Objections to Astrology: A Statement by 186 Leading Scientists," *The Humanist*, September/October 1975, 4.

[39] 2/7/1978 letter from Betty Scott to *The Humanist* Editor Paul Kurtz.

[40] 4/15/1982 letter from Betty Scott to Daniel Cohen of Fort Jervis, NY.

[41] 7/13/1982 letter from Betty Scott to James Randi of Rumson, NJ.

[42] 8/12/82 letter from Betty Scott to Bell Laboratories Stan Willie.

Betty worked hard to stay on top of the field. For example, in 1969, she received a letter from Egon Pearson at University College, London. He had just published tables of the percentage points of the non-central chi square probability distribution.[43] Pearson sent Betty four copies of the paper, and declared that she was "the first person I have heard [sic] of as using the tables!" He also thanked Betty and Neyman for their gift of two bottles of "that powerful Russian (?) [sic] spirit." Betty and Neyman liked to give gifts to their professional friends and collaborators across the globe.[44]

7.5   Bioscience and Health

Betty's first work in the area of bioscience and health was in 1956. It was the result of a three-year collaboration with University of Chicago Ecology Professor Thomas Park involving a Tribolium (a type of flour beetle) model, appearing in the *Proceedings of the Third Berkeley Symposium of Mathematical Statistics and Probability*. The article was actually in two parts. Part II, Betty and Neyman's contribution, was the statistical theory contribution; specifically, it was a formulation of a stochastic process model of the biological aspects of the problem presented by Park in Part I. A paper followed the next year that examined populations as "conglomerations of clusters"; it was "intimately connected" with the previous work of Betty and others on the distribution of galaxies in space. In Park's obituary, the Tribolium research was called Park's landmark study, and Park was said to be "one of the first to apply quantitative methods to the study of environmental populations."[45] It is curious that no reference was made to Betty and Neyman.

A third paper, on stochastic models of population dynamics, analyzed the two distinct stochastic models developed in the first two papers in terms of "elementary chance mechanisms." Betty and Neyman presented this paper in 1958 at the annual meeting of the American Association for the Advancement of Science (AAAS) in Washington, DC, and won the AAAS Newcomb Cleveland Prize. In Betty's and Neyman's words: "It was presented at the invitation of the Society for General Systems Research, the principal aim of which is to encourage the development of theoretical systems applicable to more than one of the traditional departments of knowledge." This award-winning interdisciplinary sciences paper was subsequently published in *Science*.[46]

---

[43]N.L. Johnson and E.S. Pearson, "Tables of percentage points of noncentral Chi," *Biometrika* 56, (1969), 255.

[44]11/23/1969 handwritten letter from University College London Professor of Statistics Egon Pearson to Betty Scott. Note: Pearson did not indicate where Betty used the tables.

[45]Kenan Heise, "U. of C. Ecology Expert Thomas Park," *Chicago Tribune*, April 1, 1992.

[46]J. Neyman and E.L. Scott, "Stochastic Models of Population Dynamics." *Science* 130, no. 3371 (1959), 303. Note: The annual meeting took place on December 29, 1958.

Next Betty took on epidemics, and then Parkinsonism. In the early 1960s, Betty and Neyman discovered a book on the mathematics of epidemics by Norman T.J. Bailey.[47] It inspired them to develop a stochastic model of epidemics, taking into account susceptibles, infections, and movements about a habitat. Betty also worked with a clinical team on an applied study of the stereotaxic surgical technique to relieve symptoms of Parkinsonism.

Then came about a dozen cancer studies, the first in 1967. Betty credited Neyman with introducing her to "an active study of carcinogenesis." The first paper was with Neyman in the *Proceedings of the Fifth Berkeley Symposium on Mathematical Statistics and Probability.*

In 1969, the Nixon administration announced it would build a set of supersonic transports, flying at 1,700 miles per hour, as a way of ensuring US leadership to the world in the area of air transport.[48] There was considerable environmental concern about the project, especially about the potential that high altitude flights would cause ozone layer depletion, which in turn could have effects on human health, including increasing skin cancer. In 1971, both the US Senate and House of Representatives disapproved funds for the project.[49] Betty was invited to serve on the newly formed (organized in 1972) Committee on National Statistics of the National Academies; she was the only woman among the ten members.[50,51] She subsequently was invited to serve on a panel focused on the environmental effects of supersonic transports, especially ozone depletion, and how it could increase skin cancer. Betty volunteered to chair this Panel to Review Statistics on Skin Cancer.[52] She involved her graduate students in this effort by forming a skin cancer workshop in the statistics department at Berkeley that she led every Tuesday.[53] According to some of Betty's closest colleagues:

> *She attacked the problem with her usual energy and determination, becoming an expert on skin cancer and a frequent visitor to the melanoma clinic at UCSF. Predicting how much skin cancer would increase was a formidable task. Betty set out to collect data on each and every relevant*

---

[47]N.T.J. Bailey, *The Mathematical Theory of Epidemics* (New York: Hafner, 1957).

[48]Richard Nixon, "Remarks Announcing Decision To Continue Development of the Supersonic Transport," *The American Presidency Project,* September 23, 1969, accessed February 2, 2014, http://www.presidency.ucsb.edu/ws/?pid=2240.

[49]Richard Nixon, "Statement about Senate Action Disapproving Funds for the Supersonic Transport Program," *The American Presidency Project,* March 24, 1971, accessed February 27, 2014, http://www.presidency.ucsb.edu/ws/?pid=2949.

[50][n.d.] listing, Committee on National Statistics, Division of Mathematical Sciences, 136. Note: William Kruskal built and chaired the committee with a mission to provide "policy guidance for the application of statistics to problems of national interest which arise either within or outside the NRC."

[51]5/29/1978 letter from Betty Scott to University of Chicago Statistics Professor and Social Sciences Dean William Kruskal. Note: Betty wrote to Kruskal, "You have made a terrific contribution to statistics in building this committee and I congratulate you!"

[52]Note: Other members of the Panel to Review Statistics in Skin Cancer included Peter Bloomfield, Philip Cole, Richard A. Craig, Harley A. Hanes, and Richard B. Setlow.

[53]1977–78 Jerzy Neyman's Calendar, Academic Year 1977–78.

factor and devised some very complex statistical models, together with some simpler ones: They all were in general agreement, but the complex models better illustrated the uncertainties in the predictions. The models have been used for many years to evaluate threats to the ozone layer.[54]

Betty had some personal cancer scares, which surely contributed to her interest in cancer research. She had a third scare in 1977 that put her in the hospital for many tests.[55]

Early efforts "reviewed the evidence for ultraviolet radiation as a cause of cancer."[56] In 1977, Betty explained that her efforts were toward "estimating the increase in skin cancer that will result from the increase in ultraviolet radiation due to depletion of stratospheric ozone."[57,58] The next year Betty wrote: "The depletion of the stratospheric ozone by man-made pollutants, such as exaust [sic] from high-flying airplanes or the chlorofluoromethanes used as propellants in spray cans and in refrigerants, will reduce the stratospheric ozone which provides partial protection from ultra-violet radiation. Ultra-violet radiation is a major cause of skin cancer, the most common of all cancers in the United States. Thus, depleting stratospheric ozone will increase skin cancer but how large will this increase be?"[59]

Also in 1977, Betty further expressed her statistical thinking about skin cancer in a brief abstract:

> *The estimate of the increase in skin cancer that will result from increasing ultraviolet radiation can be done by two methods: (1) Estimating the relationship between the occurrence of skin cancer at different localities and the intensity of ultraviolet radiation at the same localities, and (2) Studying the increase of skin cancer with age at each of several different localities. There are difficulties in estimating the relationships: (a) the observations of skin cancer, and also of ultraviolet radiation, are uncertain and incomplete, (b) the carcinogenesis of skin cancer is unclear and possibly complicated by complex ultraviolet repair mechanisms, (c) there are other associations with skin cancer besides ultraviolet radiation and these may be interacting, (d) melanoma rates are increasing with time (all over the world) especially in younger persons and at sites which were in years past protected by sunlight, such as women's legs and men's backs,*

---

[54]Blackwell, David; Colson, Elizabeth; Ervin-Tripp, Susan; Le Cam, Lucien; Lehmann, Erich; and Nader, Laura. Elizabeth Leonard Scott, Statistics: Berkeley. University of California: In Memorium. Accessed on 2/17/2014 at http://content.cdlib.org/view?docId=hb4t1nb2bd&doc.view=frames&chunk.id=div00061&toc.depth=1&toc.id=.

[55]1/2/1977 letter from Betty Scott to UC-Berkeley Stat Lab Assistant Marcella.

[56]E.L. Scott, "Epidemiology of skin cancer under increasing ultraviolet radiation," in *Indo-US Workshop on Global Ozone Problem*, ed. A.P. Mitra (New Delhi: INSDOC, 1984), 235.

[57][?1977] "Brief Sketch of Activities" by Betty Scott.

[58]3/5/1977 Elizabeth L. Scott. In: New Directions for Women: Portraits of 21 Women in Science. Participants in a conference at Yale University.

[59]6/3/1978 letter from Betty Scott to Sonoma State College Mathematics Professor Rick Luttmann.

presumably due to changing life styles (if so, nonmelanoma, which occurs in older ages than melanoma, on the average, can be expected to increase sharply also), (d) [sic] there are certain statistical biases in the standard methods of estimation, tending to underestimate the estimated increases in skin cancer and (e) competing risks of death.[60]

Here Betty worried about the different methods of estimation, and the various factors that could bias the estimation.

In 1976, the interdisciplinary cancer research that Betty led for the NAS was recognized in various UC presses. It had already been argued by other scientists that human factors such as supersonic jet exhausts, compounds in aerosol containers and fertilizers, space shuttles, and nuclear fallout, could cause ozone depletion. In March, the UC-Berkeley *Monday Paper* and the UC *Clip Sheet* both reported from Betty's latest panel study that if 10% of the ozone layer were destroyed by human factors and thereby exposed people to more of the sun's ultraviolet rays, then there could be a 30% or higher increase in skin cancer cases. This estimate was noteworthy because it was "significantly higher" than Betty's team reported in the previous year's NAS study. Put more plainly, Betty's new work indicated cancer from ozone depletion was a bigger threat than previously thought. Betty's new work included adjustment for underreporting of non-melanoma cases, and took into account more factors, including differences among individuals in types of clothing, kinds of occupations, and susceptibilities to sunburn.[61,62]

Betty thought there were many interesting problems in the area of skin cancer. "Partly because it is so common with human beings, is easy to study on the back of a mouse, and also easy to keep alive in a test tube, there is a great deal of literature and lots of interesting experiments at the cellular level." She wanted to continue her studies "from this point of view."[63]

Betty continued this research into the 1980s, focusing on it during a spring semester sabbatical in 1981. She analyzed data on melanoma and non-melanoma [sic], "with emphasis on the carcinogenesis, tumor growth, environmental effects, and changes with treatment." In addition to writing her own research papers, she supervised theses in this area.[64]

---

[60]9/30/1977 letter from Betty Scott to Purdue University Statistics Professor Gregory Campbell.

[61]3/12/1976 newspaper article, "Danger of less ozone: Scientists predict rise in skin cancer," *Monday Paper* (University of California, Berkeley), March 12, 1976.

[62]"Study on skin cancer-ozone link," *UC Clip Sheet* (University of California), March 16, 1976.

[63]3/30/1978 letter from Betty Scott to University of Chicago Statistics Professor and Social Sciences Dean William Kruskal.

[64]5/21/1984 report from Betty Scott, "Report on Sabbatical Leave, January 1 to June 30, 1981."

7.6 Symposia, Panels, and Talks

The Berkeley Symposia on Mathematical Statistics and Probability were a Neyman project, a vehicle to bring together some of the best talent in statistics and where he could control what was presented and published. There were six symposia in all, with five years between symposia, spanning 25 years. Betty participated in all but the first symposium, contributing a total of 11 scientific papers to the series.[65],[66] Betty co-edited the proceedings of the sixth symposium, along with Neyman and Lucien LeCam. This was the last in the series and totaled over 3,000 pages.

Betty recognized that Berkeley could be intellectually isolating and she needed to be part of larger (national or international) conversations of experts. This would ensure she was exposed to diverse arguments about a scientific issue, and then she could decide whether she agreed or not. "It is an interesting but confusing experience to be set down in a group of experts," she observed.[67] And so Betty served on a number of panels of experts over the course of her career in addition to the NAS skin cancer panel. For example, in 1978, she was one of three members on a contraceptive drug study advisory committee to the National Institute of Child Health and Human Development.[68]

Betty spent a lot of time on the road communicating her science. She was in demand as a research speaker, in a variety of settings both on campus and off. Take for example the first three weeks of May 1977. This was her travel schedule for her research talks:

> *May 3, University of Pacific, Stockton: "Does Cloud Seeding Decrease Rainfall?"*
>
> *May 3, University of Pacific, Stockton: "The Status of Women in Higher Education – The Facts of the Matter"*
>
> *May 4, San Diego State University: "The Expanding Universe"*
>
> *May 13 & 14, UC-Berkeley Lawrence Hall of Science: "Consider the Sciences! An Invitation to Minority and Women Students"*
>
> *May 17, UC-Davis: "Skin Cancer and Ozone"*

---

[65]Neyman, Jerzy, Hodges, Joseph L. Jr., Le Cam, Lucien, Lehmann, Erich L., and Scott, Elizabeth L. Evelyn Fix (1904–1965). UC-Berkeley In Memorium. Accessed on 2/16/2014 at statistics.berkeley.edu/people/evelyn-fix-1904-1965. Note: The dates given in the official UC-Berkeley Memorium are used here and differ by one year from the dates given in Constance Reid, *Neyman: From Life*, p. 176, 1982.

[66]Erich L. Lehmann, *Reminiscences of a Statistician: The Company I Kept* (New York: Springer, 2008), 33.

[67]8/18/1981 letter from Betty Scott to Temple University School of Medicine, Skin and Cancer Hospital, Center for Photobiology, Frederick Urbach.

[68]3/29/1978 letter from The Permanente Medical Group, Contraceptive Drug Study, Director Savitri Ramcharan to Betty Scott. Note: The other two members were from Johns Hopkins University in Baltimore and the Royal College of General Practitioners (England).

> *May 20, Skyline Community College: "Does Cloud Seeding*
> *Decrease Rainfall?"*[69]

Betty was not shy about being interviewed by the media. She and Neyman were so connected that Neyman kept Betty's travel days on his own schedule.[70]

At all times, including when they traveled to speak, Betty and Neyman were scrupulously honest. For example, when they reserved a hotel in relationship with a speaking engagement at the University of British Columbia in the summer of 1974, they were told each of their rooms would cost $18.00 a night. However, they were charged only $8.93 including tax each night. They had Neyman's secretary write to the clerk at the university to ask him to please check his records one more time and "clear up this misunderstanding."[71]

7.7 Managing Neyman

Some of Betty and Neyman's astronomy research had focused on differences in radial velocities from absorption and emission lines. Thornton Page at NASA had a new manuscript on radial velocities in which he attributed differences between absorption and emission lines to Morton Roberts at the National Radio Astronomy Observatory. In June of 1969, Page asked Roberts to review and comment on the manuscript. Roberts disclosed that, although prompted by his own findings and suggestions, the differences between absorption and emission lines were due to Betty and Neyman; therefore Page should cite Neyman and Betty in addition to himself. Roberts suggested Page check with Betty and Neyman: They "have only very tentative plans as to how to present these findings and how to explain them." Neyman, as usual, looked to Betty for advice. He scribbled a note to her: "What should we do?"[72]

Research collaborators often corresponded with Betty and Neyman as if they were one. Betty often handled the return correspondence. For example, when she and Neyman were sent some interesting photographs from C.R. Rao (1920–), Betty told him how pleased they all were to see them and included a specific message from Neyman.[73] Betty also often tidied up Neyman's affairs. For example, Neyman was overcharged for airline travel to India. When many other attempts by others to rectify the overcharge failed, it was Betty who

---

[69] 1977 travel schedule of Betty Scott, "Travel: end of April – end of May, 1977."

[70] 1977–78 Jerzy Neyman's Calendar, Academic Year 1977–78.

[71] 8/30/1974 letter from UC-Berkeley Secretary (to Jerzy Neyman) Kay Kewley to The University of British Columbia Totem Park Convention Center.

[72] 7/1/1969 letter from Morton Roberts to Thornton Page. Note: Betty's response was not recorded.

[73] 2/7/1975 letter from Betty Scott to Indian Statistical Institute Professor C.R. Rao.

wrote a lengthy letter to Air India trying to get the airline to issue a refund to the Indian Statistical Institute.[74]

It is interesting that Betty and Neyman's research collaborations did not extend to gender equity research, because they worked together in many other areas, especially problems in mathematical statistics, astronomy, and weather modification. In these other areas, Betty followed through for Neyman, often interfacing with the statistical assistant to make sure his large and small ideas were correctly implemented in addition to her own. Betty's follow through was done ably and efficiently because the work was a shared endeavor and she understood all of its nuances.[75] She was Neyman's unofficial manager in addition to his collaborator, and she had affection for their Stat Lab assistants. For example, she wrote to Marcella and Sheryl: "With best regards and many thanks, and love to all, Betty."[76]

---

[74]3/13/1975 letter from Betty Scott to Air India, Sales Manager for the US and Canada, F.G. Martin.

[75]1/2/1977 letter from Betty Scott to [UC-Berkeley Stat Lab Assistant] Marcella.

[76]8/28/1976 handwritten letter from Betty Scott to [UC-Berkeley Stat Lab Assistants] Marcella and Sheryl.

8

The Case of Cloud Seeding (1950–1985)

This is slippery stuff being handled by even [more] slippery people. . .
-James Wallis

Betty worked on weather modification for 35 years, starting in 1950, only four years after initial attempts by cloud seeders.[1] At first her work focused on design and evaluation of weather modification in California, specifically Santa Barbara. Then there was a six-year gap in Betty's published work on the subject that culminated in four papers by her and Neyman in the *Fifth Berkeley Symposium on Mathematical Statistics and Probability*. These papers swallowed up over 75 pages in the proceedings. Betty and Neyman continued their work on weather modification through the 1970s using both international and domestic data. There were even two straggling papers in the 1980s.

The work was controversial. By 1972, weather modification had become an important public policy issue: There was strong and increasing public pressure to try to bring precipitation to areas experiencing severe shortages of water. The problem was that weather modification had not advanced scientifically to the point where evidence indicated it was effective, including in the US. Yet there was extensive cloud seeding intended to increase precipitation or decrease hail damage, and this was taking place all over the world. Commercial rainmakers were enthusiastically selling their product, ignoring the complexities of the atmosphere. This is where Betty and Neyman came in.

8.1 Increase or Decrease Precipitation?

In 1973, University of Chicago Statistics Associate Professor K. Alexander Brownlee was working on cloud seeding issues in Colorado. As was common

---

[1]Richard A. Kerr, "Weather modification: A call for tougher tests," *Science* 202, no. 4370 (1978), 860.

with cloud seeding, the situation in Colorado was political. In this instance, it was the farmers versus the ranchers. Farmers in the San Luis Valley were attempting to prevent summer hailstorms by seeding the clouds with silver iodide. The seeding was intended to preserve the farmers' $10 million barley crop, which had recently been sold to the Coors beer company and accounted for 15% of the valley's income. Recently, weather had affected the barley crop, and Coors was threatening to buy elsewhere. Ranchers, on the other hand, did not want the cloud seeding. They claimed it was reducing the amount of rainfall and was thereby affecting their livelihood.

Brownlee wrote to Betty informing her of the situation. He asked her for help with a cloud seeding statistical problem that had been plaguing him for many years. In their paper at the International Symposium on Uncertainties in Hydrolic and Water Resource Systems, Betty and Neyman commented that a certain result was clearly not due to chance, but they hadn't conducted a statistical test. Another researcher had conducted a test and also concluded the result was not due to chance, but Brownlee wasn't familiar with the test. He wanted to know: Did the test have underlying assumptions and, if so, were they met? Rothamsted Experimental Station Statistician Frank Yates (1902–1994), in the United Kingdom, looked at the same illustration and thought the difference was due to chance. Now Brownlee was troubled, as the "most prestigious statisticians" (Betty and Neyman versus Yates) were stating different conclusions. He hoped Betty would conduct an appropriate statistical test to help settle the issue. Brownlee wrote, "...resolution of this emotional conflict would be a great joy!"

Cloud seeding gained national attention due to projected severe droughts in the western US. In 1973, a Colorado member of the US House of Representatives introduced a bill to put federal controls on weather modification.[2] The drought continued into 1974–75 when Betty took a sabbatical. Her weather modification study results were intriguing to her administration, and they were becoming increasingly important to the public.[3,4]

In 1976, Betty summarized her work on cloud seeding in this way:

> *Commercial operators will tell you that cloud seeding, such as putting a little silver iodide smoke into a cloud, allows them to modify the weather as they wish: to decrease hail in Colorado, to increase rainfall in Arizona, and so forth. The facts are that every single experiment in the United States has given negative results: more hail and bigger hail stones instead of less, less rain instead of more and not just in the target area – the decrease stretches more than a hundred miles downwind. We have been active in the design and analysis of these weather modification ex-*

---

[2]4/9/1973 letter from University of Chicago Statistics Associate Professor K. Alexander Brownlee to Betty Scott.

[3]6/3/1978 letter from Betty Scott to Sonoma State College Mathematics Professor Rick Luttmann.

[4]2/17/1977 letter from UC-Berkeley Dean of Physical Sciences Leonard V. Kuhi to Betty Scott.

periments for many years now and have learned a lot but the problems are [far] from solved.[5]

At the beginning of 1977, in the middle of the worst drought in California's history, Betty was concerned whether "our efforts to modify the clouds actually reduce the amount of rain reaching the ground."[6,7] In other words, does cloud seeding increase precipitation, or does it decrease it? Cloud seeding operations in California were many and increasing. Taxpayers were paying the bills. Betty worried whether peoples' dollars were being put to good use or wasted. She raised important questions. Was cloud seeding providing more precipitation than would have happened naturally? Was cloud seeding affecting the precipitation downwind from the target? One would be a good use of tax dollars, the other not.

8.2   An Emotional Issue

Already by 1975, Betty and Neyman were frustrated by what they saw to be lies to the public by the cloud seeding industry. Neyman came back from a conference in the Rockies thinking: "... I learned quite a lot – not so much about hail suppression as about these people. While I am thinking of the discussions, an old Ukranian story comes to my mind including the statement that 'THERE(!) they make liars lick red hot pans.' " He planned to call them out at an appropriate time.[8]

Wallace E. Howell (1914–1999) used to be a commercial cloud seeder, but in 1976 went to work for the federal government in the Division of Atmospheric Water Resources Management at the US Department of the Interior Bureau of Reclamation Engineering and Research Center. Betty found the work of Howell to be particularly substandard, based on the fact that he had a track record of making "very slippery claims (based on tossing out some storms, leaving out some stations, difficulties with arithmetic, etc., in his favor, so that the claims cannot be verified)."[9] So substandard that when she was presented with a paper by Howell in 1976 she didn't feel it was even worthy of comment – it "turned us off." She assumed it "would die of its own accord."

Over the next few months, Betty and Howell exchanged sharp criticisms. Betty accused Howell of launching "... a completely uncalled for and non

---

[5]Scott, Elizabeth L. Panel Talk, Proceedings of the Conference on Educating Women for Science: A Continuous Spectrum, Mills College, Oakland, CA, April 24, 1976, pp. 32-24.

[6][?1977] Brief Sketch of Activities by Betty Scott.

[7]3/5/1977 Elizabeth L. Scott. In: New Directions for Women: Portraits of 21 Women in Science. Participants in a conference at Yale University.

[8]9/30/1975 letter from Jerzy Neyman to US Department of Agriculture, Southwest Watershed Research Center Investigator Herbert B. Osborn.

[9]10/12/1976 letter from Betty Scott to IBM Researcher James Wallis.

sequitur attack..." on one of her cloud seeding research colleagues.[10] Howell retorted: "...I yield at once to your superior expertise in the matter of non-sequitur attacks..."[11]

IBM Researcher James Wallis, chair of the National Academy of Sciences (NAS) Panel on Water and Climate, was trying to find out more about cloud seeding in the southwest and telephoned Betty. She responded with a lengthy letter describing the research and its limitations. With usual humility, she concluded: "I am afraid that I am not providing the information you really need."[12] But, of course, the information was exactly what Wallis needed: "Thanks very much you have been most helpful. I will not quote your letter although I will most certainly use it."[13] A few days later, after more thought and investigation, Wallis wrote to Betty again: "This is slippery stuff being handled by even [more] slippery people..." He worried to Betty about how hard it was going to be to write a report, saying "it would appear that there may be political interference from NAS luminaries."[14]

In her letter to Wallis, Betty had mentioned Howell by name and described his "very slippery claims." The letter was now in the public domain, as it had been submitted to a committee of the NAS.

The letter got into Howell's hands. He was livid. He told Betty he considered her statements to be "personally libelous and professionally outrageous," adding: "The insinuation is unavoidable that I have used biased selection of data and disingenuous arithmetical 'errors' as means of falsifying my representation of the results of weather modification operations I have conducted or with which I have been connected." He asked if her statements were a charge of research misconduct: "Coming from a person of your academic stature and reputation, on the letterhead of the bastion where unbias is supposedly the most zealously guarded virtue, and addressed to an exceptionally powerful and prestigious if limited public, the charge is one I must answer if I am to behave with any self-respect at all." He demanded Betty back up her charge with specific, explicit details that he could attempt to rebut.[15]

Betty's letter then got into the hands of Archie M. Kahan, chief of the Division of Atmospheric Water Resources Management at the USDI Bureau of Reclamation. Kahan's lengthy response, sent to Wallis, took on a number of Betty's statements, which was no surprise given Kahan's own admission that his letter was researched and drafted by Howell.[16] Then Betty's letter got into

---

[10]8/4/1976 unpublished comments by Betty Scott, "Comments on Paper by Howell."

[11]8/12/1976 letter from US Department of the Interior, Division of Atmospheric Water Resources Management, Assistant to the Chief Wallace E. Howell to Betty Scott.

[12]10/12/1976 letter from Betty Scott to IBM Researcher James Wallis.

[13]10/20/1976 handwritten note from IBM Researcher James Wallis to Betty Scott.

[14]10/22/1976 handwritten letter IBM Researcher James Wallis to Betty Scott.

[15]1/6/1977 letter from Certified Consulting Meteorologist Wallace E. Howell to Betty Scott.

[16]1/11/1977 letter from the US Department of the Interior, Bureau of Reclamation Engineering and Research Center, Division of Atmospheric Water Resources Management Chief Archie M. Kahan.

the hands of another commercial cloud seeder, the North American Weather Consultants President Robert D. Elliott in southern California. Elliott, too, was mentioned specifically by Betty and took offense to her claims that he redid his analyses until they came out the way he wanted them to. He also asked Betty to provide specific details about her statements.[17]

Wallis was surprised. He never intended that any of the panel's working documents would be released to the public, and certainly "such papers were never intended to be bandied about in an indiscriminate manner," he told Kahan.[18] Wallis apologized profusely to Betty: "I regret that your helpful and forthright letter to me should have resulted in vitriolic attacks upon your honesty and integrity. Unfortunately, weather modification is an emotional, rather than a scientific issue... One does not expect to find unbiased scientific evaluation in special interest groups, and the fact if not the manner and frequency of their attacks on those who suggest it is predictable."[19] He believed his report, which he expected would upset the Bureau of Reclamation, would take some of the heat off Betty.[20]

8.3   Radio Broadcast

Betty and Neyman seriously doubted the efficacy of cloud seeding. On July 7, 1977, they wrote a summary of their work for the KPFA Radio Station in Berkeley. Here is the full text of their broadcast:

> *Mr. David Perlman's article in the San Francisco Chronicle of July 4th informed the readers of the current apparently multi-million dollar effort in California to bring down some rain, of the deep conviction of some persons that the effort will be successful and of serious doubts of some others. We belong to this latter category.*
>
> *Since 1950, both of us, working at the Statistical Laboratory, U.C. Berkeley, have been in almost continuous contact with weather modification. In that year 1950, the then Division of Water Resources, Department of Public Works of the State of California, requested the Laboratory to investigate the claims of successes made by the several commercial cloud seeding companies. The question was whether cloud seeding could*

---

[17]1/21/1977 letter from North American Weather Consultants President Robert D. Elliott to Betty Scott.

[18]1/21/1977 letter from IBM Researcher James Wallis to US Department of the Interior, Bureau of Reclamation Engineering and Research Center, Division of Atmospheric Water Resources Management Chief Archie M. Kahan.

[19]1/21/1977 letter from IBM Researcher James Wallis to Betty Scott.

[20]1/30/1977 letter (with handwritten note to Betty) from IBM Researcher James Wallis to NAS Geophysics Research Board member Donald Shapiro.

be realistically considered as a source of water, then badly needed in the southern part of the state.

We did what we could, with negative results. It appeared that the enthusiastic claims of success of commercial cloud seeders were manifestations of fancy salesmanship. To our regret, the Report of the President Eisenhower's Advisory Committee on Weather Control, published in 1957, appeared to us unreliable. On the one side, there were overoptimistic opinions and, on the other, contradictory statements, including misrepresentations.

This about summarizes our weather modification experience during the first decade of familiarity with the subject. In 1960, we gave up a supporting grant of the National Science Foundation and returned to statistical studies in other fields, particularly astronomy.

Then, in 1964, we unexpectedly received seven volumes of annual reports on the hail prevention experiment performed on the southern slopes of the Alps. Under the influence of our earlier experience we examined the reports with some skepticism. However, with a degree of pleasant surprise, we saw distinct signs of serious research, with no signs of any cover-ups. While the experiment was concerned with hail, it also provided detailed information on rainfall, something which has been the subject of our principal concern. We did not quite subscribe to the method of evaluation used by the Swiss statistician (a Dr. P. Schmid, our one-year visitor in Berkeley) and performed one of our own. We found that, indeed, the seven-year-long experimentation with silver oxide cloud seeding showed large increases in rainfall not only in the target, Canton Ticino, but also in two localities on the other side of the Alps, one near Zurich, about 80 miles north of the intended target, the other near Neuchatel, some 120 miles north-northwest from Ticino. In particular, when Zurich was downwind from the target the apparent effect of Ticino cloud seeding on rain near Zurich was a 100% increase in rain and our calculations showed that this could hardly be due to chance variation. On the other hand, no such increases were found on days when Zurich was upwind.

This experience changed our attitudes completely. We became intensely interested in the question as to whether any such phenomena occurred in other experiments, particularly in the United States.

Analysis of cloud seeding experiments required personnel and computing facilities. We were lucky to obtain the needed resources from the Office of Naval Research and since about 1965 became deeply involved in cloud seeding studies. Our primary efforts were directed towards answering the question whether the cloud seeding performed over a customary sized target, say some 25 to 50 miles across, can affect the precipitation in far away localities when those localities are downwind.

Briefly and roughly, the examination of a number of experiments showed that most of them had certain marks of non-reliability: signs of cover-ups and of defects of "randomization." However, we found one ex-

periment performed in Arizona, which appeared reasonably reliable. As is well known, Arizona is marked by rather dry climate (long range average annual rainfall of about 6 inches, compared to about 20 inches in Berkeley). Also, the population in Arizona is sparce [sic] and so are the raingauges [sic], and without raingauges we could not check on the far away effects of seeding. As a result of some search we found that the U.S. Department of Agriculture maintains a dense network of raingauges in a small locality called Walnut Gulch, some 65 miles away from the target of the Arizona seven-years-long cloud seeding experiment. The network was (and still is) in charge of a nice gentleman, Dr. Herbert B. Osborn. After some correspondence Dr. Osborn became interested and undertook a rather extensive job to collect data from 26 reliable recording gauges in Walnut Gulch for all the 212 experimental days of the Arizona experiments. The results of the evaluation are published in a paper authored jointly by Dr. Osborn and one of us. [In fact, Betty and Neyman were both co-authors on this paper.] It was published in the Proceedings of the National Academy of Sciences. The results were: On days when Walnut Gulch was downwind, the seeded days precipitation was substantially smaller than on days without seeding, with only a minute chance of this being due to chance variation. No such results were found for days when Walnut Gulch was upwind.

While these results confirmed one aspect of our findings for the Swiss experiment, namely the phenomenon of far away effects of seeding, there was a most unpleasant difference as regards the other aspect. While in Switzerland the far away effects of seeding were "positive," that is increases in rainfall ascribable to seeding, the far away apparent effects of seeding in Arizona were negative, largely losses of rainfall. True, this applied only to one locality, Walnut Gulch. But what about other localities in Arizona and in the neighboring New Mexico??

It took us substantial amount of time and effort to locate the necessary raingauges and the necessary data. But, as a result of such long studies, our findings were that Walnut Gulch was no exception. The general summary is that average rainfall in far away localities on days when those localities were downwind, the apparent effect of seeding at the Arizona experiment was something like 50% loss of rain with a minute probability of this being due to chance variation.

All of the above refers to days when the particular localities were downwind. But, naturally, each such locality was occasionally upwind. Obviously, the fully relevant question refers to the grand average apparent effect of seeding, that is to the effects of seeding averaged over all the localities for which we had data and over all experimental days. This grand average appeared to be of the order of 20% apparent loss of rain ascribable to seeding.

Our next step was to invent a mechanism of the apparent losses we found. Such hypothetical mechanism was in fact constructed, based on findings of meteorologists, primarily by Drs. Joanne Simpson and Arnett Dennis. These findings are symbolized by the terms "precipitation break" and "orphan anvil." But these details are rather technical.

Returning to the present disastrous drought in California, we have to state that if anything, it is more comparable to the situation in Arizona than in Switzerland. Thus, we rather expect that the many "emergency" cloud seeding operations now planned or in progress will lead to losses of rain. Unfortunately, the planned operations will not be randomized, that is they will not be real experiments and no realistic appraisal will be possible.

Hopefully, as was the case in biblical times (seven fat years followed by seven lean years), the current drought will somehow end. Actually, a short time ago there was a day or so when Los Angeles experienced a very substantial rainfall. Who knows, maybe this forthcoming weekend will brind [sic] something similar to the San Francisco Bay Area! But our findings just described lead us to expect that the financing of emergency cloud seeding operation will result in waste of public funds and in an increase in the severity of drought.[21]

The radio broadcast is a story of Betty's character. She was not afraid to take a stand, criticize work of others, or even call others out, because her positions were based on well-articulated questions, massive amounts of hard data, and solid statistical reasoning. She was willing to expend a tremendous amount of time and effort toward answering questions. She was not afraid to change her mind when the data indicated a conclusion contrary to her original position.

8.4 Legislative Testimony

By the end of 1977, over 50 proposals having to do with the drought had been introduced to the California legislature. One third eventually became law.[22] A model citizen, Betty thought it was important to emphasize to the general public that the effects of cloud seeding were not yet understood. She did not want the public to be mislead.[23]

In the spring of 1978, Betty and Neyman each testified to the California Legislature, specifically to the Assembly Committee on Water, Parks, and Wildlife, about the state's oversight of the modification of weather through cloud seeding. A law was passed in 1951 that enabled an individual to get

---

[21]Betty Scott and Jerzy Neyman, "For KPFA Radio Station, Berkeley, Thursday, July 7, 1977" (radio broadcast, KPFA Berkeley, 1977).

[22]State of California Department of Water Resources, "The 1976–1977 California Drought: A Review," 1978.

[23]6/17/1977 letter from Betty Scott to Solar Energy Research Institute Representative Barbara C. Farhar.

a license to modify the weather by merely completing a form and paying a $50 fee. The only requirement was to notify the state of any activity; the state did not have the authority to control the activity in any way. The state articulated about a dozen questions for the public hearing on March 31 at the state capitol, two of which were of special interest to Betty: "Are areas downwind of cloud seeding operations deprived of precipitation?" and "Are short-term drought operations effective?"[24,25]

Neyman testified before Betty. He pointed out that the Stat Lab had been involved in cloud seeding studies "for more than a quarter of a century," beginning with a contract from the California Department of Public Works Division of Water Resources. After a first report in 1951, he had collaborated on 48 more papers in national and international technical journals. He explained it this way: "The problem proved inspiring..." His basic message was that, given the current state of knowledge based on factual experimental studies together with the complexity of atmospheric phenomena, it could not be said that cloud seeding alleviates drought. Instead, there was evidence it might actually have the opposite effect and decrease precipitation. Therefore, "budgetary provisions for operational cloud seeding are likely to be a waste of public funds." Neyman felt it was his duty as a citizen to make this clear to the state.[26]

Betty's testimony came next. She pointed out the aim of her and Neyman's studies: "We have concentrated on trying to design and analyze weather modification experiments in order to find out: Does cloud seeding have any effect on the amount of water reaching the ground? If so, is precipitation increased or decreased, and for how many miles and for how many hours?" She emphasized that many answers were still needed:

> ...*there is strong evidence that cloud seeding often accomplishes the* _opposite_ *of what was intended; precipitation will be* _decreased_ *when an* _increase_ *was desired, hail will be* _increased_ *under seeding when the operator wants to decrease hail. And the effects will extend for many miles – indeed, has been observed to extend for more than 100 miles downwind. The effects persist even after seeding itself has stopped; they can last for hours.*

Betty then described the details of their cloud seeding findings from experiments in Arizona, Missouri, Colorado, Washington, and Oregon. She also described the difficulties conducting experiments in California due to "the large number of commercial operations" in the state. Betty strongly urged the state to regulate cloud seeding. Always in the role of scientist, she concluded her

---

[24]3/31/1978 hearing agenda on weather modification (cloud seeding) from the Assembly Committee on Water, Parks, and Wildlife of the California Legislature.

[25]3/31/1978 notice of public hearing on weather modification (cloud seeding) from the Assembly Committee on Water, Parks, and Wildlife of the California Legislature.

[26]3/31/1978 testimony by Jerzy Neyman, "Introductory Statement Before the Water, Parks, and Wildlife Committee Assembly, California Legislature," to the Assembly Committee on Water, Parks, and Wildlife of the California Legislature.

testimony by urging the committee to "arrange an extensive experiment to study the effects in the target and also downwind and to the side (both in California and in Nevada) of seeding in selected parts of California."[27]

Betty and Neyman followed the progress of the committee and its draft legislation. The latter concerned them. No one knew better than they that California needed valid and reliable information about the effects of cloud seeding. What they saw was language that would make it difficult for interdisciplinary atmospheric physics researchers at universities to obtain licenses to conduct experiments on cloud seeding. Betty told the chair of the committee: "...the main effect of these alternative minimum qualifications will be to blanket in essentially all of the present commercial Cloud seeders. In my opinion this would be a mistake. On the other hand, a specially trained person with expertise in cloud physics...would not be able to obtain a license...sufficient flexibility should be put into your bill...to encourage research and experimentation."[28] She worried that credible university research on cloud seeding "could be stifled," and that California should encourage such research "before it wastes any more tax money on cloud seeding operations..."[29] Neyman told the committee chair: "In order to have a reasonable chance of advancement [in atmospheric physics], the studies must be conduced by mature scholars endowed with vision and initiative." Presumably to make his case stronger, he made it clear he wasn't talking about himself or Betty, but about a well-published full professor with full-time specialty in atmospheric science. Neyman critiqued the wording of the bill and offered specific suggestions for change to the chair of the assembly committee.[30] He also sent a copy of the suggested changes to the UC president, stating plainly that the current draft bill "...is adverse to the status of universities as centers of research on societal problems."[31]

8.5 Professional Association Leadership

While all this research was going on, Betty was getting involved in leadership of the statistics profession. Shaffer recalls: "Betty did a lot of public work with professional associations. This is not easy to do well."[32] Here is Betty's

[27] 3/31/1978 testimony by Betty Scott, "Statement of Elizabeth L. Scott, Professor of Statistics, University of California, Berkeley 94720," to the Assembly Committee on Water, Parks, and Wildlife of the California Legislature.

[28] 5/7/1978 letter from Betty Scott to the California Legislative Assembly Committee on Water, Parks, and Wildlife Chair The Honorable Eugene T. Gualco.

[29] 5/10/1978 letter from Betty Scott to UC Government Relations representative Lowell Paige.

[30] 5/8/1978 letter from Jerzy Neyman to the California Legislative Assembly Committee on Water, Parks, and Wildlife Chair The Honorable Eugene T. Gualco.

[31] 5/18/1978 letter from Jerzy Neyman to UC President David S. Saxon.

[32] Juliet Shaffer, e-mail to author, August 5, 2014.

list of association memberships for 1978: IMS (Fellow), International Statistical Institute, Bernoulli Society (Council member), International Astronomical Union, American Astronomical Society (AAS), Astronomical Society of the Pacific, Biometric Society, Royal Statistical Society (Fellow), Sigma Xi, AAAS (Fellow), and Caucus for Women in Statistics.[33]

Betty was elected by her mathematical statistics peers to the prestigious position of president of the IMS. She served as president in 1977–78, the first woman to do so, all 42 previous presidents having been men. Betty must have done well, because she went on to become vice-president of the International Statistical Institute (ISI) (1981–1983) and the first woman president of the Bernoulli Society for Mathematical Statistics and Probability (1983–1985). She was well respected across the globe, and her appointment to these positions is the evidence.

As a leader, Betty tried to persuade statisticians to survey themselves. She wasn't successful, even with the shortest and "simplest sort of survey." She conjectured they must be afraid of what they might find out. She said that statisticians "...readily admit that statisticians know less about themselves than any other profession but they are almost proud of this and start quoting maxims about shoemaker's children."[34]

Betty advocated for close contact between statisticians and other scientists. She thought the statistical societies could and should promote such contact. She had experience with the International Astronomical Union, and borrowing from their protocols, she suggested to the ISI in 1976 that there be special meetings to connect statisticians and other scientists, "either immediately before or immediately after the big ISI meetings." Betty thought there were "several areas of world wide impact" where the use of statistics was "deplorable, even scandalous." Among these were "the effects of man on the environment and the corresponding effects of pollution on health," and the "estimation of natural resources." Interestingly, she added "I might have added changes in climate but for the fact that I am ignorant of the work in this domain."[35]

Even in 1977 during her year of service to the IMS as president-elect, Betty considered using her platform to advance the quality of research in weather modification. She drafted a letter to Neyman that was to be placed on IMS letterhead, although she acknowledged she was not able to speak on behalf of the IMS. The letter pledged to do whatever she could as IMS president to stimulate weather modification studies that would develop new statistical techniques and involve participation of selected statistical centers.[36]

---

[33] [?1978] Math and Science Network Questionnaire completed by Betty Scott.

[34] 9/16/1978 letter from Betty Scott to UC-Berkeley Anthropology Visiting Professor Silvia Forman.

[35] 7/16/1976 memorandum from Betty Scott to members of the Committee on Statistics in the Physical Sciences of the Bernoulli Society.

[36] 11/7/1977 handwritten letter from Betty Scott to Jerzy Neyman.

The political climate was right for Betty's pledge. Around this time, the US Congress mandated a Weather Modification Advisory Board, which in turn appointed a Statistical Task Force to critically review the existing research. The conclusion of the task force was that researchers didn't always adequately take into account the complexities of the atmosphere, and there was "a well-founded need for completely anchored conclusions."[37]

In the spring of 1980, the US Department of the Interior announced a large-scale decade-long cloud seeding experiment around Lake Tahoe, California. It was an $11 million effort designed to increase the snow pack in the Sierra mountains. By this time, there were a few other statisticians working in the area of weather modification. Most were not allies; they had sharply criticized Betty and Neyman for their work. Neyman, also reflecting Betty's position, was quoted in the *San Francisco Chronicle* as saying, "There's a racket going on. To study cloud seeding – yes! To expect it to work – no!"[38,39]

8.6   Still Cloud Seeding

Given the concerns about climate change today, it is interesting to note that Betty was interested in climate change and its effects way back in 1981. That year she suggested a good conference topic would be "the association between stratospheric temperature and stratospheric ozone and temperature on the earth." She had seen some of the research in this area. It was done in a deterministic framework, whereas she knew the problem was stochastic. Some of the topics she supported for the conference were effects of chlorofluorocarbons, climate prediction, weather modification, and climactic cycles.[40]

In 1984, Betty was working hard on hail modification. It was a large seven-year experiment in Switzerland that involved Soviet, French, Italian, Swiss, and American methods, equipment, and scientists. Betty was interested in conducting "a joint analysis as multivariate observations" in which she would carefully employ new methods for data analysis: "We are interested in possible downwind effects which opens up a messy stochastic process that we are trying to explore," she explained.[41]

In 2005, Wyoming spent $14M dollars on a ten-year experiment that aimed to answer the question: Will cloud seeding result in extra snow from winter

---

[37]Richard A. Kerr, "Weather modification: A call for tougher tests," *Science* 202, no. 4370 (1978), 860.

[38]10/30/1980 memorandum from Betty Scott to UC-Berkeley Statistics Professor David Brillinger.

[39]David Perlman, "Long-term U.S. experiment: Vast Sierra cloud seeding," *San Francisco Chronicle*, May 6, 1980.

[40]2/23/1981 letter from Betty Scott to ISI Bernoulli Society program planner and Princeton University Professor Geoffrey S. Watson.

[41]6/27/1984 letter from Betty Scott to USSR Vilnius Director of Institute of Mathematics and Cybernetics Professor V. Statulevicius.

storm clouds? The results were "thin:" "There was a 3 percent increase in precipitation, but a 28 percent probability that the cloud seeding had nothing to do with it." The US Bureau of Reclamation declared that "the 'proof' the scientific community has been seeking for many decades is still not in hand."

Many elected officials today are refusing to accept the existence of climate change, and some are continuing to spend money on cloud seeding without convincing proof that it works.[42] In 2014, there were cloud seeding operations in about ten states: Arizona, California, Colorado, Idaho, Kansas, Nevada, North Dakota, Texas, Utah, and Wyoming.[43]

---

[42] Allen Best, "Blue sky dreams: Cloud seeding is a work in progress with thin results," *Missoula Independent* (Missoula, MT), April 2, 2015.

[43] Elizabeth Daigneau, "Dried Up and Maxed Out, California Tries to Make it Snow," *Governing Magazine*, January 2014.

9

Almost Alone in Statistics (1955–1988)

Betty could be a holy terror, but her love of the department was clear and fully returned.
-Peter Bickel

Betty became a faculty member during a period in America that has been called the Golden Age of Science. More dollars were being spent, more people trained, more jobs created, and more articles published. Yet women did not benefit equally with men in this age. Women in science remained rare, and senior women on science faculties were especially rare. In 1960, only about 5% of tenure-track mathematics faculty members at 20 leading universities in the US were women, and less than 0.5% of full professors in mathematics were women.[1]

Betty beat the odds against women and became a faculty member in mathematics and then statistics at UC-Berkeley, and then she beat more odds and rose to senior status in the faculty. She fared better than most women in science nationally at the time, partly because she had a foundation with Jerzy Neyman based on their war work and his budding interest in astronomy. On the one hand, he was an equal opportunity guy, and on the other hand, she met his high standards. Also, Neyman considered those who worked for him to be family, and the trusting nature of his relationship with Betty ensured he didn't exclude her out of fear she would get married and then leave, which was a common bias experienced by women at the time. The deep intellectual collaborations between Betty and Neyman ensured she wasn't completely marginalized in the academy as were other women.

Yet outside of the Neyman family relationship, Betty persevered in the academy in the presence of a climate of difficulty for and discrimination against women. This catalyzed her to devote so much of herself to gender equity work in the later part of her career. Betty's overall perseverance as an academic

---

[1] Margaret W. Rossiter, *Women Scientists in America Volume 2: Before Affirmative Action 1940–1972* (Baltimore: The Johns Hopkins University Press, 1995), 129.

scientist was a noteworthy pillar of her success and was memorialized in a poem in which a colleague called her "the indefatigable Miss Scott."[2]

Betty had to be indefatigable, in part because she was almost alone as a woman in statistics at UC-Berkeley. Betty was a faculty member in the statistics department from the day it was born in 1955 until the day she died in 1988. Over those years more and more women pursued degrees in statistics, and the department more than doubled in size, but the representation of women on the faculty remained constant. Betty was either the only woman, or one of only two or three, depending on whether you counted lecturers or visiting professors in addition to those on the tenure ladder ranks. For 22 of her 33 years in the department, i.e., between 1965 and 1987, Betty was the only woman in a tenure ladder position.

In fact, only four women overlapped with Betty in all of her 33 years in the department. Evelyn Fix was a tenure ladder faculty member at UC-Berkeley from 1951 until she passed away in 1965. Fix completed her dissertation in mathematical statistics under Neyman at Berkeley and had a career trajectory similar to Betty's, beginning as a lecturer in the mathematics department and ending as a professor in the statistics department. She is known for work in power calculation, discriminant analysis, risk assessment, and statistical problems in biology and health.[3] F.N. David (1909–1993) was a visiting professor at UC-Berkeley in the late 1960s and early 1970s. "FND" completed her dissertation in mathematical statistics under Neyman at University College London and taught there for years in various ranks. Later she became founding chair of the biostatistics and statistics departments at UC-Riverside. She had over 100 publications, plus two monographs and nine books.[4] Juliet Popper Shaffer (discussed in this chapter) was on the lecturer track beginning in the 1970s. Deborah Nolan overlapped with Betty in the last year and a half of Betty's life (between 1987 and 1988).

9.1   New Statistics Department (1955)

Mathematics Department Chair Griffith Evans had approved appointments to the faculty of Neyman as a professor in 1938, and then Betty as a lec-

---

[2]Constance Reid, *Neyman: From Life* (New York: Springer, 1982), 282. Note: The colleague was Russian number theorist, probability theorist, and mathematical statistician Yuri Vladimirovich Linnik (1915–1972).

[3]Neyman J., Hodges Jr. J.L., Le Cam L., Lehmann E.L., and Scott E.L. Evelyn Fix: In Memorium. Accessed on 5/11/2015 at http://statistics.berkeley.edu/people/evelyn-fix-1904-1965.

[4]Garber M.J., Gokhale D.V., Utts J.M., and Beaver R.J. Florence Nightingale David, Statistics: Riverside. University of California: In Memorium. Accessed on 2/16/2014 at http://texts.cdlib.org/view?docId=hb238nb0fs;NAAN=13030&doc.view=frames&chunk.id =div00019&toc.depth=1&toc.id=&brand=calisphere.

turer ten years later in 1948. But Evans and Neyman had different visions for statistics. Evans' was for a broad-based department of mathematical sciences that included statistics. Neyman's was for his own statistics department. As time went on, Neyman, an "ambitious and dynamic administrator," pressed Evans for more and more independence of the Stat Lab from the mathematics department, until the lab was essentially autonomous and part of the department in name only.[5] Soon after Betty joined the faculty, Evans handed the chairmanship over to vice chairman Morrey, who was supportive of a separate statistics department. In 1955, statistics and the Stat Lab were moved to a freestanding department.[6]

Neyman became the founding department chair and continued to serve as Stat Lab director. Betty and nine others were appointed as tenure ladder faculty members. The Major in Mathematical Statistics was retitled The Major in Statistics. Forty-two courses were moved: two lower division, 12 upper division, and 28 graduate. In addition to statistics theory and methods courses, there were courses on areas of application, including actuarial science, city planning, engineering, demography, economics, and production/quality control. Almost all of the applications courses were already in place in the mathematics department.[7]

The year 1955 was also when UC-Berkeley began offering degrees in biostatistics. Neyman and School of Public Health Professor Jacob Yerushalmy (1904–1976), a biostatistician and child health and development expert, made an agreement to offer the MA and PhD degrees in biostatistics jointly between their two units. The proposal was a model of collaboration.[8] Betty was involved from the beginning as a member of the executive committee in charge.[9] In 1968, it was Betty who proposed that biostatistics should have a separate listing in the university catalog.[10] She continued to co-chair the biostatistics group for many years, until 1982 (a year after Neyman's death) when the program was no longer concurrently listed in the statistics section of the university catalog.[11]

Enrollment in statistics courses up to the point of the creation of the new statistics department has been described as "small and erratic," especially at the undergraduate level. It took only a year for this to change. Two tracks were

---

[5] E.L. Lehmann, "The Creation and Early History of the Berkeley Statistics Department," *Statistics, Probability and Game Theory* 30, (1966), 139.

[6] Terry Speed, Jim Pitman, and John Rice, "A Brief History of the Statistics Department of the University of California at Berkeley," *Cornell University Library*, January 31, 2012, accessed February 2, 2014, http://arxiv.org/ftp/arxiv/papers/1201/1201.6450.pdf.

[7] "Statistics," *UC-Berkeley Course Catalog, 1955–56*, 370.

[8] 12/3/1954 memorandum from Jerzy Neyman and UC-Berkeley School of Public Health Professor Jacob Yerushalmy to UC-Berkeley Chemical Engineering Professor and Graduate Council Subcommittee Chairman Charles R. Wilke.

[9] 9/12/1955. letter from Jerzy Neyman to UC-Berkeley Statistics Professor Joseph Hodges, Jr. and Betty.

[10] 8/15/1968 Minutes of Group in Biostatistics Meeting. UC-Berkeley. Attended by Neyman, Yerushalmy, Scott, Brand, and Chiang. Le Cam was not present.

[11] "Statistics," *UC-Berkeley Course Catalog, 1982–83*, 185.

introduced to the undergraduate major, one emphasizing theory, the other applications.[12] Soon statistics saw annual increases in enrollment of 20–25%. This pattern persisted for a decade. Ten years later in 1965, an undergraduate honors program was added, as well as a thesis option. By this time there was a wide spectrum of courses in theoretical statistics, applied statistics, and probability.[13] Between 1968 and 1972, 55 PhD degrees were granted[14] (see also Lehmann, 1996;[15] Speed, Pitman, and Rice, 2012;[16] and Department of Statistics, 2015).[17]

In 1968, Betty spoke as statistics department chair at the groundbreaking ceremony for Evans Hall. She took her turn right after the mathematics department chair, who bragged about his department's productivity since 1934. Mightily protective of the statistics department, Betty said this: "If we want to quote ratio of productivity, the department of statistics is far ahead [of mathematics], since the measure is infinity. There was no department of statistics in 1934."[18]

In 1978, Betty gave a commencement speech to the statistics graduates. She included a description of the space allocated to the department:

> *Statistics has moved ten times in its 40 year history, from one desk in a crowded room 420 on the fourth floor of Wheeler to several rooms in the old building of the Unitarian School of Religion (upstairs was the School of Social Welfare; cut in stone above the entrance was non ministrari sed ministrare [not to be served but to serve], then to our first permanent home, the New Classroom Building – now Union Hall, then came World War II and we were evicted to make space for the Radiation Laboratory – We were moved to RSB into part of the space of Botany and of Zoology. After the war, back to the New Classroom Building – by then called Durant Hall. One lovely afternoon in June, an organ started playing in the street outside – right in the middle of some meeting, AAAS of statistics and of hepatalogists. We couldn't stop the racket – Durant Hall was being rededicated, The School of Optometry Bldg, so as to retain a small bequest. We were pushed to one end and then to our second permanent home in Dwinelle Hall, then our third permanent home in Campbell and now here. What will the future bring to our offices and to our graduates??*[19]

---

[12] "Statistics," *UC-Berkeley Course Catalog, 1957–58*, 384.

[13] "Statistics," *UC-Berkeley Course Catalog, 1965–66*, 599.

[14] "PhD Alumni," *University of California, Berkeley Department of Statistics*, accessed July 14, 2015, http://statistics.berkeley.edu/people/alumni/phd.

[15] E.L. Lehmann, "The Creation and Early History of the Berkeley Statistics Department" (Lecture Notes – Monograph Series 30, Institute of Mathematical Statistics, 1996), 139.

[16] Terry Speed, Jim Pitman, and John Rice, "A Brief History of the Statistics Department of the University of California at Berkeley, 2012," accessed July 14, 2015, http://arxiv.org/ftp/arxiv/papers/1201/1201.6450.pdf.

[17] "History," *University of California, Berkeley Department of Statistics*, accessed July 14, 2015, http://statistics.berkeley.edu/history.

[18] 9/25/1968. Evans Hall site dedication. UC-Berkeley (sound recording).

[19] 6/9/1978 handwritten Commencement Speech by Betty Scott.

This detail underlines the importance Betty placed on the adequacy of space for people to work.

9.2 Teaching

In 1955–56, the first year of the statistics department, Betty taught courses at all levels. She co-taught the second semester of the lower division Elements of Probability and Statistics, along with David Blackwell and two lecturers. This course was for both majors and general education students, and it included examples from genetics, bacteriology, industrial sampling, and public health. Betty also taught the upper division Statistical Inference and was in charge of the associated lab course. This course was for students in natural or social sciences, and it emphasized concepts and applications rather than mathematical proofs of theorems. Finally, Betty taught the graduate Probability Models of Natural Phenomena. This was essentially a modeling course (structural, deterministic, static stochastic, dynamic stochastic models) that focused on the "relationship between natural phenomena and their mathematical theory," and it included examples from astronomy, general biology, and economics.

The variety of courses Betty taught over the many years she was on the faculty is evidence of her expertise. We know which courses she was responsible for in the years when the university catalog published names of instructors along with course descriptions (1955–1975 and 1985–88). At the lower division level, Betty taught an introductory course for statistics majors and other students with mathematics backgrounds who wanted basic concepts of statistics for general education. At the upper division level, she taught courses in theory of probability and statistics, descriptive statistics, statistical inference, and sampling surveys. At the graduate level, she taught courses in advanced introductory probability and statistics, probability models of natural phenomena, theory of probability and statistics, advanced statistical inference, linear models, and analysis of discrete observations. She taught most several times. She also taught seminars in current literature or statistics research.[20]

Like many instructors, Betty sometimes thought teaching was a struggle. For example, many times introductory statistics students try to rely on their intuition to solve basic probability problems. In Betty's words:

> *One-fourth of my students tell me that because some Republican husbands have wives who are Democrats, therefore the two properties: (a) husband is a Republican [and] (b) wife is a Democrat are independent. Another fourth tell me that since some Republican husbands have wives who are*

---

[20]"Statistics," *UC-Berkeley Course Catalog, 1956–75 and 1985–88.*

Republicans, therefore, these two properties are dependent. Only half the students wrote down a few numbers to do the problem correctly.[21]

Students need to suspend their basic intuitions and use established definitions and rules in order to correctly determine probabilities. But they also need to use their basic intuitions to determine if the answer makes sense.

The course Betty taught most often in her career was the graduate Analysis of Discrete Observations; no other course came close to this in frequency. The course entered the curriculum in 1954 when Fix taught it; the next year William Kruskal taught it as a visitor; thereafter, Betty taught it some 30 times between 1956 and 1988. She only missed two years when she was on leaves of absence.[22] Discrete observations was Betty's signature course.

In 1956–57, the first year Betty taught the course, the course description read:

> *Discrete models. Discrete distributions and their limits. Transformations. Chi-square and other best asymptotically normal methods.*

The prerequisite was one year of upper division Statistical Inference, or one year of upper division Theory of Probability and Statistics.[23]

Over a 33-year period, Betty continued to develop and expand the course based on modern developments. For example, Betty was on sabbatical in spring semester 1981, when in addition to her research on carcinogenesis and distribution of galaxies, she revised this course. The revision included new theories of generalized methods and associated software. It expected graduate students from a broader range of applied fields, especially economics, to enroll; these would be students who already knew of discrete observations and wanted to know more.[24]

By 1978–79, Betty was distributing 67 pages of typed notes on contingency tables that students would find useful for many years in their work as statisticians. She also recommended two published textbooks: Cox's *Analysis of Binary Data*[25] and Bishop, Fienberg, and Holland's *Discrete Multivariate Analysis*.[26] There were homework problems, lab assignments, and exams. Betty liked to draw from the real world for her homeworks and labs. She frequently crafted these activities from problems she was working on in science, like melanoma and Actinic Keratosis, or ultraviolet radiation and skin cancer. She would provide a one- to two-page detailed description of the problem, carefully laying out the background and science, then framing the questions in

---

[21] 11/5/1951 letter from Betty Scott to Mt. Holyoke College Professor Grace E. Bates.

[22] Note: The UC-Berkeley course catalogs list instructors for each course, except for years 1875–1985. It is assumed that Betty taught Analysis of Discrete Observations during that period, given that it was a regular course of hers from 1956–1975 and 1985–1988.

[23] "Statistics," *UC-Berkeley Course Catalog, 1956–57*, 375.

[24] 5/21/1984 report from Betty Scott, "Report on Sabbatical Leave, January 1 to June 30, 1981."

[25] D.R. Cox, *Analysis of Binary Data* (London: Chapman & Hall, 1970).

[26] Yvonne M.M. Bishop, Stephen E. Fienberg, and Paul W. Holland, Discrete, *Multivariate Analysis: Theory and Practice* (Cambridge: The M.I.T. Press, 1975).

real world language. This approach fit her coming to statistics from a science background. In the second semester, she asked students to work on a discrete data problem of their choosing. The author of this book chose to work on a problem in human fertility that led to her master's thesis in statistics under Betty's direction, which was published in the journal *Demography*.[27] A carefully taken set of lecture notes, saved for over 35 years, indicated Betty was a meticulous lecturer; and students in the classroom referred to her as "Miss Scott."

In 1987–88, the last year Betty taught the course, the description read:

> *Discrete stochastic models, generating functions, birth-death processes. Contingency tables: sources, models, sampling schemes, analysis, exact tests. Linear, log linear, logistic models. Search for models. Power. Chi-square. Quantal response; probit, logit. Asymptotics. Cluster analysis.*

The prerequisites were one year of either upper division Introduction to the Theory of Statistics or graduate Introduction to Probability and Statistics at an Advanced Level.[28]

Betty also liked to draw on real world problems for her examinations. As she wanted to evaluate her students' reasoning, Betty wrote in capital letters at the top of her tests: "SHOW ALL WORK. EXPLAIN WHAT YOU ARE DOING. GIVE REASONS." Here is a question from her first semester final examination in fall 1978:

> *Under court orders, the City of Berkeley has removed the traffic barrier at the corner of Tunnel Road and The Uplands. Some persons living on The Uplands did not want the barrier removed because they claim it reduced traffic on their street (since drivers could not get off The Uplands due to the barrier, some drivers learned not to enter Uplands). If the driver does not enter Uplands, he can be expected to enter an alternate street (see map), El Camino Real, and some of the residents of El Camino wanted the barrier removed therefore. Ignoring the purple prose about how narrow and crooked the two streets are, we want to compare the traffic U on The Uplands now and V the traffic on El Camino Real now with their former values when the barrier was up. U and V are both random variables with expectations depending on the day of the week, the time of day, the weather, etc. The data before removal of the barrier were observed last year.*
>
> *(a) What sort of data would you request to make the comparison (what days, hours, etc.)?*
>
> *(b) Use the data on the next page to set up a likelihood ratio test. What is the hypothesis to be tested in your version of the problem?*
>
> *(c) What is the class of admissible hypotheses? What is the distribution*

---

[27] Amanda L. Golbeck, "A probability mixture model of completed parity," *Demography* 18, (1981), 645-658.

[28] "Statistics," *UC-Berkeley Course Catalog, 1987–88*, 341.

of the observations under these hypotheses? You will need to make some assumption about how the data were collected – Be clear about this [sic]

(d) Now what is the distribution under the hypothesis tested?

(e) Are you now able to write down the likelihood ratio and get the test? Remember that you have no guarantee that your answer will be simple.

(f) Suppose you observed the value lambda = 0.1 for testing your hypothesis, what would you conclude? What if your value were 10.0? Remember to explain what you are doing.

(g) Now set up the problem as a chi-square test. Show all work but do not do arithmetic.[29]

Betty was thinking analytically, mathematically, and statistically about the community issues in her neighborhood.

9.3   Administrator and Professor

Before 1970, administrative appointments in the US for women in science were so unusual that Cornell University History of Science Professor Margaret W. Rossiter (1944–) in her book on women scientists in America between 1940 and 1972 mentions Betty by name along with only four others.[30] At a time when most women in science who managed to get positions in prestigious schools were isolated and marginal in status and power, Betty managed to rise to the ranks of assistant dean and department chair.

Between 1965 and 1967, Betty served as assistant dean in the UC-Berkeley College of Letters and Science where the statistics department was located. Her duties probably focused on student affairs. The next year, Betty went on to serve as the fifth chair of the statistics department, following Neyman, David Blackwell, Lucien Le Cam, and Henry Scheffé. How was Betty selected to become chair? She did have administrative experience as an assistant dean, and Le Cam gives us another clue: "I know that he [Neyman] had a reputation for being bossy. That is how I became Chairman after Blackwell's term. Nobody else wanted to take the job as long as Neyman was around." Neyman had a big presence, even though he gave up the chair position a year after the formation of the department. Le Cam explains it this way: "Neyman was a big shot and the *pater familias*. He rarely ordered you to do anything, but you felt compelled to do it... Generally, I found it very easy to get along with

---

[29] Fall 1978. Final Examination. Statistics 236A: Analysis of Discrete Observations. Elizabeth L. Scott, UC-Berkeley.

[30] Margaret W. Rossiter, *Women Scientists in America Volume 2: Before Affirmative Action 1940–1972* (Baltimore: The Johns Hopkins University Press, 1995), 144.

him. It hurts me that, after Neyman died [in 1981], his colleagues at Berkeley essentially revolted against the authority of *pater familias*."[31]

Betty remained chair for five years until 1973, except for spring of 1970 when she was on leave and asked Erich Lehmann to fill in for her.[32] The assignment was half her workload. UC-Berkeley Statistics Emeritus Professor Kjell Doksum was an assistant professor at the time. He remembered Betty as having many get-togethers at her house and recalled that: "Betty was really good at bringing the faculty together socially. She made sure everybody got along. In fact, she created a friendly situation where we all could get along. So we were a unified department when she was chair."[33]

Betty counted the number of women in chair positions at UC-Berkeley. In 1971–72, she was one of only four. The next year, no additional women had been appointed to chair positions, and there were only two left of the four: One left to become a divisional dean, one left the university when her department (demography) was disestablished, one was chair of the design department which was scheduled for disestablishment on June 30, 1974, and then there was Betty. She wrote in parenthesis: "In case anyone is paranoid about disappearance of departments with women chairmen, Statistics Dept. will have man chairman from July 1, 1973." The following year, in 1973–74, there were also no new women appointed to chair positions.

Betty counted other administrators. There were few women in the picture of organizational leadership postilions in 1972–73. The chancellor, provost, vice-chancellor, 12 deans, and two acting deans were all men. Three of four divisional deans were men. All six associate deans were men; and eight of ten assistant deans were men. Twenty-three of 24 directors were men; and one of three associate directors was a man.[34]

In all the years Betty was at Berkeley, she was the only woman to serve as chair of the statistics department. When she was chair, all faculty members in, and hired into, tenure ladder positions were men.[35] Betty did hire one woman into a visiting position, her friend and colleague F.N. David.

Betty was deeply involved in affirmative action research during the five years she was chair. She worked on many statistical reports that showed the underrepresentation of women on the faculty, and she advocated for individual

---

[31]Grace L. Yang, "A Conversation with Lucien Le Cam," *Statistical Science* 14, no. 2 (1999), 223.

[32]E.L. Lehmann, *Reminiscences of a Statistician: The Company I Kept* (New York: Springer, 2008), 108.

[33]Kjell Doksum, e-mail to author, July 2, 2014.

[34]n.d. handwritten notes by Betty Scott titled "Chairmen."

[35]Note: Fix passed away in 1965; she did not live long enough to see Betty rise to the position of statistics department chair. The ten men hired were: Professors David R. Brillinger (1970), Jacob Feldman (1970 joint with math); Associate Professor Richard E. Barlow (1970 joint with operations research); and Assistant Professors Rudolph J. Beran (1969), Robert G. Azencott (1970), Myron L. Straf (1970), Charles E. Antoniak (1971), Louis A. Jaeckel (1972), Norman L. Kaplan (1972), and Jean-Francois Mertens (1972). In addition, Albert H. Bowker had been appointed Chancellor of the University of California in 1971, and he was given an appointment as a professor of statistics in 1973.

and groups of women in other departments both locally and nationally. She talked freely about this work to others in the department. Doksum remembers Betty talking to him about it: "She talked to me about how she met opposition even though her statistics showed discrimination. She would correct for age, productivity, and so forth, did the regressions, and found that men were paid more than women. Other people came back with strange arguments, like 'but the slopes are the same,' as if this mattered. But the intercepts were different and that's what really mattered. She expressed her frustration to me."[36]

Betty knew it was the departments that could improve the representation of women. Why, then, were no women hired into tenure ladder positions in the statistics department while Betty was chair? Generally, people were hired through professional networks. Seven of the hires were at the assistant professor level. Based on the fact that the proportion of PhDs in statistics at UC-Berkeley going to women between 1967 and 1973 was only 13.7%, one could expect the networks to produce names of men. There were just not many women in the availability pool.

When asked why no women were hired into tenure ladder positions while Betty was chair, UC-Berkeley Statistics Professor Emeritus Peter Bickel cited cultural blindness on the part of males and the low percentage of women PhDs: "I had the sense that Betty had an aversion to affirmative action in the department although... she otherwise fought for many womens causes... Betty could be a holy terror but her love of the department was clear and fully returned."[37] It has recently been suggested that critical mass is important. In the words of one department chair: "The single female was not able to convince the faculty to hire more females. When a 2nd female came... that completely changed the climate and we were able to increase our numbers."[38]

Betty was promoted to associate professor in 1958 and professor five years later.[39,40] Promotion to professor was a remarkable achievement for Betty, as most women in science in tenure ladder positions at the time were given large teaching loads of introductory courses which stalled them at the associate professor rank. In fact in 1960, just three years before Betty's promotion to professor, there were no women whatsoever at the professor rank in mathematics and statistics at 19 leading universities in the US. In the same universities, there were only 13 women mathematicians and statisticians, all employed at only three of the universities (Illinois, Berkeley, and Minnesota). In other words, 16 of the 19 universities had no women faculty whatsoever in either math or stat.[41]

---

[36] Kjell Doksum, e-mail to author, July 2, 2014.

[37] Peter Bickel, e-mail to author, April 12, 2014.

[38] M. Gumpertz and J. Hugher-Oliver, "The many facets of leadership," in *Leadership and Women in Statistics*, ed. A.L. Golbeck, I. Olkin, and Y.R. Gel (London: Chapman & Hall/CRC Press, 2015), 353.

[39] "Mathematics," *UC-Berkeley Course Catalog, 1945–54.*

[40] "Statistics," *UC-Berkeley Course Catalog, 1955–88.*

[41] Margaret W. Rossiter, *Women Scientists in America Volume 2: Before Affirmative Action 1940–1972* (Baltimore: The Johns Hopkins University Press, 1995).

9.4 Colleague Juliet Popper Shaffer

Juliet Shaffer was younger than Betty, having completed her PhD in psychology at Stanford University in 1957. Like Betty, Fix, and FND, Shaffer was a successful academic woman. She obtained a faculty appointment as an assistant professor of psychology at the University of Kansas in 1959. Her position went part time when she had her children, and that time did not count toward her tenure. Shaffer didn't have any mentoring when she started as an assistant professor: "No one told me anything...that I would be dropped if I didn't get tenure. I had a child, and I didn't get tenure the first time I went up. The chair told me they didn't think I was that serious, that they would put off the decision for a year. Then I got a few more pubs and I got tenure. In a way it was good I didn't know, because I was much more relaxed, as long as it worked out,[42,43] which it did." In 1974, 15 years after her initial appointment, Shaffer achieved the rank of professor.

Along the way, Shaffer become more and more interested in statistics and published a number of articles on statistical methods. She was Statistics Editor of the journal *Computer Studies in the Humanities and Verbal Behavior* from 1970–74, taught all of the undergraduate and graduate statistics courses in the University of Kansas (KU) psychology department, had extensive statistical consulting experience, and published in the areas of mathematical learning theory, multiplicity, and analysis of multidimensional contingency tables, including a paper on log-linear models that predated textbooks on the subject.[44,45]

In the academic year 1973–74, Shaffer decided to spend a sabbatical year in the department of statistics at UC-Berkeley to take her statistical skills and knowledge to the next level.[46] She went to Berkeley with her husband, KU economics faculty member Harry G. Shaffer, and his two younger children from a previous relationship. She did not do any teaching, but attended classes. According to Shaffer, "Betty was chair when I came to Berkeley the first time. She was always very good in helping women. But I felt I was a rival to her. She never did anything bad to me. But she wasn't overly friendly either."

Shaffer's life took a major turn in that year. She was well aware of the problems for women of having nepotism rules in academe, having published

---

[42]Faculty Listings. In About: History of the Department of Psychology, University of Kansas. Accessed on 2/24/2014 at http://psych.ku.edu/about/history/.

[43]Juliet Shaffer, e-mail to author, August 5, 2014.

[44]Dan Robinson, "Profiles in Research: Juliet Popper Shaffer," *Journal of Educational and Behavioral Statistics* 30, no. 1 (2005), 93.

[45]Juliet Popper Shaffer, "An Unorthodox Journey to Statistics: Equity Issues, Remarks on Multiplicity," in *Past, Present and Future of Statistical Science*, ed. X. Lin, C. Genest, D.L. Banks, G. Molenberghs, D.W. Scott, and J. Wang (London: Chapman & Hall/CRC Press, 2014), 49.

[46]E.L. Lehmann, *Reminiscences of a Statistician: The Company I Kept* (New York: Springer, 2008), 211.

an article in 1966 with her husband called "Job Discrimination against Faculty Wives: Restrictive Employment Practices in Colleges and Universities."[47] Yet she somehow couldn't escape the problem in her personal life. In her own words: "At the end of the sabbatical, Erich [Lehmann] and I were desperately in love. We were both married [to others]. It was a difficult situation – what should we do? We talked to a psychologist, who told us to go back to our lives for a year. We did that and then both got a divorce."[48] And so in 1977, Shaffer left her position as a full professor at KU[49] to move to Berkeley and marry Lehmann. It was a costly move, financially as well as emotionally. She had to pay back half the cost of her sabbatical: "They [KU] put me on a payment plan. At Kansas the sabbatical contract said you had to come back for two years."

According to Shaffer, "When I came back [to Berkeley] the second time, Erich [Lehmann] was chair and thought there would be a conflict of interest if he hired me, so I went to UC-Davis." This was a visiting position in the mathematics department. Shaffer recalls:

> *They had Gertrude Cox there to give advice about statistics, and she recommended putting together an interdisciplinary program; she and I went out to dinner once. During that year I commuted every day [65 miles each way] from Berkeley. I was in a Monday-Wednesday-Friday carpool and a Tuesday-Thursday carpool... Most people didn't commute every day... The next year Erich was no longer chair [Peter Bickel had become chair] and so I was able to get hired [as a lecturer] at Berkeley [in the statistics department]. My degree was in psych and so it was not easy to get a job in statistics. I had a lot of experience with practical work. I talked to Bickel and he said, why don't we try this [doing a consulting service]... I had a lot of experience in consulting, and for ten years I did that, and then others started doing it ... "[50]*

And so Shaffer ended up with a lecturer appointment in statistics at UC-Berkeley, not at all like the tenured full professor appointment she had at Kansas in psychology. In her own words:

> *...I looked for a job in the Berkeley area. Psychology enrollments were decreasing in the 1970s, so there were few jobs available in that field... I became a lecturer in the stat dept at Berkeley. Because my mathematical background was relatively weak I was not fully qualified as a mathematical statistician, but because of my extensive empirical research background, I could supplement the offerings in the department on the applied side,*

---

[47]Harry G. Shaffer and Juliet P. Shaffer, "Job Discrimination against Faculty Wives: Restrictive Employment Practices in Colleges and Universities," *The Journal of Higher Education* 37, no. 1 (1966), 10.

[48]Juliet Shaffer, e-mail to author, August 5, 2014.

[49]Faculty Listings. In About: History of the Department of Psychology, University of Kansas. Accessed on 2/24/2014 at http://psych.ku.edu/about/history/.

[50]Juliet Shaffer, e-mail to author, August 5, 2014.

which was relatively weak then. A few years later, I was appointed to Senior Lecturer, a position that had security of employment.[51,52]

Thus, because she had a talent for statistical consulting and extensive consulting experience in her KU psychology professor position, Shaffer was given the responsibility for the UC-Berkeley statistics department's consulting service. She regularly taught a seminar in short-term consulting that was introduced into the statistics curriculum when she arrived at UC-Berkeley in 1977. The idea was that enrolled students would serve as consultants in the department's drop in statistical consulting service, and then get together in a seminar format to discuss the consulting problems, how to handle them, and how to work with clients to make sure their questions were answered.[53] In 1982, Statistical Consulting became a regular course.[54] Shaffer explained:

I wanted anyone to be able to drop in, from school or community. Each student had to be available for two hours a week. I had them work in pairs. I got the graduate division to give them a little money in recognition. For about ten years I was the only one who ran the consulting service. After that others did consulting... I got a big variety of problems. I enjoyed it. I like variety.[55]

Over the years, more than 2,000 clients were served under Shaffer's supervision.[56] The consulting service was half of Shaffer's position. She also taught many regular statistics courses, both undergraduate and graduate.

Betty indicated on a questionnaire that she was more than happy to participate in the consulting service when the department received requests for help, either paid or unpaid. But the problem had to be something she found to be interesting; it had to come when she had some available time; and she preferred to work in the physical and biological sciences, and also affirmative action. Betty was more than willing to supervise and encourage student consultants, even when they were being paid and she was not.[57]

---

[51] Dan Robinson, "Profiles in Research: Juliet Popper Shaffer," *Journal of Educational and Behavioral Statistics* 30, no. 1 (2005), 93.

[52] Statistics. Various UC-Berkeley Course Catalogs, 1982–1989. Note: Juliet Shaffer is mentioned for the first time in the 1982 catalog.

[53] Statistics. *UC-Berkeley Course Catalog, 1977–78*, pp. 207–209.

[54] "Statistics," *UC-Berkeley Course Catalog, 1982–83*, 185.

[55] Juliet Shaffer, e-mail to author, August 5, 2014.

[56] Juliet Popper Shaffer, "An Unorthodox Journey to Statistics: Equity Issues, Remarks on Multiplicity," in *Past, Present and Future of Statistical Science*, ed. X. Lin, C. Genest, D.L. Banks, G. Molenberghs, D.W. Scott, and J. Wang (London: Chapman & Hall/CRC Press, 2014), 49.

[57] 10/11/1978 questionnaire from UC-Berkeley Statistics Department Consulting Committee Members David Brillinger and Juliet Shaffer to Statistics Faculty and Visitors.

9.5   Flexibility and Resilience

Bank of America Senior Vice President and Risk Modeling Executive William Kahn took Shaffer's consulting class when he was a graduate student at Berkeley. He really enjoyed and benefited from it: "I almost have to say that might have been one of the most influential, most important classes I took, because what I actually ended up doing in my career for 35 years now is being a consulting statistician, and so I got my first formal training in it from Julie." Kahn noted that the research of people like Shaffer are "what gives Berkeley its prestige." [58] Over the years, on the other hand, many others saw Shaffer as a person in a staff position, and as Lehman's wife, rather than the accomplished academic she was. Former students like University of Wisconsin Statistics Professor and Chair Brian Yandell remember Shaffer as a well-organized and solid academic, and as a kind person who always had time for students. Students felt they got a lot from her and "lapped up what she had to say." At the same time, Yandell felt some students saw her as being somewhat uneasy, perhaps because she was not considered to fit the Research I mold for the field of statistics at the time and therefore not considered an equal in the statistics department. Yandell went further to say that Shaffer seemed to accept the secondary role that put her into the background of the department with its "sense of elitism." [59]

What was it like to go from a tenured to a lecturer position? Shaffer said this:

> *Every year I had to be renewed. Often the renewal came very late. Sometimes my insurance was canceled. This was a problem, because I had children. It was terrible. Sometimes I didn't even know it [that the insurance was canceled]. Then I became a senior lecturer and had security of employment. You can't get as much summer salary as a senior lecturer [as you can in a ladder position]. It is very unpleasant to be in a non-ladder position. They let you know you are different. You deal with it by trying to forget you are different. I kept a quote by Eleanor Roosevelt on my desk: "No one can make you feel inferior without your consent." It bothered me. If I hadn't already been on the regular faculty [at KU], maybe it wouldn't have bothered me as much. I found out another senior lecturer became a professor. I made inquiries. Everyone said, no, etc. I don't like being rejected. I gave up the idea. Now people call me Professor Emerita. I used to correct them and say I am Senior Lecturer Emerita. But now I no longer correct them. I call myself Senior Lecturer Emerita, but now I no longer correct them if they call me Professor Emerita. Some people say, they are happy just having the pay. But I would like to have the title and not just the pay. Erich [Lehmann] and I thought of going somewhere*

---

[58] Bill Kahn, e-mail to author, October 9, 2014.
[59] Brian S. Yandell, e-mail to author, June 24, 2014.

else, and we got an offer from a university in Florida, but my offer wasn't that great anyway, so we stayed at Berkeley.

Shaffer said she didn't really feel discriminated against, because her PhD was in psychology, not statistics, and her background included less mathematics in comparison with others in the department. "So I didn't blame the department," she asserted, "it's just that the position came with many demeaning aspects. In spite of that, I loved the stimulation of that wonderful department." Recently, Shaffer's title was changed to Teaching Professor Emerita, which "makes things much more comfortable for me."

Sadly, many highly talented and accomplished women like Shaffer found themselves in the situation of giving up their career tracks and advantages to follow their husbands and had to accept their situations as best they could. Shaffer coped by keeping the poignant reminder close at hand, that "No one can make you feel inferior without your consent." Women like Shaffer have had to be exceedingly flexible about life and mightily resilient, while keeping their longings for their aborted career tracks mostly locked in the secret chambers of their hearts.

Without a doubt, Betty had a more comfortable situation in her lifetime than most women in science in the US. Yet Shaffer offered: "I'm not sure how comfortable Betty really was."

10

Students and Memories (1948–1988)

...15 minutes with Betty were as invaluable as an hour, because she was so concentrated in her attention.
-Dennis K. Pearl

F.N. David recalled Betty "was helpful and supportive to all graduate students, paying particular attention to young women going to their first employment." The author of this book experienced this firsthand when Betty came to her dissertation defense at UC-Berkeley even though Betty's handbag had just been stolen. FND continued: "To visitors to the statistics department she was unfailingly kind in trying to make them feel at home both on and off the campus."[1] The author of this book experienced this too when, during a two-year leave of absence from her graduate program, she returned to Berkeley to discuss with Betty a plan to resume her studies; Betty assembled a group of women graduate students and took them all to lunch at The Faculty Club.

Betty did more than her fair share of student advising. She was a devoted, designated faculty advisor for either the statistics or biostatistics programs at Berkeley for some 20 years. First Betty served as the sole major advisor to statistics students from 1951 to 1954 while an assistant professor of mathematics. Next she was the founding statistics departmental major adviser from 1955 to 1957.[2] Then she served as co-advisor to the biostatistics students from 1967 through 1982.[3,4,5] Betty liked to work with people, especially young people: "The drawbacks are that [it] takes a lot of time and energy."[6] Betty's view was that the student's preparation needs to be diverse because the topics statisticians research and roles they adopt are diverse. She advised the student to

---

[1]Florence N. David, "Obituary: Elizabeth Scott, 1917–88," *Journal of the Royal Statistical Society: Series A (Statistics in Society)* 153, no. 1 (1990), 100.

[2]Statistics. Various UC-Berkeley Course Catalogs, 1945-1988.

[3]11/6/1967 handwritten notes by Betty Scott regarding strained relationships with the school of public health in connection with biostatistics.

[4]1/16/1968 "Program of Biostatistics in the Statistics Department," Department of Statistics.

[5]"Statistics," *UC-Berkeley Course Catalog, 1982-83*, 185.

[6]6/10/1978 letter from Ms. Mina Edelston, attendee at the UC-Berkeley Conference for Women in Engineering and Computer Science on 5/13/1978, to Betty Scott with attached questionnaire completed by Betty.

take courses in mathematics, probability, and statistics, but also in subjects where the student was likely to use statistics.[7]

Eight students are known to have completed their PhD degrees under Betty, with research areas in mathematical statistics, biostatistics, demography, or education. George Washington University Statistics and Economics Professor Joseph Gastwirth conjectured about why Betty didn't have more doctoral students: "Some of Betty's topics were on the more specialized side. This may have been a factor... Or it may have had something to do with the courses she taught. I don't know if many of her students did discrimination research. If not, it may be because legal statistics was new at that time."[8]

Betty's first PhD student was Mary I. Hanania who completed in 1957. Her dissertation was "Some Statistical Tests of Hypotheses in Learning Theory." Hans Konrad Ury completed "On Efficient Rank Tests for Comparing the Effects of Two or More Treatments" in 1971. Robert Keremis Sembiring completed "Demographic Treatment of Kinshi" in 1978. Virginia Foard Flack completed "Testing Hypotheses When There Could Be an Initial Effect" in 1983. Dennis Keith Pearl completed "A Stochastic Cancer Model Involving the Environment of Transforming Cells with Application to Skin Cancer" in 1984. Marisa Yadlin de Weintraub completed "Development of a Model for Probabilistic Discrete Decisions" in 1985. Francesca Irene Rizzardi completed "Some Asymptotic Properties of Robbins-Monro Type Estimators with Applications to Estimating Medians from Quantal Response," also in 1985. Betty's last PhD student was Susan Elizabeth Leroy (now Stewart) who completed "Robust Estimation of the Mean of a Retransformed Variable" in 1987.[9]

What became of these students? The answer is known for only some. Ury became a statistician at the Permanente Medical Group in Oakland. Pearl became a statistics professor at The Ohio State University. Yadlin-Weintraub became a statistics professor at Catholic University of Chile; she was the first woman in Chile to earn a PhD in statistics and be the director of the Statistics Department at Catholic University in Chile.[10] Rizzardi became a principal statistician and managing partner at Piscataqua Capital Management in Portsmouth, New Hampshire.[11] At the time this book was written, Leroy Stewart was a biostatistics associate professor at UC-Davis.[12]

---

[7]Elizabeth L. Scott, "Proceedings of the Conference on Educating Women for Science: A Continuous Spectrum" (Oakland: Mills College, 1976), 35.

[8]Joseph Gastwirth, e-mail to author, August 11, 2014.

[9]University of California, Berkeley Department of Statistics PhD Alumni. Accessed on 2/3/2014 at http://statistics.berkeley.edu/people/alumni/phd.

[10]Iglesias, Pilar and Icaza, Gloria. Leadership Research in Teaching and Learning Statistics in Chile. International Statistical Institute, 53rd Session 2001. Accessed on 2/3/2014 at http://iase-web.org/documents/papers/isi53/716.pdf.

[11]"Frances Rizzardi," *Piscataqua Capital Management*, accessed July 14, 2015, http://www.piscataquacapital.com/team-2/.

[12]Susan Leroy Stewart. Accessed on 2/18/2014 at http://biostats.ucdavis.edu/Faculty Profile.php?id=745.

10.1 Remembrances

Dennis Pearl was one of Betty's successful doctoral students. At the time this book was written, Pearl was a professor at Penn State in the Department of Statistics, after being a professor at Ohio State for 32 years. His research interests included the statistical analysis of nucleotide sequences and phylogenetic trees, simulation-based estimation, statistical education, modeling biological phenomena, and problems in clinical and translational research.[13] Like Betty, Pearl was both an undergraduate and graduate student at UC-Berkeley. He spent over 16 years there as a student, some of which were part time while he raised a family. Betty was his mentor the entire time. Pearl remembered Betty as a "wonderful woman" and "a very good teacher."

Brian Yandell was one of Betty's successful mentees. At the time this book was written, Yandell was a professor and chair of the Department of Statistics at the University of Wisconsin-Madison. His research interests included "statistical genomics, with the goal of unraveling the complex relationships between observable traits (such as flowering time or clinical signs of diabetes) and molecular signals (mRNA expression, protein and metabolite levels, etc.),... [and] development of informatics platforms that allow biologists and data analysts to share data and algorithms."[14] Yandell recalled that Betty was his first contact at UC-Berkeley while he was still an undergraduate at the California Institute of Technology (Caltech) and applying to graduate school. She arranged to get him a University Fellowship for his first year of graduate school. After that "a fair amount" of his employment was as a graduate assistant for either her or Jerzy Neyman. Yandell earned his PhD in biostatistics from UC-Berkeley, completing in 1981. He took a number of graduate courses from Betty and, throughout his graduate studies, Betty was one of his primary mentors. Yandell had "fond memories" of Betty, and described her as a "great mentor."

David Draper described himself as having been "rather close" to Neyman as a student and remembered Betty from that vantage point. At the time this book was written, Draper was professor of applied mathematics and statistics at UC-Santa Cruz. His research interests concerned "... methodological developments in Bayesian statistics, with particular emphasis on hierarchical modeling, Bayesian nonparametric methods, model specification and model uncertainty, quality assessment, risk assessment, and applications in the environmental, medical, and social sciences."[15] Draper met Betty early in his studies at UC-Berkeley. He recalled attending many of Neyman's seminars and classes and drinking many martinis with Neyman after the seminars at The

---

[13]Dennis K. Pearl. Statistics Department, Ohio State University. Accessed on 2/3/2014 at http://www.stat.osu.edu/ dkp/.

[14]Brian S. Yandell. Accessed on 6/24/2014 at http://www.stat.wisc.edu/ yandell/.

[15]David Draper, faculty bio, UC-Santa Cruz. Accessed on 12/28/2014 at http://ams.soe.ucsc.edu/people/draper.

Faculty Club; he described himself as being among Neyman's favored graduate students and was in the hospital room several days before Neyman died. He saw Betty a few times after he left Berkeley at which time they caught up on what each other was doing. Draper remembered Betty very fondly.

From their diverse vantage points, Pearl, Yandell, and Draper shared some of their memories about Betty.[16,17,18] Former UC-Berkeley students Deborah Bennett, Bill Kahn, Albert Lo, Vijay Nair, and Lucy Sells also weighed in, as did long-time UC-Berkeley department faculty colleagues Kjell Doksum and Juliet Shaffer, and external faculty colleagues Joseph Gastwirth and Ingram Olkin. At the time this book was written, Bennett was retired as a biostatistician and group leader from the Lawrence Livermore National Laboratory; Kahn was a senior vice president and risk modeling executive at the Bank of America; Lo was a professor and chair of information systems, business statistics, and operations management at Hong Kong University of Science and Technology; Nair was a professor of statistics at The University of Michigan; and Lucy Sells was a volunteer to the Democratic Party.[19]

Only some who knew Betty were findable and willing to share their memories for this chapter; perhaps individuals who were on her good side or did well in the profession were more willing to speak out. Also, memories are sometimes inaccurate or change over time. These are potential issues of selection, recall, and implicit bias. As a scientist with a specialty in measurement, Betty would have acknowledged these potential biases as part of her story.

10.2　On Mentoring

Betty wasn't on Yandell's dissertation committee, but he, "like so many others," sought career advising from her. Yandell recalled:

> She was a great mentor, a great help to me, and a friend. I saw her every few weeks. I would always just go to her office and talk to her. She would give me advice about life as well as about scholarship. I felt I could talk to her about anything. I had some times in graduate school when I wasn't very productive. Sometimes I just didn't know what to do. I felt I could still go and talk to her. She would still have encouraging things to say. Betty gave me extensive help throughout my graduate career. She even administered my French language exam, giving me an article to translate, while Neyman was in the hospital dying. I don't know why I didn't ask

---

[16]Dennis K. Pearl, telephone interview with author, February 17, 2014.

[17]Brian S. Yandell, telephone interview with author, February 20, 2014.

[18]Brian S. Yandell, telephone interview with author, June 24, 2014.

[19]Various personal communications: With Bennett on 8/30/2014 and 7/8/2015, Kahn on 10/9/2014, Nair on 10/6/2014, Lo on 7/3/2014, Doksum on 7/2/2014, Shaffer on 8/5/2014, Gastwirth on 8/11/2014, and Olkin on 9/30/2013.

her to be on my dissertation committee. I think it was just one of those things that happens when you are a student. I had too many people: I think I may have been talking more to Michael Klass at the time. When I went to Betty's office and told her I wasn't going to include her on my committee, she had a startle reaction but then quickly composed herself.

Yandell described it as a major regret that he didn't include her.

Betty also gave solid and encouraging job-seeking advice. Yandell recalled at first casting a wide net when looking for PhD level employment. Betty advised him to cast a narrower net. She didn't feel he should waste his, or the employer's, time on positions at less prestigious institutions. She had confidence he would be offered a job at a Research I university, and this is what happened. Betty also taught students about professionalism. Yandell recalled mentioning some things that were less positive about Berkeley in one of his interviews. Betty didn't ignore it, but made it into a teaching moment: "People should be circumspect when interviewing because things get back."

Yandell recalled Betty was highly regarded by the students for her studies of the equality of women in academe. "Women gravitated toward her. She was one of the few role models for women," he remarked.

10.3 On Generosity

Betty gave her life to the UC-Berkeley statistics department. She didn't just give her efforts: She also was a financial donor to the department.[20]

Betty is particularly remembered for giving that directly involved students. Doksum recalled there were lunches with Betty, Neyman, a few faculty members, and some students. At one of these lunches, Betty overheard Doksum and a student talking about basketball. Doksum remembered:

We decided we needed some exercise, went over to the gym, saw the campus had a basketball league, decided to put up notices in the department, and recruited a departmental basketball team; but we needed a basketball because the one we could check out at the gym wasn't always available. And so Betty went out and bought us a basketball. The team wasn't very good – we lost every game! Then we decided to form a soccer team, and we won every game. Later, there was a dinner in Betty's honor, perhaps for her 60th birthday. People made statements, and I decided to make up this story in my statement: The true story of our success in soccer is that Betty gave us a basketball, which we used to practice soccer; a basketball is much heavier than a soccer ball, so that's why we were so good at soccer!

Betty's act of generosity became a lifelong memory for Doksum and others.

---

[20] "Grateful Thanks to Our Friends," source unknown.

Pearl and Yandell recalled Betty's extraordinary generosity with the same two examples. One is that she supplied the baked goods or cakes for the hard-working people who came to the office on weekends. Betty would put the free Saturday pastries in the statistics department lounge. Yandell remembered the pastries as an incentive to put in extra work hours, but he also felt the reason she did it was to be kind, and she knew people needed to eat. According to Shaffer, Betty bought the cakes and brought them into the office herself.

The other example is more personal. Betty formed a skin cancer workshop in the department, where a handful of students regularly discussed the issue of ozone depletion and skin cancer. Pearl was heavily involved in the workshop for seven or eight years: He was employed as a research assistant and eventually developed his dissertation research out of the workshop. There were rules at the university about the payment of benefits. As a cost savings, most students were hired at less than 50% time because then they were not eligible to receive benefits. Pearl and his wife were expecting a baby and didn't have health insurance to cover the costs of prenatal care and delivery. Betty bumped Pearl up to over 50% so the costs would be covered, and she paid for his benefits out of her research grant. Later, when Pearl was employed by another statistics faculty member, Betty continued to pay the overage salary and health insurance. Pearl offered: "I think her grant was over. I don't know how she managed to pay. This shows how much she cared about students. She affected me a lot with her generosity."

Skin cancer became Betty's avocation. It was personal when one of the statistics graduate students died of melanoma. Yandell recalled this was around 1976, and the student had arrived to Berkeley from Germany only about six months earlier. He had a lesion on his arm and had put off going to the doctor. Betty would talk about that. According to Pearl:

> This [when the student died] was one of the two times I ever saw Betty cry. Betty said something like, 'We know so little, we can't even keep one of our own students from dying...' The only other time I ever saw her cry was when she told me that Mr. Neyman had died.

Betty could care deeply for students.

There are a number of stories about Betty continuing to find support for students when projects they were employed on were winding down. She somehow found money to keep them employed. Doksum recalled that "When students ran out of money, they would talk to her, and she would actually pay them out of her own pocket. My student Albert Lo... told me he was one of the ones, but he told me he wasn't the only one." Lo said this:

> In the 70s, the Berkeley Statistics Department separated the graduate students into four ranks, and only those in the top two ranks received department financial supports that were enough to cover expenses for a year; those who were third rank might have a "quarter-time" TA [teaching assistantship] and would need to find some part time jobs to supplement

> *their income. If you were ranked fourth, you were entirely on your own. (We used to marvel at a graduate student friend from Africa who had many kids (5?) and who supported his study and his family by pumping gas day and night at Berkeley gas stations. He survived and completed his Ph.D. degree.) It took me five years to finish and I was ranked third or fourth (mostly fourth) in my first three years of study, and I also had to scrub to survive. (Luckily I was single then.) I was searching for jobs in '75 (or '74), and somebody told me that Betty Scott (whom I did not know) was known to help students. I approached Ms Scott and she said that she needed some help with programing work in an applied research project, that the pay would be by hours worked, and that the duration of the work is a month (or 2, exact days are forgotten). Specifically my job was to go over, i.e., double check, the computer codes of a completed Fortran program of her project. I accepted, with the understanding that Ms Scott had put me on her grant and my work would be paid by research funds. At the end of my appointment, I reported the hours I had worked on the project, and was astonished when Ms Scott just pulled out a checkbook and wrote me a personal check on the spot, with no other documentations involved. It has since occurred to me that this was just her way to help out a student in need.*

According to Yandell, Betty reputedly gave people money "right and left."

It is unlikely that students or others took advantage of Betty's generosity. Betty wasn't the kind of person you would approach with a frivolous request.

10.4   On Personality and Professionalism

Pearl described Betty's personality as folksy. She liked to use sayings, such as "Give me the four parameters and I can fit an elephant." Some of the sayings originated from her and some were borrowed. But Betty's main story was that women were not able to use the big telescopes at Mount Wilson.[21]

Yandell remembered:

> *Betty had a way of complaining about things. She would start on a rant. Then she would turn it around. She was blunt, but tactful. She would make sure you knew she was aware if you weren't working hard or weren't on track. She had a way of bending rules to get what she wanted. Generally what she wanted was sensible, but unorthodox.*

---

[21] A.L. Golbeck, "Four Leadership Principles for Statisticians: A Note on Elizabeth L. Scott," in *Leadership and Women in Statistics*, ed. A.L. Golbeck, I. Olkin, and Y.R. Gel (London: Chapman & Hall/CRC Press, 2015), 31.

Former sociology student and gender equity collaborator Lucy Sells confessed that Betty could be scary.

Betty took care of her mother at home. Yandell met Betty's mother once and recalled she wasn't easy to hang out with: She had strong opinions, and always seemed to need things. Betty gave her mother a lot of attention and always made sure her mother's needs were met.

Whereas Betty took care of her mother at home, she took care of Neyman in the office. According to Pearl, if a fire alarm went off, Neyman would become concerned even if others around him were not. Betty would go to his classroom, explain to him what was happening and that everything was fine. She would always make sure Neyman was okay. "I'm not sure if she was daughter or mother during those times," explained Pearl. Yandell recalled it was often hard to think of Neyman and Scott separately.

Doksum remembered Betty as being a very good and responsible colleague:

> *She could sometimes be brisk. If I did something wrong, she would tell me. But I didn't take it badly at all, because she was treating Neyman the same way! I figured if she treated Neyman that way, it was fine with me if she treated me that way.*

Doksum recalled that all of the faculty respected her.

The UC-Berkeley and Stanford University statistics departments' joint colloquium used to meet nine times each year, alternating locations. Ingram Olkin remembered Betty being a big part of these colloquium events, bringing cakes and always being in the background making preparations. Betty had a powerful presence and voice, both intellectually and socially.

When asked about Betty's professionalism from the viewpoint of a visitor, Gastwirth had this to say: "What's there to dislike? She was very competent, very productive, very close to Neyman, did very solid work, was very well respected in the profession. Everyone thought she was very dedicated. I don't think it is easy to be the only minority in a department or group, but she did a very good job." He also said this about her organizing abilities: "I remember one thing that was unusual. There was a Western Regional Meeting in Berkeley. I went there. Betty was on the program and the local committees. She cooked the whole buffet herself – she maybe had some help, but it wasn't catered. She was quite capable in terms of organizing."

10.5   On Concentration

Betty organized her office hours by preparing a daily sign up sheet with 15-minute slots; she would post this sheet on the wall outside her crowded office. You had to get there early in the day to get a slot because they usually filled up by 10 am. Betty discouraged people from signing up for more than one consecutive slot. Pearl reflected on his time as a doctoral candidate: "It was like your 15 minutes of fame. You had to come prepared and motivated. And even though it was not very much time, 15 minutes with Betty were as invaluable as an hour, because she was so concentrated in her attention."

Then, when you were in her office or a seminar with her, Betty would close her eyes and tilt her head back. She would seemingly fall asleep, but then wake up with the key question. Yandell remembered:

> *Her head would go back, and she sometimes snored. The talk would end, and she would snap back and ask the most insightful question. She was paying attention even though she looked like she was asleep!*

Pearl reflected:

> *I used to feel funny and think, is she still listening? Should I walk out and let her sleep? But then she would suddenly come back, jump out, and ask a detailed question, or say 'shouldn't it be...?' And it would impress you about how deeply she had been concentrating on the discussion at hand. Betty worked very hard, and she got her rest in while paying attention! She was always paying attention. In fact, she always did a great job redirecting my energies in a more interesting direction, bringing me back to the big picture. Her input was always valuable.*

Betty was always present.

10.6   On Political Acumen

Pearl recalled Betty as a person with high political acumen. An example involved the tall Evans Hall right after it opened in 1971. The design of the building called for two banks of elevators. These had new technology: They were among the first elevators driven by computers. Betty used probability models to make sure they worked with maximal efficiency.

The problem was that the construction of Evans Hall suffered from inflation and cost overruns. The shafts were built, the first bank of elevators was put in, but there was no money to put in the second bank. After the building opened, many people on the upper floors became irritated after having to wait long periods of time to get to their floors.

Here is what Betty did. She was in a meeting with the UC-Berkeley chancellor and got him to come with her to her office to get something. He and Betty became stuck waiting for an elevator. The elevator was built very soon after that. Pearl said: "Betty had a way of doing that kind of thing. It was neat!"

When Evans Hall was being designed, statistics and mathematics were among the programs to be housed in the building. Yandell added that Betty worked with the architects:

> *She would climb up the scaffolding with a hard hat on and inspect the building. She noticed the women's restrooms would only be on every other floor. She kept making a fuss about this, all the way up to the chancellor, letting him know this wasn't acceptable. She had a vision that the climate for women in statistics and mathematics would improve over time and there would be more and more women in these fields. Because of her, every floor in the building has a women's restroom.*

Betty's influence on the design of Evans Hall was recognized by the chancellor at the groundbreaking ceremony in 1968:

> *As many of you know, the planning of a building such as this involves a great deal of faculty participation in the definition of the program and then working with the chancellor's office, office of engineers, architect, with the state, in the final definition of the program. And playing a very important role throughout that whole lengthy process in connection with this building is the next person whom I am to introduce, Professor Elizabeth Scott. She for almost ten years was a member of the planning committee, several times as chairman. And a more vigorous defender of the proposition that buildings ought to be useful to the people who are going to use them cannot be found. It is a pleasure for me to introduce Professor Elizabeth Scott who is chairman of the department of statistics and a valiant fighter for its present design.*

Betty had a generosity of spirit and used her speaking time to acknowledge and thank the many individuals who worked with her on the building project.[22]

Betty's political acumen did not temper her natural proclivity to be gruff. What did the administration think of Betty? Shaffer recalled: "She was probably regarded as somewhat of a pest to the administration. I imagine they didn't find her easy to deal with." What did staff think of Betty? Shaffer recalled: "One day we were both in the basement [of Evans Hall] waiting for an elevator. She chewed out some workers about something she didn't like. They were looking like, who is this old lady telling us what to do?!"

Shaffer also recalled:

> *I had an assistant who came into my office in tears. Betty was harsh to the person, and I told the person Betty doesn't distinguish between those above and below her: She will chew out a dean and a student, she didn't distinguish between them. She didn't operate on status, and she didn't use her status. She could be rough and kind at the same time... She was brusk,*

---

[22]9/25/1968. Evans Hall site dedication. University of California, Berkeley (sound recording).

but not unfriendly either. She was matter of fact. She was equally brusk (egalitarian) with those above and below her.

Betty was authentic.

Doksum also remembered that "Betty and Neyman had political power with respect to the administration. That was my sense of it. The two of them would agree on something, and then it would probably happen."

10.7 On Approach to Science

Pearl described Betty as a researcher who was "straightforward about policy matters," driven by research questions that were formulated from real world problems rather than those that a particular set of data could answer.

Yandell also remembered Betty as a scientist who let the questions emerge from the science and then used mathematics to try to answer them:

> *She didn't act like a mathematician. She came across as a scientist who would say, 'here's the problem,' and then ask, 'now what should we do about this?' She was taking the science questions she encountered, framing them, and asking statistical questions... Statisticians coming from a mathematics background would approach a problem the other way around, namely by coming up with a mathematical method and then finding an application for it. Coming from a background at Caltech where I was connecting mathematics, new computing technologies, and biology, I was really impressed with her approach. I was already on that path, but she really enforced it.*

Betty was a quintessential woman in science.

Draper talked about the Neyman-Scott collaborations and expressed admiration for their scientific work together. He specifically talked about a Neyman-Scott problem in mathematical statistics:

> *[It] became a celebrated example of an inconsistent maximum likelihood estimator. You create a two way analysis of variance problem where each group is allowed to have a different variance, and you let the sample size grow in one direction and not the other, and in the end there are just too many parameters for ML to do well in the problem, and so it ends up in simple settings converging with a lot of data but to the wrong answer... this is a rather embarrassing situation for ML. They were the people who noticed that.*

Draper still calls Neyman "Mr. Neyman" and Betty "Miss Scott."

When asked his impression of how Betty and Neyman did science together, Draper said this:

> *He was always the theoretical strong force in their relationship... My impression was that Betty was always a very strong applied statistician and that she stimulated Neyman by bringing important applied problems to him, and getting him interested in them, and then he would see how to apply his own existing theories and methods, or if necessary invent something new. I think that was the dynamic of their work relationship together. Betty was the one who brought applied problems to their joint collaboration, and then I'm sure she pulled her weight on the actual work involved in helping to solve the problem, and if there had been any computing that needed to be done, it would be Betty that did it or Betty supervising somebody else to do it.*

Draper remembered Neyman working on several papers in his hospital bed the day he died:

> *... and papers would be spread out all over Mr. Neyman's figure underneath the sheets and blankets, and he and Betty would be talking animatedly about how this section wasn't done yet because they had forgotten to talk about bla bla bla, and this other section was looking pretty good, and she was essentially making sure that she understood his desires on what should be in those papers after he died. Everybody knew the last month of Mr. Neyman's life that it was over. And so he worked very effectively with Betty during that last month to get as much of his thinking into the final papers they were working on.*

Betty was always there for Neyman.

10.8   On the Other Side

Not everyone saw the same side of Betty. Bennett recalled this experience:

> *A year or two after receiving my master's in biostatistics in 1978 [at UC-Berkeley], Elizabeth Scott was offering a course at UC-Davis. I was very excited as I had not had the opportunity to take any of her courses while working on my master's at Berkeley and she was such a big name... [on the] first day of the class, Dr. Scott addressed the class by saying how pleased she was to see all of us there, although she expected the girls were there just to sit next to the boys. I was crushed and so, well, I guess offended. How could a woman in her profession put down other women striving to also be in her same profession? Maybe she was trying to be funny, maybe I was being over-sensitive, but I had encountered enough of that attitude at work that it stung. It formed my opinion of her and that was unfortunate.*

Later, Bennett considered: "Perhaps it was a very bad joke? Maybe it's time I cut her some slack..."

Also, Shaffer and Betty "never got really close":

> *I never had a feeling that Betty promoted me in any way... I sometimes felt rivalry, but I have no basis for it. But one time she attacked a student assuming he was a student in my consulting class, which he wasn't. It was Bill Kahn, he was a graduate student, who had the run-in. He was a very independent, bright, nice guy who worked for various banks. He wanted to do a project for his master's degree and would come to my office to talk about it, and Betty wasn't in favor of him doing a project. He railed about Betty to me. He wasn't going to take anything.*

Kahn had earned a bachelor's degree in physics at UC-Berkeley, worked for awhile, and then wandered into the field of statistics because he was doing data analysis jobs and loved to work with data.

Kahn recalled the incident:

> *Betty was the master's advisor. I wanted to do the thesis option. By Berkeley standards I was nothing to write home about in terms of theoretical statistics. But it turns out that in my element [of data analysis] I was actually pretty good. For example, I took David Brillinger's time series class. For my midterm practical project, I had just gotten a A++ on from David Brillinger doing time series analysis. I had done well in Julie's ANOVA class – I think I may have been the top student – when I took the second quarter of the class. Julie then asked me to be the reader for the class, which you know is a small honor to take the class but then have the responsibility to be grading the other students' homeworks. So that was a small degree of endorsement for me having some bit of promise. Anyway, I'm in the elevator in Evans Hall – I don't want to say it was a jammed elevator, I'd say there were six or seven people in it – and apropos nothing, Betty turns to me and starts speaking to me in the elevator, and she says, "Young man, you have an inadequate mathematical background," with aggressiveness. And I was just stunned, and everyone in the elevator just stared straight ahead, not wanting to get involved. And I don't know, I was in my mid 20s, and I didn't understand that different people have different psychological and chemical balances, and if they are out of balance that day, they say funny things. But Berkeley back then was an easy place to be insecure about one's intellect, and I was certainly not immune from that. I felt pretty good about my ability to do practical data analysis but all of this other measure theory and statistical asymptotics and such was quite beyond me. So there was no need for Betty to lash out to me like that. If I have an inadequate mathematical background, there isn't anything I am going to do about it, we all have our natural skills. It was completely inappropriate to have said that, ever. She might have talked with me about how I might remediate my weakness during an*

*advising consultation, but to lash out at me in public like that, it was just
really horrible. Given my strengths, I wanted to do a thesis. I actually did
end up finding an advisor to do my thesis with.*

Shaffer conjectured that Betty may have "flared up at the student [Kahn]"
because she didn't approve of Shaffer's consulting service: "Neyman and Betty
may have resented that I was doing something different." In any case, it was
an experience that left a lasting impression on Kahn, who summarized Betty
as coming "within an inch of destroying my fledgling career. . ." But instead,
Kahn earned his MA in statistics at UC-Berkeley and his PhD in statistics
at Yale and went on to have a highly successful career in industrial statistics
and financial services, working at the executive or senior executive levels for
Capital One, Fannie Mae, Travelers Insurance, AIG, and Bank of America.

Upon hearing Shaffer's comment indicating he might have been, in his
words, "an irrelevant pawn in a slightly larger game," Kahn remarked:

*I've always had the philosophy, rare events happen rarely, which basically
means every time you see a data point and you say, that seems strange,
what you're really saying is, I don't understand the story that is generating
this data, but there is some story, and it's generating related data. So what
I have tried to understand is, why would this senior tenured professor
turn to a measly graduate student and clearly actively mistreat him. . . now
you've introduced the idea that there were other dynamics at play. . . Maybe
after I read [this] book I will be able to see Betty as a multidimensional
person and not the evil witch of my youth.*

Then, with regards to this book, Kahn said: "Oh, thank you!"

10.9   On Gender

At the time Pearl, Yandell, Draper, and the author of this book were graduate
students at UC-Berkeley, Betty was the only woman in a tenure ladder position
in the statistics department. She had been one of Neyman's students. Draper
recalled:

*I heard rumblings about this from time to time – people who felt that
she should not have been hired, that Neyman was using nepotism and
privilege to get her to be hired into the department, and it is true that,
by the stringent, narrow standards that the Berkeley faculty liked to use
in those days for just exactly who are the kind of people that we want,
under those criteria Betty would never have had a faculty position in
that department. At most she would have been something like what they
did to Julie [Shaffer], making her this thing called senior lecturer with
security of employment. Maybe it is better now, with Deb Nolan and people*

> like that, but there was a lot of sexism in the department when I was
> there. Betty was the only female figure you ever saw around except for
> the woman [Jeanne Lovasch] who did the computing and then also Julie
> Shaffer coming on the scene in the mid to late 70s as well. That was
> pretty much about it. I don't think they were in the habit of bringing in
> any female grad students at the time, or if they were, they were quite
> small in number. It makes sense Betty would work on this [women's]
> issue, because she herself was facing substantial headwind in even being
> there, and you would have a natural desire to try to make things better
> for people who came after you in that situation, and I admire her again
> for that.

When asked why no women were hired into the department when Betty was
department chair, Draper continued:

> I would guarantee you that it was not for lack of trying on her part.
> But I would also make the guess that every single man in the department
> claimed that all female candidates at that time were sufficiently weak, were
> not strong enough for the rigors of receiving tenure at Berkeley, so that
> it was not something that got off the ground, and that is my reasonably
> educated guess.

When asked why Shaffer wasn't hired into a tenure ladder position, Draper
offered that this was all about power. Neyman was the founding chair, whereas
Lehmann was the student of the founding chair. Neyman had a card to play:
Because he was the founding department chair, he was entitled to get what
he wanted. So it was a story of headwind, sexism, and power.

10.10   Summing It Up

Olkin called Betty the First Lady of the department. Nair, who Shaffer remembers as being her first consulting student, described Betty as "the one
who took care of all of us grad students and faculty." Yandell summed up
Betty's legacy this way: "Betty wanted to make sure the department was relevant. It has now evolved into a well-balanced department. I think she had
influence on that."

11

Letters to Jerry
(1954–1955)

*Anybody with a pair of eyes and any knowledge of what it means
for a warm, intimate look to pass from one human to another
could not have failed to see that they were close.*
-David Draper

In 1953–54, Betty taught upper division Statistical Inference for the second time, co-taught with Neyman the graduate Probability Models of Natural Phenomena, and taught the lab associated with Neyman's graduate Advanced Topics in Probability and Statistics. It was the first time Betty and Neyman taught together, and it brought them closer.[1]

UC-Berkeley had a nepotism rule at that time that prohibited more than one relative from working in the same department. The rule had to be of particular interest to Betty. She could easily imagine how it could have terminally affected her career had she been married to co-worker Jerzy Neyman, or "Jerry" as she called him. Betty had a professional relationship with Neyman, one that by the mid 1950s had become personal. It happened in Paris.

The year was 1954, 15 years after Betty met Neyman. Betty had accomplished much professionally and was now in her third year as a mathematics assistant professor. Betty was Neyman's student; worked for him at the Stat Lab and alongside him as faculty member in the mathematics department; co-authored five publications with him (two in mathematical statistics and three in astrophysics); and taught a statistics course with him. Neyman wanted to work with Betty, and Betty wanted to work with Neyman: Betty didn't want anything more for her life and career than to maximize her ability to contribute to Neyman-Scott collaborations. To that end, Betty thought it would be helpful to take a leave of absence from UC-Berkeley during the next academic year, 1954–55. Her objective would be to strengthen her background by getting additional training from leaders in the field of mathematics.

Most would judge Betty to have been an excellent student of mathematics. Betty's UC-Berkeley transcript indicates, as an undergraduate student in astronomy, she had 33 semester units of mathematics, with mathematics grade point average of 3.455; as a graduate student in astronomy, she had 41

---

[1] "Mathematics," *UC-Berkeley Course Catalog, 1953-54*, 217.

semester units of graduate-level mathematics, with mathematics grade point average of 3.8. This was an enviable record of scholastic achievement, but now she felt she needed to know even more mathematics to excel in her research. New mathematical disciplines were rapidly emerging that were influencing development of the theory of statistics. Betty felt deficient in them. She felt the deficiency was impeding her research.

At first Betty tried to teach herself, but she quickly ran into limitations of this approach. The leave of absence would give her the opportunity to fill gaps in her mathematics education. She wanted to spend the year "in systematic studies from the bottom, so to speak."[2] She hoped that with the additional training, her level as a mathematical statistics researcher would improve. She undoubtedly also thought her additional training would enhance her chances of earning tenure in the Berkeley mathematics department.

Betty reasoned she could get the additional training she needed within the US, even at Berkeley. But if she stayed in the US, she would be distracted. She had, in addition to her own research and the exclusive Neyman-Scott collaborations, several significant collaborations going on with eminent scientists. One was with C.D. Shane at Lick Observatory on the distribution and clustering of galaxies; another was with R.R. Reynolds at the California State Department of Water Resources on weather modification; yet another was with Professor Thomas Park at the University of Chicago on the Tribolium model in ecology. While Betty was heavily invested in all of these collaborations, she needed some space to accomplish her goals in mathematical statistics. She also needed some space from her responsibilities as a teacher, consultant, and professional association volunteer. To really get away from everything, Betty reasoned, she needed to leave the country, or else she could easily get dragged back into things. But where should she go to maximize her experience?

Neyman was foreign born, had lived in a number of countries, and was widely traveled. One of the places he had a special affinity for was Paris. He spent the 1925–26 academic year at University College London on a Rockefeller fellowship. There he met Egon Pearson who was a lecturer in statistics. Then Neyman spent the 1926–27 academic year in Paris, also on the Rockefeller fellowship. There he studied pure mathematics under renowned mathematicians Emile Borel (University of Paris) and Jacques Hadamard and Henri Lebesgue (College de France). While in Paris, Neyman received a visit from Pearson. It was on that visit that they began their work on statistical hypothesis testing,[3] work that had solidified both of their places in the history of statistics. Neyman also went to Paris in 1936, delivering his then-new ideas about confidence intervals.[4] Obviously Paris had been a special and productive place for Neyman. Betty surely had this in her mind as she contemplated

---

[2]n.d. [?1953]. Proposal Plan of Study (E.L. Scott).

[3]C.L. Chiang, "Jerzy Neyman 1894–1981," *Statisticians in History*, accessed June 11, 2012, http://www.amstat.org/about/statisticiansinhistory/index.cfm?fuseaction=biosinfo &BioID=11.

[4]Constance Reid, *Neyman: From Life* (Berlin: Springer-Verlag, 1982), 207.

where to spend her leave of absence, wondering if she too would have such a special and productive experience.

Betty had been advised by some to go to Nancy, and she almost went there. In her words:

> *In thinking of studying abroad I discussed the matter with my colleagues here. Also, I wrote for advice to Professor Paul Lévy,[5] whom I know, and to Professor G. Darmois[6] in Paris. After these discussions I decided that I am likely to profit most by going to Nancy, which is in the center of the modern French school of mathematics connected with the name of the "Bourbaki."[7] There I propose to attend courses given by Delsarte, Godement and Serre.[8]*

But Betty ultimately decided to go to Paris, to the Institut Henri Poincaré.

This was an institute of mathematics at the University of Paris. In 1923 John D. Rockefeller, Jr. donated a huge sum of money to promote and advance education throughout the world. His vehicle was the International Education Board (IEB). The IEB had already given substantial funding to a number of European universities, but little to universities in France. And so it decided to give to a new institute, the Institut Henri Poincaré. "Noting the importance of the French Mathematical School," the IEB reasoned that "... helping mathematics in France was perhaps one of the best ways to help science all over the world."[9] The Institut was accordingly set up in 1928 as a center for teaching and research on mathematical physics and probability.

There was so much going on at the Institut. Already it was "the privileged place where the excellence and the dynamism of French mathematics [were] embodied, under the double sign of hospitality and sharing. Every year, this 'House of Mathematics and Theoretical Physics' [hosted] hundreds of visiting scientists, from the entire planet, and thousands of visitors who [came] to share their scientific expertise."[10] Yes, Betty would go to Paris. She wouldn't do actual research while she was there. She would write papers, but only on research she had already completed. She would study and learn, so she could do better research after she got home. She would attend and present papers at some congresses. Probably at the International Congress of Mathematicians,

---

[5]Note: Paul Pierre Lévy (1886–1971) was an eminent mathematician who worked at the École des Mines. He was especially active in probability theory and has many mathematical ideas named after him.

[6]Note: Georges Darmois (1888–1960) was a mathematician and statistician who had been involved with the Institut from its beginning and who had been influenced by the work of statistician R.A. Fisher. Darmois was appointed head of the Institute of Statistics at the University of Paris after World War II.

[7]Note: Nicolas Bourbaki was a pseudonym for a group of mathematicians who wrote a series of books seeking to base all of mathematics on set theory.

[8]n.d. [?1953]. Proposal Plan of Study (E.L. Scott).

[9]M. Fréchet, "The Inauguration of the Institute of Henri Poincaré in Paris," *Bulletin of the American Mathematical Society* 35, no. 2 (1929), 198.

[10]Institut Henri Poincaré, accessed March 22, 2012, http://www.ihp.fr/en/instituteIHP /presentation.

scheduled for September 1954 in Amsterdam; and probably at the meeting of the International Astronomical Union, which was to be held in summer 1955.

In 1954, Neyman was 60 years old and had been married to Olga Solodovnikova for 34 years; she was a Russian girl he had met on a trip to the Crimea where he had gone to recover from tuberculosis. Now the couple had grown apart, their son Mike was 18 and had left the nest, and they set up separate residences.[11] Neyman was available, and so was Betty. She was 37 years old, had never been married, and was still living with her parents. Betty was almost half of Neyman's age, but the age difference didn't matter.

In fall 1954, Betty and Neyman traveled together to Paris.[12] Neyman returned to Berkeley at the end of October. The morning after he left, Betty moved out of the hotel and into the Fondation des Etats-Unis, a private student residence hall that was part of the Cité Internationale Universitaire de Paris and located across the street from the 37-acre Parc Montsouris.[13] Established in 1930, the Fondation continues to offer housing to over 250 international students. Now in a beautiful building with tall windows and manicured courtyard, in Betty's time it was institution-like: "Black and run-down." There was a metro station across the street, which was convenient, but when the train came it shook the entire building. In room 442, Betty was on the street side of the top (5th) floor, in the women's half of the building. Betty would be pleased that her floor is now devoted to the fine arts, since at the beginning of her college life she was interested in art as well as astronomy.

Life was not easy for Betty at the Fondation. The elevator in her half of the building was broken; she had to walk up and down many flights of stairs to get to her room. Betty had a washing machine at home, but at the Fondation she had to wash her clothes by hand in a tub. There were no bathtubs, and the showers were being repaired on every floor except the first, so she had to compete to use the showers. Betty had a keyed closet in her room, but the key to one of the halves was gone; she worked hard to try to get into the locked half of her closet. Other problems at the Fondation were easier to solve. There was no ceiling lamp, just a bedside lamp. Betty could buy another lamp. In the meantime, she could move the bedside lamp to the desk when she needed to work, and then move it back to the bedside at night when she was done. The meals were reasonably priced (about 100 francs per meal), but they did not include many green vegetables or much fruit. Betty figured she could purchase supplemental vegetables and fruit and still spend about $100 a month ($801.73 in 2010 dollars) on living expenses (rent, food, soap, etc.). At least she hoped so.

---

[11] Constance Reid, *Neyman: From Life* (Berlin: Springer-Verlag, 1982), 236.

[12] Note: Betty and Neyman probably flew commercially, but on propeller aircraft, as this was before the beginning of the commercial aviation "jet age." In 1955, there was only one jet airliner put into service – the British de Havilland Comet.

[13] Note: The address was Cité Universitaire, Rm. 442, Fondation des Etats-Unis 15, Blvd. Jourdan, Paris 14.

There was one possibility of an alternative living arrangement. It was a room in a private home. Betty met the landlord, who spoke good English, over tea. She was proud of herself that she got there at the appointed time, "on the dot." The room was pleasant and had its own bathroom. But Betty decided it was too expensive and way too far from the Institut to go by car.

Betty decided to remain at the Fondation. Then, being the end of October, it started to get cold and rainy. She worried about whether the management would soon turn on the heat. Her room had a tiny balcony (most of the others did not), but it was on the north side and on the boulevard, which meant it was too cold and noisy to use. Betty longed for spring to come as the trucks roared down the boulevard outside her window.

Betty began to get to know some of the neighborhoods. She walked all around and as she got to know her environment, she found that her French was causing her trouble. Betty had studied French for her first two years in college, but it was not her strongest suit (she got Bs and Cs in French, but mostly As and Bs in everything else). More French language instruction was the answer. Betty talked to people at the Fondation, and they recommended the Alliance Francaise. But that wouldn't work: The classes were five days a week in two-hour blocks, and the schedule interfered with her mathematics. She would look into the Sorbonne. Their language instruction was only three days a week. She wasn't sure if it would be too advanced for her.

11.1   October 1954: Paris

On October 29, 1954, Betty wrote a four-page letter to Neyman telling him all about the Fondation, her room, and her French. "The pictures from Spain came! Some of them are very nice. The Alhambra is not so red as I remember it. I'll label them and mail them soon... Best wishes and love to all. I miss you and envy you in Berkeley."[14] Betty added a footnote in French that contained corrections she wanted to get to the author of a manuscript she was reviewing. She also added a page in French that consisted of an outline of a manuscript by Neyman on the probabilistic theory of clusters of galaxies.

The next day, Betty wrote to Neyman again.[15] She started the letter by sharing some technical ideas about their consultations for statistician Donald B. Owen at Sandia Corporation. Some 20 years later, Betty would publish an article on statistics in astronomy in the US in a book that Owen edited on the history of statistics.[16] But back in October 1954, Betty worried to Neyman that Owen could "grind out some Bs and some B*'s because it cer-

---

[14] 10/29/1954 handwritten letter from Betty Scott to Jerzy Neyman.

[15] 10/30/1954 handwritten letter from Betty Scott to Jerzy Neyman.

[16] "Statistics in astronomy in the United States," in *On the History of Statistics and Probability*, ed. D.B. Owen (New York: Dekker, 1976), 318.

tainly would be exciting to have the quasi-correlations with and without expansion. . . perhaps he could try the higher order integrals (we called them As). . . " She worried to Neyman about what she promised to do for the December Berkeley Symposium on Mathematical Statistics and Probability.

Betty was pretty well unpacked and had gotten most of her clothing washed, but she worried about the many errands she still had to run in order to get her living situation settled. Like getting an electrical adaptor for the ironing room so she could begin to iron all the clothing she had washed. Two men worked throughout the day to try to get her closet unlocked. They finally got it open and, lo and behold, it was jammed full of books. Betty was relieved to find that the heat was turned on. But she was dismayed that no one at all was working to fix the elevator, and that the work to fix the showers on her floor was happening at a snail's pace.

Neyman had gone home, and it was hard for Betty to be left alone in Paris. She had been unwell in Spain and was now feeling better, but she worried about whether she would relapse and have to see a doctor. But mainly she was unhappy because she had no one to talk to. She complained to Neyman: "No one even tells you what to do." She wished the cashier at the cafeteria would say more to her than merely "Vetre carte" (Your card).

As an astronomer, Betty was a trained observer, and it showed in her descriptions to Neyman of her environment:

> *I do wish you could have been here with your camera the morning I left the hotel (the morning after you left). You would have gotten a magnificent picture: Matched horses – dark brown with black manes and tails or dappled greys with black manes and tails parading along spiritedly to the music of a band (also riding) in front. Blue uniforms with red trimming, plenty of braid, sabers jammed into the right arm, fancy red and white cushions in front of and behind the red saddle, blue and gold bicrown [sic] hat with black plume streaming out of horse's mane. All on the Quai Voltaire – what a sight! And the sun was shining. . . Bringing up the rear was a policeman on a little bicycle.*"

Referring to Mount Hamilton and Lick Observatory, Betty closed her October 30 letter to Neyman with the words: "Please say hello to the Mountain for me – ," echoing the words "Please remember me most kindly to the mountain" that her Aunt Phoebe delivered to William Campbell some 35 years earlier. Betty remembered the 365 turns to get from the bottom to the top of the mountain. She reminisced about the work she had done there. This mountain was as special to Betty as it had been to her aunt.

There were many long letters and postcards from Betty to Neyman. These communications were filled to the brim with information about Betty's collaborations with Neyman, her instructions to him about next steps in their research, and detailed stories about her travels. On one postcard, she told Neyman that she liked her hotel in London, and she especially liked the breakfast: "grapefruit, eggs, bacon (ham like), real toast, butter, real coffee!" She continued, using every corner of the postcard, to describe a manuscript she had just drafted and mailed to Berkeley about "change in apparent magnitude due to Doppler type shifts in energy." She called Neyman's attention to a disagreement between astronomers "Hubble + Humason (really Tolman) and de Sitter and McCrea (and McVittie or conversely)." Betty had her own opinions, and she asked Neyman to show the paper and discuss the matter with one of them. "We need this stuff right away yesterday."[17] She warned Neyman about a specific detail that could be an issue in one of his upcoming talks. Her intellect and love for research literally jumped off the 4x6 inch card.

It turned out that what Betty wrote in the draft manuscript was, by her own final assessment, "very naïve,"[18] and it caused Neyman some "difficulties and/or embarrassment" with the astronomer Betty suggested he consult. But, as she explained to Neyman, she drafted the manuscript because she was anxious to get something computed in the Stat Lab back at Berkeley: "After all, you have had the formulae for a long time and we have computed nothing... [the astronomer] himself asked me to compute the Δ_m. But, that's it [French trans]." She went on to say she thought one of [the astronomer's] remarks about the mathematics was "nonsense." "I am not too convinced that [the astronomer] knows which end is up. What do you think? But he is very sensitive... I am glad Chandra didn't shoot you. On second thought, shooting is not in character probably."[19]

11.2 January 1955: Paris

January 31, 1955 was a good day for Betty. All the astronomers talked to her in French, and she was excited. She considered it a sign of acceptance and familiarity that they roared when she answered with "wrong" French.

Life Magazine was a weekly news magazine that emphasized photojournalism. The January 24 issue of the international edition caught Betty's eye. The articles were nice, but the illustrations were "magnificent." They were of "the star-studded reaches of measureless space." Struve's stars were on the cover. "Inside you'll note the touch of Baade, Struve, Humason, Bondi, Gamon et al explaining stars and galaxies and expanding universe rather nicely and clearly, especially the drawings," she told Neyman in her letter. She and Neyman could use these kinds of drawings in a future lecture.

Betty was working on a February 25 symposium talk. It was too long (18 pages in Neyman's size of handwriting), and Betty had the additional

---

[17] 12/26/[?1955] handwritten postcard from Betty Scott to Jerzy Neyman.

[18] 1/31/1955 handwritten letter from Betty Scott to Jerzy Neyman.

[19] Note: This was probably a reference to astrophysicist Subrahmanyan Chandrasekhar (1910–1995) who shared the 1983 Nobel Prize for Physics for discoveries on the evolutionary stages of massive stars.

challenge of writing in a foreign language, so she sought some professional help. As luck would have it, a Berkeley mathematics and statistics colleague, Michel Loève, had been on sabbatical in Paris in 1953–54, and he was still there. Loève corrected the first five pages of Betty's talk in terms of both grammar and style. Then Betty met with the student secretary for the seminar at the Institute for Astrophysique. He graciously corrected the whole talk, but only in terms of grammar. Betty wasn't doing too badly: He found only one or two corrections (wrong word, or wrong article) for each page.

While waiting for the student secretary in the library, Betty noticed the November 1954 issue of the *Astrophysics Journal.* "I haven't seen anything like that since I left Berkeley – the libraries are difficult here." There were several articles of interest to her and Neyman, and she pointed them out to him. One was the abstract of their clumpiness paper! Another had to do with reddening versus distance or apparent velocity.

> *Is it a straight line? Does it exist? Is it really reddening excess over Doppler or are there other explanations such as selection, number of galaxies in cluster, etc., etc.? This leads to what we need and is clearly connected to the paper I sent that led to grief with [the astronomer]. Also, part of this is what I did before... but didn't do anything with after I talked at L.A. because you & I had a discussion disagreement? [inserted] about the estimation part. It might be very interesting to get this out again. You remember that this is a case of "both variables subject to error."*

She cited some references for this. A third article of interest had to do with magnitudes of bright galaxies. "Now we can redo the test of the expansion of the Coma Cluster." This cluster was just then being thoroughly studied at Mount Palomar and becoming understood to its full extent.

In her January 31 letter to Neyman, Betty mentioned she "received the letter about plans! This is very exciting... I shall try to write nicely." She also mentioned that she got vaccinated. "So now I can come home! But don't worry, not today." She was worried about the figures for her February 25 symposium talk, and asked if they would be ready in time.

11.3   February 1955: Paris

Half of Betty's next letter to Neyman, on February 4,[20] was packed with business. Neyman sent her a draft of the manuscript that she, Neyman, and Shane were preparing to publish in the *Proceedings of the Third Berkeley Symposium.* Neyman had gone through some trouble transferring most of Betty's corrections to another draft and had it typed. He now wanted her to review

---

[20] 2/4/1955 handwritten letter from Betty Scott to Jerzy Neyman.

the latest draft, and she did. She outlined her suggested changes, making the more trivial corrections right on the typed manuscript. She detailed the more important, complicated, or controversial corrections in the letter: incomplete; change notation; confusing as written; not true as typed; clumsy; redundant; incomplete captions; non sequitur. At the same time, Betty was apologetic for all of these corrections and changes, because she knew it would take work to get them all put into the manuscript back in Berkeley. "Sometimes I am not in complete accord with that which resulted but I tried (and try) to be agreeable (or 'stubborn together')."

Betty's former French teacher also corrected the February 25 talk and listened to her deliver it, correcting her pronunciation. Betty now considered the talk to be finalized. In Betty's case, the astronomers suggested making copies of her entire talk, rather than the usual two pages, and handing it out before she talked rather than after. Betty welcomed the suggestion. It would make it easier for her to get through the material, which she knew was too much. She knew the suggestion was a reflection of the quality of her French language skills, but she also thought it was a reflection of the strangeness of the subject matter to the audience: galaxies, and probability, and statistics (even though, she admitted, there actually wasn't much probability or statistics in the talk).

Betty was always grateful for Neyman's advice, suggestions, and help. She missed him. Part of her was still the student, and he the mentor. Given their age and experience differences, it was hard for her to feel intellectually grown up and independent. Betty trusted Neyman completely, and so it wasn't hard for her to confide these feeling to him. "I wish I could talk to you!"

At the same time, Betty was feeling some push back from the Paris astronomers about the merits of statistics in astronomy:

> ∃ *[there are] astronomers in Paris who feel the need of a statistician (and [Georges] Darmois does not supply this need) and who want to learn some statistics themselves. But these are a small minority. There is another, strong, older, more vocal group who say that all that modern statistics can supply is tests of significance and since these always give the result "not significant" then why use statistics? (When ignorance is bliss, 'tis folly to be wise). I can sense this hostility (I think, but I am still terribly insecure in Paris so it may be partly my imagination plus their reaction to a stranger) and I have tried to behave very "nicely" at all times.*

Betty was asked by astronomer J. Delhaye to give another talk in French on a statistical topic: the properties and applicability of least squares estimation versus maximum likelihood estimation. He wanted her to focus on astronomy examples, rather than theory. "This does seem to be an extremely delicate topic, pretty much treading on peoples' toes." She promised to think about it, even though it would be quite a bit of work, as she had left most of her examples back in Berkeley.

Betty asked Loève to look at her sentence: "C'est un honneur pour moi et, en meme temps, un plaisir de... [It is an honor for me and, at the same time, a pleasure of...]" She wanted this politeness to be absolutely perfect. Loève thought it was fine as written. He then offered to do a dress rehearsal with her a few days before the seminar. Betty accepted his kind offer. Loève even offered to go to the seminar to, in his words, "defend" Betty. "He feels that the underlying theme of building a model, etc., will be antagonistic. This does alarm me. Honestly, I don't expect anyone to jump on me," she told Neyman.

11.4   April 1955: Paris

Betty's many letters to Neyman, even those that were mostly business, contained personal notes, but there was a definite change of tone after her February 4, 1955 letter. The rest of her letters from Paris made it clear that the two had become lovers.[21] There were declarations of love. "Oh Jerry!" There were expressions of sadness at being so far apart. Longing to be together in Paris, Berkeley, or anywhere. Desire to wake up in the same bedroom. Yearning to go home a week earlier. Betty sent Neyman a thousand kisses, and she did it in French, with the valediction "Ton aimante" [your lover].

At the beginning of May, Betty's time left in Paris was getting short. To get home, she planned to fly on June 3 to Lisbon on TAP [Transports Aeriens Portugais, or Air Portugal], lay over for a day in Lisbon, fly to Bermuda on AC [Avianca], lay over in Bermuda for a day, fly to New York on PAA [Pan American Airways], lay over in New York and Philadelphia, and then fly on June 9 to Los Angeles on United. She asked Neyman if he would be in Los Angeles when she arrived. It would be only a month more before she would see him!

11.5   May 1955: Paris

May 1955 was Betty's last month in Paris on the one-year leave of absence she had taken from her Berkeley faculty position to study advanced mathematics. The 1954–55 academic year was Betty's first long international stay. By the time May came around, she longed to be back home. She made a plan to leave Paris on June 3 and arrive in Los Angeles on June 9. This was a little earlier than originally planned. She hoped Neyman wouldn't disapprove.

Betty had committed to giving a series of talks in England the second week of May before heading to Los Angeles. The first one was at the University of

---

[21]4/12[?1955] handwritten letter from Betty Scott to Jerzy Neyman.

Oxford on May 11. The second was at University College – London on May 12. The third was at Cambridge University on May 17.[22] She would be talking in front of some of the world's greatest scholars in statistics and mathematics. Betty was still a young and relatively inexperienced academic: She had only just completed her third year as an assistant professor. But she was already well on her way to becoming a scholar with an international reputation.

Betty would go on to attend many international conferences in her career. She would take a sabbatical at Cambridge University with Neyman (1957–58), be elected to the ISI (1966), be elected an honorary fellow of the Royal Statistical Society (1981), and go on to take important roles in international professional associations. There are many examples: She would become program chairman of the IMS (1958, 1960, 1965); a member of Commission 28 (Extra-galactic research) of the International Astronomical Union (1957); the scientific secretary for the International Association for Statistics in Physical Sciences (1960–1972); a member of the council of the Biometric Society (1970–1973 and 1978–1981); a member (1978–1981) and chair (1980–1981) of the committee on the selection of administrative officers for the society; and more. Like Neyman, Betty would become a global statistical leader of a number of international statistical societies (see Chapter 8).

But back in May 1955, before all of these significant international roles, Betty was wanting to make Neyman proud of how she conducted herself on her leave of absence. She was writing to Neyman several times a week. Like her many earlier letters, these letters still contained the usual information about her European activities and her thoughts on their collaborative research projects. But now these letters contained more personal thoughts. For example, Betty often mentioned her hair. Betty talked about having her hair cut. She talked about taking a bath and washing her hair. She talked about buying shampoo. She talked about her hair being "fluffy" and "completely kempt." In these times, women had to be preoccupied with their hair, because this was before wash-and-go hairstyles.[23] But Betty also talked about her hair in an intimate way, to her lover; she talked affectionately to Neyman about her hair, about how she wished he could be with her to mess it up and brush it for her. As Betty's letters to Neyman became more personal, Betty expressed herself more often in the French language.

---

[22]Note: This was at the invitation of Henry Ellis Daniels (1912–2000). Daniels helped to establish the Cambridge Statistical Laboratory.

[23]Note: Vidal Sassoon is credited with creating the first wash-and-go hairstyles in the early 1960s. Women no longer had to wear curlers to bed, and they no longer had to go to the beauty salons each week.

11.6   Thursday, May 5, 1955: Paris[24]

Betty was planning her trip to England. Her English hosts had sent an itinerary for her upcoming visit. She told Neyman she was pleased with it, and wrote at length about their research. "I will ask about M. Oort's constants. I'm thinking also of an RH diagram in color. In this case, there is the nonlinear regression that you noticed in the past, plus $X = Y - Z$ versus Y. We are looking for the relation between ζ and η. Oh, la, la!"[25,26,27]

Being thoughtful and generous by nature, Betty considered what gifts she would bring to colleagues in England. She would bring 1/12 Remy Martin VSOP cognac and 1/5 of wine for F.N. David; and she would bring 1/5 wine and 2 dolls for the Chapman family.[28] All of this was 1/5 more than she could import without paying duty. But the duty wasn't so expensive as to deter her from her plans for giving.

Neyman had sent Betty a copy of their Berkeley Symposium article on galaxies. She wasn't sure when he sent it and asked him about the postmark. Betty wanted to keep up with all of Neyman's work and asked him to send her another of his articles. She would pay for it. But she hadn't had time to read all the articles she had already accumulated in Paris. She was so close to going home now. Betty proclaimed to Neyman that the time had come when she would have to burn the articles: "Too bad!"

There was more business in Betty's letter to Neyman. The next session of the International Astronomical Union (IAU) would be in Moscow in 1958. She had talked about this to her colleagues in Paris who told her she needed to get an invitation from Russia in order to get a Russian visa. They told Betty there would be special trains, hotels, and all kinds of arrangements from which there could be no deviation, no freedom. The IAU meeting was three years away and maybe things would ease up. Would Neyman like to make a plan with her to go to Russia?

It had been raining constantly, and Betty was tired of it. Neyman had given her a camera. She took a lot of pictures in spite of the rain – over three rolls of film recently. Photography was giving Betty a lot of pleasure. She expressed her gratitude to Neyman for the gift. Betty was glad she didn't have to go

---

[24]5/5[?1955] handwritten letter from Betty Scott to Jerzy Neyman.

[25]Note: Oort constants are parameters that characterize the local rotational properties of the Milky Way.

[26]Note: Probably a Hertzspring-Russell diagram, which plots star brightness against temperature.

[27]Evilina Batery, e-mail to author, June 26, 2012. Note: Betty sometimes used the expression Ah, la, la!, and other times she used Oh, la, la! According to translator Evelina Batery, Ah, la, la is an "American misinterpretation" of Oh, la, la.

[28]Note: Douglas Chapman (1920–1996) was a mathematical statistician and expert in wildlife statistics who earned his PhD under Neyman and could have been on sabbatical at Oxford.

to England on May 5. The storms were so strong the Queen Mary (Reine Marie, in French), a premier ocean liner and primary mode of transatlantic passenger service, could not make the trip. Betty wrote to Neyman: "Not the Queen Mary, not me neither."

Betty longed for details about news back home. She wanted to know more about what they could do to help a difficult political situation in the department at Berkeley. Betty had heard a woman was found dead in front of her house in Berkeley, but neither Neyman nor her parents had told her about this. She wanted to know about Evelyn Fix. Betty gently chided Neyman for not giving her more news from back home.

11.7 Sunday, May 8, 1955: Dieppe, Newhaven, Winchester[29]

Betty had a car when she was in Paris. To give the talks in England, she had to get from Paris to England, and she decided to drive. This required she put her car on the ferry to cross the English Channel.

It was close to the time Betty would return to the US and she didn't want to bring the car back with her. The person she planned to sell the car to would be there when she was in London, but selling the car in England wasn't allowed. This meant Betty would need to take the car back across the English Channel to Paris after her talks, and the person would need to travel from England to Paris to get the car. Inconvenient.

Betty found getting to England to be complicated. On May 7, she tried to take the evening ferry from Le Havre to Portsmouth. She had to cross the Seine River to get to the Le Havre port, but she got there too late to cross the river. The last ferry had already left, so she stayed in a hotel near Caen. The next morning she got up at 5:30 am, left her room at 6:15, arrived at the ferry port in Dieppe at 9:30, "ah, la, la!" Betty contracted with an English automobile club to get her car transported on the ferry. She gave them the car and her car papers. They put the car on the ferry using a big crane and took it off the same way. "I did the rest." The cost for this service was 250 francs. Then Betty waited for customs to open so that she, too, could depart for Newhaven.

As Betty waited for the ferry to depart from Dieppe, she worked for an hour and a half on the Tribolium research. She finished making some charts. She worked on her talks. She found some misprints that she hadn't noticed before in the draft manuscript and logged them onto a piece of paper to send to Neyman. After she got on the ferry, she kept working on the Tribolium research. One of the assumptions they were making in the paper seemed un-

---

[29]5/8[?1955] handwritten letter from Betty Scott to Jerzy Neyman.

comfortably strong to her. She wanted to weaken it, but then "things" got very messy, and otherwise the estimates got very biased. "Oh dear!"

The crossing went well, but there was a problem upon landing. Betty's car had Dutch license plates on it. When she arrived in Newhaven, a British automobile club agent made her buy British plates, even though she was only going to be in Britain for one week! She didn't get them fastened on. Instead, she put them under the hood, loose. Then she talked to the customs agent and he said she didn't need the British plates. "What is going on?" she asked. But the money she spent on the new plates was already gone. It was not to be seen again. The fact that customs didn't charge her extra duty for the gifts she was carrying was small consolation.

So many things had gone wrong getting to England that Betty started to worry about things going wrong with her talks. She wanted to know more about the Symposium talks back at Berkeley.

Betty's May 8 letter contained more of her observations of the environment:

> *Last night I nearly froze to death in a very nice hotel right on the channel, right on the beach... Tonight I am in a lovely town [Winchester] with an old school, old cathedral, etc. the former capital of England – William the Conqueror was crowned here. This hotel has a huge garden and chintz drawing rooms (I am having coffee in one)... It is just spring here – jonquils and peach trees; there were pink apple blossoms in Normandy. By going north each week, I can have perpetual spring!*

Betty hoped Jerry was okay. She missed him. She hoped he knew it.

11.8   Thursday, May 12, 1955: Oxford[30]

Betty was excited whenever she received a letter from Neyman. This time, on May 11, Neyman included in his letter to Betty a reprint of a published article that was of special interest to her, and she was grateful for it. Neyman also wanted to be sure Betty received the proofs of a manuscript he had sent to her earlier. She had already taken care of the proofs. Betty was surprised Neyman had included a table that had three significant figures. She thought that was more than needed. She was also surprised he left in a remark that some of these figures were not accurate. The fact that the computing was not perfect distressed her.

Betty's research talk in Oxford the day before had gone well. "Everyone was very complimentary." She also had to talk the next day in London. She was worried. She thought this talk would be harder because she didn't think

---

[30]5/12[?1955] handwritten letter from Betty Scott to Jerzy Neyman.

what she had prepared was very unified. Then, the next week she had to give a talk in Cambridge on the Tribolium research.

People normally get nervous before giving academic talks, and Betty was no exception. Betty confided in Neyman: "As you know, I am not a good talker and all this is a terrible strain on me. – I just cannot exaggerate how uncomfortable I feel at the moment."

In Oxford, the Chapmans had Betty as their guest for dinner at the local faculty club. Betty thought the Chapmans were very nice. Others came after dinner, and she had a very pleasant time. Betty mentioned the Kendalls[31] and the Hammersleys[32] specifically to Neyman as being very nice. Betty also mentioned the kindness of F.N. David. Betty was staying at FND's apartment in London, and FND had insisted that Betty eat breakfast in bed. But all of these kindnesses could not make up for the fact that Betty missed Neyman. She hoped he wasn't angry that she had decided to come home earlier than planned. She would see him in less than a month.

11.9   Thursday Night, May 12, 1955: London[33]

Betty's second talk was on May 12 at University College – London. It took place at 4:30 pm, after tea. Betty delivered it as Neyman had advised her. Afterwards, FND took Betty to dinner at La Suisse, a place that Betty and Neyman had gone to together previously. Did he remember it? And then Betty and FND went to a movie. Betty really appreciated FND's niceness and cordiality. They got back very late, 11:30 pm. Betty was very tired, but she missed Neyman, wished she could be with him, and wrote another letter to him before going to bed, even though she had written to him earlier the same day. "Oh, oh, Jerry!" This letter was in French, with the exception of a footnote in English.

University College – London was the academic home of Egon Pearson,[34] Neyman's famous London collaborator. Betty saw Pearson in the hallway, but he didn't know Betty. Up until two days prior, he had been down with German measles, and he was suffering from rheumatism. "Too bad [in French]!" He didn't make it to Betty's talk. Betty offered Neyman other gossip about

---

[31]Note: David George Kendall (1918–2007), professor at Oxford, was a highly decorated mathematical statistician.

[32]Note: John Michael Hammersley (1920–2004), professor at Oxford, was a highly decorated mathematician.

[33]Blackwell, David. "An Oral History with David Blackwell," an oral history conducted by Nadine Wilmot, Oral History Center, The Bancroft Library, University of California, Berkeley, 2003.

[34]Note: Egon Sharpe Pearson (1895–1980) was a mathematical statistician who co-authored the Neyman-Pearson theory of testing statistical hypotheses.

people in the department. Finney[35] was awarded a Fellow of the Royal Statistical Society, "but not Mr. Pearson because Mr. Fisher[36] has a veto. Alas [in French]!" Given his heritage, Neyman was always interested in the advancement of Polish intellectuals. Betty noticed there wasn't a single Pole at London, "this is politics." She was told politics was a reason why one of Neyman's countrymen[37] didn't receive the recently vacant readership at Oxford.

On the evening of May 12, Betty sat down to collect her thoughts. It had been a long day. She thought about writing another academic paper, but she knew she wouldn't have time to do it before returning to Berkeley. After all, when she got back to Paris, she would have only 16 days left. "Besides my calculations have to be verified with the method r (i.e., to divide the stars by the apparent distance but not exactly)," she wrote in French.

Only 16 days before seeing Jerry again! Betty wrote to Jerry that she adored him. She sent him a thousand kisses.

In a postscript to this letter to Neyman, Betty mentioned she had purchased a French lamp so that she could read and study her French verbs while in bed. It was a medium-priced lamp that she got from her parents for her birthday. There was no point in bringing it home, because the electrical wiring wouldn't work in the US. She thought about leaving it for the French girls at the Fondation, but they would go home in June, and so "it would be more bother than good to them." Betty asked Neyman if she could give the lamp to his "niece" [sic]. The two hadn't met, but Betty thought his niece might like the lamp and be able to use it. Betty could deliver it on her last day in Paris, or she could mail it. "But I wouldn't want to hurt her feelings" by giving her a used lamp. She asked Neyman what he thought.

11.10 Tuesday, May 17, 1955: Cambridge[38]

It was 9 am on May 17. Betty was in Cambridge and thinking about Neyman. She was about to give her talk. The stress of all of the conference activity was taking its toll. Betty hadn't had trouble with headaches for some time. But three days earlier, and yesterday, she had terrible headaches. "The cords and veins in the back of my neck and head were all swollen and tight and sticking

---

[35] Note: David John Finney (1917–) held professor positions at several universities in the United Kingdom, lastly at the University of Edinburgh. He is best known for his work in the area of drug safety.

[36] Note: Sir Ronald Aylmer Fisher (1890–1962) is best known for pioneering the principles for the design of experiments. It is well known that Fisher and Egon Pearson had an acronymous relationship.

[37] Note: Betty named Kammersky. A Reader in the United Kingdom is a rank equivalent to a full Professor in the US. A person appointed to a Reader position will have a distinguished international research reputation.

[38] 5/17[/1955], handwritten letter from Betty Scott to Jerzy Neyman.

up literally in knots." Aspirin wouldn't work. She needed something stronger, and she didn't have it. She would have gone to the doctor (it was that bad), but it was Sunday and so it wasn't possible to get help. Betty knew the pain was stress-related.

On this Tuesday Betty was recovered from the headaches. She washed her hair and was ready to begin the new day. But she was tired and run down after the days of headaches. She wanted to write to Neyman the day before, but she couldn't with her headache. Betty's status left room for self-doubt. She told Neyman she didn't agree with him that she was more grown up after her academic year in Paris. "I feel most insecure even now in Paris, although England is much better – But, as you can see, I didn't stand up very well. Oh, dear – "

Betty apologized to Neyman that her letters, including this one, had been so chaotic. Betty answered Neyman's questions about the remarks she made in earlier letters about their research. She referred to publishing the work with Oort's constants. He was right that the charts for the Tribolium research were enlargements and/or revisions of figures in their paper. David Kendall had some suggestions about plotting the stochastic variability of the Tribolium.

Betty admitted to Neyman that most of the gossip she included in her May 12 evening letter was incorrect. R.A. Fisher wasn't going to retire in June: Cambridge changed the retirement age from 65 to 67, so he would work for two more years. Pearson wasn't going to retire after one more year: He would turn 60 on August 16 and would work for 5 or 7 more years, depending on whether London would also change the retirement age from 65 to 67. Fix planned to go to England next year to be with FND. Betty had other news for Neyman. Mrs. Loève was somewhere in the US and would qualify as a US resident. So the Loèves could pay for her trip and still make money on their income tax. FND seemed to want to stay at University College.

"Jerry, I can count the days now! But I can't wait."

11.11 Wednesday, May 18, 1955: Paris[39]

Betty got back to Paris the night before. As it turned out, the travel was a real adventure, leaving her somewhat exhausted. There had been "ghastly" weather getting out of England: hard rain and strong winds. By the time Betty got to Canterbury, her engine got so wet that when she needed to start it at 8 am to get to Dover for the ferry, it wouldn't start. She needed to get the car to a garage, but they didn't open until 8:30, and then the mechanic didn't arrive until 9:05. By that time she had gotten the car started herself, but she was obligated to pay the mechanic ("for nothing").

---

[39] 5/18/1955 handwritten letter from Betty Scott to Jerzy Neyman.

Then the pouring rain confused Betty and she headed in the wrong direction! She went back through Canterbury twice. It was slow going through town. After she got out of town, she began to make good time and managed to get to Dover by 9:35. But she was five minutes too late for the Dover-Boulogne ferry. Fortunately there were two places left on the Dover-Calais ferry. It was scheduled to leave at 11 am. Betty had time to mail a package containing her heavy Swiss shoes, a heavy Dutch skirt, and some papers. The nice and efficient attendant commented he could see she was a little excited, "but don't worry, we'll hold the ferry for you!" The package was "all done up in sealing wax, British fashion." Then she got on the ferry. "What a life!"

In England, Betty had purchased some sweaters for her brother and a tie for Neyman. She worried that the tie was "not exactly right, I am sorry to say." She had wanted it to be perfect for her Jerry.

11.12   Thursday Morning, May 19, 1955: Paris (continuation of the previous letter)[40]

Betty got back to France on Tuesday evening, and the storm was even worse in France than was in England. "The Queen Mary couldn't get into Southampton, but hid behind the Isle of Wright. I wish I had done the same." But eventually she got back to Paris, with the experience of a glimpse of sunlight along the way and with the setting sun illuminating a lovely drive along the Seine River.

Betty was excited to find letters from Neyman waiting for her. As usual, his letters included details to further their research collaborations. It also contained some welcome news about astronomer Fritz Zwicky. This time, however, Neyman also wrote about a serious political issue.

Several years earlier, Erich Lehmann was offered the editorship of the *Annals of Mathematical Statistics*, and Neyman strongly encouraged Lehmann to accept. At that time, Neyman also agreed to provide a secretary to Lehmann, as well as suitable office space for the secretary. When, in 1954, Lehmann was asked to remain the editor for a second term, Neyman agreed to continue their earlier arrangements, which included the secretary and space with a window. Now, in 1955, there was a change of circumstances. Neyman needed to find funds to publish the *Proceedings of the Berkeley Symposium on Mathematical Statistics and Probability*. The *Proceedings* was Neyman's personal publishing project. He was not going to see funding issues limit its production. One of the places to find some funds was Lehmann's secretarial position. Neyman notified Lehmann that there would be no more funding for a secretary for the *Annals*.

---

[40]5/19/1955 handwritten letter from Betty Scott to Jerzy Neyman.

Betty was distressed at Neyman's stories about the situation, and she told him so. She suggested he write "a nice, formal letter" to the Institute of Mathematical Statistics (IMS) and ask for some overhead to cover the secretary. She had the impression that Neyman had "more or less promised" Lehmann a half-time secretary "if possible." Betty knew this was a bad situation, because if Lehmann didn't get the secretary, then the work would fall on the department office, and they were already totally overburdened. "Oh, Jerry, what a hard life you lead! You sound even more excited than I but you have real problems."

This day was Ascension Day, a holiday. Betty told Neyman she had "seen 1,000 (x 10?) paintings of Mary being sucked up in a cloud!"

Betty guessed that her talks had gone well. "Not world-shaking, but every one was very polite and cordial and etc. I'll tell you more when I see you. I have so many things to tell you, I'll probably explode; and you'll have to say: "Get angry! Ohe [sic], Jerry! [in French]."

Betty gave Jerry some advice about an article he was writing for the *Annales de l'Institut Henri Poincaré*. This time she wrote her advice in French. But then she was interrupted because some Swedish girls showed up. Betty was scheduled to have lunch with them. After lunch she planned to work on a talk that she was scheduled to give on Monday.

Betty decided her French was "now terrible." In her opinion, phonetics wasn't helping. She always had to go back to how something was written, not how it was spoken. "What a pity!" Her handwriting at this point may have reflected her fatigue, but it also was larger and had some strikeouts and write-overs that reflected her lack of certainty about the language.

11.13   n.d: Paris[41]

This letter from Betty to Neyman was entirely in French, except for the postscript, which she wrote in English. Betty decided she wanted to speak to Neyman in French. "Now I'm thinking that this language is a little special for you and me. Can you feel what I'm feeling?"

At the same time, she knew her French was deteriorating "fast, fast, fast." She was particularly lost without a dictionary, when she had to guess words and their gender. Her difficulties got worse when she was tired, as she was that Friday after having a class and three seminars. Maybe she and Neyman could speak only in French at lunch on one day each week? "But, I don't speak French, of course there is my biggest problem."

Like most of us, Betty had favorite words in English. Betty also had her favorites in French. Her favorite French word used to be "charmant [charming]."

---

[41]n.d. handwritten letter from Betty Scott to Jerzy Neyman.

Then, when she got to Paris, it was "formidable." Now, it was "impeccable." "We have to use current expressions, don't we?"

It had been a good lunch with the Swedish girls. Later they did some sightseeing and went to Saint-Jean aux Bois and the big Compiégne forest, north of Senlis. She saw J.J. Rousseau's tomb. Betty found all of it to be "very pretty and interesting." Lilies of the valley [in English – she didn't know the French word]! Betty loved "these charming flowers a lot." She happily brought a big bouquet of them back from her excursion. She didn't have a vase, and so she put them in the base of her little Italian coffee machine. No more espresso, but the beautiful lilies of the valley were worth the sacrifice. Eventually she would make some coffee, but for now the coffee machine was a vase.

Betty had more issues with customs. Neyman had sent her some vitamins and dried fruit, and she was "greatly" appreciative. At least some of the fruit was for Loéve, who declared it to be "very good, delicious." But it looked like Betty was going to have to pay a huge tax on the vitamins and fruit. The value of the two items was $11 ($88.54 in 2010 dollars), and the tax was set at a whopping 46%. Before, when the vitamins were sent separately, she didn't have to pay any tax because the value of the package was less than $2. And this time she could have done without the vitamins. She only needed two weeks of them. Most of the vitamins would go back with her to the states. She considered sending the vitamins back to Neyman, "return to sender," but this wouldn't have worked given how Neyman addressed the package ("Director, US Foundation"). "Oh, la, la!" Betty talked with everyone in customs and at the SNCF train platform. Then she got lucky and was able to talk with the customs director. He thought $6.85 was too much for dried fruit and vitamins, most of which would go back to the states. He decided that Betty would not have to pay any taxes whatsoever. "Oh, la, la!"

Betty's thoughts turned to the camera Neyman had sent her. She hadn't paid any taxes on it either, and decided she had been lucky in that instance as well. She had just taken pictures of the residence halls at the university. "I think that it is interesting to show the various huge buildings displaying a national identity."

Her thoughts also turned to the cold weather. Daytime temperatures were only 42–55 degrees Fahrenheit. There was strong wind and heavy rain. The Fondation had turned their heat off on May 1. So had the Sorbonne. Betty lamented having sent her heavy shirt and shoes home. "Oh, la, la!"

Betty was too tired to write any more in French. She told Neyman she missed him and loved him, sent him her kisses, and told him she was counting the days until she could see him. Betty's brother Dick lived in Los Angeles and had offered for her to stay with him and his family when she returned from Europe to Los Angeles. But "I prefer to stay with you if it's at all possible. We could see them Sunday and maybe other days."

In a postscript, Betty mentioned that she just received a letter Neyman had written to her on May 18 at 7 am. First she chided him for writing that early. "Too early! Your life is most distressing. Oh, Jerry – "

Then Betty encouraged Neyman to arbitrate his differences with Lehmann. "This is the wrong spirit – and is most disheartening. It is disheartening for you because public quarrels just make every one bitter." Betty again encouraged Neyman, or Lehmann, or the two of them, to ask the IMS to provide some moneys for the *Annals* secretary for Lehmann. She thought they could make a good case because Berkeley had already paid for the secretary for three years and now IMS could take a turn. "Because it certainly is true that Erich needs a secretary and having the Editorship does add to the prestige of the Lab, etc. What do you think?"

11.14 Back in Paris

Back in Paris, 1955, there were more letters from Betty to Neyman, several in French. By this point, Betty had an academic proficiency of French that was more than adequate. But she didn't always use verb tenses correctly, as is common today among domestic students in their third year of French.

A translator had this to say about Betty's use of French in her letters to Neyman:

> *The impressive thing is that she wrote in French at all, and out of love. We must give her a great deal of credit for that. Also, Betty certainly didn't understand the nuances of the idiom which made sometimes for a few hilarious misunderstandings that I did not translate because they would have made her a great deal more licentious than she intended (or perhaps not!). For example she wrote to Jerry 'Je te baise' (in current French it means: I will f… you) instead of 'Je t'embrasse' (I will kiss you).*[42]

One wonders what all was lost in translation.

11.15 Monday Midnight [most likely May 23, 1955]: Paris[43]

This letter didn't have a date on it. But whatever day it was, Betty was having a bad day. She wrote to Neyman: "I am all distressed by so many things that I should not write to you, but yet I must. I cannot imagine a day when more things were worse."

---

[42]Evelina Badery, e-mail to author, March 29, 2011.
[43]5/23/1955 handwritten letter from Betty Scott to Jerzy Neyman.

Betty was having a bad day because she realized she gave Neyman some incorrect professional advice in a previous letter. She had told him to change the remarks about the "picture" of the universe in one of their draft manuscripts. But she realized at 1:30 am that there was no reason to make any changes. She didn't feel she had a good excuse. It didn't help that the assistant told her something that was wrong, or that the Swedish girls came to pick her up an hour early. "But it was foolish, and I am sorry." When she realized her error, Betty "reached down [sic] the typed manuscript from the shelf over my head and yes, what I had written was nonsense. About 10 minutes later, 3 books came down on me and a bunch on the floor. They hurt, and what a crash! The shelf is about 6 ft. from the floor." First thing in the morning, she sent Neyman a telegram telling him she had been wrong.

A second reason Betty was having a bad day was that she lost her umbrella. She looked for it in her room and car. No luck. Then she had an idea where she left it. She wrote to the place, sending them 100 francs to mail it back to her. It wasn't her favorite umbrella because it wasn't black. But she still wanted it back.

A third reason for Betty's bad day was because she had given Neyman some premature long-distance counsel, having to do with the problem he was having with Lehmann. Betty had now received copies from Neyman of some of his letters about the problem. Earlier, she had given him some guidance. She had urged him to arbitrate his differences with Lehmann. After reading the letters, however, Betty understood the situation differently. She told Neyman she would have given him different advice had she seen the letters first. In any case, Betty's reaction was that it was an "awful mess and it is a shame it gets publicity. And you and Erich sound so bitter." Now she wrote: "I guess the horse is really dead." She went on to say this: "I wish so much I could help you – I have been so unhappy about this all day. This is the most depressing thing that could happen. Can't the money be found somehow? I feel very guilty because I have been getting money and what do I do – ?"

A fourth reason that it was a bad day was that Betty's French teacher criticized her and she wasn't up for the criticism. In fact, she almost "came to blows" with him. The teacher had announced at the previous session that students should work at home on certain sentences and come to the next session prepared to record them. But when he came to the session, he decided to work on the sounds "oi, oui, etc.," which was something completely different. He asked the students to speak "some messy sentences" that contained these sounds, like "Louis lui louait" [English: he rented]. Then he put out a sentence that started with "Moi" and called on Betty. From her standpoint, "m" was supposed to be nasal, and she pronounced "Moi" accordingly. Before she could get out a second word, he jumped all over her about her pronunciation of the first word, telling Betty she was being "nasal."

This wasn't the first time the teacher had given this criticism to Betty. Instead, it was a common criticism, and Betty was sick-and-tired of it. After all, she talked just like her parents did. Betty was weary of being told by the teacher that she was wrong "...every day the same thing..." She didn't confront him, but she did say flat out in French "...that it would be better not to speak of nasal again since this wasn't helping anything, there never was any progress..." Betty had less than ten days left in Paris. She told the teacher "it would seem sensible to work on something else."

The teacher's reaction was to look in Betty's mouth, after which he proclaimed that she had a normal palate. She just needed to open her mouth more and project her voice forward. This was a hard thing to correct. Keep practicing! Betty was really unhappy with the criticism. Growing up, she had had rubber bands attached to her eyeteeth in order to fasten her jaws together. She had endured this discomfort for 15 long years. She reasoned that her teacher would have talked in the back of his mouth, too, under these circumstances...with his mouth shut tight...with the sound coming out of his nose... "Where else can it go?" She was pretty sure of one thing: The French teacher's system of nagging was not going to get her to change. His system surely hadn't helped up to this point. Betty was plenty discouraged.

That day, Betty gave the last talk that she would give during her extended stay abroad. She beat herself up. The talk was too complicated, technical, theoretical, and long. Her previous talk had gone really nicely, but she knew this one wasn't going to be good before she got up to give it, and she was displeased with it when she was finished. She had tried to make it better but couldn't get it fixed in time. The material simply didn't lend itself to a talk. She was unhappy with the talk all the way around.

The people attending Betty's seminar were astronomers. There wasn't time during the seminar for her to present an example to illustrate her method.

> *I think that in many ways astronomers, although generally intelligent, learn only from examples – They and their science, especially astrophysics, are in the same stage as medicine. Every time [an eminent astronomer] sees a peculiar shape of a line in a star, he publishes another paper. Then, if he finds this in another star, he publishes still another paper. And if he doesn't find in the other star, he publishes another paper anyway. I am exaggerating, but as far as max. lik.[44] [sic] is concerned, I think that this is the level of the astronomers in Paris.*

"Ah, la – what a horrible day." Betty wished many things were different: that it wouldn't rain and hail every day; that her room and the Sorbonne had some heat; and that the British Automobile Association and the Women's Home Industries hadn't gypped her. If tomorrow were the same as today, she would go home tomorrow.

Betty closed her curtains. The traffic was thinning and her room was quieter. But the artificial light in her room was harsh. It reflected off of the copper pitcher on her desk and onto the yellow Scotch broom flowers in the pitcher.

---

[44]Note: Betty was probably referring to maximum likelihood, a statistical technique for using a sample of data to estimate factors in a statistical model. She didn't have data on all astronomers, but was making a generalization based on her experience with astronomy.

The flowers looked cheerful, but they were starting to droop. Betty reflected on picking them. She was behaving like the English and the French, picking wildflowers along the road; but being American, she felt like she was stealing them. She imagined being fined $25 for each red clover she picked. By the next evening she would have a stuffed up head and headache. Perhaps it was an allergic reaction to something, maybe the wild flowers. Within 24 hours, she would end up throwing all of the beautiful wild flowers away.

Betty wished Jerry were there. Then she wouldn't be complaining. She would be happy! She would put her head on his shoulder. She would whisper, "I love you," into his ear. "Can you hear me?" she wrote to Neyman.

11.16   Tuesday [May 24, 1955]: Paris[45]

The next morning came, and Betty continued her previous letter. It was already five handwritten pages. She was very depressed and had more to say to her Jerry. She poured her young but tired heart out to the one she completely trusted.

> *The ONR[46] has spent a lot of money on me and I really can see no benefit: I am much more unsure of myself than before, more worrying, I can't even talk French nor give a proper seminar at the right level for the audience, who knows whether I learned any mathematics, I don't have any real friendly contacts with either French mathematicians or astronomers: I feel that French mathematicians think that statisticians, ipso facto, are ignoramuses; and that the French astronomers think we are interested only in technicalities and fault finding and live in an abstract world that they are afraid of and want no part of. Now that it is time to go home I am so distressed – Paris is interesting but I spend all the days and nights locked up in my room – and I can only listen to the chants for $\frac{1}{2}$ hour because I have to worry about my seminar (time wasted) and sit in the car while Virginia goes in – If I could learn not to worry then I would give a better seminar because the worrying means no sleep and headaches. And etc. – But I think I am worse than I used to be – certainly I was yesterday – And I thought I was getting better last time in France and the times in England. Oh, dear Jerry, I wish I were a different person – .*

On this particular morning Betty was feeling desolate. But this lowest-of-the-low wouldn't last. She would very soon regain all of her considerable strength. After all, she had gone places and done things that most people of her time could not have imagined. She had met and communed with many of the world's leading statisticians, mathematicians, and astronomers. She had been invited

---

[45]5/24/1955 handwritten letter from Betty Scott to Jerzy Neyman.
[46]Note: The ONR had funded her activities in her extended stay abroad.

to communicate her research at some of the world's leading universities, and she had interested many in her work. Maybe the French language would never be her strong point. But she would leave Paris as an emerging world-class academic.

11.17   May 26, 1955, 7:20 am: Paris[47]

Betty felt bad about her previous letter to Jerry – it wasn't a "nice" letter – and she was determined to write him a cheerful letter this time. So she started out talking about food.

She had gone out with one of the Swedish girls and a friend the evening before to a three star restaurant called the Lapérousse[48] to celebrate a birthday. They didn't go on to the theater because the dinner took too long. Everything was the best ever: soup; Ragons Veaux (veal kidneys); Champingons [sic] (mushrooms) in a white sauce with cheese, ham and wine; crepes with nuts inside and carmelized [sic] sugar; red wine and coffee. She told Jerry that the dinner was the best she had ever had. But it was expensive. "Ah la I wish you had been there!"

Betty reported to Neyman the latest news about their research. She had gone to the Gauthier-Villars publishing firm about proofs of one of their manuscripts. Betty found her way to the "chief" of the firm and talked to him and one of the assistants who she had talked to before. She learned it wasn't too late for the publishers to remove a remark about the lack of accuracy of some of the figures that Betty told Neyman should be removed in her May 12 letter. They were "very pleasant" about the whole matter.

At this point Betty was homesick. She desperately wanted to see Neyman. She imagined how their reunion would be. "I dream of it!" She had only a week left in Paris before she would fly to Lisbon on the first leg of her trip back to the US. She would land in Los Angeles on June 9, and Neyman would be waiting for her.

Betty's brother in Los Angeles also wanted to meet Betty when she landed, and Betty wanted to see him too, but she told him no thanks, she would telephone him when she arrived and then see him on Sunday June 12 (three days later). She claimed she had to go to the Observatory. Betty didn't want to hurt her brother's feelings.

Betty was deeply sympathetic to Neyman's hard academic life. He was just then creating the statistics department at Berkeley, with all of the associated political headaches. "...you have so many problems and some people are so

---

[47]5/26/1955 handwritten letter from Betty Scott to Jerzy Neyman.

[48]Note: This was probably La Pe'rouse, a historic restaurant from 1766 situated in a beautiful mansion on the Seine that has been called the monument of French gourmet cuisine (http://www.idealgourmet.fr/1252-restaurant-laperouse-presentation-gb.html).

terrible. . . " Betty wrote. In his letters, Neyman confided all of his difficulties, and she imagined that he must have been "all worn out." She hoped he hadn't been hurt by what she had written to him in her previous letter.

In Neyman's last letter, he asked Betty for the names of the hotels she would be staying at on her trip back to the US. He wanted to be able to write to her at each place. In Lisbon, it would be the Hotel Borges on the romantic side of the city. In Bermuda it would be Hotel Oxford House (Betty couldn't name the town it was in). In New York it would be The Biltmore. The Biltmore was a luxury hotel, and Betty made sure Neyman knew she had the "college rate." Of course, she wouldn't be able to go into the bar at the Biltmore, because the Men's Bar, as it was named, didn't admit women. A decade or two later, Betty would surely have noted with interest the part that The Biltmore played in the women's movement, when the New York City Human Rights Commission ordered the hotel's bar to admit women.

Betty told Neyman he didn't need to write letters to her on her way home. He was busy, and she would see him very soon and then she would kiss him a huge number of times. . . without limits. . . an infinite number of times.

Then Betty's mind went back to their research. She wanted to know about David Blackwell's appointment: Was it for one year, or was it permanent? She told Neyman how much she liked the Tribolium paper. She had talked with Kendall and Daniels about the paper when she was in England, and they liked it too. But one thing Betty didn't like about the paper was the assumptions. They "are gross," she told Neyman. She thought of some changes to the methods, but the assumptions would end up remaining the same, at least for now, as a problem to be solved.

Neyman suggested he and Betty should visit with an astronomy professor at Caltech on June 11 while they were in Los Angeles. Betty agreed to meet with the professor, but she had her own point of view about him:

> *You have seen [the professor's] papers and heard him talk. His reputation as a big wind with little real information is well deserved. Also, his pig-headedness. From our point of view, it might be better to have the data published separately, without interpretation. . . Then we could use and discuss the data ourselves with a certain amount of freedom (but never complete, of course).*

With these thoughts in mind, Betty suggested perhaps she and Neyman should also meet with some other astronomers. Then she wrote about the astronomy paper they were preparing for the *Berkeley Symposium*. She had read the most recent version before giving her talk at University College – London. She was "struck" by the edits Neyman had made to the paper that were "against" the CalTech professor.

Betty had a tempering influence on Neyman. She advised him: "Many of these are new inserts, and thus caught my eye, but I still think that you are rather heavy in stepping on him. Not so heavy that he can scream, but heavy enough to make him very hurt and angry. Perhaps some of the inserts are too aggressive? I am not sure, and this is a matter of taste and policy, but you might want to reconsider."

Next Betty's mind turned to how she was going to get all of her things back home. She had already mailed two packages from England. One contained gifts, the other clothing and papers. She planned to send the rest of her things from an express parcel office in the Metro station across the street from the Fondation. Convenient in terms of location, but complicated: People told her it was normal to have to repack at least three times in order to meet the shipping requirements. Betty hadn't gotten to the parcel office yet because she still had academic work to finish. She had to read a thesis. She had missed classes while she was in England, so she had to make up work in mathematics and French. She felt she had forgotten a lot of her French. She also had to write a summary of the last talk she had given. It was an astronomy talk. The summary would be different from the talk. It would be better. Or at least she hoped so. "No more seminars, Je suis d'accord!! [I agree!!]."

11.18   Le 26 Mai, 17 hr: Paris[49]

This letter was brief (one page) and written entirely in French. It was both a love letter and a letter lamenting Neyman's current difficulties with Lehmann, which she had underestimated. She had tried to give Neyman advice but realized her advice may have been mistaken because she didn't have all of the information. She had written to Lehmann and realized that perhaps this was a mistake from Neyman's point of view. "I'm sorry, but all I want to do is to help you... It is too long and you are too far from me!"

11.19   29 Mai 1955: Paris[50]

Today Betty was happy and hoped Neyman was, too. She was feeling upbeat and confident, and so everything she wrote on this day was again in French. Betty wanted Neyman to know she was with him even when he was "up against a wall."

In general, it isn't always possible to immediately tell how your academic talk went or how it was received. Your first reaction about your own talk could be completely false. You often misread the reactions of academics in your audience. Betty thought her talk to the astronomers on Monday had not gone well. But afterwards she found out otherwise: The astronomers agreed

---

[49]5/26/1955 handwritten letter from Betty Scott to Jerzy Neyman.
[50]5/29/1955 handwritten letter from Betty Scott to Jerzy Neyman.

with her positions. She still intended to write a summary that was different, and hopefully better, than her talk. She had been given a deadline of Tuesday at 4:45 pm, before Pieter Oosterhoff's [51] presentation.

In two days Betty expected the person who planned to buy her car[52] to show up. She had grown attached to the car and would really miss it. The countryside was beautiful, and Betty longed to travel in it with Neyman. She had been invited to go to the historical town of Avignon for a long weekend in the southeast of France. But Betty had a lot of things to do to get ready to go back to the US. She had work to do and a lot of errands to run.

Betty purchased perfumes for the girls in the Stat Lab. They were expensive, but she still considered them to be a bargain... "the only bargain I can find in Paris!" Betty also purchased some perfume for herself. She hoped Neyman would like it and sent him a thousand kisses, asking if he would like all of them when they were once again together on June 9.

11.20   June 1st: Paris [1955][53]

There were two letters in the post from Neyman the day before, a day which Betty noted was Pentecost, and there was another letter today. With these three letters in hand, Betty considered herself to be rich. Plus, she was calming down. With the new calm, she wrote again entirely in French, this time with particularly handsome handwriting.

Betty ended her stay in Paris with good feelings about the astronomers. She said goodbye to them the day before. "... they are all very polite, very agreeable." She raised the question to them about whether to publish a notification and correction of a mistake having to do with Oort constants. Betty and the astronomers had some interest in publishing the notification, but they questioned whether it would be a good idea. This is because the Lick Observatory was expected soon to unveil a new method that was supposedly significantly more precise.

The Englishman who was to purchase Betty's car showed up as expected. "It was sad [to have to sell the car] but that's life." The process of transfer took a while, because the buyer didn't have the documents prepared ahead of time and didn't speak much French. Betty spoke to him that evening. The English may have imposed a tax on the car, but Betty decided not to ask him any more about it.

Neyman planned that he and Betty would stay in the Los Angeles area for three weeks. That was fine with Betty. It would be convenient for their

---

[51] Note: Pieter Oosterhoff (1904–1978) was co-administrator of the Leiden Observatory in the Netherlands.

[52] Note: There is no information about the type of car.

[53] 6/1/1955 handwritten letter from Betty Scott to Jerzy Neyman.

research agenda to be in Pasadena. But Betty was still living with her parents, and she didn't know how they would react to such a long stay. In terms of housing arrangements in Pasadena, she didn't care what Neyman came up with. She would be thrilled to be with him no matter what. Also, with the extended stay in the Los Angeles area, Betty would have to strategize what to bring back with her in her luggage as opposed to sending in the mail. Clothing that would fit the weather in June in Los Angeles... she would have to guess how nice the weather would actually be. In any case, she would have to repack. She was limited to 20 kilos [about 44 pounds]. Altogether, her clothing, an iron, and the gifts would have to be under that weight. Speaking of weight, Betty had "luckily" lost some weight over the first half of her year abroad. Her weight had stayed the same since December, even though she was eating a lot. Her clothes didn't fit her new weight. Betty was still enjoying the dried fruit Neyman had sent to her. Every day she ate some of it, a tangible reminder of Neyman.

Betty had just read some articles in *The New Yorker* on customs, and they were depressing. She was also reading an article on how the Bible was written. She was finding it to be very interesting. It explained the power that Jesus had in his time, she told Neyman.

11.21   Saturday, June 4: Lisbon[54]

Thursday, June 2, was Betty's last full day in Paris. She said good-bye to Loève, who had just returned to Paris from Stockholm. Loève sent his thanks to Neyman, Fix, and Lucien LeCam for the dried fruit and vitamins. He was having trouble with his leg, but the doctor had taken him off Vitamin B. Betty described her last days in Paris as "ghastly." She rushed around like mad all day every day, "never breathing it seems," but these last days were the worst. She was exhausted and prepared to take an afternoon nap.

Betty was unhappy about how the car transfer had played out. The buyer had paid almost $5,000 for the car, but he had taken advantage of her. It was important to her to part on good terms with him, and so she fed him lunch. She didn't ask him to take her to the airport or to pay for her merchandise passports, which transferred to him (he saved $20–$35 on the new international customs papers – over $200 in 2010 dollars). She paid half of the check charges, where he should have known that there would be a hold on a personal check for $600 on a foreign bank. Betty was ashamed of herself, but she "didn't say boo, of course," and she and the buyer parted as friends. Betty fretted about a document she signed about the car with Pacific Motors before she left home. The document said that the car company would sue Betty if

---

[54]6/4/1955 handwritten letter from Betty Scott to Jerzy Neyman.

she didn't return the merchandise passports to them. But these transferred to the car buyer. And so all Betty could do was hope for the best with Pacific Motors.

There were about nineteen pieces of unfinished business. One of them had to do with a rebate that British Rail was to send Betty. She hadn't received it yet. Perhaps the delay had to do with them being on strike. So she couldn't blame them. Perhaps they could pay the rebate to her in dollars at her Berkeley address. Another piece of unfinished business had to do with her travel expenses to give her talk at the University of Cambridge. The university paid her on a London bank, which meant she couldn't cash the check in Paris. So she would have to ask Cambridge to have a transfer made to the Bank of Berkeley. The charges, she thought, might be more than the check was worth. All in all, it was just another thing to attend to. She would never have made it out of Paris if a friend hadn't come over to help her.

But then at last Betty was on her way home. On Friday, June 4, she flew the first leg of her trip, which was from Paris to Lisbon. The flight was rough. There were clouds, and the plane couldn't get above them. The plane wasn't pressurized, and so it flew around to the west of the Pyrenees Mountains that divide France and Spain. Also, the plane had engine trouble. They had a forced landing in Porto, Portugal, which was about 170 miles north of Lisbon. They got to Lisbon two hours late. At least the stewardess and airport people were pleasant. They spoke perfect English and helped Betty.

On her one-day layover in Lisbon, Betty wrote to Neyman asking him to bring a few things to Pasadena. One was a coffee pot. She didn't have one with her. The other was a folder from a room in their suite of offices. The folder was labeled "3rd approximation." It contained some intermediate formulas that could be useful to them in Pasadena when discussing results. She had the final formula with her, so if their assistant Jeanne couldn't find the folder, then he shouldn't worry about it. Perhaps Neyman could also bring some office supplies that they might need, like an eraser.

> *I am sitting in the shade of one of the myriad old walls of St. George's Castle. It is magnificent – on a steep hill looking down on all sides on the city and the broad river. There are white peacocks everywhere and black ravens in the old Moorish bakery, and there are storks and storks' nests too, and baby peacocks no bigger than baby chicks. There are Portugese [sic] soldiers who explain what you are looking down on (in English) – very pleasant – but I long to see you –*
> *. . . the peacocks squawk. . .*
> *The peacocks are very friendly. One is watching me – not 3 feet away. Walking carefully along the old stone steps, with his tail high behind him and his big feet reaching down, he looks kind of ludicrous. Now the soldier who explained the high towers just came by. There is a nice breeze and everything seems calm and relaxed up here. But not in the city – there is bustle and chatter everywhere. Women with big baskets of fish or fruit*

(cherries, strawberries, bananas) or flowers on their head coming up from the waterfront...

The soldier says I am sitting in the original gate to the castle. It was built by Moors in 997 and captured by the Portugese [sic] in 1147. There is a bust of the Capt. Maximillan (it sounds like), who led the capture, over my head.

Jerry, I long to see you – I guess this is the last time I can write.

It would be a very short time before Betty and Neyman would be reunited. Five days later, Betty would land in Los Angeles.

Betty may have left Paris, but she had an impact there. In the words of W. Edwards Deming, "...I know that you did statisticians there a lot of good."[55] In fact, her impact extended beyond Paris. Betty was awarded the Medal of the University of Liege in 1955.

11.22 Soul Mates

Betty and Neyman would remain inseparable partners their entire lives. Betty could be daughter, significant other, or mother to Neyman, depending on the time and circumstances. But Betty and Neyman were not to marry. In the assessment of David Draper, Neyman's wife – having brought up Catholic and being angry, stubborn, and wanting to spite him – refused to give him a divorce; Neyman would end up never publicly recognizing his separation from his wife, but he would refer to her in a toast:

Then at the very end of [his] graduate seminars, he insisted we accompany him every week to The Faculty Club, where he bought all of us drinks, and then aways run the tab and didn't close it until everybody was done. And so he didn't just buy everybody one drink, he bought everybody two or three drinks. Because he liked to take a drink himself, and so he enjoyed martinis, I can't remember if they were vodka or gin, and so some of us learned to drink martinis as well to be along with Neyman, and he always insisted on giving the first toast every time we went, and the first toast was aimed directly at his wife, because she didn't give him the divorce. His toast was always the following: To all the ladies present, and some of those absent. And of course what he meant by that was, I do not wish well at all to my wife, she is the one I'm talking about when I say some of the ladies absent. He was extremely courtly and courteous with everyone he was with... and the one person in his life that I can tell he truly hated

---

[55][unkn.] letter from New York University Professor W. Edwards Deming to Betty Scott.

was his wife, and so he made a point to insult her every time he raised his glass.[56,57]

There was also the nepotism rule. If Betty and Neyman could have married, the rule would have prevented their employment in the same department.

And so Betty and Neyman failed to disclose the extent of their relationship, even to their closest students and colleagues. According to Draper, "Anybody with a pair of eyes and any knowledge of what it means for a warm, intimate look to pass from one human to another could not have failed to see that they [Betty and Neyman] were close. The students who don't mention this just weren't very perceptive. And they [Betty and Neyman] were really sweet with each other." According to Juliet Shaffer, "Everyone in the department speculated wildly about [whether Neyman and Scott were a couple]."

While Betty and Neyman were alive, many could only wonder about the nature of their relationship. But Neyman saved the letters that Betty wrote to him from Paris. He kept them in a file in his office. Did he keep them there instead of at home because he was separating from his wife? Or did he keep the letters there because he eventually wanted the world to know of the love between himself and Betty?

---

[56] David Draper, telephone interview with author, October 31, 2014.

[57] Note: Brian Yandell recalls that Neyman liked a vodka gimlet with an olive. Brian Yandell, e-mail to author, September 21, 2015.

12

Civil Rights Advocacy (1950–1953, 1963–1968)

Political advocacy... has become one of the important but subsidiary activities on an intellectually active campus.
-UC-Berkeley Academic Senate

Betty grew up in a family with mixed political views. When she first registered as a voter in 1940, she registered as a Republican like her mother. That same year, one of Betty's younger brothers registered as a Democrat like his father.[1] "Women's rights" at the time was a Republican issue, and Betty and her mother were in line with Republican women who tended to support the Equal Rights Amendment. Because the Democrats had fought hard for laws protecting women against excessive physical demands, sexual harassment in the workplace, and the like, Democrats tended to oppose the ERA because it would have eliminated such protections.[2] One can only imagine the political discussions around the Scott family dinner table. Betty had the opportunity within such a family to learn to think for herself and advocate for her own views. This exposure to debate within her formative years would have helped to shape the liberal nature of her spirit toward the views and actions of others.

12.1 UC Loyalty Oath

Betty earned her PhD at Berkeley some three years after the start of the Cold War between the US and USSR. This was the same year the UC Board of Regents proposed that all employees should sign a Loyalty Oath saying they were not Communists. Employees were accustomed to signing a Constitution Oath: "I do solemnly swear (or affirm, as the case may be) that I will support

---

[1]Miles, Josephine. "An Interview with Josephine Miles," an oral history conducted by Suzanne B. Riess, in "The Women's Faculty Club of the University of California, Berkeley, 1919-1982," Oral History Center, The Bancroft Library, University of California, Berkeley, 1983.

[2]G. Collins, *When Everything Changed: The Amazing Journey of American Women from 1960 to the Present* (New York: Little, Brown and Company, 2009), 70.

the Constitution of the United States and the Constitution of the State of
California, and that I will faithfully discharge the duties of my office according
to the best of my ability." The additional Loyalty Oath, passed by the UC
Regents on April 12, 1950, was something else altogether:

> *Having taken the constitutional oath of the office required by the State of*
> *California, I hereby formally acknowledge my acceptance of the position*
> *and salary named, and also state that I am not a member of the Commu-*
> *nist Party or another organization which advocates the overthrow of the*
> *Government by force or violence, and that I have no commitments in con-*
> *flict with my responsibilities with respect to impartial scholarship and free*
> *pursuit of truth. I understand that the foregoing statement is a condition*
> *of my employment and a consideration of payment of my salary.*

The Loyalty Oath was like a dark cloud that hung over the UC for some ten
years. Betty and many other employees felt they should not be required to
sign a loyalty oath: They weren't in the same category as, say, deputy sheriffs,
who were clearly officers of the state and held a public trust. UC employees
felt they should have been consulted about a loyalty oath, as would have been
customary within traditions of shared governance and principles of academic
freedom. In 1956 the AAUP censured the UC administration over this issue.[3]

Although just a new PhD, Betty wrote a forceful letter to the governor of
California advising him of her strong views. It turned out the governor was
in sympathy with Betty and sent her his public statement indicating he had
cast a minority vote against the requirement.[4,5] Some university employees,
including 19 at Berkeley and 12 at other campuses, took the extreme step of
actually refusing to sign the Loyalty Oath. They were dismissed from their
positions, not because they were charged with any disloyalty whatsoever, and
not because they were found to be professionally unfit, but merely because
they would not sign. One of the non-signers was one of Betty's colleagues,
UC-Berkeley Mathematics Associate Professor Pauline Sperry.[6]

---

[3] *Bulletin of the American Association of University Professors*, Appendix A, 42, no. 1 (1956), 100.

[4] 3/13/1950 letter from California Governor Earl Warren to Betty Scott.

[5] 2/28/1950 press release – Statement from California Governor Earl Warren.

[6] Note: By 1950 Sperry had worked for the University for 33 years. As she put it, "The greatest gift to mankind – the freedom of the mind – is in great peril. If we lost that we lose everything. The universities are its greatest bulwark. They are the first to be attacked. The battle is only just begun." Sperry and others took the Loyalty Oath requirement to appellate court. The court declared the oath to be unconstitutional. But Sperry's career at Berkeley was already over.

12.2   IMS and Racial Segregation

Around the same time, momentum for racial integration in the US was building. Academics were getting into the dialog, and issues involving segregation were emerging within academic professional organizations like the IMS. Betty was more than just a IMS member: In 1951 she was elected a Fellow "in recognition of his [sic] contributions to the development, dissemination, and application of Mathematical Statistics."[7]

In 1953, a group of IMS members including Betty and Neyman decided to send a petition to the elected governing council requesting an amendment to the IMS by-laws that would explicitly prohibit racial segregation.[8] They reasoned: How could they stand by and watch any persons be subjected to segregation, including their brilliant friend and colleague David Blackwell, because of the color of their skin?

The IMS council received the petition. Betty and Neyman heard a loud and clear response from one of the council members, who felt the IMS should not get involved in "an entirely unstatistical matter." He saw the specification of non-segregation as equivalent to the specification of other "desiderate" for meetings like indoor plumbing and said it would be impossible to add all of these things into the by-laws. He "deplored" the fact that the petition would split the society on "an issue that is completely irrelevant to the goals" of the IMS.[9] He wrote: "It will irritate a minority, and it is barely possible that this minority will take the attitude that the Institute has gotten into the hands of a group of wild-eyed reformers who are perverting its purpose."[10] The council member expressed a "faint hope" that the petitioners did not feel strongly, could be swayed by his arguments, and would reconsider. He even told the petitioners how exactly they should indicate they were withdrawing (what they should say, to whom, and using what medium).[11]

But the council member did not know with whom he was dealing. Betty responded by writing to the IMS President, as indicated. But she did not withdraw. Instead, she asked the president for reassurance that she and others would be allowed to vote on the proposed non-segregation by-law (to make sure there was no question, she specified that it was "Bylaw Article 4 combined with Constitution Article 12."), and she furthermore asked if the secretary would send the amendment to all members in the next mailing. She wrote: "I may say that I feel somewhat absurd and even insulted to have to write this letter and I hope that you will not have the same feelings." She furthermore asked if the

---

[7]Note: 1951 certificate to Betty Scott upon her election as an IMS Fellow.

[8]Note: Blackwell accepted a permanent position as Professor of Statistics at UC-Berkeley in 1954. In 1955 he served as IMS President. In 1965 he became the first African American to be named to the NAS.

[9]10/8/1953 memorandum from IMS Council Member Alex Mood to Council.

[10]10/14/1953 memorandum from IMS Council Member Alex Mood to Council.

[11]10/29/1953 letter from IMS Council Member Alex Mood to Ladies and Gentlemen.

amendment would appear on the ballot so that all members could vote by mail; if the answer was "no," then how would a member vote by mail, and how would such mail ballots be counted? "But I have been reliably informed by several members of the Council and other members of the Institute that you have ruled 'no.' It seems unbelievable that you would rule the disenfranchisement of the great majority of the Membership unable to attend the meeting in person!"[12]

On November 7, 1953, Betty and Neyman, together with a group of 17 of their associates and students at Berkeley and Stanford, sent a response to the council member's memoranda. They said they wished to inform the council of IMS that they were not withdrawing their names from the petition. That was that. And their firm convictions seem to have contributed to a profound effect: Blackwell was elected president of the IMS only three years later.

12.3   Civil Rights Solicitations

Fast-forward ten years to 1963. This was a turbulent year toward the advancement of civil rights in the US. Neyman witnessed some of the issues for blacks firsthand in spring when he was an invited lecturer at several colleges and universities across the country, some of which were in the South.[13] When he returned to Berkeley, Neyman decided to take action and organized a group of 11 faculty members who were equally "disgusted" by the overall atmosphere for blacks and recent developments in the South and felt "compelled to do something."[14,15] Betty was the one woman among them. The group cited arrests, school suspensions, school dismissals of large numbers of youth, and threats by a governor in the South to "personally" block a few black students from passing through the entrance of a university. But their main thrust was the same as Martin Luther King Jr's, namely, voter registration.[16] Neyman's faculty group wanted to raise money to help young blacks "in their struggle

---

[12]11/5/1953 letter from Betty Scott to IMS President Morris E. Hansen.

[13]Note: One of Neyman's lectures was in Mississippi at an all black college. While there, he accepted an invitation to attend a party at a black faculty member's home. He was then informed of a state law that "forbids white persons from being the guests of Negroes." He was informed that he was breaking this law by attending the party.

[14]Note: The group included David Blackwell, Physics Nobel Laureate Owen Chamberlain, Milton Chernin, Lucien LeCam, Benson Mates, Jerzy Neyman, Herbert Robbins, R.M. Robinson, Betty Scott, Kenneth Stampp, and George Stewart.

[15]"Committee Raises Funds for Negroes," *The Daily Californian*, June 28, 1963, 5.

[16]6/19/1963 letter from Jerzy Neyman to Jean Brockell. Note: In the words of King: "We are making the main thrust of our work in the area of Voter Registration; the effort to achieve universal adult suffrage and truly representative government in the South. Despite the bombing of churches, police harassment, jailings and attempted assassinations of Voter Registration workers, S.C.L.C. has assisted local communities in getting more than 40,000 Negro voters added to the registration rolls, from Virginia to Mississippi [underline sic]."

to attain the status of citizens." They wanted to raise the money for bonds, legal expenses, and the like. . . moneys that could supplement the scarce monetary resources of communities in the South. And so at the end of May, the group of 11 sent a letter, personally signed by each of them, to every faculty member on the Berkeley campus and colleagues on other campuses asking for contributions: They should make their checks out to King's organization, the Southern Christian Leadership Conference (SCLC), and send them to Neyman at his home address.[17] By the end of June, the faculty group had 250 contributors among faculty and students and raised $3,200 ($22,538 in 2010 dollars). They continued to raise moneys.[18]

Betty assumed the role of preparing the moneys and sending them to the SCLC. King responded to Betty with a letter of "deep appreciation" for the "generous contribution," saying:

> *Without your moral support we would be caught in a dungeon of despair without knowing that many people all over the nation are supporting us in our struggle. By aiding us in this significant way, you are telling the world that the rights of Negroes cannot be trampled in any community without impairing the rights of every other American. Thank you again for making our financial problems a little less burdensome. I am confident that if we continue to gain this type of support, this sweltering summer of discontent can be transformed into an invigorating autumn of justice and freedom for all people.*

Betty saved this treasure of a letter.[19]

What else could academics do to promote civil rights and the Civil Rights Bill? They could put out some strategic advertisements! The paid ads, placed in the university student newspaper, *The Daily Californian*, urged people to express their support to their congressmen. It was a matter of "liberty and justice for all." Neyman assembled an initial group of 12 that included Betty. Let's place some ads! Let's not delay! Large and small! Let's keep placing ads! But the group needed funds for the ads, so they sent out another solicitation to colleagues asking specifically for contributions.[20] Three days later the group of 12 expanded to 28, with Neyman and Scott as two of six leaders,[21] calling themselves the Faculty Reminder Group because the ads would be written to remind people to support various civil rights organizations.[22] By the beginning of December, the group had grown to 39.

---

[17]5/24/1963 letter from 11 faculty members to Dear Colleague.

[18]"Committee Raises Funds for Negroes," *The Daily Californian*, June 28, 1963, 5.

[19]8/20/1963 letter from Martin Luther King, Jr. to Betty Scott.

[20]11/6/1963 letter from list to Colleague. Note: The initial list of people who solicited gifts for paid advertisements to support civil rights included David Blackwell, Robert Cogburn, Jacob Feldman, Henry Helson, Leon A. Henkin, Lucien M. LeCam, Michel Loève, Jerzy Neyman, Henry Scheffé, Betty Scott, Kenneth M. Stampp, and John V. Wehausen.

[21]11/9/1963 memorandum from Henkin, LeCam, Loève, Neyman, Scheffé, and Scott to 22 people.

[22]12/1963 memorandum to the Faculty Reminder Group from Secretary M. Darland.

The first ads were basic, calling on people to transform their "beliefs into action" by contributing to "organizations fighting for civil rights." Later ads read like newspaper articles, explaining the issues, calling on people to help defeat certain pieces of proposed anti-civil rights legislation, and calling on people to vote for pro-civil rights candidates.[23] Both early and later ads included lists of the contributors, at first members of The Faculty Reminder Group, and later members of a group that included more than just faculty called the Scholars Committee. By the end of October 1964 there were over 135 contributor-signers. The group planned to put out one large and four small ads each month. Betty, Neyman, and two colleagues formulated a large ad for February 1965, the front end of which read:

> *In recent weeks*
> *Dr. Martin Luther King, President of SCLC, was awarded the Nobel Peace Prize in Oslo.*
> *In Selma, Alabama, he was slapped and kicked a few weeks later, and then arrested.*
> *The Civil Rights Bill created the illusion that the civil rights struggle is won. But voting rights are still denied by legal trickery, while protesters against this trickery are arrested. The struggle for civil rights needs your support more than ever.*[24]

Betty remained a consistent contributor-signer of all of the ads.

12.4   Saving Aquatic Park

At the same time, Betty was in the thick of a movement to save Aquatic Park. This park was comprised of both land and lake with a Yacht Harbour. It was built in the 1930s as part of a Federal Work Projects Administration endeavor to be an open and clean air space for recreation within the city of Berkeley.[25] Even though some saw the park as Berkeley's "finest recreational resource," the park suffered from neglect. This gave an opening for entrepreneurs to bring proposals forward to buy the park and convert it to "Industrial Use."

Betty worked hard to save Aquatic Park. She asked: Why sell this fine resource that contributes vitally to the "whole picture" of Berkeley, when it could be "gradually put in order" and then maintained?[26] She wanted to know why the city manager failed to include in his budget proposal any repairs for

---

[23]10/30/1964 advertisement "Vote AGAINST EXTREMISM" placed in *The Daily Californian* by The Scholars Committee.

[24]2/8/1965 letter from Jerzy Neyman to Dear Colleague. Note: The two colleagues were H. Helson and E.H. Huntington.

[25]Note: It is highly visible, being at the University avenue entrance to Berkeley from the Bay. It is quite large for a city park, covering 97.8 acres.

[26]n.d. ?1963. Anonymous. White Paper: Aquatic Park.

Aquatic Park which were sorely needed for the road, to open a vent, and to repair a sidewalk. Betty even cost these out, but none were written into his proposal. She thought it was "ridiculous to have to fence off part of the Park – to keep children out – because the sidewalk needs repair, but that is what is happening. Let the children use the park, do not let it go to ruin. I realize that it would take more money (although much less than the Recreation Secretary imagines) to put Aquatic Park in a completely manicured shape but this is not what is needed. Berkeley needs a place where Berkeley children (and adults) can sail."[27]

Betty was concerned about Aquatic Park right down to the water valves. She heard that industrial dumping had clogged two of the valves, and the clogging was getting worse. This retarded the circulation of water, which caused sediment to build up, which in turn caused the water to get warmer and more shallow, too shallow and marshy for boating. Betty wanted the city engineer to find out who was doing the dumping: Was it the Berkeley City Dump, or did the state fail to close the valves when they built the University Avenue overpass, or what? Betty wanted the city to identify the perpetrating organization and hold it responsible. She wanted the valves to be fixed and then properly maintained so the park could resume good sailing conditions.[28]

At this point Betty started to bring her statistical expertise into her public advocacy. The sale of Aquatic Park was discussed at the February 13, 1963 meeting of the Berkeley City Council. Betty was there and heard a statement that surprised her, namely, that Aquatic Park was "not used." This was contrary to Betty's impression, which was that the park was used for all kinds of recreational activities including, for example, walking, dog tracking, boating, and water sports. It was also her impression that people went to the park to just look out at the water.

Betty wanted to know if her impression was correct, or whether the park was not used as had been alleged. She decided to collect some data and conduct a statistical analysis. She knew that good research designs have control groups, so she decided to compare utilization of Aquatic Park with utilization of two control park areas: one was Codornices Park and the adjacent Rose Garden, which was located in northeast Berkeley and "expensively landscaped"; the other was the south end of Tilden Regional Park, which was a jewel of the East Bay park system. Betty reasoned that Codornices and Tilden would be suitable for comparison because both had about the same land area, and therefore similar maintenance costs, as Aquatic Park.

Betty wanted to test the null hypothesis that Aquatic Park had the same pattern of use as the other two parks. Her measurements were counts of people in the parks. She expected counts to vary with different conditions and arranged to have students in Advanced Statistical Inference class collect data at

---

[27]n.d. ?1963. first handwritten draft of a letter from Betty Scott to the Berkeley City Council.

[28]n.d. ?1963. second handwritten draft of a letter from Betty Scott to the Berkeley City Council.

different conditions for each of the three park areas. They made sure no special events were taking place in the park at the time of data collection. Data included date, hour, park area (Aquatic, Codornices, or Tilden), weather (e.g., terrible, clear, etc.), and number of people using the park. Betty displayed these data in a frequency table titled "Observed Use of Parks." One student also conducted a telephone survey, selecting 100 numbers at random from the residential telephone directory. Of these, 32 were Berkeley residents who were asked, "If you had the time and if the weather is good this coming weekend, which one of the two parks, Aquatic or Codornices with the Rose Garden, would you prefer to go to? These results were also displayed in a frequency table.

The next step was to conduct a statistical analysis. Betty assigned this to the students because they were studying the kinds of nonstandard statistical techniques that the Aquatic Park problem would require. They found the hypothesis was not rejected and concluded Aquatic Park had the same pattern of use as the other two parks, observed differences being "not more than can be expected with the fluctuations with time caused by people coming and going in groups." They also did not have enough statistical evidence from the small numbers in the telephone survey to conclude an unequal preference for Aquatic and Codornices parks.

Betty prepared three pages of written testimony, plus the two frequency tables, and presented it to the City Council on March 3, 1963. She explained that a definitive assessment would require a larger sample size over the course of a whole year. However, on the basis of her more limited assessment over a 19-day period, she found "no support for the statement that Aquatic Park is 'not used,' judged by the number of persons actually using it or relative to the use of other parks."[29]

Two years later, Betty and others were still trying to save Aquatic Park. One of the others was Josephine Miles (1911–1985), a colleague of Betty's at UC-Berkeley. Miles was a trailblazer at Berkeley: She was the first woman to be tenured in the English department, in 1947; and she would become the first woman to be awarded the coveted title of University Professor in retirement. Miles was a poet and, like Betty, brought her specific expertise to her Aquatic Park advocacy, writing a poem about the efforts to save the park. Called "Saving the Bay," it was published in the *Atlantic Monthly* and has been listed among Miles' primary works. The poem is lengthy: It has 21 stanzas, one of which (the 16th) describes Betty's study and testimony to the City Council. Here is how Miles converted Betty's frequency table on the "Observed Use of Parks" into poetry:

> *And a letter from a statistician, fond of the facts,*
> *Compares the use of Aquatic Park with the Rose Garden: the same pattern;*

---

[29]3/3/1963. Testimony from Betty Scott to the Berkeley City Council: Comparative Usage of Aquatic Park.

Fewest people, about five each, on a Friday of terrible weather,
Next, about fifty, on a warm Wednesday afternoon,
Most, a hundred and fifty on a clear windy Saturday. Signed, sincerely.

Miles sent Betty a signed copy of the poem, with a note: "For Elizabeth Scott, with thanks!"[30]

Betty was thrilled to have poetry written about her statistical work. She wrote to Miles: "It pleases me greatly to be in a poem; I never imagined that such a thing would happen."[31] Betty was also grateful that Miles had written the poem because she thought it could influence the city council members to Save the Bay. Betty forwarded the poem to a professor in the city planning department on the Berkeley Campus who at the time was also an elected member of the Berkeley City Council, saying: "It is possible that if the members of the City Council are informed of the poem and read it, they will be influenced in the right direction." She then proudly wrote in a P.S.: "I learned of the poem because Miss Miles sent me a copy indicating that, among other things, she described my own performance as a statistician!"[32]

12.5 Free Speech Movement

Betty's colleagues in the statistics department recognized her penchant for advocacy. When asked who was involved with the Free Speech Movement, Blackwell said, "I don't think most of the people in the statistics department were involved one way or another... My guess is that Betty Scott would have been, and very much on the side of the students."[33]

The precipitating incident for the Free Speech Movement began on December 2, 1964. Students on the Berkeley campus thought the rules of the administration about assembling, speaking, and raising funds were too restrictive. 768 students chose to occupy the Sproul Hall administration building in a climactic protest. They were trying to exercise their First Amendment Constitutional rights to freedom of speech. When the students refused to leave Sproul by the next day, they all were arrested. The charges were trespassing, unlawful assembly, and resisting arrest. They were thrown into jail under the jurisdiction of the municipal court for the Berkeley-Albany Judicial District, County of Alameda, State of California. The arrested students faced Judge Rupert Crittenden who found them not guilty of unlawful assembly. But he

---

[30] J. Miles, "Saving the Bay," originally in *Atlantic Monthly*, reprinted San Francisco: Open Space, 1967.

[31] 1/15/1965 letter from Betty Scott to Professor Josephine Miles.

[32] 1/15/1965 letter from Betty Scott to Professor T.J. Kent.

[33] Blackwell, David. "An Oral History with David Blackwell," an oral history conducted by Nadine Wilmot, Oral History Center, The Bancroft Library, University of California, Berkeley, 2003.

did find many of them guilty of trespassing and resisting arrest; these students were given jail sentences, with periods of probation and fines up to $300 (over $2,000 in 2010 dollars). For students who appealed, the judge set bail at $550 ($3,760 in 2010 dollars) or more. This was a huge amount of money for a college student. No students were released on their own recognizance.[34]

So many students were thrown into jail that defense attorneys brought a class suit on behalf of the group. Students faced a mountain of legal fees, including what the defense attorneys called "excessive and therefore illegal bail." It was looking to the students and others like bail was being used as a punishment. It was a serious situation: Students were faced with a choice of posting this "excessive" and "illegal" bail or else losing their right to appeal their convictions forever, unless they chose to remain in jail pending appeal. As a group, their appeal bonds would cost over $350,000 ($2.3 million in 2010 dollars). If they couldn't come up with the large sums, they would have to pay premiums of over $35,000 ($230,000 in 2010 dollars). The costs were staggering, and the students had no time for a hearing and appeal before they were faced with "irreparable damage." The class action suit asked to restrain Judge Crittenden "from further sentencing until he can appear before the Supreme Court to show just cause why he should not grant [the petitioner's] request" for "release without bail pending further appeal of the conviction."[35]

Betty and many others on the faculty had deep concern for the students and did not condemn them, thinking the matter of whether they violated the law should be decided within the legal system and not by the faculty. They thought it should be decided with all the facts on the table and with competent legal representation. So Betty and others formed an Independent Faculty-Student Legal Fund.[36] Neyman went out front and led the formation of an initial group of professors and students to organize the fund. First there were four trustees, and then seven including Betty. They wanted to make sure the arrested students had good legal help. Some attorneys had generously offered their time without charge. But there were going to be other legal expenses that would add up, such as the costs of obtaining records, preparing documents, secretarial help, expediting court cases, paying fines, or purchasing bail bonds. The group set out to raise moneys to help with these legal expenses. They were willing to take donations from "anyone and everyone."[37] Neyman wrote: "It is urgent that these students receive the full protection of the law."[38]

---

[34]n.d. ?1965. In the Supreme Court of the State of California: David Noble, Judy Peters, Kenneth Baker, and All Other Similarly Situated, Petitioners vs. Municipal Court for the Berkeley-Albany Judicial District County of Alameda, State of California, Respondent. Petition for a Writ of Mandate; in the Alternative for an Alternative Writ of Prohibition; Points and Authorities.

[35]8/2/1965. Sit-inners ask writ against Crittenden. *Berkeley Daily Gazette*.

[36]2/25/1965 letter from Betty Scott to Professor Margaret T. Hodgen.

[37]n.d. ?1964 memorandum to unspecified from The Independent Faculty-Student Legal Fund Trustees (Henry Nash Smith, Louis Simpson, Jerzy Neyman, and Girard Pessis).

[38]n.d. ?1964. letter to unspecified from The Independent Faculty-Student Legal Fund, Jerzy Neyman, Chairman.

The trustees organized a meeting of interested parties at The Faculty Club. They talked about how to proceed. A path was clear. They should try to help the students without putting money on the table. Betty grabbed a piece of faculty club stationery. She wrote: "We wish to ask Judge Crittenden to release students to professors, rather than requiring them to post bond of $550 (or more) to appeal their cases. No money would be involved; faculty members would vouch for students they know. We are preparing a list by majors. Please sign if you agree in principle." Fifteen people signed, about half from statistics or mathematics, the rest from other departments across the campus.[39]

The group wasn't immediately successful in getting the judge to release the arrested students to professors. When the method was tried for five students on July 27, it simply didn't work. Maybe, they thought, it didn't work that day because their requests were made on behalf of students who were leaders of the movement, and so they kept trying. They remained hopeful they could get at least a few students released to professors. They got a glimpse of encouragement on July 28. On that day an arrested student requested a two-day delay of his sentencing so that he would have time to pull together letters requesting his release without bail. The judge granted the request. So there remained hope among Betty and Neyman that this strategy would work: "... real hope that with a reasonable backing of students by the faculty, many unnecessary days in prison may be saved and/or the returnable sums paid for bail bonds may be substantially decreased."

Betty and Neyman proceeded to write a letter to their colleagues that included a list of students from the colleague's department who were arrested and hadn't yet been sentenced. The idea was that it would be easier for Betty and Neyman to ask their professor colleagues for a letter than it would be for the student to ask the professor on his or her own behalf; the expectation was that the professor would only write a letter for the student if the professor knew the student well. So Betty and Neyman's letter asked if the colleague would be willing to vouch for one or more of the students on the list. If so, would the colleague please be willing to write a letter to Judge Crittenden, so that the letter could be used in court at the time of sentencing? To make it easy for the faculty member, Betty and Neyman included a sample letter prepared by one of the defense attorneys that asked if the student could be released without bail on his or her own recognizance.[40]

How could they help out more than just a few students? The answer again was a letter addressed to Colleagues. It was from 18 Berkeley professors and students, including Betty and Neyman, requesting a gift for the Independent Faculty-Student Legal Fund. The fund would pay the legal costs for students who decided to appeal. Bail would be expensive, although it would be recoverable; a bail bond would cost a fee, and it would not be recoverable. Without

---

[39] 1965. Memorandum to Judge Crittendon from 15 faculty members.
[40] 7/28/1965 letter to Colleague from Jerzy Neyman and Betty Scott.

bail or a bail bond, the student would go to jail. Students' financial situations should not keep them from appealing, they thought. Time was of the essence. Because students were being sentenced now, moneys were urgently needed now. The appeal was fervent: "These are OUR [sic] students."[41]

Michael Rossman (1940–2008) was a legendary planner and leader of the Free Speech Movement. He felt that Berkeley's reputation of being a very liberal institution, one highly supportive of free speech, was a myth. He developed a report that exposed how the administration had been dealing with political issues. Rossman gained a reputation from this report. It landed him on the executive and steering committees of the Free Speech Movement.[42,43] Rossman was among those arrested in December of 1964.

It so happened that Rossman was a teaching and research assistant in the statistics department. Betty knew him personally for about two years before he was arrested. Seeing that Rossman was on the list of people arrested, Betty got to work. She was willing to vouch for Rossman. Betty told Judge Crittenden: "I know from personal observation and from spontaneous discussions with his students that he is a very good teacher, very dependable, conscientious and does what he says he will do. I believe him to be an honest and trustworthy citizen... I know Mr. Rossman to be a respected member of this community and to have a good reputation for reliability and honesty. I therefore wish to vouch for him in connection with the matter now pending before the court and to request that he be released upon his own recognizance, and to the extent permissible, upon my recognizance, during the period of his appeal from the judgment entered against him in this matter."[44]

By the end of July 1965, over 100 faculty members from across the campus had either written or agreed to write letters on behalf of their students. Betty and Neyman were the ones who led the effort. They "respectfully transmitted" to Judge Crittenden the names of the faculty members, hoping he would be overwhelmed by the volume of faculty support and would "see fit" to release the arrested students on their own recognizance. They hoped he would be moved by the fact that all of these faculty members were willing to "stand behind" their students.[45] They hoped if the judge wouldn't eliminate bail, then he would at least reduce it.

Once again, Betty and Neyman included with their letter to colleagues the sample letter that had been prepared by a defense attorney. The sample was like the letter Betty wrote for Rossman. The intention was to provide a general format to use when drafting their letters. The colleagues were asked

[41]7/28/1965 letter to Colleague from 18 Berkeley professors and students including Betty Scott and Jerzy Neyman.

[42]M. Taylor, "'60s activist Michael Rossman dies in Berkeley," *SFGate*, last modified May 17, 2008, http://www.sfgate.com/cgi-bin/article.cgi? f=/c/a/2008/05/16/BASB10O523.DTL.

[43]*Free Speech Movement Archives*, accessed December 24, 2014, www.fsm-a.org.

[44]7/29/1965 letter to The Honorable Rupert J. Crittenden from Betty Scott.

[45]7/30/1965 letter to The Honorable Rupert J. Crittenden from Jerzy Neyman and Betty Scott.

to write letters "similar to the general format attached." But it turned out that many, including Betty, thought their letters had to be "substantially of this form," that the specific "phraseology" was necessary since it came from an attorney. Consequently, many of the faculty members, wanting to be as careful and helpful as possible, used this phraseology in their letters for their students, and the municipal judges received many identical or similar letters.

The judges didn't like this at all. They accused the faculty of writing "form letters." Since Betty and Neyman had been the ones to "respectfully transmit" the names of the faculty member letter writers to the judges, they were the ones who were contacted by the judges about this matter. It was an uncomfortable situation. Betty ended up talking with, and then writing a two-page handwritten letter to, one of the judges about it. Her motives, and those of the faculty members, were student-centered and noble, and she wanted the judges to know that. She thought it would help clear things up if the judges had exact copies of the letters she and Neyman had written to their colleagues, and so she attached those to her letter. She explained the motivation for the faculty letters and the rationale for the nature of her and Neyman's request. She assured the judge that the faculty members were not writing form letters, that they were "using what we thought to be the necessary phraseology in this legal matter." Humbly, and hoping to extinguish the fire, she closed her letter by saying she hoped that the judge "will take into consideration the fact that I have indeed misinformed my colleagues."[46]

Campus support for the arrested students continued to grow. On August 2, a letter signed by about 120 faculty members was introduced in court requesting that the students be released on their own recognizance, "especially if they are vouched for by a professor, rather than requiring each student to post bond in order to appeal his case." It referred to the over 100 faculty members that Betty and Neyman had rallied to write letters on behalf of individual students.[47] The letter was accepted by the court, but not entered into the official record. Nevertheless, Betty, Neyman, and the faculty had succeeded in putting pressure on the court, and Judge Crittenden said he would consider the proposal.[48]

Next Betty teamed up with Philosophy Professor John Searle (1932–) who four years earlier had been "flabbergasted" when the administration "trampled on" his own free speech by canceling his talk, a critique of an anti-communist propaganda film, at the last minute.[49] Now Betty and Searle led a group of 80 faculty members to prepare statements that vouched for the good character of several arrested students. The two hoped the judge would consider the weight of the voucher statements and lower the bail for these students. Unfortunately, Judge Crittenden would not budge. Betty was unde-

---

[46] 8/16/1965 handwritten letter to Judge Talbott from Betty Scott.

[47] "Warrants issued for 16 sit-in defendents," *Berkeley Daily Gazette*, August 3, 1965.

[48] n.d. ?1965. FSM defendants fight bail. Source unkn.

[49] "His speech muzzled before, professor stood with students," *The Daily Californian*, October 8, 2004.

terred. She said she would continue to explore all kinds of avenues with the defense attorneys to help the students. She continued to maintain the bail was excessively high, especially since the students were residing locally. The prosecution didn't agree: They referred to the "wanderlust" of the students and maintained the students would be hard to find when their appeals came up, which would be "in a year or so." The next idea was that the faculty would set up "property bonds." The thought was that these would carry a lot more weight than simple voucher statements. The judge said he would consider, but reserved judgment, calling them "very cumbersome."[50]

The attorneys for the defendants drafted a legal document for the faculty members to sign. In the document, the faculty member pledged to pay the bail if the student failed to perform any conditions of judgment. Two faculty members would sign to be responsible for each student. For example, Betty and Physics Professor Sherwood Parker signed to be responsible for the $550 bail for defendant Robert Greenberg.[51] The attorneys also drafted an affidavit for each faculty member to sign declaring the amount of their personal worth and names of defendants the faculty member would be responsible for. Betty declared her worth to be $4,840 (just over $33,000 in 2010 dollars). She was one of two faculty members who pledged to pay bail for the following ten students studying in a variety of departments: at $550 bail each, these included Mary Lloyd (English), Jonathan King (Political Science), Richard Nanas (Sociology), Peggy Hallum (English), Ellen Wedum (Chemistry), Jon McDaniel, (History), Lawrence Shapiro (English), and Robert Greenberg (English); and at $220 bail each, these included Andrew Ross (Humanities) and Rafaella Del Bourgo[52] (Anthropology).

12.6 A Changed University

The university administration slowly backed down on their regulations having to do with political activity on campus. In spring 1965, the academic senate took a lot of credit for solving the crisis. Teaching, learning, and scholarship eventually returned to normal, but with a changed environment. An Emergency Executive Committee of the academic senate described it this way: "... there is ... unprecedented discussion about the objectives and techniques of education. Students and professors are talking with each other more than ever before – in classrooms, studies, restaurants and professors' homes. Polit-

---

[50] "'No contest' sit-ins given light terms," publication unknown, n.d.?1965.

[51] n.d. 1965 SAMPLE legal document from Attorneys for Defendant (Malcolm Burnstein, Richard M. Buxbaum, Henry M. Elson, Stanley P. Golde, Douglas J. Hill, and Norman Leonard), People of the State of California, Plaintiff, vs Robert Greenberg, Defendant.

[52] n.d. 1965 SAMPLE affadavit from Attorneys for Defendant (Malcolm Burnstein, Richard M. Buxbaum, Henry M. Elson, Stanley P. Golde, Douglas J. Hill, and Norman Leonard), People of the State of California, Plaintiff, vs Robert Greenberg, Defendant.

ical advocacy – the key issue in the fall controversy – has become one of the important but subsidiary activities on an intellectually active campus. Problems continue to arise, but they are handled by rational discussion and mutual adjustment in a context of candor rather than by rigid legalisms and hostile confrontations in an aura of suspicion."[53]

Betty received a number of letters of thanks from the students she co-sponsored. On November 2, 1965, Del Bourgo wrote this to her:

> *This is just a short note to thank you for going my surety and to assure you that I am a live and breathing person and not just a complicated name.*
>
> *I am a senior in anthro but plan to get my M.A. in social welfare, probably here. I am sort of putting myself through school and so it was really a relief to know that the bail money I put up is applicable to my fine (if convicted).*
>
> *I'm taking 17 units this semester & must get c. [sic] a 3.5 so I'm studying very hard & not socializing very much – however if I do get a free hour (& can keep my conscience clear long enuf to stay away from the books) perhaps we can have coffee together and chat.*
>
> *Frankly, I was beginning to feel a victim of "the machine" or "the combine" or "Catch 22" or whatever you want to call it – Knowing that university people put themselves on the stand for us really made me feel better.*

On November 14, Hallum wrote to Betty:

> *Your generosity in acting as surety for me in the appeal of the Free Speech case is deeply appreciated. Without your help it would have been impossible to make the appeal.*
>
> *It is a good feeling to know that faculty and community people still stand behind the students in this case. December 2 was a long time ago and to most of those not directly involved the sit-in has become merely an event of the past that is now done with. It was exciting to collect bail in the atmosphere of urgency that obtained on the campus December 2 and 3. To those of you who are willing to support our case even without the dramatic happenings of those days we must extend our sincere thanks.*

Wedum thanked Betty for her time and trouble, hoping Betty didn't have to go to court multiple times to provide Wedum's surety.

Betty's colleagues were also generally grateful for her and Neyman's efforts on behalf of those who had been arrested as part of the Free Speech Movement. A journalism professor, who had written a letter on behalf of two students, wrote to Betty wishing to congratulate her and Neyman "on the action you've taken to help these students this way, and to help everyone negotiate this sorry business."[54]

---

[53] 3/1/1965 memorandum from the UC Academic Senate to Members of the Academic Senate, Berkeley Division.

[54] n.d. Friday ?1965 memorandum to Betty Scott from Pete Steffens.

Betty's work with the arrested students crept into her academic correspondence. In December 1966, she submitted a book review to the ISI president. She told him she thought her review sounded "cross," she had a sore throat, and he should "Please... not publish the review if you find it unpleasant." But the interesting thing about her letter was that all of this was in the postscript. The main body talked about the hectic life in Berkeley: "Now 8000 students are striking against the University administration. A Dutch professor in Civil Engineering here says that students in Holland never strike or demonstrate. Ha! But life is difficult [sic] for them and for us (and the administration)."[55] The ISI president wasn't phased by the tone of Betty's review and said he thought criticism makes reviews worthwhile. He had more to say about the demonstrations than he did about her review, offering: "I appreciate your difficulties with the students on strike."[56] Across the globe, colleagues were interested in what was happening at Berkeley.

By August 1967, Betty was the official designated secretary for what was then being called the Independent Faculty-Student Surety Fund. The Free Speech Movement cases had concluded and been disposed of. Betty reported that the fund paid fines to the court whenever a defendant had difficulty getting prompt payments to the court. She also reported that, at the end of the process, "there were no expenses to any of the sureties." Now it was time to close the books.[57] In August 1968, Betty polled the Trustees of the Fund asking them their choice for disposition of the remaining funds, indicating that if there was agreement, the fund could be terminated.[58] The Free Speech chapter of Betty's life had come to an end.

In the 1960s at Berkeley, Betty had been involved in many high profile liberal political issues. She had been an advocate for academic and political freedom... for voting rights... for her community... for her students. Betty was well prepared to become a lead advocate for academic women's rights.

---

[55] 12/4/1966 letter from Betty Scott to ISI President Bart Lunenberg.

[56] 12/12/1966 letter from ISI President Bart Lunenberg to Betty Scott.

[57] 8/23/1968 memo from Betty Scott to Trustees of the Independent Faculty-Student Legal Defense Fund concerning Disposition of Remaining Fund.

[58] Note: Jerzy Neyman, one of the Trustees, responded to the poll that he mildly preferred to turn the moneys over to the ACLU. The final dispensation of the fund is not known.

Part IV

The Status of Academic Women at Berkeley

13

A Disgraceful Situation (January – September 1969)

...I am not allowed to join the Faculty Club...I am not going to encourage anyone to join a Faculty Club that practices discrimination. I hope that you will take some action.
-Elizabeth L. Scott

13.1 Two Faculty Clubs

Most campuses don't have faculty clubs, but UC-Berkeley has two. One is named The Faculty Club, the other the Women's Faculty Club (WFC).

The Faculty Club was dedicated in 1902 and for many years the membership was only open to men. Hence it has been, and sometimes still is, informally called the men's faculty club, although the official name is The Faculty Club. This club, located in the heart of the campus, looks like a men's club, definitely masculine with the look of a hunting lodge. There is a lot of dark wood, including wainscoting. There are animal faces carved into the ends of large ceiling beams and mounted animal parts that include a moose head, deer heads, and antlers. There is a tavern in addition to food and hotel services, and a gift shop where official Berkeley ties are sold.

The WFC was conceived in 1919 as a reaction to the exclusion of women from The Faculty Club. Between 1902 and 1923 it was very hard for academic women to walk by The Faculty Club and watch men going into their exclusive male-only meetings and gatherings, and so in 1919 the women decided to do something about it. They didn't have the influence to open the men's club to women, but they could build a club of their own, and they did. The women had fund-raisers and built the WFC a stone's throw away from The Faculty Club. It was the first, and for many years only, faculty club for women in the US. The WFC looks like a European pension, with an international atmosphere. No beams with carved animal faces. No mounted animal parts. No tavern. But bright rooms, healthy food, and good hotel services.

In all the 14 years Betty was a Berkeley student, she didn't have a woman professor. But she did have a few women substitute teachers, one of them being mathematician Pauline Sperry. Remembered as being "militantly anti-male," Sperry believed that "women owed something to each other as separate from what they owed to humanity."[1] Sperry was always recruiting for the WFC.[2] In fact, it was Sperry who invited Betty, along with a few other graduate students in the physical sciences, to lunch in the club's dining room and encouraged her to join, and Betty joined in 1941.[3]

In 1969, over 65 years after its founding, The Faculty Club was still open only to men. The club president decided at the turn of the New Year that it was time to solve the worst of the club's problems, which was that young faculty members were underrepresented within their membership. What he meant was that young male faculty members were underrepresented within their male membership. He accordingly wrote a letter asking department chairs to help him by reminding their wives about an upcoming party, "which should be a humdinger" with cash bar, optional dinner, and dancing. The club mailed out invitations to all the young men they could identify as eligible for membership, including assistant professors and certain research personnel. The idea was that wives could telephone their networks and thereby increase the party turnout. The president asserted that "...this is a way both to help the Club and make your department more friendly and convivial. A few minutes of your time could make a big difference for the Club."[4]

Betty received the president's letter as statistics department chair. She didn't have a wife to remind about an upcoming club dinner, and she was livid the president of The Faculty Club would ask her and other women to help recruit for a male-only organization. She fearlessly wrote back to him: "You should know that I am not allowed to join the Faculty Club. I want to tell you that this is a disgraceful situation. I am not going to encourage anyone to join a Faculty Club that practices discrimination. I hope that you will take some action." She copied WFC President Margaret Thal-Larsen.[5] Betty as a woman wasn't just barred from joining The Faculty Club, she had also been banned in the past from attending meetings there. Reports are that Jerzy Neyman used to sneak her in through the window.[6,7]

---

[1] Colson, Elizabeth. "Anthropology and a Lifetime of Observation," an oral history conducted by Suzanne B. Riess, Oral History Center, The Bancroft Library, University of California, Berkeley, 2002.

[2] Note: Sperry was an officer in the WFC who brilliantly co-handled its finances leading to an early payoff of the club's mortgage. She also organized a monthly dinner group at the club where research papers were presented and discussed.

[3] 1941 invitation to The WFC Annual Dinner to be held Saturday, November the first at half past six o'clock.

[4] 2/27/1969 letter from UC-Berkeley Physics Professor and The Faculty Club President John Reynolds to UC-Berkeley Department Chairmen.

[5] 3/3/1969 letter from Betty Scott to UC-Berkeley Physics Professor and The Faculty Club President John Reynolds.

[6] Brian S. Yandell, telephone interview with author, June 24, 2014.

[7] Kjell Doksum, e-mail to author, December 4, 2014.

Betty's letter, it seemed, had an immediate and profound effect. At their next meeting, the all-male board of directors of The Faculty Club acknowledged that, in spite of a recently established exchange of dining privileges, having separate clubs for men and women led to problems. They favored a "Federated Faculty Clubs of Berkeley" that integrated program privileges between the clubs. The president wrote in his 'Notes from the Board,' "...as a first step, unless the minority who still favor the old idea of the Faculty Club as a masculine retreat comes on terribly strong against our plan, the Board proposed that restrictions against ladies in the Great Hall, and in the Member's Lounge be dropped." He offered the disclaimer that having two clubs was "an accident of history that your writer knows little about" and referred to faculty women as "the fair sex" or "the ladies."[8] So there was still some distance to be traveled, but it did seem that Betty's letter caused a measurable elevation in consciousness among the men on the board.

Thal-Larsen asked Betty: "Do you suppose your communication sparked the discussion noted in the attached record of the men's last meeting of their board?"[9] Thal-Larsen's question had to be rhetorical. The action came within one month after Betty wrote her letter of protest.

13.2 Berkeley Academic Senate Subcommittee

By April, women on campus who taught, either within regular or temporary positions, began to organize themselves as the *Berkeley Faculty Women's Group*, later the *Women's Faculty Group* (WFG). Attaching to the WFCSC goals, the WFG was concerned with the status of women on campus, including problems with their recruitment to and advancement within the faculty, and their admission to graduate school. There was no obvious source of funding for their efforts, but it did not deter the group from moving ahead.

The WFG planned for regular meetings every two weeks over lunch at the WFC. A steering committee was appointed that overlapped the WFCSC: Diaz as chairman, Kay as vice chairman, Betty and Sociology graduate student Arlie Hochschild as members at large, and Social Scientist Ellen Gumperz as secretary.[10] Hochschild would later become a professor and celebrated author at Berkeley.

The WFG steering committee got right to work setting up committees

---

[8]n.d. 'Notes from the Board' of The Faculty Club written by UC-Berkeley Physics Professor and The Faculty Club President John Reynolds.

[9]4/7/1969 memorandum, UC-Berkeley Institute of Industrial Relations, Industrial Research Specialist, Margaret Thal-Larsen to Betty Scott re "Notes from the Board" of The Faculty Club.

[10]Eleanor Swift, "Oral History of Herma Hill Kay," *California Legal History* 8, (2013), 1.

and defining the charge.[11] Betty's handwritten notes indicate the group had interests in organizing graduate women to help orient undergraduate student women; helping to orient new women staff members; and connecting with national meetings, collecting case histories, and maintaining master files relating to their goals.

It was decided Betty would obtain statistics of women in academic departments across the campus.

Lucy Sells (1933–2014),[12] who in 1967 was a sociology graduate student working on a dissertation involving related issues, was asked to help with acquisition of dropout rates for women versus men students; as well as proportions of women among the staff, graduate students, undergraduate students, and researchers.[13] The Sociology graduate students had just founded the Women's Sociology Caucus, and members were beginning to build friendships, provide education, and share problems among themselves. They were beginning to realize sex discrimination was so pervasive they themselves often didn't know that it existed, and problems all women faced had social rather than psychological roots. They would take action over the next year by organizing women at the national American Sociological Association convention, organizing a woman's conference on the Berkeley campus, and negotiating with the UC-Berkeley Sociology Department to hire more women faculty. At the time, there were 25 men on the sociology faculty, and no women.[14]

Importantly, Kay took the action she promised the WFCSC in 1968: She went to the senate policy committee of the Berkeley division of the academic senate and said, "Why don't you guys do something about women faculty?" Under the leadership of Law Professor Sanford (Sandy) Kadish (1921–2014),[15]

---

[11]Scott, Elizabeth L. "An Interview with Elizabeth Scott," an oral history conducted by Suzanne B. Riess, in "The Women's Faculty Club of the University of California, Berkeley, 1919-1982," Oral History Center, The Bancroft Library, University of California, Berkeley, 1983.

[12]A.G. Jennings, "Lifelong Berkeley resident and activist dies at 81," *The Daily Californian* (Berkeley, CA), February 24, 2014. Note: Lucy Sells was a political activist who had a significant effect on the gender equity movement. She granted an interview for this book and passed away as it was being written. Her obituary states: "In the 1970s, Sells coined the phrase 'critical filter' to describe the significant role that high school mathematics plays in preparing women and minorities for college-level mathematics and, correspondingly, careers in science – and mathematics – based fields. In her dissertation research, Sells found that women and minorities were vastly misrepresented in high school science and math classes and were therefore grossly underprepared for basic calculus classes at UC Berkeley, in comparison to first-year male students. Sells' research helped launch a national response that included funding for additional research on women in math and science. 'She was this master of taking a well-chosen statistic and trying to use it to mobilize people to take action,' said Elizabeth Stage, director of the Lawrence Hall of Science and Sells' former colleague. 'Lucy was indomitable and relentless.'"

[13]4/22/1969 handwritten notes by Betty Scott re Meeting of Steering Committee.

[14]"Women: Breaking the Shackles," *The Every Other Weekly* (Berkeley, CA), January 27, 1970, 3.

[15]Note: Sanford Kadish was a criminal law theorist. He granted an interview for this book and passed away as it was being written. He served as dean of the UC-Berkeley Boalt School of Law from 1975–1982.

the committee affirmatively formed a subcommittee to thoroughly investigate academic women's issues.

The policy committee presented its 1969 state-of-the-campus message to the academic senate on May 6. Citing differential treatment of women at Berkeley, it announced the appointment of the *Subcommittee on the Status of Women* (Policy CSAW):

> *It is surprising that so few women – only 15 at the present time – achieve the rank of full professor at Berkeley. A relatively small number of women are enrolled in graduate schools on this campus and elsewhere. All too frequently women who intend to pursue academic careers have been forced to adapt themselves to the uninterrupted training and apprenticeship patterns established by men with consequent loss to themselves as women. The recognition of this choice has itself discouraged many able women from seeking academic careers with the consequent loss to the world of scholarship.*

Asked 40 years later why his committee hadn't previously considered equality of women, Kadish recalled members were so busy thinking about discrimination against ethnic minorities that they just didn't think about women.[16]

It was no surprise that Betty was chosen for the Policy CSAW. The senate policy committee was aware of Betty, as she served on the committee just two years prior (May 1966–January 1967). Also, Betty had a relationship with Kay through both the WFCSC and steering committee of the WFG. Policy CSAW membership included both men and women. Anthropology Professor Elizabeth Colson (1917–) was appointed chairman [sic]. Sociology Profesor Herbert Blumer (1900–1987), Speech Associate Professor Susan Ervin-Tripp (1927–), and Law Professor Frank Newman (1917–1996) were also appointed. It was around this time that Newman introduced the emerging field of human rights law to the UC-Berkeley curriculum. Kay served as liaison between the Policy CSAW and the parent committee on senate policy.[17]

Betty received a letter from Kadish at the end of April confirming her agreement to serve and suggesting the Policy CSAW concern itself with recruitment and advancement of women faculty, and admission of women to graduate school. The letter then offered the subcommittee carte blanche by encouraging it to adopt its own priorities and concerns. It set a time line so that the senate policy committee could include the subcommittee report in their state of the campus message in the early fall of 1969.[18] That was only six months time. Betty responded she would be happy to serve, as she had come to realize over the past few years that "a difficult problem does indeed

---

[16]Sandy Kadish, interview with author, June 17, 2010.

[17]5/6/1969 State of the Campus Message from the Committee on Senate Policy to the Berkeley Division of the Academic Senate.

[18]4/30/1969 letter from UC-Berkeley Law Professor Sanford H. Kadish to Betty Scott.

exist." She told Kadish she thought the Policy CSAW could work with the short time line, as the work had already been started within the WFG.[19]

The first meeting took place on May 19th in Colson's office.[20] The subcommittee brainstormed about ideas. They should look into problems involving the timing of children, concerns of graduate students, and issues around age at school. The subcommittee brainstormed about information sources. Try the UC-Berkeley vice chancellor of academic affairs (VCAA); dean of the graduate division; former dean of the Harvard Law School who was a member of the Carnegie Commission, an influential higher education think tank; and professional societies. Look at materials from the UN Commission on the Status of Women and Women's Bureau in the Department of Labor. Look at the California Constitution: It has provisions against discrimination of women.[21]

13.3   Subcommittee Data Collection

Betty telephoned the VCAA's office the next morning, asking for information about the relative advancement, including speed, of women versus men at UC-Berkeley. The secretary said they didn't collect this information directly; there were complications; a comparative study would be needed; and there were confidentiality issues. She told Betty they were willing to cooperate, but Betty should first telephone the chair of the academic senate budget committee to see what information they could make available. The chair responded to Betty that he didn't know of any study on the relative advancement of women versus men on the Berkeley campus, but he would take up the matter with his committee when they met in a few days.[22]

Sells consulted the centers for law and society, and survey research, about collecting data from women professionals on campus for the report. She was advised it would be "financially, temporally, and methodologically impossible to design a valid, reliable questionnaire, get it reproduced, collated, addressed mailed [sic], returned, coded, key-punched, analyzed, and summarized by June 30." In those times, computing was expensive and slow. The time-cost estimates given to Sells were prohibitive.

---

[19]5/5/1969 letter from Betty Scott to UC-Berkeley Law Professor Sanford H. Kadish.

[20]n.d. (5/1969) memorandum from UC-Berkeley Anthropology Professor Elizabeth Colson to members of the Policy CSAW re meeting of the committee.

[21]n.d. handwritten notes by Betty Scott re first meeting of the Policy CSAW. Note: The VCAA was William J. Bouwsma (1923–2004); the dean of the UC-Berkeley Graduate Division was Sanford S, Elberg (1913–2011); and the former dean of the Harvard Law School was Pat Harris (dates unknown). Betty would later be asked by the Carnegie Commission to do research involving academic women nationwide.

[22]5/20/1969 handwritten notes by Betty Scott re telephone calls to VCAA William J. Bouwsma's office and to UC-Berkeley budget committee chair John Raleigh.

Sells accordingly recommended asking professional women for more targeted information about their work history, specific examples of discrimination against women, and special needs of women that should be addressed. She suggested the campus' official bio-bibliography form be sent to all professional women (what she counted to be 44 tenure ladder faculty, 233 lower level teaching faculty, and 234 professional researchers). She suggested a separate sheet be used to ask open-ended, exploratory questions about graduate school and professional life, whether the woman had any suggestions for structural changes in professional life, and whether the woman had any strong feelings about an issue and would like to be interviewed. "We can learn more about the relevant dimensions of 'discrimination' from fifteen or twenty case studies than we can from 500 questionnaires in which we are not yet clear what the right questions to ask are... we can learn more from a few people with creative ideas on the subject than we can from 500 'don't know's.'" Sells also suggested the Policy CSAW tap into the wealth of existing data "... and it should be exploited to the hilt." She saw this as the area in which she could be most useful, for example, in accessing the NAS database of Berkeley PhDs, and the Berkeley Graduate Division database.[23]

It turned out the budget committee had already started a study that could be useful to the subcommittee. They divided women by "department area" and were doing a record study to see if the progress of the women looked "normal." The chair offered his suspicion that they would find women advancing more slowly than men due to two factors: "prejudice against women" and women "produce less." He gave Betty figures on assistant professors, associate professors, and professors so she could calculate the total full-time equivalent (FTE) number for each gender. Betty's handwriting got significantly larger as she got to the percentage of total FTE for women. It was only 3%! This was a smaller percentage of women faculty members than UC-Berkeley had over much of the century. There were 1,721 full time equivalent tenure ladder faculty members on the campus, and out of them only 60 were women![24]

The Policy CSAW took Sells' advice and sent a letter to the women professionals in June. The letter started out with the 3% statistic and explained that while the UC-Berkeley faculty grew rapidly over the past decade, the number of women dropped; the proportion of women dropped dramatically; and this was a matter of much concern to the senate policy committee. The letter informed the women of the existence of the Policy CSAW and its mission to improve opportunities for qualified women to have academic careers. Attached to the cover letter was the bio-bibliography form, and a page with the open-ended, exploratory questions.

The bio-bibliography forms were going to be placed in a central file and used to examine career patterns over time, even beyond the deadline for the

---

[23] 5/27/1969 memo from UC-Berkeley Sociology Graduate Student Lucy W. Sells to the Policy CSAW.

[24] 11/4/1969 memo from Ellen Gumperz to Herma Kay, Hanna Pitkin, and Elizabeth Scott re Minutes, WFG Luncheon, 10/31/1969.

Policy CSAW report. This resource would help promote equal opportunities for women in promotion and tenure, and locate points of difficulty in the professional ranks. The one-page questionnaire was meant to suggest issues for possible elaboration, either by letter or interview: "Any advice you can give us on matters which you think pertinent to the inquiry is welcome... You may know of instances where you feel that you or other women have been treated unfairly either in working for a graduate degree at Berkeley or in obtaining appointment or promotion. You may know of informal arrangements within your professional society or in the structure of your profession which make it difficult for a woman to build a reputation and acquire recognition. You may have suggestions for changes in academic routines which would help meet special needs of women academics. We invite you to write as fully as possible." The Policy CSAW gave several assurances of confidentiality.[25]

Also in the summer of 1969, the VCAA solicited comments from all deans, directors, and department chairs on behalf of the Policy CSAW. He asked for their views on the disadvantages and advantages of having women colleagues and their suggestions on how to improve the status of academic women:

> *The small proportion of women who receive academic recognition at Berkeley could be interpreted as an indication of the poor training which Berkeley and other major universities are providing for women students, or as due to a lack of interest among women in academic careers. It may also reflect selective admission to graduate studies and discriminatory hiring policy for academic posts. These are matters largely controlled by departments since they admit students and initiate hirings and promotions. It would be of great interest to the special committee, therefore, to know whether you encourage your able women majors to go on to graduate school? Do you help your able women graduate students to compete for jobs at major universities where they could become visible in the academic world? In your search for new faculty do you ask colleagues to submit the names of their able women students along with the names of their able men? Some departments seem to appoint women only to lectureships on an annual basis or in non-tenure positions. In following such practices, departments no doubt have their reasons. It is in the interests of the academic community that these should be made explicit so that they can be subject to examination and the test of research.[26]*

A similar follow-up letter would be sent a few months later by the person who had been the budget committee chair and was now the new VPAA.[27]

---

[25] 6/1969 letter from members of the Policy CSAW to unspecified women professionals on the Berkeley campus.

[26] 6/4/1969 memorandum from VCAA William J. Bouwsma to All UC-Berkeley Deans, Directors, and Department Chairmen re their views on the advantages/disadvantages of having women colleagues and their suggestions for improving the situation.

[27] 10/8/1969 memorandum from VCAA John Henry Raleigh to All UC-Berkeley Deans, Directors, and Department Chairmen re their views on the advantages/disadvantages of having women colleagues and their suggestions for improving the situation.

A second part of the Policy CSAW's charge had to do with admitting women to graduate school. Here they took advantage of the invitation given them to adopt their own concerns. They were concerned with more of the pathway than just the entrance. They were also interested in women's progress through graduate school and recruitment into academic positions.

Betty and the Policy CSAW had their hands full, and they quickly got to work. They held discussions with women graduate students in several departments to get ideas for a survey. Betty prepared a three-page pilot survey based on an earlier survey of women graduate students in political science and sent it to departments with which she had the strongest experience and connections: astronomy, mathematics, and statistics. Betty had the impression that women were experiencing different difficulties in different academic areas.

Betty was extremely concerned with, and careful about, confidentiality issues. At the bottom of the survey, Betty indicated:

> *If there are some issues about which you have strong feelings, I hope that you will contact me or otherwise provide more detailed information. If you wish you may send me an anonymous statement of any particular incidents you may know of.*

In her cover letter to the pilot survey, she indicated:

> *... you may wish to provide a confidential answer on a separate sheet. Anything that you wish to have held confidential will be. These data will be used only as they apply to the entire area... unless there is some help or special difficulty on which you want the committee's assistance... we will be sure that no individual woman can be hurt by our report.*

Betty also offered to meet with any women who were interested or willing, either over lunch or coffee, or in her office.[28]

In June, the VCAA offered Betty reappointment as chairman [sic] of the statistics department. As a humble servant, she said: "I accept and will try to do my best." Betty had established a good working relationship with the VCAA on behalf of women, and she told him she admired his position "on the issues that beset us."[29]

Colson headed to Minnesota in mid July. Before leaving, she wrote to Betty reporting on Policy CSAW activities: questionnaire returns, interviews, attempts to get into departments' files, and graduate assistant help for accessing these files. Colson included a handwritten note, saying they could write their report praising the cooperative stance of departments, with the exception of one department "which shudders at the mention of women." Colson concluded by saying she had told helpers Lucy Sells and Lucy Stout to report to Betty, and thanked Betty for taking over the Policy CSAW work for

---

[28]6/6/1969 letter from Betty Scott to UC-Berkeley women graduate students in Astronomy, Mathematics, and Statistics.
[29]6/25/1969 letter from Betty Scott to VCAA William J.Bouwsma.

the summer.[30] Colson later asked that everything pertaining to the Policy CSAW be sent to Betty. Colson's parting words to Betty were: "I hope all goes well."[31]

Betty immediately met with Sells and Stout. Betty already knew Sells and was very interested in the information she was collecting, thinking it would be useful. It had to do with the progression of women, compared with men, up the ranks in selected departments. Betty advised Sells to include temporary appointments among her numbers, thinking these would be one of the biggest differences between the genders. Stout made a positive first impression. Betty found her to be "full of steam" and having "lots of ideas." They made plans for Stout and a work-study student from the statistics department to get student names and addresses from files of departments across campus. They had to work especially hard to get student addresses from the departments. Betty wanted to make some changes in the questionnaire that would "further encourage replies" before they were sent out to students.[32]

The pilot survey was turned into a five-page survey for women graduate students and recent PhDs in other selected academic areas. It was thorough. It asked women graduate students for all kinds of information. Where and when she was born. Where and when she graduated from college. Whether, by whom, and where, she was advised to apply to graduate school. Where and when she was enrolled in graduate work. When she entered into graduate work at Berkeley. If she transferred to Berkeley, why she transferred. Financial support she received in graduate school, including if and when she was a teaching assistant or research assistant, and if and when she received a fellowship and what kind. Her degree goal and career objectives at and after graduate school admission. Degree, major, and dates of any graduate degrees she received. Dates she was admitted to any Qualifying Exam. Whether and with whom she discussed the possibility of a future academic position. Whether she was married and the marriage dates. Any help she had in maintaining the home. Whether she had children and the birth dates. If she had children, how they were cared for while she was on campus.

The survey also used open-ended questions to ask the women for their opinions and suggestions. Betty was especially interested in answers to these questions. Details of any unusual difficulties the woman had in being admitted to the Berkeley graduate school. Any special advice she received about future possible academic employment. Details of any handicaps she perceived in being a woman and getting financial support. Reason for any decisions to postpone or interrupt and then begin or resume graduate work, length of time of the postponement or interruption, and details of any difficulties she had

---

[30] 7/8/1969 letter from UC-Berkeley Anthropology Professor Elizabeth Colson to Betty Scott.

[31] 7/23/1969 letter from UC-Berkeley Anthropology Professor Elizabeth Colson to Betty Scott.

[32] 7/24/1969 letter from Betty Scott to UC-Berkeley Anthropology Professor Elizabeth Colson.

readjusting to academic routine. Details of whether she received any advice or suggestion about her progress or program as a graduate student that appeared to be predicated upon the fact that she was a woman. Any restrictions on the kind of position she would accept as a woman. Details of how marriage affected her academic progress or career goals. Details of how graduate study affected her marriage. Details on any additional work or time she needed to spend at home because she was a woman. How motherhood has affected her academic progress or career goals. Any suggestions for structural changes in the academic program, in admissions, in administrative policies and services, or other ways, which would help meet the special needs of graduate women. How she would like to use her graduate training, and how she planned to use it. Whether she had any other items or problems to report to the Policy CSAW beyond what was asked in the survey.[33]

Sixteen academic departments were asked to submit names of women who had entered within a window that extended two more years back than the typical time to obtain the doctorate in that department. Sells and Stout took on the job of mailing out questionnaires to the women graduate students in the departments. Some departments were easier to work with than others. Across the pilot and full surveys, a total of 1,208 women with working addresses were surveyed; 566 surveys were completed: 345 in a first mailing and 221 in a second. Betty hoped the 53% who responded were representative of the entire population. By the middle of August, Sells had collected enough data that she was ready to use the mainframe computer to make some data tables. Scott either met with or wrote to several of the students who wanted to talk with her.[34,35]

The women who were surveyed almost uniformly provided well-defined and articulated responses to the survey, with little whining. But the responses were so varied that Betty and others had difficulty summarizing them. Betty tried to summarize them in one draft paragraph:

> *...for both present (current) and past students... many were given positive advice to apply to graduate school, some received no advice on this subject while very few received negative advice. Almost no students thought that they had admission problems... few students think that being a woman made obtaining financial aid more difficult although quite a few said they did not try (usually because they thought there was no hope but sometimes because they did not need aid). Many students postponed or*

---

[33] n.d. (1969) survey of UC-Berkeley women graduate students in selected areas. Note: These areas included a sample of science, language and humanities departments, the School of Librarianship, and the School of Social Welfare.

[34] "Number of Women Married While in Graduate School and Number of Women with Children While in Graduate School" (data table, UC-Berkeley, 1969). Note: The departments were Astronomy, Bacteriology/Immunology, Botany, Engineering, English, French, History, Journalism, Librarianship, Mathematics, Music, Physiology/Anatomy, Social Welfare, Spanish/Portuguese, Statistics, and Zoology.

[35] 8/17/1969 letter from Betty Scott to UC-Berkeley Professor of Anthropology Elizabeth Colson.

interrupted graduate study, usually with readjustment problems following. The figures as to advice on progress predicated on being a woman show that almost no woman received positive advice while there were [sic] quite a lot of negative advice with the most answers being no advice at all. The difficulty with restrictions on the kind of job they will accept is pronounced: half the women do make restrictions, usually based on husband or children; women in professional schools make fewer restrictions which they explain by the fact that the positions they are training for are widely available.[36]

There were attempts by at least three others to write summaries of the survey responses (with 4-, 6-, and 13-page attempts). These focused more on the negative comments, and were looking for evidence of discrimination. Betty's brief summary was characteristically more balanced. Rather than approach the data from a viewpoint, she let the data speak for themselves. This set the tone for balance in the Policy CSAW's official report.

Betty was also interested in the status of academic women within the UC Retirement System. She reviewed their booklet, "Basic and Supplemental Plans" and, seeing indications of potential gender discrimination, wrote in August to the chair of the system's governing board asking for information. It appeared there were differences in benefits for widows and widowers; the tables indicated a female co-annuitant, and it wasn't clear what the situation would be if the co-annuitant was male; also, it wasn't clear what would happen when both spouses were members. Betty was uncovering gender inequalities and asked for some answers, with tact: "It is possible that the booklet is not clear."[37] The chair confirmed different actuarial tables were being used to determine pay for widows and widowers. He reported that adjustments were made for male co-annuitants to reflect the greater life expectancy of females.[38]

Betty decided to take the survivor benefits issues to the committee on university welfare, asking them to look into possible discrimination against women. She saw the inequities in the death benefits as archaic and hoped the committee could help effect a change. She saw the differences in rates as arguable: If the employee was viewed as purchasing something, this could justify higher rates for women; however, if the university was viewed as "spreading the risk," the higher rates did not appear to be justified.[39]

Betty also hoped the Policy CSAW would look into short-term disability income replacement benefits. While it is true that such benefits paid to women are greater than to men, it is largely because the woman takes sick leave when a child is sick. Since women are known to be generally healthier than men, she did not expect this pattern to hold after, say, 30 to 60 days.

---

[36]n.d. (1969) Survey Summary by Betty Scott.

[37]8/17/1969 letter from Betty Scott to UC Retirement System Governing Board.

[38]8/22/1969 letter UC Retirement System Governing Board Chairman Robert R. Headley to Betty Scott.

[39]12/18/1969 letter from Betty Scott to UC-Berkeley Law Professor Adrian A. Kragen.

Betty later volunteered to look into possible discrimination against women in insurance policies. She found no evidence in employee life, accidental death and dismemberment, or dependent life insurance policies. Nor did she find evidence in the regents' death benefit allowance. Betty did, however, find inequities "against women (attributed to traditional family patterns)" in the retirement system's death benefit schedule. She also found that "the actuarial tables used in the Retirement System and the Short Term Disability Plan depended on sex, with women paying more."

14

Making Visible
(October – December 1969)

I am afraid that we asked too many questions!...I hope that what
we are outlining is not too much work for your busy committee.
-Elizabeth L. Scott

Institutional research became critical toward mandatory reporting of affirmative action progress to the federal government. Betty was deeply involved. She began this work in 1969 when asked to be a member of a UC-Berkeley academic senate subcommittee to study the status of academic women. She continued this work for the rest of her career without an official administrative position. She approached institutional research from within the faculty as a service toward the greater good while respecting administrative channels. At the time this book was written, the campus institutional research office with a staff of 13 had a mission to help make campus decisions better.[1] In 1969, Betty with a staff of less than one embarked on a personal institutional research mission to improve campus decisions. She did this within a climate of partial gender segregation within the faculty.

14.1   Subcommittee Research

At the end of summer 1969, Betty was getting ready to be out for a month. She took a large carton, placed it on a chair in the corner of her office farthest from the door, and created folders for each department. Betty put all the files related to the Policy CSAW work in the same corner, behind the telephone on the windowsill. She arranged for her assistant to look at her mail and file incoming completed questionnaires in the folders.

When she returned, Betty approached the budget committee office. They suggested she articulate specific questions the Policy CSAW was interested in. Betty offered the following:

---

[1] "Directory," *UC-Berkeley Office of Planning and Analysis*, accessed October 27, 2013, https://calnet.calnet.berkeley.edu/directory/details.pl?uid=93688.

1. *Advancement*. Do women advance through the ranks at the same rate as men? There are examples of women who were promoted to the professorship (or associate professorship) the year before retirement. This seems atypical. Is it relatively restricted to women?

2. *Employment policy*. Do some departments have a policy of not considering women for tenure positions or even for ladder positions? Do some departments have a quota limiting the number of women to be considered? Do some departments, such as Design and Nutritional Sciences, have a policy of preferring men appointments in order to "improve" their image? We note that on the list of new appointees for the year 1968–70 the great majority of women are lecturers or instructors whereas few men are at these ranks. Further, looking at all positions held by women on the academic staff in 1968–69, it appears that about half are lecturers or instructors; the proportion of men holding such non-ladder appointments is tiny.

3. *Research*. Do women publish at the same rate as men at the same rank in the same field? Does the quality of women's publications appear to be about the same as men's in the same field at the same rank?

4. *Teaching*. The quality of women's teaching? Are women less likely to be assigned to graduate teaching?

5. *Committees*. Do women serve on important Senate committees and important administrative committees at the same rate as men of the same rank? When a woman's ad hoc committee is appointed, does this committee have at least one woman member? We are informed that there once was such a rule. Do women have the same load of small administrative duties such as, for example, serving on departmental fellowship committees?

6. *Permanence*. A question which is raised very often and which your Committee may be able to answer, at least partially, is concerned with the relative permanence of women faculty members. We presume this should be answered separately for different ranks. Thus, for a given rank and a given field, do women faculty members remain at Berkeley for shorter or longer periods than men faculty members? We know of examples of women who resigned to "follow" their husbands (who also resigned) and of examples of men faculty members who resigned for reasons connected with the health of his wife or divorce from his wife, etc. Women faculty members are often told that they were not even sent offers to go to another university because it is assumed that they will not consider leaving Berkeley where their husbands are. (We suspect that this censoring of offers makes it more difficult for women to obtain increases, especially jumps, in salary).

7. _Discrimination_. *Are there specific cases of discrimination against individual women faculty members at Berkeley? If so, is there anything that the Subcommittee for the Status of Women should report in an effort to alleviate and indeed eliminate discrimination?*

Betty asked the committee to partner with the Policy CSAW to refine this set of questions, knowing they had just undertaken a study of academic staff that investigated the relative progress of men and women in the same fields. This experience could be helpful to the Policy CSAW in shaping their questions.[2]

Lucy Sells pursued information on PhDs. She obtained a computer tape from the graduate division that contained data submitted to the NAS-NRC's "Survey of Earned Doctorates." The data had been de-identified, so there were no concerns about confidentiality. Sells sent a letter to "Miss Scott" informing her about what was on the tape.[3] There were lots of variables. The aim was to determine whether there were differences between men and women on these variables. Sells accordingly provided to Betty a number of useful two-way tables. There was a table for each of the following variables cross-tabulated by gender: field, citizenship, number of dependents, marital status, type of high school, employment status, employment prospects, type of doctoral employment, work experience, function in postdoctoral job, father's education, mother's education, years of full-time undergraduate work, years of part-time undergraduate work, years time out after the BA, years full-time graduate, years out of graduate school, and decade of birth.

Sells asked what additional analyses she could do for Betty. From the data on the tape, they produced a table: "Number of Doctorates Awarded to Men and to Women by Decade and Field." They also produced two graphs of the "Distribution of Number of Years to Doctoral Degree" by gender. One was for 1957–62, the other 1963–68; they counted each part-time graduate student as one half of a full-time student. The table and graphs would appear later in the Policy CSAW's official report.

Betty got into her own hands the data tape Sells was using. She created a printout, and it was huge. She discovered the data set prior to 1957 was too sparse to be useful. It contained only year of degree, institution that granted the degree, and gender. Betty planned to make a new tape that contained the subset of degrees granted in or after 1957, where there was enough information to be useful to the subcommittee.[4]

There was a growing professional literature on the status of women in higher education, and Betty was devouring it. She had seen a chapter and an appendix from a forthcoming book, *Human Resources and Higher Education*, that particularly interested her.[5] She was especially struck by the very large

---

[2] 10/13/1969 letter from Betty Scott to UC-Berkeley Chemistry Professor David H. Templeton.

[3] 10/23/1969 letter from UC-Berkeley Sociology Graduate Student Lucy W. Sells to Betty Scott.

[4] 1/6/1970 letter from Betty Scott to members of the Policy CSAW.

[5] J.K. Folger, H.S. Astin, A.E. Bayer, *Human Resources and Higher Education: Staff Re-*

percentage of women who continued to be employed some seven years after earning their doctorates. She didn't expect the percentage to be that high.

Betty also observed the book contained some linear regression analyses of salary, using factors such as sex, age, institution, and some measure of quality (e.g., number of citations in the citation index). Betty didn't have access to these chapters or the data, but she was keenly interested in learning more, because she suspected some of the factors used in the models interacted. She was referred to Helen Astin (1932–), a faculty member in higher education at UCLA who had done additional analyses and reportedly would soon publish them.[6] This was the beginning of Betty's involvement with regression analyses of salary, an involvement that would continue for the rest of her career.

14.2   Women's Faculty Group Debriefing

The WFG was still going strong. Twenty-three women attended the lunch meeting at the beginning of November. The group decided to add two new members to the steering committee; they should be from the physical or natural sciences, if possible, because May Diaz was on leave and Arlie Hochschild had taken a position at Santa Cruz. Over the next few months, Political Science Associate Professor Hanna Pitkin and Public Health Clinical Assistant Professor Mary Murai were added to the steering committee.

Colson gave a progress report on the work of the Policy CSAW. She reviewed the summer data collections and indicated the final report would deal with women wanting to compete equally with men. The report would emphasize safeguards for all women, full time and part time. For all women, there needed to be maternity and pregnancy leaves, and access to university childcare centers. For those women who chose to work full time, there needed to be relief from the nepotism rule. For those who chose to work part time, there needed to be regular personnel reviews. The report would emphasize that women researchers needed to be able to be principal investigators and get credit for their own work: At present, only members of the academic senate were allowed to assume the role of principal investigator; many women researchers were not members of the academic senate and so were required to work under a nominal principal investigator who was a member of the senate; and this requirement handicapped these women from receiving appropriate recognition for their research and contributions to the university's reputation. Colson emphasized that women needed to be taught strategies in order to compete successfully. The report accordingly would "urge women faculty to

port of the Commission on Human Resources and Advanced Education (New York: Russell Sage Foundation, 1970).

[6]H.S. Astin, *The Woman Doctorate in America* (New York: Russell Sage Foundation, 1969).

discourage the overly simple idea in women graduate students that simple merit gets rewards." Betty reported on the Policy CSAW's efforts to discern problems felt by graduate women. Based on the returns received, Betty was able to cite some major problems: need for childcare, sense of isolation, and resentment of attitudes of significant others.

Other items were discussed: women's caucus activities connected with professional associations; whether the WFG should support proposals for campus childcare centers; possibilities of small membership dues; improving employment benefits; and a protest on the policies of The Faculty Club. They also discussed relationships with graduate women's groups, reactivation of committees, and establishment of co-educational swimming hours in the winter.[7]

14.3 More Research

Betty had seen an article, "Unequal Peers: The Situation of Researchers at Berkeley,"[8] published the year before. She and the Policy CSAW viewed the situation of women researchers to be particularly unequal.[9] Colson wrote a letter to the UC-Berkeley Vice Chancellor for Research (VCR) on behalf of the Policy CSAW requesting that qualified women researchers be permitted to be principal investigators. She observed the nepotism rule often prevented fully qualified women from being employed in regular faculty positions.[10]

Betty investigated in detail the nepotism rules. She spoke to the campus research office and learned they merely flagged cases of suspected nepotism before referring them to the chancellor's office. Betty then spoke with the VCAA's office and learned nepotism rules were put into place during the depression. The UC system president delegated authority to make exceptions to the campus chancellor in July 1966. The nepotism policy applied equally to all researchers, whether or not they were members of the academic senate. The procedures for exceptions were pretty basic. The unit head (chair, director, etc.) needed to request the exception and include an explanation, to which the campus chancellor would either approve or disapprove.

The assistant in the VCAA's office asked Betty if she thought the nepotism rules discriminated against women. Betty thought they did. But, Betty went on in her characteristically analytical and even-handed fashion to say that:

---

[7]11/4/1969 memo from Ellen Gumperz to Herma Kay, Hanna Pitkin, and Betty Scott re Minutes, WFG Luncheon 10/31/1969 with attachment of committee memberships from spring 1968.

[8]C.E. Kruytbosch and S.L. Messsinger, "Unequal Peers: The Situation of Researchers at Berkeley," *The American Behavioral Scientist* 11, (1968), 39.

[9]11/5/1969 letter from Betty Scott to UC-Berkeley Anthropology Professor Elizabeth Colson.

[10]n.d. (11/1969?) letter from UC-Berkeley Anthropology Professor Elizabeth Colson to VC L. Sammet.

It is not at all clear that the University itself discriminates against women rather than just against one of the two related persons. If the University is going to force either the husband or the wife to work without salary or to work in a less secure position, such as a research institute, our social order is such that the husband will take the well-paid secure position and let the wife manage with no job or no salary or, at best, a year-to-year appointment.[11]

Betty had seen first-hand evidence of this type of discrimination.

Betty heard back from the budget committee chair. He addressed three of her seven questions.

Regarding advancement in faculty ranks, the chair said there was so much variability among both men and women that it would not be meaningful to compute an average rate of advancement. Nevertheless, they did compare women and men full professors, ignoring department affiliation, and found on average: On the basis of age, women have one step lower salaries; and on the basis of date of PhD, one half step lower. He did not know of examples of women who were promoted the year before retirement, and he did not have an easy way to search the emeritus professor files for such cases.

Regarding employment policy, the chair wasn't aware of any departmental policies that excluded women for tenure ladder positions, with the exception of physical education. He was aware some departments had never appointed women and some had made frequent, but temporary, appointments to women. He then offered that there was a system policy, not a department one, which discriminated against women. It was the nepotism rule. The rule didn't distinguish men and women, but in practice it worked against women:

In the opinion of the Budget Committee it is an obsolete rule which results in a waste of talent and which works an unnecessary hardship in an age when marriages between professional colleagues are increasingly frequent. This rule has been the direct reason for termination or exclusion of some women from our faculty. It has been a source of difficulty even in some cases where exceptions have eventually been approved.

The Policy CSAW and budget committee were aligned in supporting repeal.

Regarding committees, the chair said they had no reliable statistics on committee service, nor were they aware of any rules about women on ad hoc committees. The impression was that women were being treated fairly:

We have the impression that the lack of women in the tenure ranks results in part from a small supply of good candidates. Since we never see the dossiers of persons not recommended by department chairmen [sic], we have no basis for judging to what extent prejudice or disinterest causes women not to be proposed for appointment. We do see cases of women who would be expected to advance through the tenure ranks but who resign

---

[11]n.d. [11/6/1969?] memo from Betty Scott to the file.

> *to follow their husbands elsewhere or to cope with family problems. Men also leave for personal reasons, but we have the impression that a smaller fraction of men than women do so.*

He recognized these feelings needed to be backed by reliable data.[12]

Susan Ervin-Tripp thought a study of the "durability" of women academics was needed. She was impressed with the "frequency with which people mention the temporariness of women teachers." This was a primary argument used by some to rationalize why women shouldn't be hired into permanent positions. But was it true? A study would tell the true story. Ervin-Tripp's idea was to match individual female faculty members with a male at the same rank and in the same department; then the length of stay of all the women could be compared with that of all the matched men. The matching would make for a valid comparison as controlled by matching factors. Ervin-Tripp wanted this study to be part of the Policy CSAW's report.[13]

Ervin-Trip used data from the General Catalogues of the university, accessing them in the Main Library for a 17-year period (1951 through 1967–68). She tried to identify all new women assistant professors, then selected a group of males from the same department as each woman; for the women and men, she extracted the starting year, the last year as assistant professor, and whether or not the person was promoted to associate professor. She tried to do the same for associate and full professors. Ervin-Tripp was able to compute basic measures of differences between the men and women for duration, promotion, and years in rank. A limitation was that the data in the catalogues weren't very accurate. They didn't reflect all appointments or promotions. Furthermore, there was no information about ranks and steps, salaries, or ages. Betty, Ervin-Tripp, and other Policy CSAW members hoped they could get more accurate and complete information from the budget committee.[14]

Betty now understood the budget committee could provide information on individuals with assistant professor rank and above, but she would have to look elsewhere, perhaps to the deans, to get information on lecturers and acting instructors. The Policy CSAW was particularly interested in the issue of advancement that Betty had outlined earlier to the chair, as well as Ervin-Tripp's issue of durability. Betty didn't just want to work with simple unadjusted averages, she wanted data on individuals over time. She approached the chair with tact, saying that his letter was very helpful to the Policy CSAW: "I am afraid that we asked too many questions!...I hope that what we are outlining is not too much work for your busy committee."

Study bias was a significant concern for Betty. Like most statisticians, she wanted to avoid anything that could be construed as systematic deviation

---

[12] 11/1/1969 letter from UC-Berkeley Chemistry Professor David H. Templeton to Betty Scott.

[13] n.d. (8/1969?) letter from UC-Berkeley Rhetoric Professor Susan Ervin-Tripp to Betty Scott.

[14] 12/8/1969 letter from Betty Scott to UC- Berkeley Chemistry Professor David H. Templeton.

from the truth.[15] In this case, Betty was concerned with bias in how faculty members were selected from the database. She knew it wouldn't be sufficient to study only those who were at Berkeley at the time. These were "successes." She knew she also needed to study those who entered teaching but had left. These were "failures." She wanted to observe faculty members from some fixed date, as early as possible in the budget committee records.

Betty gave more thought to study designs using matching that could estimate differences in advancement or duration due to sex: She already knew there were differences between departments in how women were treated, so a man matched to a woman would have to be in her department. The man should also have started at Berkeley around the same time as the woman, at the same rank and step within rank; and he should have been around the same age and with around the same number of years since the PhD. Betty recognized in many cases it would be hard to get a good match. In such cases, she thought they should take several men who could provide a balance among rank-step, age, and time since PhD. This would produce a group who were similar to the woman with respect to the balancing variables. The women could then be compared with the men to see if there was evidence that, on average, men continued teaching at Berkeley longer than women, or men advanced in rank-steps at a more rapid rate than women. Betty also hoped to get information from the budget committee about reasons for leaving.

Betty continued to be sensitive about data confidentiality issues. She thought, since there were so few women on the faculty, she would be able to identify some of them from the raw data, and she didn't want to chance such a breach of confidence. She therefore proposed the budget committee give her only a table of differences, rather than all the raw data. The table would have a row of information for each woman, but would not include the woman's name or her department. The difference measures (male minus female) for each woman would include: years duration at Berkeley; whether promoted to associate professor ('1') or not ('0'); if promoted to associate professor, years from assistant professor step II to associate professor; number of steps from start to present; and increase in salary over the previous six years.

Betty wasn't afraid to put her own case forward as an example. She wanted the committee to be perfectly clear about what she was asking. Her example showed her difference measures relative to five men who became statistics assistant professors around the same time as she. It indicated she had been at the university 5.7 years longer than the average of the five men; she had been promoted to associate professor but only three of the five men had been promoted; the men had advanced an average of 2.3 more steps than she; and the average increase in salary among the men over the previous six years exceeded hers by $1,067 (over $6,000 in 2010 dollars).

---

[15]David L. Sackett, "Bias in Analytic Research," *Journal of Chronic Diseases* 32, (1979), 51.

14.4   Nearing End of Data Collection

By late November, the Policy CSAW had collected much of the information they thought they needed for their report.[16] They had surveys of women graduate students, recent PhDs, faculty members, and senior research staff members. They had data from the graduate school, as well as information from department chairs and deans solicited through the chancellor's office. Additionally, they had all kinds of reports on the status of women prepared by students, studies at other universities, and in specific professions. The Policy CSAW was busy analyzing all this information. They did their work without extra funding. The $250,000 that the Ford Foundation gave to the UC to support the urban crisis initiative was distributed in December. None of the funds to the Berkeley campus went to initiatives involving women.

The information showed serious problems with the pathway for women from undergraduate student to graduate student to faculty status. For many departments, the decrease in proportions of women at the three stages was striking, with the proportion of women on the regular faculty being "very tiny, usually zero." This just raised more questions. Why the steady and strong decrease? Do women not apply to graduate school? Are women refused admission to graduate school? Do women decide not to continue for a PhD after their Master's? Are those who continue less successful at obtaining the doctoral degree? Do women PhDs decide to enter nonacademic fields? Are they less qualified to be on the UC faculty? Is there pressure to drop out? Are the decisions made freely?[17]

The WFG decided to hold their December meeting on the first Friday over lunch at the WFC. Herma Kay volunteered to begin drafting by-laws for the group. They heard a proposal from the graduate women's sociology caucus for an all-university meeting on women in January. They heard reports from the steering committee and four standing committees (faculty appointments, graduate students, orientation for undergraduates, advanced training).[18]

Betty chaired the faculty appointments committee, which had a mission to ensure women faculty members had the same opportunities for advancement as men. It was developing a database on all women who were teaching, including information on special problems they faced.

The steering committee was working on recommendations regarding women researchers. A few months earlier, deans asked department chairs and research unit directors to review and comment on proposed policy changes for the administration of externally funded grants and contracts. These changes were expected to be discussed by the academic senate soon. The committee

---

[16] 11/21/1969 memo from UC-Berkeley Anthropology Professor Elizabeth Colson to Members of the Policy CSAW re progress on their work.

[17] n.d. (?1969) Manuscript notes, unattributed (?Betty Scott).

[18] 11/26/1969 memo from Ellen M. Gumperz to women members of the faculty re agenda for the December 5 meeting of the WFG.

was interested in the effect the changes would have on the status of women researchers and faculty. The most significant change for women would be that submission of a research proposal would be "only by a staff member who will personally serve as the Principal Investigator"; the role of principal investigator would have to be a substantial one in all cases. This would affect all women who planned, directed, and carried out their own research. The steering committee decided to make a formal statement about these changes, as they would affect a large proportion of women PhDs on the campus. They wanted to present a formal resolution on the status of women in research, as well as on the nepotism rule, to the academic senate before the end of the year. They wanted to "...use all these devices to make ourselves visible to the University community as a defined group."[19]

The Policy CSAW was still waiting for a more detailed report from the budget committee. What Betty really wanted was to be able to compare the progress of women and men appointed in the same department at the same time. Right after Christmas, Betty learned from the chair that the budget committee was unable to give her the information she requested for the advancement and durability studies, at least not in time for the Policy CSAW report. The budget committee didn't have the departmental roster data from previous years that she needed.

Rather than give up, Betty figured she might be able to get some of this information for the College of Letters and Science by going directly to them.[20] Success! They were willing to give her what she needed to answer "quite a few questions" for their regular appointees. On the plus side, the college had a roster for each department, but on the negative, it did not list women separately and only went back 10 years. Because it showed the step data, which the Policy CSAW needed, the college couldn't give the roster to Betty. But the person who was in charge of regular appointments liked to work on this kind of thing, and she would see what she could do.

To get information on temporary appointments, the college would have to contact another person in their office. But it was Betty's impression that these records of temporary appointments were less complete and more difficult to interpret than records of regular appointments. The office would give Betty specific information grouped by departments, but without identifying them, and Betty promised not to guess.[21,22]

As the year came to an end, Betty obtained a letter from the office of the UC Vice President for Academic Affairs (VPAA). The letter was written in October and addressed to the Berkeley campus personnel manager. Betty assumed it wasn't confidential and forwarded it to Kay and the women on

---

[19] 12/11/1969 minutes of the WFG Steering Committee.

[20] Note: It is unclear whether she approached other colleges.

[21] 12/26/1969 letter from Betty Scott to College of Letters and Science Assistant Constance J. Wilson.

[22] 1/6/1970 letter from Betty Scott to Members of the Policy CSAW.

the Policy CSAW.[23] The manager had asked about the university's policy on non-discrimination in employment based on sex,[24] indicated there was no evidence the university had ever explicitly publicly spoken on the subject, and quoted the university policy on general non-discrimination in employment. He had summed up "that it is the University's policy not to make distinctions on the basis of sex in academic hiring except in connection with bona fide operational requirements of the positions to be filled." At year's end, Betty again questioned whether there should be restrictions on discrimination in academic employment.

---

[23] 12/26/1969 memo from Betty Scott to UC-Berkeley Professors Herma Kay, Elizabeth Colson, and Susan Ervin-Tripp.

[24] 10/31/1969 letter from UC-Berkeley Professor of Mathematics Angus E. Tayor to UC-Berkeley Campus Personnel Manager John Wagner.

15

Not a Good Time
(January – April 1970)

...women feel that women students are discouraged in the system
and that things are generally made harder for them.
-Betty Neely

At the beginning of 1970, Elizabeth Colson headed to Caltech as a visiting professor for the winter quarter to teach and gain experience with a time series software program using her Gwembe Tonga of Zambia data.[1] This meant the chair of the Policy CSAW was absent from Berkeley. It also meant Betty's role on the subcommittee, and on women's issues at Berkeley in general, was about to grow.

Drafts of parts of the Policy CSAW report were materializing. Colson left Betty with a partial draft of the body of the report. Lucy Sells had a draft of part of the appendix. Betty circulated these to the subcommittee.[2]

Betty looked for even more data. She approached the graduate division associate dean, being very specific about what she wanted that would allow the Policy CSAW to form conclusions about graduate student admissions, fellowships, dropouts, and completions. She wanted the data by department, so the subcommittee could discern possible problems for women among the departments. She asserted they would keep the data confidential and arrive at conclusions based on the data. Betty proposed that the associate dean review the Policy CSAW's conclusions prior to their entry into the draft report.[3]

The associate dean agreed to let Betty use more graduate division data. Two weeks earlier, Betty had obtained the data tape containing information on UC-Berkeley graduate students who obtained their PhDs either at UC-Berkeley or elsewhere. The new data set would be much more complete and consistent over time and much more useful to the Policy CSAW.

Betty's efforts, focusing on the collection of hard data and presentation of facts, were having an effect at the chancellor's level. An internal memo-

---

[1] Colson, Elizabeth. "Anthropology and a Lifetime of Observation," an oral history conducted by Suzanne B. Riess, Oral History Center, The Bancroft Library, University of California, Berkeley, 2002.

[2] 1/6/1970 letter from Betty Scott to Members of the Policy CSAW.

[3] 1/5/1970 letter from Betty Scott for the Policy CSAW to UC-Berkeley Anthropology Professor and Graduate Division Associate Dean Eugene A. Hammel.

randum marked confidential and signed with four sets of initials was sent to the chancellor and vice chancellors (VC), showing the distribution of faculty and staff by sex in four pages of tables. The authors thought the top administration might be interested in these hard data, "...the gathering of which resulted partly from a request from Professor Elizabeth Scott for a study she is doing...We thought it would be interesting to see what the facts were."[4]

15.1   One Faculty Club

Betty was an early proponent of having one faculty club on the Berkeley campus. In fact, the board of directors of the women's club (the WFC) was busy working on a proposition to combine the two clubs under the facilities of the men's club (The Faculty Club). The idea was this amalgamation would serve the needs of the university over the long term better than separate clubs.

It was proposed that the combined clubs be on the premises of The Faculty Club because the facilities were in better shape than those of the women's. The campus architect estimated it would take $234,000 to repair and renovate the women's club to bring it up to a minimum safety standard, or $340,000 to bring it up to an acceptable operating standard for residence and dining. The latter is $1.9 million in 2010 dollars, not a small amount of money. The women's club had only $18,000 of reserves that were the remains of hard fund-raising by a group of members over a three-year period. The women thought they simply didn't have the resources to repair and renovate, even to the minimum safety standard. On the other hand, it was looking like the men's club would have the funds to renovate to a standard that would be acceptable to the membership of both clubs. Combining dining services would reduce operating costs. The women's club building could be used for other university needs that did not include residence and dining. A Joint Committee on Merger of the Faculty Clubs was formed and began working on a plan based on requirements specified by the women's club.[5]

In spite of the apparent interest and activity related to combining the clubs, some things remained business as usual. Betty and Neyman housed an out-of-town guest at The Faculty Club in March. When the person tried to check out, the front desk couldn't tell him how much was owed, because meals hadn't yet been figured into the bill. So the guest paid an approximate amount, and the front desk asked him to leave the name of a staff member to whom they could report the exact amount. The guest tried to leave Betty's name, but the front desk refused it because she was a woman.[6] Betty held the

---

[4] 2/10/1970 confidential memorandum from NL, SM, LT, and AW to RWH, REC, RLJ, JHR, and EWM [names not identified] re Distribution of Faculty and Staff by Sex.

[5] 1/15/1970 Proposition by the Board of Directors, Women's Faculty Club, Inc.

[6] 3/1970 handwritten letter from unknown to Betty Scott.

rank of tenured Professor, the highest faculty rank at the university, but the club would not allow her to be responsible for the bill of her visitor.

Betty had thought the faculty club situation at UC-Berkeley was unique, where there were separate and exclusive faculty clubs for men and for women. But then she learned this was also the case at the University of Pennsylvania. She wanted to know more, and so she wrote to Elizabeth Kirk Rose (1902– 2008), president of the women's faculty club at Penn. Betty explained the situation, that the women's club at UC-Berkeley was formed 51 years prior in reaction to discrimination against women at the men's club. Women who tried to eat at the men's club had been forcibly ejected. Betty had experienced this herself. Women had not been allowed to stay overnight in the guest rooms. Women had not been allowed to become members. Betty also reported that, in the past five years, the discrimination against women had lessened somewhat. She could now eat at the men's club without being forcibly ejected, but she still couldn't stay overnight or become a member. Betty described the ongoing efforts to merge the two Berkeley clubs. It was her understanding that the interest in a merger was due almost entirely to financial difficulties of the two clubs. However, the merger that was being discussed would leave two distinct clubs. Betty told Rose the Policy CSAW would be putting forward a recommendation that the boards of directors of the two clubs be asked to form one single club open to both women and men.[7]

15.2 Interpreting Biases

Betty was emerging as a counselor to women experiencing campus issues. An example is a woman graduate student who was having difficulties being posted to a teaching assistant position because her husband had such a position in the same department. The woman noted her situation on Betty's campus survey. Betty immediately followed up, suggesting the woman either bring the 1966 relaxation of the nepotism rules to the attention of her chair, or else authorize Betty to approach the chair on her behalf.[8]

The VCA was trying to figure out what the Women's Liberation Group meant when they demanded the end to "the systematic discouragement and cooling out of female students which occurs throughout the University," their number one demand of their summer resolution. Betty Neely, the UC-Berkeley dean of women and associate dean of students, admitted she did not fully understand the meaning of the demands, but nonetheless tried to explain to the VCA that this was not a good time to be a woman student: "Our

---

[7]4/12/1970 letter from Betty Scott to University of Pennsylvania Medicine faculty member Elizabeth Kirk Rose.

[8]1/15/1970 letter from Betty Scott to student Susan Kishler.

interpretation is [that] these women feel that women students are discouraged in the system and that things are generally made harder for them."

Neely offered an example of bias against women in the classroom: Sorority girls were unable to wear their sorority pins on campus, because some instructors seeing girls with pins would refer to them as "flighty sorority girls."[9] Betty herself had been a member of a sorority for a brief period of time when she was an undergraduate, and she may have experienced this personally: On January 19, 1936, Betty was initiated into the Alpha Iota Chapter of the Beta Sigma Omicron Sorority;[10] however, she became inactive after only one year, because the Monday night meetings conflicted with one of her astronomy observing courses.[11] Those who knew Betty in her later years would have had a hard time imagining Betty wearing her pin and being referred to as a flighty sorority girl.

Neely offered other examples where there was bias against women. In terms of policies and procedures, some women felt they needed to offer detailed explanations in order to request exceptions and get action. Relating to seeking career advice, some departments reportedly actually discouraged women from seeking graduate school or job placement information or suggestions. In department offices, some women felt men students got more cooperation from the women clerks. With respect to women who were homemakers, some had trouble getting permission to take reduced loads, because homemaking was not considered equivalent to a job. Concerning residency requirements for tuition purposes, some women who were separated from their husbands had trouble establishing California as their place of residence, because the legal residence of the wife was taken to be the legal residence of the husband.[12]

Male administrators weren't totally on the same page with women. The graduate division dean felt comfortable enough with the VCA to refer to women as "militant suffragettes." But then he presented applicant and admissions percentages, which showed some imbalance in favor of men; the degree completion statistics were, in his words, "dismal and grim," being around eight to ten men completers to every one woman who was a completer.[13]

Let's look more closely at "Jane" as an example of why it was not a good time to be a woman student. Jane was a graduate student and married to another student in the same field. Due to the nepotism rule, she wasn't able

---

[9] 1/16/1970 letter from UC-Berkeley Dean of Women and Associate Dean of Students Betty H. Neely to VCA Robert L. Johnson.

[10] 1/19/36 certificate of initiation of Elizabeth Leonard Scott to the Beta Sigma Omicron Sorority. Note: Beta Sigma Omicron was founded in 1888 and merged with Zeta Tau Alpha sorority in 1964.

[11] Scott, Elizabeth L. "An Interview with Elizabeth Scott," an oral history conducted by Suzanne B. Riess, in "The Women's Faculty Club of the University of California, Berkeley, 1919-1982," Oral History Center, The Bancroft Library, University of California, Berkeley, 1983.

[12] 1/16/1970 letter from UC-Berkeley Dean of Women and Associate Dean of Students Betty H. Neely to VCA Robert L. Johnson.

[13] 1/19/1970 memo from UC-Berkeley Graduate Division Dean Sanford S. Elberg to VCA Robert L. Johnson re Graduate Division statistics.

to get an appointment as a teaching assistant because her husband had such an appointment. Jane and her husband had a child, but she didn't have access to adequate childcare. Childcare was expensive, and there were long waiting lists (at the time there were some 5,000 children of UC-Berkeley students). Her childcare wasn't tax deductible. Jane applied for a fellowship, but she didn't get one: Over a third of the graduate students in her department were women, but almost all fellowships went to men, even when women were equally qualified. Even if she had gotten a fellowship, it would have offered no relief for childcare because her husband was employed. Jane was in one of the six most criticized fields for its treatment of women.[14] She was discouraged by her male advisor, who told her she wouldn't be able to get a job, and if she did she would be a poor wife and mother. Because her professors were all men, Jane didn't have any women role models to emulate. Even though she was gifted, her scholarship wasn't taken seriously. Her male classmates often ridiculed her.

There was a two in three chance that Jane would drop out of graduate school and could not look forward to a normal professional career. She was totally outside of the academic prestige system. She was unlikely to be hired in an academic job, especially at UC-Berkeley, where members of her department felt that hiring a woman would lower the department's prestige. So even if she finished her PhD, Jane planned for a nonacademic job. The data showed the proportion of women on the UC-Berkeley faculty had been steadily declining back to what it was in the 1870s, namely, 3.5%. Women were routinely discriminated against within the hiring processes. If lucky enough to be hired, they were likely to be hired in the lower ranks. This meant lower pay, heavier teaching loads, and slower promotions. It also meant no tenure, and no ability to apply for research grants.

How could this be fixed? Susan Ervin-Tripp proposed that adding more women to the faculty immediately would have the greatest effect, that putting "men of good will on search committees in every department could alter the system now."[15] Betty and her collaborators understood they needed the support of the power structure to make changes.

Betty's advocacy on behalf of students was not limited to women. She was deeply interested in the quality of selection of all graduate students. When one of their best graduate students, a Kansas farm boy who had won an NSF fellowship in national competition, decided to transfer to Harvard, Betty took the matter directly to the chancellor.

A major factor in the student's decision to transfer was that he was faced with a police record without cause. Betty described the incident in detail to the chancellor. The student and several friends were on campus, going from a movie to their car. The university police stopped them. There was no

---

[14]Note: The most criticized were French and Music. Also heavily criticized were English, Botany, Librarianship, and History.

[15]Susan Ervin-Tripp, "The Academic Mythology," *The Every Other Weekly*, January 27, 1970, 3.

apparent reason, and no reason was given. The students were asked for their identification. The police transmitted the students' identifications by walkie-talkie so that they could be checked. The police said the information would be recorded into a folder in Sproul Hall, the university's central administration building, indicating the students had been stopped by the police and then released. The student asked whether the record would state they were stopped for no apparent reason. He was told the record would not state this.

When Betty conferred with her statistics department colleagues and students, she found out this was not a unique occurrence. All were appalled by "what seemed to us to be unnecessary and unfortunate setting up of files which are likely to harm the students for many years to come." They understood the police were nervous because it was a time of fires and explosions, but they did not see why they should stop and set up files on ordinary students who were not "far-out hippies," just because they are walking across campus. Betty asked the chancellor to stop this practice of indiscriminate police recording of students and to see that the files of innocent students were eliminated.[16]

15.3   Problems in Zoology and Chemistry

At the end of January, the chair of the UC-Berkeley zoology department sent the agenda for their next meeting in a confidential memorandum to his faculty. The first item was an eyesore:

> *We should address ourselves to the question of shift in sex ratio of the graduate students. My view is that an increasingly disproportionate portion of our resources is being invested in women. We should address ourselves to the question, should a limit or ceiling be put on [the] proportion of applicants for 1970–71 who are women?*

He then presented numbers of admitted students entering over the past three years (men/women were 16/11 in 1967, 23/10 in 1968, and 18/18 in 1969), and number of applicants for 1970 (44/11), concluding that: "It appears that more women than men accepted by us come to UC-Berkeley; i.e., competition for men students is keener – we will certainly not get four men to every woman in the lot we draw in September 1970." The zoology department had 34 faculty members, none women. It wasn't that qualified women weren't available: It was estimated there should have been at least five on the zoology faculty given the number of UC-Berkeley PhDs granted over the previous 20-year period.[17,18]

---

[16]4/16/1970 letter from Betty Scott to UC-Berkeley Chancellor Roger W. Heyns.

[17]"Sex Ratios in Faculty Hiring, Berkeley Campus: Selected Letters and Science Departments" (data table, UC-Berkeley), 1970.

[18]1/26/1970 memorandum from UC-Berkeley Zoology Professor and Chair Frank A.

Someone intentionally sent a copy of the chair's memo to the women students in the department. They in turn sent the memo to the UC-Berkeley chancellor, telling him it required his immediate attention: The proposed zoology department quota system for women admits was not only clear sex discrimination, it was illegal. A long list of women signed the letter, from law and other fields in addition to zoology women, calling themselves the Coalition of Campus Women (CCW). They specifically cited Article IX, Section 9 of the California State Constitution (as amended November 5, 1918), that "... no person shall be debarred admission to any department of the university on account of sex." They also specifically included a table of sex ratios in faculty hiring for selected departments in the college of letters and science which indicated that any sex discrimination was not confined to zoology. "We feel it is imperative that you, as Chancellor, exercise the full powers of your office to bring all units of the Berkeley Campus into deliberate and immediate compliance with the spirit and the letter of state and federal laws," they wrote.

The women wanted adequate representation of women on all department admissions committees. They wanted departments to maintain and submit admission records (all applicants, acceptances, and denials), so that a check could be made on compliance with the laws. They wanted the zoology department chair to be made aware of the illegality of his proposed quota system, even though the department had tabled the matter. They wanted the chancellor to take action, saying that: "At a minimum, we demand from [the chair] a public statement that the Zoology Department does not and will not discriminate against women." They gave the chancellor a two-week deadline for his response, or else, they threatened, they would take "further appropriate action." Betty, as representative of the Policy CSAW, was copied on the letter.[19]

Around the same time, Betty marked a handwritten note "Confidential" and put it into her file. She had just spoken with a woman graduate student who complained of discrimination by the zoology chair and some others in the department. The specific complaint was that he would not let women go on extended field trips and never allowed them to go to his field station in the Arctic. The student said it used to be that no faculty members in zoology allowed women to go to their field stations, but now some do. According to the student, the chair suggested that women should do their PhD field research in the Life Sciences Building courtyard on the campus.[20]

The CCW copied the zoology chair on their letter to the chancellor. Rather than waiting for the chancellor to act, the chair wrote a response admitting the discussion and explained the entire matter was tabled because "the im-

---

Pitelka to Zoology faculty re Agenda for staff meeting Tuesday noon, 27 January, room 1005.

[19] 3/4/1970 letter from the CCW to UC-Berkeley Chancellor Roger Heyns.

[20] 2/1970 handwritten confidential memorandum from Betty Scott to the File re a graduate student in zoology.

plications of the discussion were considered objectionable." He responded to the CCW demand with a statement that the zoology department "does not and will not discriminate against women." He said his letter could "serve as a public statement if such is needed."[21]

The VC responded to the CCW's letter, hoping the zoology chair's letter would satisfy the women's concerns, at least their immediate ones. He assured the women their concerns were also the concerns of the UC-Berkeley administration, and the administration expected all units to comply with federal and state laws. Regarding the women's specific requests, the VC said assignment of a concerned woman to department admissions committees would sometimes be impractical, for example, when only one person is assigned to evaluate applications. However, he assured the women he would bring the matter to the attention of all the units. He also said it was the responsibility of the admissions office and graduate division to maintain admissions records of undergraduates and graduates, respectively; the administration did not consider it practical to require individual units to maintain the same records. Ten days later, the VC sent to the CCW's leader, a law school student, a copy of the memorandum the graduate dean had written "in response to the visitation by militant suffragettes" containing statistics on admissions of women to the graduate school.[22]

Also in January, complaints about women in chemistry were published in a union newsletter. The article pointed out that lab technicians faced most of the same problems as clerical workers, but also faced uncertainties of employment due to uncertainties in grant awards that supplied funding for their positions. It went on to say that women in chemistry at UC-Berkeley also faced job discrimination, that the college of chemistry "takes pride in excluding women from certain research laboratories and boasts of employing very few female technicians." The author exposed that there was no tenured woman faculty member in either the chemistry or biochemistry departments.[23]

The chemistry assistant dean acknowledged there were no tenured women faculty members in chemistry, but missing the point, went on to say that "neither is there any tenured teaching person in the department of Chemistry whose birthplace was in Lapland." The assistant dean said the statement that chemistry "takes pride in excluding women from certain research laboratories and boasts of employing very few female technicians" has "no basis in fact." "The fact that there is no tenured woman teacher no more demonstrates willful discrimination against qualified women per se than does the second fact demonstrate willful discrimination against qualified candidates

---

[21]3/9/1970 letter from UC-Berkeley Zoology Professor and Chair Frank A. Pitelka to UC-Berkeley Chancellor Roger Heyns.

[22]3/30/1970 letter from VC Robert E. Connick to UC-Berkeley Law student Maria Tankenson.

[23]Gail Sheehan, "Ladies in Lab Coats," *U.C. Clerical, Technical and Professional "Employee Press,"* 4, no. 1 (1970). Note: The UC-Berkeley College of Chemistry has two departments: The Department of Chemical and Biomolecular Engineering, and the Department of Chemistry.

who happened to have been born in Lapland." He worried about the damage the union newsletter article could do to college recruiting and asked to have the allegations in the article investigated: "...if they find her allegation to be baseless, please see that appropriate official protest concerning such malicious statements is placed."[24]

15.4 Subcommittee Follow-Ups

The national recession brought new worries to California, and campus women felt these worries on top of their struggles for equal status. UC up to this point had never charged California residents for tuition. But now the board of regents was considering whether to establish a resident tuition charge in order to mitigate the effect of the budget crisis on the university.

The WFG was concerned about how this new charge would affect women students. The steering committee, with Betty and Herma Kay as the most senior members in terms of faculty rank, wrote a letter to the board of regents chair saying they had reason to believe, based on a recent campus survey conducted by the counseling center in fall 1967, that a resident tuition charge would have a greater negative impact on women than men students. Among their worries was that families strapped for funds might choose to send only their boys to college: "As faculty women, we are interested in encouraging capable women to attend the University and to prepare themselves for lives enriched by greater opportunities than many of them now enjoy."[25]

Betty continued her work on the surveys of women graduate students and recent PhDs. She worried the "Women's Liberation publicity" might generate more comments and bias the results of the second mailing. She was relieved when she didn't see any difference between mailings.[26]

In her continuing search for data for the Policy CSAW report, Betty got in contact with the gifts and endowments office. They had survey data they were willing to let her cross-tabulate. Betty looked at the list of variables and found many were of interest for the report. Support versus degree by field and sex was especially of interest, as well as measures of occupation or income, which could be used as indicators of success. Betty offered to send copies of any tables she produced back to the endowments office.[27]

Betty was also in contact with colleagues at Stanford University, who sent her a copy of their report on the status of women. Betty read the report and

---

[24]2/5/1970 letter from UC-Berkeley Chemical Engineering Professor and College of Chemistry Assistant Dean David N. Lyon to John Wagner.

[25]2/11/1970 letter from The Steering Committee of the WFG to UC Board of Regents Chairman DeWitt A. Higgs.

[26]2/21/1970 letter from Betty Scott to staff assistant Susan Smith.

[27]2/21/1970 letter from Betty Scott to UC-Berkeley Gifts and Endowments staff member Joseph R. Mixer.

concluded the problems for women at Berkeley and Stanford were similar. Betty noted the biggest problem with the UC-Berkeley report was in getting complete or equivalent data, and that the appendices were a continuing challenge. Betty promised she would send a copy of the completed report to Stanford.[28]

The subcommittee met toward the beginning of March. It was time to make plans to finish the report.[29] Colson was still away at Caltech. Frank Newman was also unable to attend; Betty followed up with him to make sure he received Colson's and Lucy Sells' drafts and suggested they set up a time to talk about what to put into the final report, especially the appendices.[30,31] At the meeting, Betty, Ervin-Tripp, and Herbert Blumer discussed what had been assembled so far, including the several hundred documents relating to the status of academic women. They discussed what evidence and data they had assembled on faculty, including: question lists and bio-bibliographies; letters from department chairs; relative proportions employed in various ranks; and comparative data on attrition and advancements. Some of the statistics were from official sources, others from the university catalogs. They also discussed what evidence and data they had assembled for graduate women, including: question lists; proportions of majors, degrees, and placements to women; and data on time to degree, financial support, and retention. There were statistics for UC-Berkeley, California, and the entire country. There were data from original questionnaires, the NAS, and the UC-Berkeley Graduate Division. They talked about including reasons advanced by men for gender imbalances "without analysis" and "without passing judgment." They discussed including recommendations on nepotism (which "is indeed difficult"), childcare, and part-time employment.[32]

Betty followed up again with the budget committee about the information she had requested for the advancement and durability study. They were sorry: They had been, and would be in the near future, too pressed for time to conduct further study. The chair told Betty they might be able to help in the summer. That would be too late to be of use for the report.[33]

---

[28]2/23/1970 letter from Betty Scott to Stanford University Psychiatry faculty member Barbara Arons.

[29]3/3/1970 three memoranda from Betty Scott to each of UC-Berkeley Speech Professor Susan M. Ervin-Tripp, UC-Berkeley Sociology Professor Herbert Blumer, and UC-Berkeley Law Professor Frank C. Newman re March 6 meeting of the Policy CSAW.

[30]3/8/1970 letter from Betty Scott to UC-Berkeley Law Professor Frank C. Newman.

[31]3/23/1970 memorandum from Betty Scott to UC-Berkeley Sociology Professor Herbert Blumer, UC-Berkeley Anthropology Professor Elizabeth Colson, UC-Berkeley Speech Professor Susan Ervin-Tripp, and UC-Berkeley Law Professor Frank Newman re confirming the date of the meeting of the Policy CSAW.

[32]3/6/1970 handwritten notes by Betty Scott from the meeting with UC-Berkeley Speech Professor Susan M. Ervin-Tripp and UC-Berkeley Sociology Professor Herbert Blumer.

[33]3/13/1970 handwritten notes from Betty Scott to The File re conversation with UC-Berkeley Professor of Chemistry and Budget Committee Chair David H. Templeton about getting budget committee study data.

The Policy CSAW was establishing a high profile, and people were bringing more gender inequities to the attention of its members. For example in April, another potential problem was identified with health insurance premiums. The situation was that, regardless of age or sex, there was a higher premium for two-party than one-party coverage; and then the increase in premiums from two- to three-or-more-party coverage was much less than the increase from one- to two-party coverage. The problem was that, when a woman was covered and the second party was a child, she was likely to be a single parent, and the premiums did not take this disadvantage into account. Betty asked the committee on university welfare to look into this problem.[34]

15.5   Problems in Mathematics

Given the CCW's letter and the emerging activity on campus around the status of women, it is no surprise that by the end of March, a number of complaints bubbled up to the chancellor's office about sex discrimination. These alleged that "various departments" were "practicing discrimination according to sex" in employment, as well as in graduate school admissions of women. The complaints prompted the VC to write a memorandum to all deans and department chairmen pointing out the illegality of such practices.

The VC especially mentioned the following:

- Title VII of the Civil Rights Act of 1964 which "prohibited employment discrimination based on race, color, religion, sex, and national origin."[35]

- Executive Order 11247 signed in 1965 which provided for "the coordination by the Attorney General of enforcement of Title VI of the Civil Rights Act of 1964" as amended by Executive Order 11375 signed in 1967 "relating to equal employment opportunity."[36]

- Article IX, Section 9 of the [1879] California State Constitution, which stated that: "No person shall be debarred admission to any of the collegiate departments of the university on account of sex."[37]

- University Policy 7.4 which stated that: "Except in connection with bona fide operational requirements of the positions to be filled, as set forth in University Regulation 7.4 Policy of Non-Discrimination, it is the policy of

---

[34]4/12/1970 letter from Betty Scott to UC-Berkeley Committee on University Welfare Chair A.A. Kragan.

[35]US Equal Employment Opportunity Commission, "Title VII of the Civil Rights Act of 1964," http://www.eeoc.gov/laws/statutes/titlevii.cfm.

[36]"Administration of Lyndon B. Johnson (1963–1969)," *National Archives*, http://www.archives.gov/federal-register/executive-orders/johnson.html.

[37]1879 California State Constitution: Article IX: Section 9.

the University to make no distinctions on the basis of sex in connection with hiring or admission."[38]

The chancellor's office knew they needed to be careful in their handling of women's issues. This time their specific concern was the WFG. Barbara Kirk, one of the original members of the WFCSC and a current member of the CCW, thought the CCW was "more politically oriented than motivated by trying to solve real problems." At least that's what an assistant chancellor reported to the chancellor, and he accordingly advised the chancellor not to act upon any matters brought forward by the CCW until hard data from two counseling office studies became available. A colleague was in the process of conducting these studies in cooperation with Kirk. In about a month, the two planned to have a report ready from a survey of 250 out of the 400 undergraduate women over age 25 in 1966–67. After that they planned to produce a study of graduate women students. They planned to send the two reports to the chancellor marked confidential, since the assistant chancellor wanted the units identified "where particular problems exist."[39]

Some departments on campus were getting defensive. Given her background, Betty was particularly aware of the situation in mathematics, although this situation wasn't at all anomalous among the departments. The mathematics department had, as she put it, zero women full professors, zero women associate professors, zero women assistant professors, and three women lecturers. The three lecturers were unusual, according to Betty, because most years there were zero women lecturers. Betty knew the last time a woman was appointed to any of these "ladder" positions was in 1953.[40]

The mathematics department chair responded in detail to the VC's memo to deans and department chairs pointing out the illegality of sex discrimination by departments in employment and admissions. The math chair wanted to make it "immediately clear" to the VC that his department "vigorously" followed a policy of nondiscrimination and that "we think we have gone out of our way to make sure that women are not discriminated against." The chair pointed out the department had hired three women as new lecturers, two full time and one part time: Karen Uhlenbeck (1942–), Lenore Blum (1942–), and Julia Robinson (1919–1985), all of whom would go on to make big names for themselves in mathematics. He also pointed out that his department's entire full-time nonacademic staff of 20 was female. He continued:

> ... *if there is any danger about sex discrimination at the level of admission to graduate school I think it may possibly be in the other direction: it*

---

[38] 3/26/1970 memorandum from VC Robert E. Connick to deans and department chairmen re complaints of sex discrimination and laws and regulations.

[39] 4/2/1970 memorandum from UC-Berkeley Botany Professor and Assistant Chancellor for Educational Development Leonard Machlis to UC-Berkeley Chancellor Roger W. Heyns re WFG matters. Note: The colleague named was Ann Stout.

[40] 4/29/1970 letter from Betty Scott to NYU Courant Institute of Mathematical Sciences Professor Cathleen Synge Morawetz.

might interest you to know that on our master list of applicants last year we had women carefully asterisked, because we were especially anxious that they should come (in part because they are not subject to the uncertainties of the draft which have plagued our attempts at orderly planning of graduate enrollments). In summary, it is a pleasure to report that the Department of Mathematics is trying to be in the forefront of the campaign to protect the rights of the great American disadvantaged majority, WOMEN [sic].[41]

The VC replied with three questions about the math department:

Do women get appointed only as lecturers – not as regular ladder members? Why are women relegated to non-academic positions? Is not preference of women graduate students over men, because of the draft, discrimination by sex?[42]

The chair provided some answers. He skirted around the first question. Rather than a yes/no answer to whether women get appointed only as lecturers, he talked about how "truly phenomenal" the competition had become for assistant professorships, they were "hopeful" that women appointed as lecturers would be promoted to assistant professors, and they were "not unproud" of their record of having hired three women as lecturers that year. He mentioned Robinson in particular. Within only five years, Robinson would have the distinction of being (in 1975) the first woman mathematician elected to the NAS; this was a major mark of excellence, as members are elected in recognition of their significant contributions to original research, and election is one of the highest honors for a scientist. But her marriage to one of the math professors, Raphael Robinson, held her back to a lectureship at UC-Berkeley. The chair said that if it were not for the nepotism rules, they "might well have appointed her to a regular faculty position long ago."[43]

The chair contended that women were not exiled to nonacademic positions: None of the 20 women in such positions had a PhD and so did not qualify for academic positions. He added the responsibilities of some of the nonacademic positions, especially assistant to the chairman and the business officer, "are certainly greater than that of some of our regular faculty positions. (If this is not accurately reflected in their rate of pay, I believe the responsibility lies with the Administration rather than with the Department.)."

Regarding the third question about preference for women graduate students because of the military draft, the chair contended that women were not given preference in admission or financial aid because of the draft. He suggested that: "even if we were not so noble in purpose, we would still have a

---

[41] 4/1/1970 letter from UC-Berkeley Mathematics Professor and Chair John West Addison Jr. to VC Robert E. Connick.

[42] 4/14/1970 letter from VC Robert E. Connick to UC-Berkeley Mathematics Professor and Chair John West Addison Jr.

[43] 4/20/1970 letter from UC-Berkeley Mathematics Professor and Chair John West Addison Jr. to VC Robert E. Connick.

built-in reason for not discriminating against women, namely, their advantage to us because of the draft situation." He suggested that a case could be made that opportunity programs for minority students are a form of discrimination by race. He asked if there would be similar programs for women students. He said he would like to hear discussion about such programs before forming a definite opinion about them.

Betty told the chair she would forward his letter of response to Colson. Betty was noticing the proportion of women who received a PhD in math at UC-Berkeley was low compared with other top ten math departments in the country. She expressed concern about this to the chair.[44]

15.6   Information Exchanges

It was mid April 1970, and Betty figured that the *Human Resources and Higher Education* book must have come out. She wrote Helen Astin asking for a copy of it and another book Astin had been working on, *The Woman Doctorate in America*. Betty was keen to get a look at them before the Policy CSAW published their report, and so she asked Astin for prepublication copies. Betty hoped Astin's books would affirm some of the Policy CSAW's data-based findings.

Betty took the opportunity to comment to Astin on the linear regression analyses of salary in the first book, saying she expected there would be a need to include interaction terms in the regression equations. Betty wrote that, if Astin hadn't already included the interactions in her expanded study, then perhaps they could collaborate on this.[45] When Betty had an interest in something, she usually offered to collaborate or help.

Betty was corresponding with people all around the country about gender equity issues. They were exchanging studies and information about circumstances. Betty informed her correspondents that at UC-Berkeley there were "difficulties" for women graduate students in certain departments, and the situation for women faculty was "rather dismal."[46]

---

[44]4/9/1970 letter from Betty Scott to UC-Berkeley Mathematics Professor and Chair John West Addison Jr.

[45]4/12/1970 letter from Betty Scott to UCLA Higher Education Professor Helen Astin.

[46]4/12/1970 letter from Betty Scott to University of Pennsylvania Medicine faculty member Elizabeth Kirk Rose.

16

Grounded in Hard Fact (May – June 1970)

We offer the report now as the most detailed and thoughtful study of the status of women on the Berkeley campus that has ever been prepared in the hope that it will serve as the basis for sustained discussions...
-Sandy Kadish

In May 1970, Betty led completion of the *Report of the Subcommittee on the Status of Academic Women on the Berkeley Campus*. Fourteen recommendations backed by 15 appendices comprised the heart of the 78-page report that took a year to prepare.[1]

The Policy CSAW considered data from a wide variety of sources. The report required "quite some effort."[2] It was "grounded in hard fact."[3] There were surveys of department chairs, faculty women, professional women in research units, and current and former women graduate students. There were reports, catalogs, and statistical summaries, some previously published, some not. There were new extractions or breakdowns of university data prepared especially for the report, mostly coming from a variety of confidential administrative sources, at a variety of levels in the hierarchy. These were extracted by staff, or sometimes even by Policy CSAW members themselves or their assistants. The Policy CSAW also listened to anyone who wanted to offer her or his impressions, reasonings, or suggestions.

The data and listening revealed substantial evidence that women were being denied equal opportunities and recognition at UC-Berkeley. This was the case even though exactly 50 years had passed since women got the vote. Put in a journalistic way: "The University of California is not using the talents of the women it helps to train..."[4] Put more simply: Men were getting better academic opportunities than women. The bottom line, backed by hard

---

[1] H. Blumer, E. Colson, S. Ervin-Tripp, F. Newman, and E.L. Scott, *Report of the Subcommittee on the Status of Academic Women on the Berkeley Campus*, May 19, 1970.

[2] 7/6/1970 letter from Betty Scott to Harvard Statistics Professor Frederick Mosteller.

[3] S. Almazon, "UC Failing on Women Teachers, Study Shows," *S.F. Examiner* (San Francisco, CA), June 5, 1970, 13.

[4] S. Almazon, "UC Failing on Women Teachers, Study Shows," *S.F. Examiner* (San Francisco, CA), June 5, 1970, 13.

data, was that discrimination against women hurt the women and the whole university, as it led to lost opportunities and productivity.

16.1   Completing the Subcommittee Report

Betty reported the overall findings to women members of the faculty at the May 1 meeting of the WFG.[5] The report was close to completion. Betty hoped in general it would result in some substantive changes at UC-Berkeley. She hoped in particular it would result in better, more detailed university data record systems, so that accurate planning, analysis, and publication could be done in areas like employment, degrees, and student status of women.[6]

On May 9, Betty and Susan Ervin-Tripp discussed some last changes to the report.[7] Two days later, Betty sent a preliminary report to Herbert Blumer and Frank Newman for their review.[8] Shortly after, Betty sent the draft report to members of the academic senate policy committee. She also sent a copy in confidence to the equal employment opportunity officer asking for her comments or suggestions.

Blumer suggested they consult with the office of educational career services about another possible "arch in the structure" of unequal opportunities facing women graduate students, namely, placement of women in the tightening job market.[9] Betty followed up with the career services office. The representative informed her that only some jobs in junior [community] and state colleges specified gender as a qualification; his office did not play a major role in placing graduates; and junior colleges were reluctant to hire PhD candidates. Betty deemed this to be "... important in assaying the 'performance' of graduate women students because these women students inform me that they are advised not to obtain the Ph.D. since it will make it difficult for them to obtain a position in a junior college. Yet at the same time they are advised that a position in a junior college is the highest they can hope to obtain."[10]

Betty also sent a copy for review to Doras Briggs, the woman who got the "status of women" initiative started two years earlier when she called together a group of women to discuss the urban crisis initiative. Briggs had been very helpful to the Policy CSAW, both with information and suggestions. Betty wrote: "You certainly made my life much easier, and, quite probably, there are some things I never would have found without your help." Indeed, Betty

---

[5]4/22/1970 memorandum from UC-Berkeley Social Scientist and WFG Secretary Ellen Gumperz to women members of the faculty re the agenda of the May 1 meeting.

[6]5/21/1970 letter from Betty Scott to UC Staff Member Doras M. Briggs.

[7]5/11/1970 letter from Betty Scott to UC-Berkeley Speech Professor Susan Ervin-Tripp.

[8]5/11/1970 letters to UC-Berkeley Sociology Professor Herbert Blumer and UC-Berkeley Law Professor Frank Newman from Betty Scott.

[9]6/1/1970 letter from UC-Berkeley Sociology Professor Herbert Blumer to Betty Scott.

[10]5/25/1970 letter from Betty Scott to UC-Berkeley Sociology Professor Herbert Blumer.

attributed her own involvement in gender equity research to Briggs and the WFCSC on the urban crisis initiative. "The problems of the urban crisis are very much with us, and, indeed, so are the smaller problems of women, but your committee did activate us to make a start in trying to solve some of the problems." Betty offered Briggs a "hearty thanks."[11] Briggs had planted a good seed.

The policy committee was looking forward to printing the report. Members Herma Kay and Sandy Kadish personally asked the chancellor for $500 ($2,775.67 in 2010 dollars) to pay for the printing. It was expected the report would be distributed in a blue cover to the whole senate in early June.

The completed Policy CSAW report was introduced to the Berkeley Division of the Academic Senate of the UC on May 19, 1970. Kadish introduced it as part of the senate policy committee's report. He presented it as a "factual investigation" and "prelude to consideration of remedial change." Betty was acknowledged as co-chair of the Policy CSAW, along with Colson. The two male members of the Policy CSAW were listed first as authors,[12] in spite of the fact that the women did most of the hands-on work.[13]

Kadish in his introduction said this:

> *The Committee on Senate Policy is not prepared at the present time either to endorse or to take exception with any of the substantive recommendations made in the subcommittee's report. We offer the report now as the most detailed and thoughtful study of the status of women on the Berkeley campus that has ever been prepared in the hope that it will serve as the basis for sustained discussions next year by the Berkeley Division and in the hope that it may serve to stimulate similar studies on other campuses.*[14]

The report focused on evidence-based "differences in opportunities faced by men and women in making their way in the academic world." It considered all categories of academic women: faculty members, researcher employees, and graduate students. The findings confirmed "the supposition that women face a large number of obstacles in obtaining recognition as members of the academic community in their own right." It concluded that "The status of women on this campus will be improved only by increasing the number of women on the faculty in a substantial manner." The Policy CSAW was "not recommending that the University should lower its standards, but rather that it should broaden its vision."[15]

---

[11]5/21/1970 letter from Betty Scott to UC Staff Member Doras M. Briggs.

[12]5/19/1970 memorandum from UC-Berkeley Law Professor and Senate Policy Committee Chair Sanford H. Kadish to The Berkeley Division re the "Report of the Committee on Senate Policy."

[13]Susan Ervin-Tripp, interview with author, June 13, 2010.

[14]5/19/1970 memorandum from UC-Berkeley Law Professor and Senate Policy Committee Chair Sanford H. Kadish to The Berkeley Division re the "Report of the Committee on Senate Policy."

[15]H. Blumer, E. Colson, S. Ervin-Tripp, F. Newman, and E.L. Scott, *Report of the Subcommittee on the Status of Academic Women on the Berkeley Campus*, May 19, 1970.

The body of the report was nine pages. Its focus was on UC-Berkeley. It provided unranked, evidence-based, targeted recommendations, 14 in all: two to the president, four to the chancellor, seven to the senate committees, and one to the faculty clubs. The recommendations "boiled down simply to asking that women be judged on their merits and records."[16] The Policy CSAW asked the senate to consider these 14 recommendations "in light of the fact that at present only 45 women are appointed to ladder positions which carry Senate membership and that the proportion of women in the Senate is less than it has been at any time since the 1920s. This fact alone warrants quick action to ensure that conditions leading to such a situation be rectified."[17]

16.2   Recommendations

Nepotism Rule: The first recommendation was to the UC system president: The nepotism rule should be abolished and procedures should be developed to prevent conflicts of interest. The rule meant that "...if a woman is trained in the same field as her husband, no matter how good she is there is no chance for her to be employed in the same department."[18] Married graduate students are often in the same department. The rule complicated the awarding of teaching assistantships to these couples. Women may be highly qualified for tenure-ladder faculty positions. The rule often relegated such women to lesser positions. These women sacrificed "salary, job security, and the hope of promotion and access to the normal encouragements for academic excellence." The Policy CSAW reported the nepotism rule "has been repeatedly singled out as a major barrier to the employment of qualified women...It is the one single practice most commonly raised by the various persons consulted as discouraging the advancement of women in academic life..." The rule was introduced in the 1930s during the Great Depression and was, simply put, archaic.

Scarcity of Women: The Policy CSAW recommended the UC-Berkeley chancellor take steps to achieve a faculty with representation of women proportional to the number trained. Specifically, he was asked to make available new positions for women faculty, and to direct departments to accept women students. He was asked to ensure female representation on search and promotion committees for women candidates. The UC-Berkeley senate budget committee was asked to instruct academic units to review women faculty members routinely and promote them as rapidly as possible. These three recommendations

---

[16]L. Spears, "No Equality for Women on Faculty," *[Oakland?] Tribune*, June 6, 1970.

[17]H. Blumer, E. Colson, S. Ervin-Tripp, F. Newman, and E.L. Scott, *Report of the Subcommittee on the Status of Academic Women on the Berkeley Campus*, May 19, 1970.

[18]S. Almazon, "UC Failing on Women Teachers, Study Shows," *S.F. Examiner* (San Francisco, CA), June 5, 1970, 13.

were a direct response to the fact that women in ladder faculty positions were scarce. Betty described the number as "minute" or "very tiny."

The statistics at UC-Berkeley were worse than at many other major universities in the US. In terms of absolute numbers, there were 1,245 individuals of senate rank (instructor, assistant professor, associate professor, or professor); almost all of these (1,200 of the 1,245) were men. Many academic units had no women faculty members at all, even though they had a sizable female graduate student population. In terms of percentages, only 2% of professors, 5% of associate professors, and 5% of assistant professors were women. Even worse, the percentages at all of these ranks had decreased steadily over the years: They were smaller than in the 1930s and 1950s; the percentage of assistant professors who were women in 1970 was only one-third of the percentage in the 1920s. The problem had become "increasingly acute at Berkeley in recent years."[19] The Policy CSAW thought this trend was "ominous for the future":

> *There is clearly a disproportionate tendency to put women into the position of lecturer or [teaching] associate. The majority of women are employed in non-tenured positions from which they have no access to research funds, sabbatical leaves, or other facilities which are vital to productive scholarly careers. Few departments on the campus have the number of faculty women that could be expected if they were appointed in proportion to the representation of women in the pool of Ph.D.s.*

Women were clustered at the low academic ranks, even among units that had proportionate gender representation. This, together with the scarcity of women professors, raised questions about promotion policies. Were men being pushed for promotion at higher rates than women? It didn't help that it was common practice for men to leverage outside offers in order to obtain promotions, and "...women faculty are less likely to obtain outside offers because of their presumed immobility ..." Furthermore, women reported being actively discouraged from seeking promotion.[20]

The situation in the psychology department was particularly shocking. There were no women among the 41 regular faculty members, but over a quarter (26.7%) of the PhDs produced were women. When a male faculty member in the department was asked why no woman had been appointed to a regular teaching position since 1924, he reportedly replied, "they don't make women like that anymore."[21] But according to Ervin-Tripp, the reality was that there were large numbers of qualified women. They were just being hired as research staff and not as regular faculty members.

---

[19]6/3/1970 memorandum from the UC Berkeley Office of Public Information to Editors & News Directors re the Policy CSAWs news conference about the report.

[20]H. Blumer, E. Colson, S. Ervin-Tripp, F. Newman, and E.L. Scott, *Report of the Subcommittee on the Status of Academic Women on the Berkeley Campus*, May 19, 1970.

[21]S. Almazon, "UC Failing on Women Teachers, Study Shows," *S.F. Examiner* (San Francisco, CA), June 5, 1970, 13.

California State Senator Mervyn M. Dymally (1926–),[22] Democrat from Los Angeles, introduced a packet of bills that would have, among other things, required the proportion of women on the faculty in the UC system match the proportion of women in the California population.[23] While the Policy CSAW agreed with this principle for the long term, they regarded it as unrealistic for the short term. They wanted to be pragmatic. The first step, they thought, should be to get some women in the faculties, period. The second step should be to require the proportion of women on the faculty to match the proportion of women in the population of PhDs. After that they could work on the long-term principle.[24]

Motherhood: The Policy CSAW recommended to the UC president that he request of the regents up to two paid maternity leaves for each woman. The budget committee was asked to prepare recommendations to the president that women be permitted to hold part-time appointments for some portion of their academic careers that would count proportionally toward promotions, tenure, and sabbatical leaves of absence. The UC-Berkeley chancellor was asked to take steps to establish centers for childcare. At the time there were none.

All three of these recommendations recognized women could be both mothers and academics. Unpaid maternity leaves and inflexible tenure ladder appointments penalized academic women who were also mothers.[25] Younger women should be recruited to the faculty "without asking them to sacrifice some of their role as women or make maternity an endurance test that it need not be..." The absence of childcare centers was recognized to be a major handicap to mothers who tried to work within the university.[26]

Women on Committees: The UC-Berkeley committee on committees, a senate committee that appoints individuals to the other senate committees, was asked to appoint women to major policy-making committees in representative numbers. The most influential senate committees are the committee on committees and those committees engaged in major policy-making. Women rarely had seats on such committees: No woman had ever been elected to the committee on committees, educational policy committee, or academic planning committee. In fact, in 1970, the senate had 28 committees and only eight of them had a woman member.

Women Graduate Students: The Policy CSAW recommended the UC-Berkeley graduate council ensure academic units meet certain standards in recruiting and supporting graduate students. Women should be included

---

[22]Note: Mervyn M. Dymally was one of the first blacks to serve in the U.S. House of Representatives. He also served as the 41st Lieutenant Governor of California.

[23]"Women Take Battle to Senate Education Group," *San Francisco Chronicle*, May 28, 1970, 27.

[24]5/20/1970 letter from Betty Scott and UC-Berkeley Anthropology Professor Elizabeth Colson to UC Assistant Administrative Analyst to the President Rebecca Mills.

[25]H. Blumer, E. Colson, S. Ervin-Tripp, F. Newman, and E.L. Scott, *Report of the Subcommittee on the Status of Academic Women on the Berkeley Campus*, May 19, 1970.

[26]11/20/1970 letter from UC-Berkeley Anthropology Professor Elizabeth Colson and Betty Scott to UC President Charles Hitch.

on graduate student admissions committees. They should be admitted and awarded fellowships and other support based on their academic qualifications, in equal competition with men, and not according to quotas or their marital status. Furthermore, "mature" women – i.e., women who have completed childbearing – should be admitted to graduate school based on their earlier academic qualifications, in equal competition with men. The university should strengthen graduate academic advising "and reduce faculty emphasis upon appeals to competitiveness in encouraging students to excel, since this last produces charges of 'aggressive, castrating females' and diminishes the chances that successful women students will be accepted on their merits..." The Policy CSAW did feel that women fared somewhat better in graduate school than they did on the faculty, but the record was unclear. While individual graduate women felt there were problems, their beliefs were only partially supported by available data.[27]

There were plenty of stories about active discouragement of women, either during the graduate application process or after admission. Most women stopped after earning their Master's degrees and did not go on for the PhD. Women reportedly were told they could not "stay the course," that academic careers were contrary to their "true 'feminine' natures," or that they wouldn't be able to find suitable jobs.[28] But discouragement per se was difficult to measure. The data did indicate clearly, however, that women were less likely to receive teaching assistantships or other non-fellowship awards during their graduate study. Some of this related to the fact that there were especially few women in the physical sciences and engineering: "Some departments have no graduate women at all or have so few that they are ridiculed by staff and students."[29] Women were more likely to be found in the humanities and social sciences where there was less financial support for graduate students.[30]

Women Researchers: The Policy CSAW recommended the budget committee take steps to improve the status of women in research units. These should include regular review of women research associates so they could be properly recognized and promoted. Particularly egregious was that women in research units were not allowed to apply for grants as principal investigators because they were not faculty members. They could write grant proposals, but they could not submit them under their own names. The Policy CSAW recommended there be procedures to recognize women who are qualified to apply for grants as principal investigators in their own right: "Given the previous reluctance of schools and departments to hire women in regular faculty positions, and given the inflexibility of the system which has encouraged women

---

[27]H. Blumer, E. Colson, S. Ervin-Tripp, F. Newman, and E.L. Scott, *Report of the Subcommittee on the Status of Academic Women on the Berkeley Campus*, May 19, 1970.

[28]5/20/1970 letter from Betty Scott and UC-Berkeley Anthropology Professor Elizabeth Colson to UC Assistant Administrative Analyst to the President Rebecca Mills.

[29]5/22/1970 letter from Betty Scott to State of California Senator and Senate Education Committee Chairman Al Rodda.

[30]5/20/1970 letter from Betty Scott and UC-Berkeley Anthropology Professor Elizabeth Colson to UC Assistant Administrative Analyst to the President Rebecca Mills.

to carry out their research through the research units since they cannot apply for grants as individuals, such a procedure is justified whether or not a similar review is extended to men research associates."

Other: The subcommittee also submitted a recommendation about the faculty clubs. There were still two clubs on campus, one men's and one women's. The boards of directors should work toward the creation of one faculty club, where women and men have equal status. "Women faculty members report that they have suffered needless humiliation in the past when they have been thrown out of official functions, held in quarters from which they were banned."

The Policy CSAW wanted to ensure its work would continue. The committee on committees was accordingly asked to appoint a standing committee on the status of women, comprised of both faculty and graduate students. Its charge would be "to report to the Senate on the annual progress of the Campus in achieving equality of opportunity for women."[31]

16.3   Appendices

Attached to the body of the report were 69 pages that contained 15 appendices. These covered a wide range of topics. Nepotism was followed by employment, promotion/attrition, insurance, senate committee membership, graduate admissions, graduate student financial support, degrees awarded, doctorates awarded in distinguished departments, number of years to obtain the doctorate, difficulties experienced, suggestions for change, and situation in research units. The richness of the appendices was a reflection of the fact that the Policy CSAW approached the report as a serious research project. This was even though the university provided no staff or funding for the research.[32] The university as a whole was researched, as well as groupings of departments and selected individual departments.

The attitude of the appendices reflected Betty's expertise. They were packed with findings, analyses, remarks, quotes, comparisons, discussions, recommendations, conclusions, data summaries, charts, tables, graphs, estimated probabilities, and references to professional literature. Data sources were carefully documented, including limitations and their effect on the findings.

New research questions invariably arise in the process of answering existing questions. Such was the case with the Policy CSAW report. A good example is Appendix V: Comparative Rates of Promotion and Attrition of Men and Women on the Berkeley Faculty, 1920–1970. The first question was: Is the

---

[31] H. Blumer, E. Colson, S. Ervin-Tripp, F. Newman, and E.L. Scott, *Report of the Subcommittee on the Status of Academic Women on the Berkeley Campus*, May 19, 1970.

[32] 7/1/1970 letter from UC-Berkeley Anthropology Professor Elizabeth Colson and Betty Scott to UC-Berkeley Physical Education Professor D.B. Van Dalen.

common claim true that "women quit?" Betty was surprised when the data indicated the claim to be false in comparison with men. This led to questions about whether there were different rates of promotion and salaries for women in comparison with men. The data were available to answer these questions and indicated there were differences. This led to questions about why differences existed. Unfortunately, the data that were able to answer the "are there differences" questions were not able to answer the "why" questions. A new study, with new data collection, would be needed.[33]

16.4   How to Proceed

The assistant administrative analyst in the office of the president asked Betty and Colson for their advice on how to proceed with improving the status of women on the Berkeley campus. The two gave comments as individuals rather than co-chairs of the Policy CSAW, but they used the Policy CSAW report to back their comments:

> *It may also be that the University needs to provide other facilities to encourage women to develop as full members of the academic community. Universities have made adjustments in the past to provide for the changing nature of its male constituency. At one time, English universities expected their faculties to be composed of celibate males. There was a major outcry about the attack upon the universities when it was decided that it might be well to recognize that many on the faculty were in fact married men with families. Until World War II, American Universities assumed that they catered primarily to the single male student. We learned to live with the fact that many men students are married men with families and adjusted to this situation. Recently we have faced the fact that minority students will need special facilities, even though this is a temporary measure. The University can no doubt absorb the fact that many of its students and employees are married women with children. A great many women reached by the Subcommittee stressed the need for child care centers. The University ought certainly to explore ways of meeting this need. Public facilities in Berkeley are inadequate and private efforts so far seem to be unable to cope with the situation. . .*[34]

Thus, Betty and Colson invoked history to support their advocacy.

Three days after the Policy CSAW report was presented to the academic senate, Betty started to think about the possibility of getting some extra-

---

[33]6/30/1970 letter from Betty Scott to University of Chicago Business Professor Harry Roberts.

[34]5/20/1970 letter from Betty Scott and UC-Berkeley Anthropology Professor Elizabeth Colson to UC Assistant Administrative Analyst to the President Rebecca Mills.

mural funding to continue work on the status of academic women. Her first idea was to approach the US Office of Education. But just then a better idea came across her desk. One of the deans sent a notice from the NSF to the chairs. Betty was a chair, and so she got the notice. The NSF program was a new one. The focus was on research about social problems and action. Betty thought work on the status of academic women would qualify. Her mind quickly scanned the possibilities. A budget committee study of why women are promoted more slowly. A Graduate Division study of why women drop out of graduate school. Departmental studies of why men and women have different post-PhD career paths. And so forth. Indeed, there were many research problems within the area of the status of academic women, and as Betty considered the possibilities, she found other NSF programs she thought might also be suitable to support their work into the future.[35]

Betty decided to go on record in support of Senator Dymally's packet of bills having to do with correcting discrimination against women in higher education, both students and employees. She prepared written testimony that contained selected findings from the Policy CSAW report and sent it to Senator Al Rodda, chair of the all-male California State Senate Education Committee.[36] Betty reasoned the packet of bills would:

> ... break the pattern of discrimination: it will provide opportunities for women to teach in university and college, it will provide models for women graduate students to see and talk to, it will help graduate women students to get a fair chance and even an extra boost which is sorely needed by many in their graduate training. I urge the Senate to pass Senator Dymally's packet of bills.[37]

Betty struggled with her advocacy letters. She generally needed to write more than one draft. She wanted her letters to have "zing yet not be emotional."[38]

It was time to also reach out to the other UC campuses. Betty obtained from the UC-Berkeley senate office the contact information for each of the other campus senate offices and telephoned the indicated individuals. Betty learned that only Davis had a committee on the status of women, where the chancellor had appointed a task force. In addition, UCLA was proposing a standing committee: Their proposal was in the academic senate and would probably be taken up at the next meeting. Betty told Colson: "No other campus has a committee. Each time the Senate Office told me 'Well, you know there are almost no women on this campus,' which is interesting in itself." There were only a few typed copies of the Policy CSAW report left.

---

[35] 5/22/1970 letter from Betty Scott to UC-Berkeley Anthropology Professor Elizabeth Colson and UC-Berkeley Speech Professor Susan Ervin-Tripp.

[36] "Women Take Battle to Senate Education Group," *San Francisco Chronicle*, May 28, 1970, 27.

[37] 5/22/1970 letter from Betty Scott to State of California Senator and Committee Chairman Al Rodda.

[38] 5/24/1970 letter from Betty Scott to UC-Berkeley Social Scientist and WFG Secretary Ellen Gumperz.

The blue copies had not yet been produced. Betty suggested to Colson that Davis and UCLA should be sent typed copies right away. She also suggested exchange of information with those campuses. The other campuses could wait for blue copies. Betty reasoned if they sent blue copies to a few people on each campus, then one would likely end up in the right hands.[39]

The report was completed, and it was time to tidy up. Colson drafted a letter to the chancellor about the report under hers and Betty's signature. Colson sent typed copies of the report to their contacts at Davis and UCLA. Betty still had a large pile of work sheets from Lucy Sells: Since men weren't named, the work sheets would probably have limited utility in future work and should be returned to Sells. The WFG was working on a letter to the chancellor about establishing childcare centers.[40]

Betty had suggested there be a watchdog committee that would have funding to carry on additional research. Colson thought this was an excellent idea. It should include women in both academic and nonacademic positions.

Colson suggested someone might like to work on combining the Policy CSAW report with content from similar studies at other universities. Perhaps combine the UC-Berkeley report with a report by AAUP for Tulane University, or one from Stanford Medical School, or one about the University of Chicago political science department. There could be an original introduction that tied the reports together.[41]

The data available for the Policy CSAW report indicated that on average both academic and research women had lower salaries and were being promoted more slowly than men. This was even though women and men who obtained their doctoral degrees in the same field at the same time seemed to have equal publishing rates. And it was in spite of the fact that women tended to be employed in universities that had higher teaching loads and less stimulating environments for research, i.e., in less prestigious universities.

Betty remained keenly interested in doing a study of the differences between women and men in terms of salary, position, performance, etc. She continued to be interested in a study that would match each woman with a man who was in the same institution and field, obtained the PhD around the same time, and was about the same age; if a good single match couldn't be found, then the woman "could be matched with the average of several men, some a little younger and some a little older."[42]

Meanwhile, local printing of the Policy CSAW report was delayed unexpectedly. There would be a special run of a few hundred copies for distribution

---

[39]Colson, Elizabeth. "Anthropology and a Lifetime of Observation," an oral history conducted by Suzanne B. Riess, Oral History Center, The Bancroft Library, University of California, Berkeley, 2002.

[40]5/24/1970 letter from Betty Scott to UC-Berkeley Anthropology Professor Elizabeth Colson.

[41]5/26/1970 letter from UC-Berkeley Anthropology Professor Elizabeth Colson to Betty Scott.

[42]C.E. Kruytbosch and S.L. Messsinger, "Unequal Peers: The Situation of Researchers at Berkeley," *The American Behavioral Scientist*, (1968), 39.

at the senate meeting on June 2. The full printing would be mailed to the senators on June 5. Betty, women members of the faculty, and others were disappointed to hear that end of the academic year senate business would dominate the last senate meeting agenda. The Policy CSAW report probably wouldn't be discussed until the senate reconvened in the fall.[43]

16.5 Perspectives

The Policy CSAW report cited the federal regulations and reminded the academic senate that these regulations required the university to correct gender discriminatory practices. Title VII of the Civil Rights Act of 1964 prohibited employers from discriminating (compensation, terms, conditions, privileges) on the basis of sex. UC-Berkeley fell under Section 703 of this Act because it accepted federal grant funding. Executive Order 11246, amended by Executive Order 11375, required the university take action to correct discrimination in employment. The Policy CSAW took this position: "In advance of a test in the courts, the Berkeley campus should ask itself if it can lag behind other employers in the fairness of its dealings, and forestall possible federal intervention by its own vigilance against inequality."

Women and minorities were in similar boats. Evidence showed that women faced many of the same barriers to academic success as members of ethnic minorities. Stereotypes and barriers led to the disappearance of women along the pathway from undergraduate student to the rank of professor. In 1970, women had to be very determined and committed to become a professor. Stereotypes drew women to the humanities, social sciences, or service work rather than the sciences. In 1970, women had to work very hard to overcome forces repelling them from the sciences. The Policy CSAW report indicated:

> *She is less likely to be judged on her own merits than as a member of a category for which there is a highly developed stereotype endowed with characteristics which run counter to academic demands. In many instances women appear to be judged by what they might do, given the stereotype, rather than by what they have done. In some instances male colleagues not only judge them in advance but decide for them what they ought to do... Men appear to accept without question that some of their number have the ability to pursue a large number of interests simultaneously. They are less willing to give a woman colleague the right to similar competence.*[44]

---

[43]5/29/1970 memorandum from UC-Berkeley Social Scientist and WFG Secretary Ellen Gumperz to women members of the faculty re last meeting of the year of the group.

[44]H. Blumer, E. Colson, S. Ervin-Tripp, F. Newman, and E.L. Scott, *Report of the Subcommittee on the Status of Academic Women on the Berkeley Campus*, May 19, 1970.

Data for the Policy CSAW report emphasized to Betty her own place of achievement within the university. Jerzy Neyman as a leader had been instrumental in affording Betty – along with other highly competent women and ethnic minorities – a pathway to success within the academy. On her end, Betty had employed all of her intellect, determination, and commitment toward success in academic life in general and the sciences in particular; and now Betty was not just a professor – which is the highest faculty rank in the university – in the sciences, but also a department chair. In 1970, there was only one other woman department chair at UC-Berkeley. There were no women vice chancellors or vice presidents. And, of course, the UC-Berkeley chancellor and the system president were men. Women did sometimes hold positions on the campus as associate or assistant deans. Betty herself had been an assistant dean in the College of Letters and Science in 1965–67. But assistant dean positions were largely focused on undergraduate students. They had a lower status than most other academic administrative positions.

Part V

Getting on the Agenda

17

A Tiny Beginning
(June – July 1970)

*Someone (guess who?) must have put some serious statistical
effort into the UC-Berkeley [report]. I was most pleased to see
that some data had actually been analyzed.*
-Fred Mosteller

The pressure was on. On June 4, 1970, Betty and the members of the Policy
CSAW held a news conference. Sanford Kadish, chair of the senate policy
committee that commissioned production of the *Report of the Subcommittee
on the Status of Academic Women on the Berkeley Campus*, also participated.
Betty and the participants explained the recommendations to upgrade the
status of academic women. They talked about the "problems of academic
women overcoming organizational rigidity and achieving equal opportunity."[1]

The next day, the report was mailed to members of the academic senate.
Feedback was generally positive. The university community would have to
digest the report and then act on it. It was expected to be a long process.[2]

The Policy CSAW had targeted 14 recommendations to specific administrators or committees. Two lawsuits were just filed against the UC and
California State College systems, charging them with violating federal laws
that prohibited gender discrimination in employment. Susan Ervin-Tripp suggested "the university begin moving on [the Policy CSAW's] recommendations
lest the courts find it guilty of discrimination against the fair sex."[3]

Some of the recommendations would require UC system policy changes.
Nepotism was one example. Betty and other Policy CSAW members were
talking with faculty members at other UC campuses to rally their support for
these changes. Data indicated Berkeley was actually doing better than other
UC campuses when it came to proportions of professors who were women.

Some changes were beginning to happen on the Berkeley campus. Betty's
May 26 letter to her chancellor on behalf of the Policy CSAW had caught his

---

[1] 6/3/1970 memorandum from the UC-Berkeley Office of Public Information to editors
and news directors re the Policy CSAW's news conference about the report.

[2] L. Spears, "No Equality for Women on Faculty," *[Oakland?] Tribune*, June 6, 1970.

[3] S. Almazon, "UC Failing on Women Teachers, Study Shows," *S.F. Examiner*, June 5,
1970, 13.

265

attention. The recommendation about increasing childcare facilities for campus constituents was something he wanted to, and could do something about. By June 16 he had taken action to form an advisory committee charged with developing proposals for one or more campus childcare centers. He established an ambitious timeline, with plans to go forward for fall semester. Committee members included faculty and staff of both genders: One was Ellen Gumperz; others were from public health, social welfare, architecture, student housing, and babysitting services.[4] Another change was also happening. At the June 1 meeting of the academic senate, a substantial number of women (15) were appointed to major committees. A woman was even appointed to the powerful budget committee.[5] These changes constituted a "tiny beginning."[6] Betty expected the very existence of the Policy CSAW report would result in some changes. But she also wondered how much long-term change there would be and was not optimistic: "...we do not expect these results to be long-lived."

The WFG held its final meeting of the academic year on June 5, the same day the Policy CSAW report was mailed to the senators. In addition to regular group business, there were important reports relating to the status of academic women: Ervin-Trip on testimony to the California legislature concerning women in higher education; Gumperz on efforts to assist the Child Care Center of the Associated Students of the University of California (ASUC); and Herma Kay on efforts to assist the WFC[7] which had been a friend in the struggles to improve the status of academic women on the Berkeley campus. The club had loaned the Policy CSAW copies of their internal reports. When Betty returned them, she reflected that the problems faced by academic women have been going on for many, many years, and they didn't seem to be getting any better. In fact, they seemed to be getting worse. She hoped the WFC would be a partner in making things better.[8]

17.1   Subcommittee Report Distributed

Betty didn't waste any time: She sent copies of the Policy CSAW report to the other UC campuses the same day that copies were sent to members of the UC-Berkeley academic senate. She recognized that some of the problems faced by academic women were common across the campuses. She issued a

---

[4]6/16/1970 letter from UC-Berkeley Chancellor Roger W. Heyns to Betty Scott.

[5]6/5/70 letter from Betty Scott to UC Riverside Psychology Professor Sally E. Sperling.

[6]L. Spears, "No Equality for Women on Faculty," *[Oakland?] Tribune*, June 6, 1970.

[7]5/29/1970 memorandum from UC-Berkeley Social Scientist and WFG Secretary Ellen Gumperz to women members of the faculty re the last meeting of the academic year.

[8]6/5/1970 letter from Betty Scott to WFC Office Manager Margaret Murdock.

plea the campuses should share information and work together to try to solve some of the problems that academic women faced.[9,10]

Betty sent a copy to Margaret Thal-Larsen, president of the WFC, who was particularly interested in improving the status of women like her in research units. She suggested identifying women principal investigators would have to be done in a timely manner given the time pressures associated with grant application processes. She told Betty: "The Committee's report is a real shocker. Words are virtually unneeded to convert those figures into a tremendous indictment of past policies and practices. I do appreciate your interest in my problems – or maybe they would better be labeled 'incredible adventures' – in this, the second half of the 20th century."[11] Betty sent a copy to a staff member at the Institute of International Studies in the UC-Berkeley Center for South Asia Studies who called it "impressive (and alarming!)."[12]

Bernice Sandler (1928–) obtained a copy of the report. Sandler was chair of the action committee for federal contract compliance of the Women's Equity Action League (WEAL).[13] Sandler gave Betty's name and contact information to others around the country, including an assistant professor at Columbia University who was a member of a committee studying women's issues at her university. Could Betty spare two or three copies? The assistant professor had several friends who she thought would make "excellent use" of them.[14] She also sent Betty a handwritten note telling of a study of women library faculty at Columbia that appeared in the *Barnard Alumni Magazine*. They wanted to look at faculty in other areas but couldn't get reliable data: "Your study is commendable from that point of view and the quotes in the back are hair-raising. The press doesn't take it very seriously and so reports little or nothing. So the university doesn't need to fear any adverse publicity," she told Betty. She wished Betty the best of luck with her efforts.[15]

---

[9] 6/5/1970 letter from Betty Scott to UC Santa Cruz Ecology and Evolutionary Biology Professor Jean H. Langenheim, UC-Santa Cruz Literature Professor Priscilla W. Shaw, and UC-Santa Cruz Anthropology Professor Adrienne Zihlman.

[10] 6/5/1970 letter from Betty Scott to UC Riverside Psychology Professor Sally E. Sperling.

[11] 6/7/1970 letter from UC-Berkeley Institute of Industrial Relations Researcher employee Margaret Thal-Larsen to Betty Scott. Note: Thal-Larsen's comments and letterhead, which had "Mrs. Herman Thal-Larsen" printed on top in bold letters, may seem incongruous when judged by today's standards.

[12] 6/9/1970 handwritten note from UC-Berkeley Institute of International Studies employee Barbara Hered to Betty Scott re the Policy CSAW report.

[13] "All About Bernice Sandler," *Bernice Sandler*, accessed January 6, 2015, http://www.bernicesandler.com/id2.htm. Note: Sandler's curriculum vitae describes her impactful work as chair (1969–71): "Conceived and implemented strategy to require the federal government to enforce existing executive orders (which prohibited organizations holding federal contracts from discriminating). Filed the first charges of sex discrimination against more than 250 universities and colleges. Played a major role in the development and passage of Title IX which prohibits sex discrimination in educational programs."

[14] 6/22/1970 letter from Columbia University Art Assistant Professor Ann Sutherland Harris to Betty Scott.

[15] [6/1970] handwritten note from Columbia University Art Assistant Professor Ann Sutherland Harris to Betty Scott re the Policy CSAW report.

The Cornell University ombudsman's office wrote to Betty asking for two copies of the report. They heard about it from one of Betty's women contacts at the Stanford University School of Medicine. They were collecting information to use in responding to frequent questions from professional women.[16]

Betty took the initiative to send a copy to the National Center for Educational Statistics (NCES). She thought they might be interested in it even though most of the information was "local." The Policy CSAW had found the national NCES gender data to be very useful and frequently quoted NCES in their report. NCES included data from UC-Berkeley, which made Betty wonder: Why didn't UC-Berkeley publish its own gender data? Betty's underlying agenda for communicating with NCES was to open a conversation about possible funding for a national study on the status of academic women. The Policy CSAW report was "only a start" toward remediation. There were also some local studies at other universities: Chicago, Stanford, Tulane, Columbia, and a few others, some of which had not yet come to the reporting stage. Some were long, some short. These also were only starts toward remediation. Betty thought a national study would be timely and warranted and let NCES know she would be happy to cooperate on such a study. She also informed them she would like to carry out a more refined study at UC-Berkeley, and she expected others would wish to do likewise at their universities. Betty wanted to do a "careful" study, using personnel records. She was confident she could carry out such a study at UC-Berkeley.

Betty also sent a copy to Alberta Siegel (1931–2001), a psychology professor at the Stanford University Medical Center.[17] Siegel's response was this:

> *I read it [the report] with interest and admiration. It is by far the best documented report I've seen. The statistical analyses, of course, exceed those in any comparable document. My late husband, Sidney Siegel, was a psychological statistician, so I am able to appreciate sophisticated and clear statistical reporting when I see it, as in your report. I appreciated also the clarity of your prose, and the thoroughness of the recommendations. Your report is being widely read at Stanford. Already I have heard many mentions of it. Copies have been distributed to the members of our new Committee on the Education and Employment of Women at Stanford... [One] member is Lincoln Moses (1921–2007),[18] Dean of the Graduate School and a colleague of yours. Both spoke admiringly of your*

---

[16]6/23/1970 letter from Cornell University Secretary to the Ombudsman Danilee G. Spano to Betty Scott.

[17]Lisa Trei, "Alberta Siegel, child development expert and first tenured female medical faculty member, dead at 70," *Stanford News Service*, last modified November 13, 2001, accessed July 15, 2012, http://news.stanford.edu/pr/01/siegel1114.html. Note: The year before, Siegel had become the first woman ever to earn tenure at the medical school.

[18]Note: Lincoln Moses founded the division of biostatistics at the Stanford University medical school. He also was head, under President Jimmy Carter, of the energy information administration in the US Department of Energy.

report, and indicated that its quality is just what they would have expected under your authorship...

Siegel had collaborated on the recent report on the status of women at the Stanford Medical School, a report she thought was less definitive, and hence less persuasive, than the one Betty had led at UC-Berkeley.[19]

The wife of the UC-Berkeley zoology department chair was a adjunct professor and research zoologist. She had participated in the Policy CSAW study by completing one of the questionnaires and had cooperated with her department chair and research unit director when asked for information needed to complete their questionnaires. She saw the press release that the study was completed and saw a copy of the report when her husband, a member of the academic senate, had brought home his copy. She read the report "with interest and admiration" and congratulated Betty for the "careful and detailed gathering of data, for the informative presentation, and for the good sense of your recommendations." But she had a major complaint: "It strikes me as ironic in the extreme that a subcommittee dedicated to a study of the problems of women on this campus and chaired by two women should altogether overlook the possibility that non-Senate women who are subjects of and contributors to the report might have a legitimate interest in reading it. The oversight is an exquisite example of the failure to take women seriously that your report is all about. Don't you agree?"[20] Betty responded: The points made about the distribution of the report, or rather the lack thereof, were well taken.[21]

The University of Chicago had a committee on university women that, like UC-Berkeley, produced a report in 1970. Statistician William Kruskal was on Chicago's committee. Not knowing Betty was on Berkeley's committee, Kruskal sent her a copy of the 122-page Chicago report thinking that "as a distinguished woman statistician," Betty might be interested. Betty noted that the Chicago and Berkeley committees experienced similar difficulties in producing their reports, and that some of the methods they used were different.

Betty then sent Kruskal a copy of the Berkeley report. Kruskal's first reaction to the Berkeley report was to notice Betty had significant "statistical influence" on it. He asked Betty to please send a copy of her report to Fred Mosteller (1916–2006),[22] who he noted was also very interested in the problems of academic women.[23] Mosteller's reaction, which he wrote to Betty on

---

[19]7/17/1970 letter from Stanford University Medical Center Psychology Professor Alberta E. Siegel to Betty Scott.

[20]7/27/1970 letter from UC-Berkeley Research Zoologist Dorothy R. Pitelka to UC-Berkeley Anthropology Professor Elizabeth Colson and Betty Scott.

[21]7/29/1912 letter from Betty Scott to UC-Berkeley Research Zoologist Dorothy R. Pitelka.

[22]Note: Frederick Mosteller founded the department of statistics at Harvard. He served as president of a number of professional associations.

[23]6/9/1970 letter from University of Chicago Statistics Professor William Kruskal to Betty Scott.

his personal stationery, was similar to Kruskal's: "Someone (guess who?) must have put some serious statistical effort into the UC-Berkeley [report]. I was most pleased to see that some data had actually been analyzed."

Mosteller was sympathetic to the cause of gender equality, and he especially had thoughts about issues surrounding pregnancy:

> *Yes, I think the Status of Women is more contraversial [sic] than WM [weather modification]. We don't necessarily start with equal values there. In WM we more or less are agreed what we wish we could do, and then the question is whether we can or should. In studying women, or under-privileged groups, (one has to be careful not to say minority groups), we need to reconsider concepts like equality, fairness, and so on, and perhaps update them. Just as the Golden Rule can be improved by changing to Do unto others as they would be done by. Partly these problems have to do with what can be forced by a united group, and partly what the society is willing to pay to share equality. I am repeatedly impressed by the in-ability of our society to face the problems of pregnant women, just as a clear problem. The report tries to talk about this, but the problem is much harder than it makes it appear. I have been a chairman, senior professor, advisor, etc. to many women over the last 25 years. Some women take pregnancy almost like the common cold – no more trouble. Others just go absolutely off the beam. The society just isn't set up to respond quickly and reasonably to the needs of these women. For example, a secretary wants to stay until the very end of her term – for very sensible financial reasons, yet she often cannot perform her duties, and may become a great drag on her group. Similarly graduate students find that requirements, even slight ones, are overwhelming when they are pregnant – requirements they would have laughed at only 6 months before. Relations between husband and wife seem to go hay-wire at this time. My point is not that anyone here is at fault, but the problem is complicated, and we just aren't set up to do any-thing proper about it. We seem to be trying in the report to do something positive toward it, but without actually recognizing how complicated this problem is. For example the sabbatical for pregnancy is all well and good, and will you promise to make them take it? How the university is to pay for it all is of course another problem. I liked your report very well indeed. It seemed to have more definitive constructive recommendations than the Chicago one [.] I think the pool of FTE's generally available to departments that could find candidates is ingenious indeed. I do not think the nepotism problem has been thought out as carefully as needed. But on the other hand nepotism rules that don't rule out nephews are scarcely appropriately named. Thanks again for your very nice report.*[24]

---

[24]6/15/1970 letter from Harvard University Statistics Professor Frederick Mosteller to Betty Scott.

Betty appreciated the thoughtful reaction of Mosteller. She and the Policy CSAW had put a tremendous amount of effort into preparing the report, but were finding it "much more difficult to find a reader who will comment on the stuff!" And Mosteller delivered. Betty replied:

> *I was taken aback by the official Berkeley attitude toward pregnant professors. They actually say, "Have your baby during the summer." The reasoning recited is that pregnancy is not an illness and not an academic endeavor so there is no way in the <u>Academic Handbook</u> to authorize a leave. A graduate student who wants to take leave or a reduced load for a quarter when her baby is expected is forced to resign her fellowship not just for that quarter but for the entire year. We claim that the rules that essentially force a girl to quit all together or come to class with her two-day old baby in a basket beside her (which is exactly what happens) need changing. I think that you are quite right in that the problem is much more complex. But allowing a maternity leave, even better, a paid maternity leave, would be an unbelievable improvement at Berkeley.*[25]

Betty was sympathetic to the issue of pregnancy even though she herself never had children.

17.2   Advocacy Letters

Betty took great trouble to correct misconceptions about women. An example, expressed by a biochemistry professor, was that women graduate students were much less likely than men to obtain the PhD. The biochemistry department upheld this false impression as a justification for imposing a quota on the number of women admits to the graduate program. Betty usually found the professor to be a very forward-looking person, a good scholar, and committed to educational excellence. She might be able to get through to him, she thought.[26] And so Betty meticulously laid out the data to him:

> *In our study of women graduate students we did not happen to make an intensive study of your department although we did talk to some women in Biochemistry and we did make a careful study of quite a few neighboring departments in the biomedical sciences. On the other hand, we collected information from university records and from the Graduate Division about men and women graduate students in Biochemistry. You will notice on page 52 of the Report that women in Biochemistry are very slightly more*

---

[25]7/6/1970 letter from Betty Scott to Harvard University Statistics Professor Frederick Mosteller.

[26]6/10/1970 letter from Betty Scott to National Center for Education Statistics Statistician Dorothy M. Gilford.

successful in obtaining the undergraduate degree, they are slightly more successful in obtaining any higher degree, they are quite a little more successful in obtaining the Master's Degree (although few Master's Degrees are awarded), and they are only slightly less successful than men in obtaining the Ph.D. Degree. In fact, you will note that Biochemistry is one of the departments closest to the line corresponding to equality of "success" for men and women. If you are correct, at least approximately, for the slightly differing goals of men and women in Biochemistry the point for Doctor's Degrees almost lies on the line corresponding to equality. The adjusted points are displayed on page 54. All these data refer to the five year period 1962/63–1966/67, which is the latest period for which all data are available. I think that there is not justification for saying that women do not obtain the Ph.D. degree, when compared to men. I would also like to refer to page 57 where the number of doctorates given at Berkeley are compared to the number of doctorates given to "the first five" and to "the first 10" departments of Biochemistry as rated on quality. It appears that the percentage of doctorates going to women at Berkeley is lower than the national average, although not much lower. Yet, at the same time, there are no women on the regular faculty in the Berkeley Biochemistry Department and there never have been any, at least not since 1920. I have heard a rumor that Mrs. Koshland will become a Professor in Biochemistry. In fact, this was told me as an example of how Mrs. Koshland was helped by being appointed as a Lecturer up until now although the speaker could not give me any reason why he should regard this as a help![27]

Meanwhile, a bill was put in front of the California legislature that aimed to reduce gender discrimination among graduate students. The university legislative office refused to support it, saying "there was no discrimination at the University of California."

Betty was also writing letters on behalf of the WFG. She was still a member of the steering committee, now with Kay, Gumperz, Colson, Hanna Pitkin, Public Health Assistant Clinical Professor Mary Murai, and Law Professor and 1970–71 Chairman of the WFG Babette Barton. By this time the charge of the group was clear. It was to promote solutions to problems and difficulties faced by academic women.

One such letter went to the VC. Graduate students had brought to the WFG's attention the need for courses, programs, and majors geared toward women's interests and values. Betty in turn brought the need to the attention of the VC. She urged him to institute study and action on courses and departments that were of particular interest to women. Betty specifically asked that the university make "special efforts" to help women achieve their educational goals. She backed her request with data from a national survey. Betty noted the percentages of women at various levels of educational attainment

---

[27]6/10/1970 letter from Betty Scott to UC-Berkeley Biochemistry Professor Charles A. Dekker.

who were employed; these were increasing "rapidly," and this trend could be expected to continue. She also noted the percentages were similar for UC-Berkeley. Betty argued that, given these statistics and trends, the UC would surely want to include in its mission to help prepare women students for employment "that is both intellectually demanding and rewarding." For various reasons, there were fields that had become known as "women's fields," ones that attracted women and where they could find employment. The university would surely want to strengthen these majors and improve advising and career counseling in these majors. Women in these, and all, fields deserved the opportunity for employment not only in low-level positions, but also in prestigious positions.

Betty singled out the design department as an example. This department was known as a "woman's field." Betty on behalf of the WFG urged the VC to promptly launch a study of, and instigate action on, the difficulties faced by design department women, including proposed changes in the department that appeared to be deleterious to women. Under the revised academic plan for UC-Berkeley, the design department would forcibly be drastically reduced: Full-time-equivalent faculty and students would be decreased by a factor of two over the next five years; and the headcount of students would decrease from 180 to 50 students. This was in spite of the fact that there were many job opportunities available for design majors. Also, there were plans to drastically change the direction of the design department. This was in spite of the will of many of the faculty and students. Appropriate committees of the academic senate should review proposed changes that are drastic, Betty argued. No senate committee had recently conducted such a review of the design department. Betty reminded the VC that the changes in the design department would affect both men and women and accordingly were of concern to all. It just happened that women outnumbered men.[28]

Betty also offered to help a secretary with analysis of survey data on women employees at Lawrence Radiation Laboratory.[29] The secretary, a founding member of the lab's women's association, sent Betty a draft questionnaire to review. Betty suggested some cross-tabulations, even if the numbers seemed to be small: "You can always throw a table away. – But very often these little tables turn out to give very interesting insight into the problems."[30]

Betty's attention shifted back and forth between the local and the national. At the national level, Betty remarkably had the ear of Martha Wright Griffiths (1912–2003), a member of the US House of Representatives between 1955 and

---

[28]6/26/1970 letter from Betty Scott to VC R.E. Connick.

[29]Note: The Lawrence Radiation Laboratory was renamed the Lawrence Berkeley Laboratory in 1959, but Betty still referred to it by its original name.

[30]6/30/1970 letter from Betty Scott to Lawrence Berkeley Laboratory Secretary Miriam Machlis.

1974.[31] Griffiths asked Betty if she could place excerpts of the Policy CSAW report into the Congressional Record. Betty was pleased. She checked with the university administration and verified there were no objections. Betty told Griffiths that "... it would please us even more if action can be taken to provide women equal opportunities for employment, in particular, academic employment in prestigious universities. We admire the steps that you have already taken and appreciate your help."[32]

Virginia Smith (1977–1986)[33] was the assistant director of the Carnegie Commission on Higher Education. Betty sent her a copy of the Policy CSAW report. Smith responded: "I was not only sympathetic to the issue but impressed by the report itself... my personal interest is great." Smith asked for half a dozen copies and asked to meet with Betty. The topic of the meeting: What the Commission might do to follow-up on the report.[34]

The urban crisis initiative was the catalyst for the Policy CSAW. Now it appeared unlikely there would be support for urban crisis programs in the next academic year. Faculty members were urged to look for external funding.[35] Kadish and Kay nonetheless obtained special funds from the chancellor to make 3,000 copies of the report. Copies were sent to each senator. Betty tried to send a copy to each person who helped prepare the report or requested a copy. She exchanged copies of reports with individuals in other institutions preparing similar reports. Betty was surprised at how many similar studies were being conducted. By the end of July, Betty and others were looking at self-publishing as a realistic option for a wide dissemination of the report. They could produce copies like the ones they distributed to the senate for 35¢ per copy. This became the plan.

17.3   First Mention of Big Telescopes

The experience on the Policy CSAW had raised Betty's consciousness.

Until I was appointed on the Berkeley Study Committee, I more or less thought that my problems were personal problems, with the possible excep-

---

[31] "Martha Wright Griffiths," *HIstory, Art & Archives: United States House of Representatives*, accessed July 13, 2012, http://history.house.gov/People/Listing/G/GRIFFITHS,-Martha-Wright-(G000471)/. Note: Griffiths became known as the Mother of the Equal Rights Amendment (ERA), having been highly instrumental in passing legislation to further women's rights; she is especially known for getting sexual discrimination included in the 1964 civil rights bill, and for getting the House to pass the ERA.

[32] 6/29/1970 letter from Betty Scott to The Honourable Martha W. Griffiths of the U.S. House of Representatives.

[33] Note: Virginia B. Smith became the first director of the Fund for the Improvement of Postsecondary Education (FIPSE) and eighth president of Vassar College.

[34] 7/17/1970 letter from Carnegie Commission on Higher Education Assistant Director Virginia B. Smith to Betty Scott.

[35] 7/30/1970 letter from International House staff member Jean S. Dobrzensky to UC-Berkeley Professor in the Graduate School, International and Area Studies Teaching Program, Edwin Epstein.

tion of no woman being allowed to use the 100" Hale telescope. During the study I was surprised, as were many other women on this campus, to learn how general many of our problems are and also how pronounced some of the difficulties are.[36]

This is the first documented instance where Betty talked about women not being permitted to use big telescopes. It was June 10, 1970. Betty's reference to women not being allowed to use big telescopes would emerge many times over the course of her career. The reference would become an iconic example of the differential treatment of professional women.

---

[36]6/10/1970 letter from Betty Scott to National Center for Education Statistics Statistician Dorothy M. Gilford.

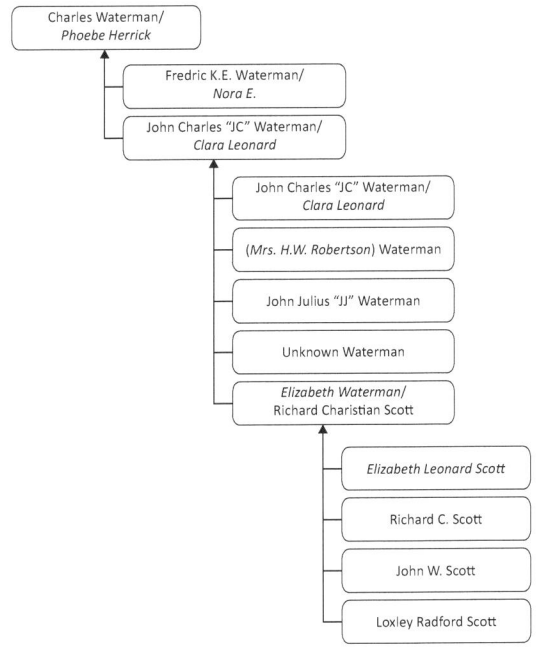

The Waterman Family Tree. Betty came from a family with a strong tradition of military service. Her grandfather John Charles "JC" Waterman, uncle John Julius "JJ" Waterman, and youngest brother Loxley Radford Scott all graduated from the United States Military Academy at West Point (1881, 1910, and 1945, respectively). Her father Richard Christian Scott graduated from the US Naval Academy at Annapolis in 1911. Betty's two other brothers also served in the military.

Maternal grandfather John Charles "JC" Waterman (1857-1939) was a Calvary officer who served at Wounded Knee and went on to have a distinguished military career in the Philippines and along the US Mexican border. (Photo courtesy of David Haas.)

Youngest brother Loxley Radford Scott (1921-1997) taught junior high school math and science after he retired from the army in 1961. (Photo courtesy of USMA West Point.)

Betty's Aunt Phoebe worked as a computer at the Carnegie Mount Wilson Solar Observatory in Southern California before pursuing her PhD at UC-Berkeley. This picture, circa 1890, shows Astronomer Edward Charles Pickering's Harvard computers at work.

Betty's maternal aunt E. Phoebe Waterman (1882-1967) (at left) was among the first two women to graduate with a PhD in astronomy from UC-Berkeley, in 1913. Pictured (left to right) are fellow graduate Sunar Einarsson who became chair of the Berkeley astronomy department when Betty pursued her PhD there; Lick Observatory astronomer William Hammond Wright who helped Phoebe "often;" and fellow graduate and close friend Anna Estelle Glancy (1883-1975) who accepted a position at the Argentine National Observatory at Cordoba with Phoebe and went on to become an optics expert with many lens-related patents. (Photo courtesy of David Haas.)

ELIZABETH SCOTT

Betty graduated in 1939 from a high school that was connected with the UC-Berkeley Education Department and had strong programs in math and science. This is her yearbook photo. (Copy courtesy of Oakland Public Library.)

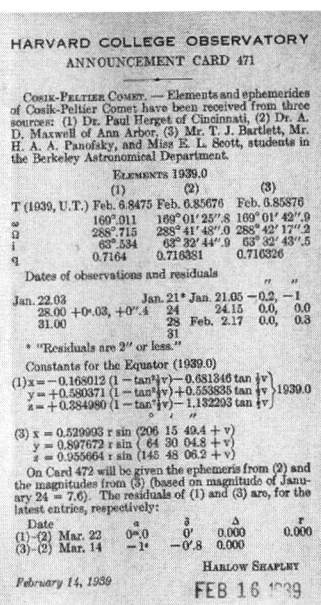

Betty achieved her first publication in 1939 while she was still an undergraduate. It was in a series managed by Harvard University to quickly announce discoveries related to comets and other objects having transient characteristics. (Copy courtesy of John G. Wolbach Library, Harvard College Observatory.)

Circa 1940s, this is a photo of Berkeley and Lick Observatory astronomy "family": (front) John Russell, D. Beard, Betty Scott, Leon Salanave, Keith Pierce, Bob Matthews, Bob Weitbrecht; (middle as indicated on back of photo) Prof. Shane, Mrs. Trumpler, Prof. Trumpler, Prof. Crawford, Prof. Einarsson, Prof. Myer, Mrs. Shane, Nancy Matthews; (back) Julie Trumpler, Margot Trumpler, Margot Weaver, Harold Weaver, Cicile Weaver. (Photo from UC-Berkeley Bancroft Library.)

The Institute of Mathematical Statistics

By action of the Council, upon nomination by
the Committee on Fellows,

Elizabeth L. Scott

was elected a FELLOW in the year 1951
in recognition of his contributions to the
development, dissemination, and application of
Mathematical Statistics.

Betty was elected to be an IMS Fellow in 1951. Her Fellow certificate read, "in recognition of his contributions." (Copy from UC-Berkeley Bancroft Library.)

Betty in 1959. (Photo courtesy of the Oral History Center, Bancroft Library.)

Betty was an active member of the International Statistical Institute. In 1961, she (second from left at the focused table) and colleagues enjoyed the gala dinner together at the meeting in Paris. Betty served as vice president of the ISI in 1981-83. Third from the left is Jerzy Neyman (1894-1981). (Copy courtesy of IMS.)

Within the Berkeley statistics department, Betty only overlapped with four women statistics faculty members over a 33-year period, two of whom are pictured here in 1962: F. N. David (1909-1993) (at left) and Evelyn Fix (1904-1965) (at right). Also pictured is David Blackwell (1919- 2010). (Photo courtesy of the UC-Berkeley Department of Statistics.)

At the top of Sather Gate at UC-Berkeley, there is a bronze sculpture with the Berkeley motto, "Fiat Lux," or "Let There Be Light." (Photo by BrokenSphere - Image:Sather_Gate_star.JPG, CC BY-SA 3.0, https://commons. wikimedia.org/w/index.php?curid=3860059)

Betty was the first woman chair of the UC-Berkeley Department of Statistics, from 1968-73. Here she is pictured with Lucien LeCam (1924-2000), who was a fellow statistics professor and close colleague in the Stat Lab. (Photo by Jostein Kåre Lillestøl.)

This is another photo taken in 1968-69 when Betty was statistics department chair. This was the same time she began her work on the status of academic women. (Photo by Jostein Kåre Lillestøl.)

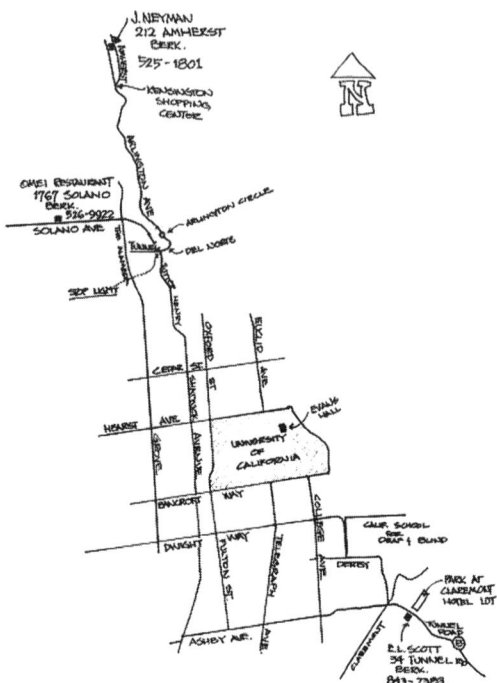

Betty and Neyman enjoyed having colleagues and students to their homes for social events. Circa 1983, this is one of the maps used by Betty or Neyman to direct guests to their houses. These maps were treasured by graduate students. (Copy from UC-Berkeley Bancroft Library.)

Betty and Jerzy Neyman were collaborators and significant others. (Reprinted from Reid, C. *Neyman from Life.* Springer, 1982.)

For over 50 years, Betty lived with her mother in a house directly across from the famed Claremont Hotel. This is her "Tunnel Road House" as it appeared in 2008. (Photo by Amanda Golbeck.)

Betty spoke to many groups around the country, inspiring emergent leaders to work on women's issues. Circa 1970-79, here Betty is speaking to the Lockheed Management Association in Sunnyvale, CA. (Photo from UC-Berkeley Bancroft Library.)

FACULTY STRUGGLE: Women Fight to Hold Gains at UC Berkeley
Trombley, William
Los Angeles Times (1923-Current File); Aug 2, 1976;
ProQuest Historical Newspapers: Los Angeles Times
pg. C1

ALONE IN STATISTICS –Elizabeth L. Scott, professor of statistics at UC Berkeley, is the only woman member of her department.

Times photo by Robert Lachman

In 1976, Betty's work on the status of academic women appeared in the *Los Angeles Times.* Betty's work on women's issues also appeared in *The New York Times* and other prominent newspapers. (Credit to Robert Lachman, Copyright © 1976. Los Angeles Times. Reprinted with Permission.)

FACULTY STRUGGLE

Women Fight to Hold Gains at UC Berkeley

BY WILLIAM TROMBLEY
Times Staff Writer

After several years of controversy, women have made few gains in their efforts to attain equality in faculty hiring and promotion at UC Berkeley, and even their slender achievements are threatened by a tightening job market and increased male resistance.

So say leaders of the women's movement on the Berkeley campus, one of the most active in the state.

"We are just holding the line for some of the gains made in the 1960s and in some cases not even that," said Arlie Hochschild, associate professor of sociology and cochairman of the faculty's committee on the status of women and minorities. Ms. Hochschild pointed to the fact that only 50 of UC Berkeley's 1,182 tenured professors—4.23% —are women, according to a report campus officials submitted to the Office for Civil Rights in the U.S. Department of Health, Education and Welfare last April.

Among newly appointed assistant professors, 33 out of 204 (16.8%) were women, but Ms. Hochschild said it is questionable how many of these ever will be promoted.

Susan Ervin-Tripp, professor of psychology, said the latest report from the campus budget committee showed that there were actually five fewer women in the faculty Academic Senate last fall than there were the year before.

Ms. Ervin-Tripp said there was a higher percentage of women on the Berkeley faculty in the 1920s and 30s than there is today.

Berkeley is not alone.

At USC, according to political scientist Judith Stiehm, there were 38 tenured women on the faculty in the 1974-75 academic year, a drop from 42 in 1971-72.

A survey by the American Assn. of University Professors found that the percentage of women on American faculties dropped from 22.5% to 21.7% last year and that the salary

Please Turn to Page 2, Col. 1

Betty participated on many panels around the country that were organized around women's issues. This 1979 panel included Bonnie G. Bogue, Betty, Belle Cole, and Anita Silvers. (Copy from UC-Berkeley Bancroft Library.)

Betty was an internationalist. She had colleagues, collaborators, and friends across the globe. Here she and Neyman are on a professional visit in Prague, Czech Republic. (Reprinted from Reid, C. *Neyman from Life.* Springer, 1982.)

This is how Betty appeared when the author knew her as a graduate student at Berkeley, during the years 1977-79 and 1981-83. (Photo courtesy of the UC-Berkeley Department of Statistics.)

Committee of Presidents of Statistical Societies
Elizabeth L. Scott Award Winners

F.N. David, 1992

For her efforts in opening the door to women in statistics; for contributions to the profession over many years; for contributions to education, science, and public service; for research contributions to combinatorics, statistical methods, applications, and understanding history; and her spirit as a lecturer and as a role model.

(Photo courtesy of the UC-Berkeley Department of Statistics.)

Committee of Presidents of Statistical Societies
Elizabeth L. Scott Award Winners

Donna Brogan, 1994

For her efforts as founder and first president of the Caucus for Women in Statistics; for serving as effective role model and mentor for graduate students and junior faculty; for promoting employment opportunities for women statisticians; and for productive scholarship in the use of statistics to serve the public health, notably in breast cancer epidemiology.

(Photo courtesy of Donna Brogan.)

Grace Wahba, 1996

For inspiring a generation of women statistical scientists through her outstanding methodological work in splines and computational methods and her leadership in interdisciplinary research; for quietly improving the image and status of women in academia by maintaining a standard of excellence in her teaching and research, and through her uniquely generous and good humored style.

(Photo courtesy of Grace Wahba.)

Ingram Olkin, 1998

For his long-standing commitment to fostering opportunities for women in statistics; for his efforts to promote the interests of women in the professional societies; for his personal service as a mentor for individual women students and junior faculty; for establishing through NSF the highly effective "Stanford Summer Program in Statistics for Women."

(Photo by Ed Souza, courtesy of *IMS Bulletin*.)

Committee of Presidents of Statistical Societies
Elizabeth L. Scott Award Winners

Nancy Flournoy, 2000

For her innovative and highly successful efforts in encouraging women to seek competitive research funding; for envisioning and supporting the pioneering Pathways to the Future Workshops; for serving as a role model and mentor for graduate students and young faculty; for her scholarship in teaching and research, and for her many contributions to the statistical sciences.

(Photo courtesy of Nancy Flournoy.)

Janet Norwood, 2002

For fostering opportunities for women through improved training and salary opportunities; for setting the standard for the advancement of women in statistical positions throughout the federal government; for correcting gender-based inequities at the Bureau of Labor Statistics; for mentoring women throughout their careers; and for serving as a role model through her dedication to professionalism and excellence.

(Photo courtesy of the Bureau of Labor Statistics.)

Gladys Reynolds, 2004

For her outstanding leadership and commitment to the field of biostatistics/epidemiology, to national and international health, and to the promotion of women and underrepresented groups to the full potential of their roles in statistics and public health management and professional society positions.

(Photo courtesy of Gladys Reynolds.)

Louise Ryan, 2006

For serving as a highly visible role model for women at Harvard and around the world; for developing an exemplary summer program for recruiting minority and female students to graduate study in biostatistics; for excellent mentoring; and for supervising numerous female PhD students and postdoctoral scholars who are now in positions to influence the next generation of women in statistics.
(Photo courtesy of Louise Ryan.)

Lynne Billard, 2008

For encouraging women statisticians as they embarked on their careers and mentoring women as they advanced; for excellent leadership to the profession, serving as a role model to the next generation of women and men in statistics; and for conducting and publishing studies to understand and end gender-based inequities in statistics.
(Photo courtesy of Lynne Billard.)

Mary E. Thompson, 2010

For outstanding contributions in research, teaching, and service that have served to inspire women statisticians; For encouraging women at all levels to seek careers in statistics; For excellence in graduate student supervision and mentorship; And for her leadership to minimize gender-based inequalities in employment.
(Photo courtesy of Mary Thompson.)

Mary Gray, 2012

For her lifelong efforts to foster opportunities in statistics for women and to further the careers of academic women; and for creating a forum for discussing the role of women in mathematics; for exposing discrimination, and for exchanging strategies, encouraging political action, and promoting affirmative action.
(Photo courtesy of Mary Gray.)

Kathryn Chaloner, 2014

For her commitment and success in developing programs to encourage and facilitate women to undertake careers in statistics; for extensive mentoring of women students and young faculty; for work to identify and remove inequities in employment for under-represented components of the profession; and for serving as a role model, balancing work and family while excelling as a teacher, researcher, and academic administrator.
(Photo by Eric Sampson/ASA, courtesy of *IMS Bulletin.*)

Amanda L. Golbeck, 2016

For her outstanding efforts in enhancing the status of women and minorities, fostering new leadership opportunities for women and men, promoting diversity at all levels, and advocating for a more inclusive, open, and supportive atmosphere in statistical sciences.
(Photo by Craig Molgaard.)

18

Persistence of Repeated Themes
(August – December 1970)

... the survey found little evidence of... directly measurable
... prejudice. The serious problems emerge more subtly...
-Ann Middleton

It was August 1970. Betty charged ahead and ordered 1,000 more copies of the Policy CSAW report. At 35¢ per copy, the cost was $350 (close to $2,000 in 2010 dollars). She ordered the copies on credit from the campus' printing department. Betty had them sent to her office, 501 Campbell Hall.

The Policy CSAW report was having a nationwide impact. Some really wanted to see this report widely disseminated so that the "entire academic community" could benefit.[1] People began offering to help pay for duplicating. Politically active UC-Berkeley Physics Professor and Nobel Laureate Owen Chamberlain (1920–2006) was the first to offer. Several other people followed by the end of the first week of August. Betty was happy to accept. Furthermore, she and Elizabeth Colson personally took responsibility for any copying costs not covered by these contributions. Betty also took on the tasks of bookkeeping and getting necessary administrative approvals.[2]

UC-Berkeley English Assistant Professor Ann Middleton wrote a review of the report for the biweekly *Berkeley Monitor*, explaining that patterns of discrimination were hard to specify and therefore would be hard to correct:

> *... the survey found little evidence of... directly measurable expression of prejudice. The serious problems emerge more subtly at other crucial points in a woman's academic career, and the persistence of a few repeated themes throughout the many kinds of data is perhaps the most valuable feature of the study.*

Middleton commended Betty et al. for not taking "refuge in [issues] from the specific 'women question,'" finding they let facts speak for themselves, which was a powerful way of attacking false myths perpetuated about academic women. She intensely expressed how profoundly damaging myths can be, and

---

[1] 9/17/1970 letter from University of Minnesota Genetics and Cell Biology Research Associate Phyllis Kahn to Betty Scott.
[2] 8/4/1970 letter from Betty Scott to Charles E. Phillips.

how distortions of images of academic women "even in the minds of women themselves" can be self-fulfilling prophesies of failure.[3]

Toward the end of August, the system administration decided to publicly support equality for academic women. The UC Vice President for Administration (VPA) said he was personally committed to upgrading the status of university women and issued a press release stating the university's concern that all women – students, faculty, and staff alike – have equal opportunities. The release acknowledged existence of gender discrimination in employment and pay; discouragement of women: "...they are frequently taught to curb their exploratory and intellectual aspirations, and are often counseled against entering some fields of study or work"; and underrepresentation of women in high-level positions. It recognized some causes, such as stereotyping "women's work," denying economic responsibilities of women, and promulgating myths about women's performance in the workplace. The release ended with the statement: "The University commends those who are calling our attention to these problems of discrimination, and pledges affirmative action to provide equal opportunity to women within the University of California."[4] Betty was without a doubt among the most deserving of this commendation.

18.1   Hard Facts about Big Telescopes

Use of big telescopes preyed on Betty as she attended the International Astronomical Union (IAU) assembly in England in late August. Two and a half months earlier, Betty had asked the NCES about funding a national study on the status of academic women, mentioning that no woman had been allowed to use the 100" Hale telescope.

Betty wanted to be sure of the facts. While in England, she spoke with a number of female astronomers. She collected details about women being allowed, or not allowed, to use the big telescopes in Southern California. The big telescopes that Betty was referring to were the 60" and 100" telescopes at Mount Wilson Observatory near Pasadena and the 200" telescope at Mount Palomar Observatory in North San Diego County.[5]

The 60" Hale, completed in 1908 nine years before Betty was born, was the world's largest functioning telescope until the 100" was completed nine years later. The 60" has been one of the world's most productive telescopes. With this telescope, Harlow Shapley showed in 1918 that our solar system is positioned halfway between the center and edge of the Milky Way Galaxy, rather

---

[3]A. Middleton, "Academic Women," *Berkeley Monitor* (Berkeley, CA), 1970, 4.

[4]8/26/1970 press release from VPA Robert L. Johnson re concern for equal opportunity for women.

[5]Note: Mount Wilson Observatory was established in 1904 and is located at 5,715 feet. Mount Palomar Observatory was established in 1928 and is located at 5,617 feet.

than in the middle; and spectroscopic analysis and parallax measurements were pioneered, as well as nebula and photometric photography.

The 100" Hooker, completed the year Betty was born in 1917, was the world's largest telescope until the 200" was completed 32 years later. The 100" is one of the foremost scientific instruments of the 20th century. A special optical astronomical interferometer was attached to it in 1919, allowing star diameters to be measured for the first time. Using this telescope, Hubble and a colleague found evidence that some nebulae were galaxies outside the Milky Way and that the universe is expanding.

The 200" Hale, completed the year Betty completed her PhD in 1949, was the world's largest telescope, until the Keck I telescope on Mauna Kea in Hawaii was completed 43 years later. The importance of the 200" cannot be overstated. Using this telescope, astronomers discovered quasars and asteroids, provided evidence of the existence of stars in distant galaxies, and advanced knowledge of synthesis of elements in the universe.

In 1928, astronomer George Ellery Hale (1868–1938) published a vision in the popular *Harper's Magazine* that big telescopes would help uncover the fundamental nature of the universe. Over the years, he took opportunities to garner public support for his vision toward building the 200" telescope. It is no wonder that young astronomers everywhere, including women astronomers, aspired to use the world's biggest telescopes.

Were women being allowed to participate in Hale's vision to uncover the fundamental nature of the universe and use the biggest telescopes? Betty had easy access to the observing schedules for both Mount Wilson and Mount Palomar, because copies went to UC-Berkeley and were posted one time each month. She was able to scrutinize the schedules. She had the hard data on who was observing and when. She drafted a long and detailed letter to Colson and copied the other Policy CSAW members, summarizing what she learned at the IAU assembly in England and from the observing schedules.

The 200 Inch at Mount Palomar: No woman had ever used it.

It used to be the case that women were forbidden to use the 200". In 1970, Betty thought this was probably no longer the case. However, one needed to have a strong scientific claim to use this telescope. This was especially true if the person was not on staff at Mount Wilson Observatory, Palomar Observatory, or Caltech (which owned the telescope). Women astronomers on these staffs would certainly not have had such a claim because, even though some had PhDs in astronomy, none had ever been given the title of "Astronomer." Also, lodging at Mount Palomar was built for men astronomers who came to do observing. It was called the Monastery, because monk-like behavior was required of all guests. No woman had ever stayed there.

The 100 Inch at Mount Wilson: Three women were known to have used it.

Dorothy Davis (1913–1999),[6] with her PhD from UC-Berkeley, was probably the first to use the 100". She used it during World War II to sustain observing programs of men who were away. While Davis always used this telescope in a man's name, she used it with the consent of the director of Mount Wilson Observatory. Betty lamented: "At the time I thought this would break the exclusion of women astronomers, but it did not."

Margaret Burbidge (1919–),[7] with her PhD from University College-London, used the 100" a number of times. Early in her career, she was rejected for a Carnegie Fellowship that would have required her to observe at Mount Wilson Observatory, because only men were allowed. Beginning in 1955, she used the 100" for her own research, but in her astronomer husband Geoffrey's name, under the guise of being his assistant. But it was well known "to everyone," including the administration, that she did all the observing.

Vera Rubin (1928–), with a PhD from Georgetown University, also used the 100" a number of times. She was the first, and up until this time the only, woman to use this telescope in her own name. Betty thought it "curious" that Rubin used the telescope when she reportedly didn't ask. It turned out that some staff members at Mount Wilson were unhappy about the treatment of women and especially unhappy that Burbidge was not allowed to observe with this telescope under her own name. Their protestations had an effect: The Mount Wilson and Palomar Observatories director invited Rubin (not Burbidge!) to use for one night of observing any telescope up to the 100". The invitation included the 48" Schmidt telescope at Palomar, which women also previously weren't allowed to use. It isn't entirely clear why the director selected Rubin. Betty thought it might have been because Rubin was employed at an observatory also connected with the Carnegie Institute. Rubin had an excellent reputation as an observer, which also had to help in the decision.

Davis was not given lodging at the Mount Wilson Monastery, nor were the Burbidges. Instead, they were assigned to one of the Cottages, which were built as summer cottages by early astronomers. Rubin, however, was treated differently: She was assigned to the Monastery under her own name.

The 60 Inch at Mount Wilson: Both female and male Caltech astronomy graduate students had used it.

From its inception, the Caltech astronomy department had a policy that allowed each graduate student to have some observing time on the telescope. The department chair brought to the attention of the Observatory Council that some of the students would be women, and that there would be "no difficulty." Betty didn't record the date that a woman first used the 60",

---

[6] Donald E. Osterbrock, "Dorothy N. Davis Locanthi (1913–1999)," *Physics Today* 53, (2000), 88. Note: Dorothy N. Davis Locanthi had the background and publications for a permanent full-time position in astrophysics, but was never given such an opportunity.

[7] Note: Eleanor Margaret Peachey Burbidge was the first director of the Center for Astronomy and Space Sciences at UCSD. She was one of the strongest advocates for equal opportunities for women in astronomy. She served as president of the American Astronomy Society and of the AAAS.

but she guessed it was after 1965, and she did see women's names on the observing schedules in 1968 and 1969. The women students were not housed in the Monastery. Instead, they were assigned special housing. Betty found it easy to identify the women on the observing schedule, because they had a footnote by their name indicating special housing.

No Simple Explanation: Women would not have had difficulty observing at the telescopes, Betty thought. For one thing, the roads up both mountains were, in her opinion, "easy," even with switchbacks. For another thing, there were staff members whose job it was to operate the machinery that electronically moved the telescopes for the observers. Observers worked at night, so Betty called these staff members "night assistants." They would just have to carry out the "fine scale guiding" at the telescope. Betty had heard second-hand the excuse that they wouldn't like to take directions from a woman: "Margaret Burbidge told me that the night assistants roared with laughter when she told them this, and replied 'All we want is an astronomer who knows what to do, and you're fine.'"

For yet another thing, in Betty's opinion lodging at Mount Wilson was plentiful. The observatories owned the Monastery and Cottages,[8] and for a long time (from 1905) there was a landmark public hotel.[9]

Betty's long and detailed letter to Colson also included these findings:

I did not find a simple explanation for the discrimination against women in using these telescopes. The explanation given is tradition and there is some tendency to blame the founder of the Observatories, George Hale. I know the discrimination was going strong in 1910 and affected women at that time and well beyond my generation (including myself). Women could and did use telescopes at other observatories. However, Mt. Wilson had by far the largest telescopes (the next largest were 36" and 40" until after World War II) and the greatest influence. Now there are other large telescopes available to women astronomers (up to the 120" at the Lick Observatory of the University of California) so the problem is less important even though it has not disappeared. A further fact. Margaret Burbidge told me that she applied for a Mt. Wilson fellowship (postdoctoral) in 1947 (as I recall the date) which she saw advertised in London. The Director wrote her that she would not be considered because these fellowships were open only to men. The advertisement had not stated this. Incidentally, Margaret Burbidge received an honorary doctorate from the University of Sussex last night, along with the Astronomer Royal, the president of the International Astronomical Union and an outstanding Russian. In the presentation there were the usual jokes about the wife and mother bit... [sic].

---

[8] 8/26/1970 draft letter from Betty Scott to UC-Berkeley Anthropology Professor Elizabeth Colson.

[9] Marvin Collins, "Benjamin's Mountain," *Regional Radio History – Los Angeles, California*, accessed July 23, 2012, http://www.oldradio.com/archives/stations/LA/mtwilson3.htm.

Betty must have had second thoughts about sending this letter. She was moving heavily into weather modification research, but continued to have interests in astronomy that could possibly have been compromised by this letter.

Betty wrote a note to herself on the top of the letter: "Never finished or sent – ELS July 1973."[10] But she apparently did send it, at least to Ervin-Tripp who still had a copy over 40 years later. Today Betty is so identified with the issue of big telescopes that Ervin-Tripp remembered the issue as being what fired Betty up about the status of professional women in the first place.[11]

18.2   Requesting Action on Recommendations

At the end of summer, Betty took notice when in August the Berkeley administration distributed a new policy on "submission for contract or grant support of research or training projects." It was related to the Policy CSAW recommendation that the status of women in research units be improved.

But Betty wasn't sure about the specific policy language. She saw two problems. One was a restriction that the person be "involved in a project in a significant manner." The other was bigger: It was the process to obtain an exception that would allow a non-senate member to be designated as a principal investigator. The process would involve up to three months of delay between request for and granting of an exception. This kind of delay didn't conform to the time pressures within the grant application process and was, in Betty's mind, "completely unrealistic," and "makes the situation impossible."

The academic senate discussed the policy, but according to Betty it hadn't finished by the time the chancellor's office distributed it. Nevertheless, it looked like it was final and already being implemented, but there were already two problems with its implementation.

Betty contacted the chancellor's office and expressed her concerns. She inquired about whether it was possible to obtain exceptions on continuing bases, in advance of submitting grant proposals, and if so, what was the procedure.

The VCR told Betty it was indeed possible to request a continuing exception. The unit head would request it for a person in the same way as if it were for a personnel appointment; or the person was welcome to ask the unit head to initiate the request. The VCR also expected deans would be asked to identify individuals worthy of such exceptions. Once the request was prepared it would be reviewed, first by the appropriate dean, and then by the chancellor's office, which would consult with the budget committee.

"He also adds that it will not be easy to obtain these exceptions!" Betty wrote to the steering committee of the WFG. "I have the feeling, and I think

---

[10]8/26/1970 draft letter from Betty Scott to UC-Berkeley Anthropology Professor Elizabeth Colson.

[11]Susan Ervin-Tripp, e-mail to author, April 17, 2012.

that there is some evidence to substantiate this, that this procedure will not turn out to allow non-Senate members to become principal investigators. It appears that the stumbling block is the budget committee. At least, they get the blame." Betty hoped applications would be submitted from some exceptionally worthy candidates. Then they could see what was really happening. "I hope that I am too pessimistic," she wrote.[12]

Betty continued to look for good scholarship on academic women. She noticed an article in *Science* titled "Women in Academe";[13] praised the author and sent her a copy of the Policy CSAW report; and described the representation of women on faculties at prestigious universities (Berkeley, Stanford, Chicago, Columbia, etc.) as going "very much down hill." The proportion of women was decreasing and would decrease more because the proportions of assistant professors were tiny. Betty found this situation most alarming.[14]

How to influence the situation? Betty felt strongly that somehow girls in the lower grades needed to be reached, both before and while they were in high school. Young girls needed to be persuaded to think about their future, and what kind of work they would like to do. Young girls should think about work in the biological or physical sciences. A scientist herself, Betty thought it would be easier for girls to get more interesting jobs in the sciences than in the social sciences or humanities. Then they could think about what courses they should take to do that work. More skills generally mean higher paying jobs. Betty thought girls should take courses in algebra and trigonometry: She was finding very few girls were taking trig. They should take courses in chemistry, even mechanical drawing: Betty was finding girls still weren't taking mechanical drawing. Girls will need these math and science courses to study the sciences in college, Betty reasoned. "I am convinced that, even with discrimination, more skills mean a better paying job."

Betty, it seemed, tried to act on every potential injustice to women that came to her attention. At the beginning of November, Betty wrote a letter to Herma Kay about a potential injustice involving California residency status. There were two situations of concern. (1) A woman was a California resident and married a man who was living in California but was not a resident. They had a home, a joint bank account, etc. in California. The wife was registered to vote in California. They left California temporarily, for less than a year, taking a leave of absence from the university. They sublet their California apartment, they maintained their California bank account, and the woman kept up her California voter registration. But when they returned, the wife was denied resident status at the university. Betty thought the wife should still be regarded as a resident student. (2) A woman was a California resident, and she

---

[12]9/21/1970 memorandum from Betty Scott to B. Barton, H. Amster, L. Gould, B. Sakitt, B. Bain, steering committee of the WFG, re principal investigator status.

[13]P.A. Graham, "Women in Academe," *Science* 169, no. 3952 (1970), 1284.

[14]10/20/1970 letter from Betty Scott to Columbia University History Professor Patricia A. Graham. Note: Graham (1935–1991) went on to become the first female dean at Harvard University.

married a man who was not living in California. When they moved to Berkeley and the wife became a student, she was denied resident status at the university. This happened even when the wife's parents were both California residents. Again Betty thought the wife should be regarded as a resident student. But! If the man instead of the woman was a California resident, he married a women who was not living in California, and the couple moved to Berkeley, then both the man and the women would be considered California residents by the university. Betty didn't think this was fair. She sought Kay's counsel about whether the regulations were being interpreted correctly by the university.[15]

In November Betty received a letter from a mathematics and statistics colleague at American University. He told Betty of an expected vacancy in his department and asked for her recommendations. Betty strongly recommended a woman statistician and enclosed for his consideration a list of other students who were coming onto the job market. Betty commended her colleague for stating in his letter that women would be good candidates. This was the first time she had seen a letter that stated this explicitly. Betty reflected on the seriousness of the problem of the status of academic women: It was "much more serious than I had ever realized." Anticipating his interest, Betty sent her colleague a copy of the Policy CSAW report.[16]

Meanwhile, Betty and Colson began to wonder: Had the recommendations of the Policy CSAW fallen onto deaf ears? Would there be any serious consideration of them? It was important for the entire Berkeley campus that the report not be shelved, and so Betty and Colson asked for some assurance of action from the incoming chair of the senate policy committee.[17] They also drafted some reminder letters to targets of the Policy CSAW recommendations requesting action toward improving the status of the academic women.

The letter to the UC president asked him to take action to abolish the nepotism rule and grant paid maternity leaves.[18]

The letter to the academic senate chair asked him to call on various senate committees to take action. The senate budget committee should assure that both women researchers and faculty be regularly reviewed for rapid promotion, and that qualified research women be given principal investigator status. This committee should also prepare a white paper for the president that would address permitting temporary reductions of percent time for tenure ladder faculty members while still allowing earned credit toward promotions, tenure, and sabbatical leaves of absence. The committee on committees should estab-

---

[15] 11/6/1970 letter from Betty Scott to UC-Berkeley Law Professor Herma Kay.

[16] 11/20/1970 letter from Betty Scott to American University Mathematics and Statistics Professor Basil P. Korin.

[17] 11/20/1970 letter from UC-Berkeley Anthropology Professor Elizabeth Colson and Betty Scott to Institute of Industrial Relations Professor Lloyd Ulman. Note: Lloyd Ulman was Professor Emeritus of Economics & Industrial Relations, Institute for Research on Labor and Employment, Berkeley, CA. He was also Director of the Center for Research in International Labor and Industrial Relations.

[18] 11/20/1970 letter from UC-Berkeley Anthropology Professor Elizabeth Colson and Betty Scott to UC President Charles J. Hitch.

lish a standing committee on the status of women. Betty and Colson's letter to the senate chair said they were pleased the budget committee recommended abolishment of the nepotism rule, but they wanted to make sure this recommendation didn't die in committee. So they nudged him: They told him they trusted that the recommendation had been forwarded to the president, along with a request that he act on it. They also told the chair how pleased they were the committee on committees had appointed women to major senate academic policy-making committees.[19]

The letter to the chancellor asked him to earmark a pool of funds for faculty positions for women, and to direct departments to achieve goals of representation of women faculty. Betty and Colson asked to have representation by 1972 and suggested positions recently detached from foreign language departments be used to advance this recommendation. They wrote to the chancellor:

> *At the time we did the research, only 45 women were in such positions at UC-Berkeley, and many departments with large numbers of women students had appointed no women to regular posts for many years. Women students are becoming increasingly resentful at this situation, and the pool of women Ph.D.s from good schools in this country does not seem to warrant keeping the UC-Berkeley faculty so distinctly a male preserve.*

They also asked the chancellor to direct appointment policies that stressed willingness to consider women candidates. These should address recruitment at national professional meetings, as well as job ads in professional newsletters. Betty and Colson further asked him to appoint a woman member to any committee established to review a woman for either appointment or promotion. Ending on a positive note, they told the chancellor how pleased they were with how quickly he acted to create a committee to explore improvement of childcare for the campus community.[20]

The letter to the graduate dean asked him to implement procedures he had recommended earlier to the Policy CSAW. Betty and Colson recognized these procedures were actually superior to ones they recommended in their report. The dean's procedures aimed to ensure admission and financial support of women, based on their merits: The graduate office would review departmental data and serve as a watchdog over decisions made by departments. Betty and Colson reminded the dean about the Policy CSAW recommendation to admit older women based on their earlier academic records.[21]

Betty went through proper channels. It was the senate policy committee that commissioned the original Policy CSAW report. Betty accordingly sent

---

[19] 11/20/1970 letter from UC-Berkeley Anthropology Professor Elizabeth Colson and Betty Scott to UC-Berkeley Academic Senate Chairman and Energy and Resources Professor Mark N. Christensen.

[20] 11/20/1970 letter from UC-Berkeley Anthropology Professor Elizabeth Colson and Betty Scott to UC-Berkeley Academic Senate Chairman and UC-Berkeley Chancellor Roger Heyns.

[21] 11/20/1970 letter from UC-Berkeley Anthropology Professor Elizabeth Colson and Betty Scott to UC-Berkeley Graduate Dean Sanford S. Elberg.

the letters forward through the chair of the senate policy committee, rather than sending them directly to the administrators.

18.3 Faculty Club Disagreements

Betty's attention was once again on the "combining" of the men and women's faculty clubs. She received the proposed ballot mailed to her and the other voting members of the two clubs. But she wasn't ready to vote.

The chancellor's ad hoc committee on the future of the two clubs resulted in the women forming a so-called "merger" or "joint" committee. This committee consisted of WFC President Mary Lou Norrie, Anne Low-Beer, Mary Ann Johnson, and economics faculty member Emily Huntington (1895–1982). Norrie and Low-Beer, both original members of the WFCSC, were co-chairs of the committee. This committee developed a proposal for the future of the clubs, which they presented first to the boards of directors of the two clubs, then to the membership of the men's club at their mid-November meeting, and finally to the membership of the women's club at their December 1 meeting.[22]

Betty could tell the committee had spent a lot of time and effort on the report, but felt there were too many questions left unanswered. She felt the report was "very unconvincing." She wanted more information. Betty talked with others about the report, and they also had questions that cried out for answers. Betty wrote to the board of directors of the WFC and asked for answers to four questions:

> *Why not have one Faculty Club without discrimination?... Why not have several estimates of the cost of minimum changes needed to bring the Women's Faculty Club building up to code?... How can the two clubs raise what added up to $600,000 when the two separately have not each been able to raise $300,000?... What is the objection to running the two clubs as they were this summer: housing and lunch in both, dinner only in the MFC [Men's Faculty Club].*

Betty had strong opinions, and she voiced them.

Betty would not support the continuation of two separate clubs with two separate membership lists and "two of everything." She wanted rules for membership in one combined club that could apply to both women and men. This would require some work. The men's club had rules that allowed for certain bank managers, which didn't appeal to some women. The women's club had rules that allowed for certain secretaries, which didn't appeal to some men. She wisely told the board this was "a problem in equal opportunities for em-

---

[22]2/1/1971 Minutes of the Annual Business Meeting of the WFC.

ployment not in equal opportunities for membership" and that membership rules could be developed to apply to both genders.

Earlier Betty had a problem with the women's club about membership rules: She and Evelyn Fix resigned their club membership because the board refused membership of a research assistant professor and a research associate professor. These were people Betty had recommended. After a long struggle, the women's club changed their rules to allow the admission of research faculty. Unfortunately one of the women died before the rules were changed, and the other remained terminally angry with the club, so angry that Betty thought she would never hear the end of it. But the bottom line was that the membership rules had been changed, and so they could be changed again.

Betty also did not support enhancements to the physical structures beyond what was minimally needed. She was inherently practical. She did not support new bathrooms, extensive remodeling, or a water sprinkler system. Rather, she limited her support to plumbing, construction, and fire safety feature changes that would either be required or needed immediately. She gave an example of such a money-saving approach they had used recently in Campbell Hall. The joint committee had offered only one cost estimate; Betty reminded the board that getting more than one would be the usual procedure.

Betty wanted a realistic plan that was less expensive, identified sources for funds, and prepared for future needs. Housing and meals were important club functions, but there wasn't enough club housing, even with two buildings, and the space was sometimes too small to accommodate everyone who wanted to eat. Betty wanted to know if rent could be raised to a level that would allow for a fund to cover future repairs. She had other questions, but these were the most burning. She apologized for the "cross" sounding letter, but felt strongly there should be only one faculty club for both genders, one that had two buildings and two restaurants that were open to all members.[23]

In follow-up, Betty and three others prepared a petition to the directors of the WFC requesting a moratorium on voting on the merger or combining of the two clubs, insisting members needed more information to make an "intelligent" vote. They asked that members be given answers to questions having to do with financing repairs, improvements, etc., and that members be given "clear" financial statements from both clubs.[24]

WFC member Josephine Smith served at the WFC formally as treasurer and informally as finance advisor, but had recently resigned from the board after serving half a term. She weighed in with a lengthy letter to the WFC board of directors that focused on legal and financial aspects of the proposed merger. Smith agreed with Betty that members hadn't been given enough information, including a rationale, to make intelligent assessments of the recommended merger. Also "it failed, entirely, deliberately or otherwise" to mention

---

[23] 12/9/1970 letter from Betty Scott to members of the board of directors of the WFC.

[24] n.d. [12/?/1970] petition from Betty Scott and four other women [handwritten signatures not clear] requesting a moratorium on the vote to merge or combine the two faculty clubs on the Berkeley campus.

any of the financial issues involved. "It was climaxed by the arrogant recommendation to close our clubhouse at the end of the winter quarter."

Obviously unhappy, Smith consulted an attorney. She learned two important things. One was that the merger would result in each club assuming the assets and liabilities of the other. In this case, the women's assets and surplus would advantage the men, and the men's debts would disadvantage the women. The assumption of liability of the women for the men's club debt would be held up in a court of law. The other thing Smith learned was it was expected that financial statements be made available prior to voting on a merger; if such statements were not available, members had every right to demand them. The statements should include a balance sheet showing the assets and liabilities, and an earnings report for a representative period. In this case, without such statements, the women would be assuming liability for an unknown amount of men's club debt. That just didn't seem smart.

Smith underlined Betty's assertion that they should not be paying for unnecessary building renovations and improvements. At one point the chair of the building committee recommended $350,000 of changes for the women's club. But Smith had no recollection that these changes had been formally approved, and now this high figure was being used as an argument to abandon the WFC building. Smith also underlined Betty's request for alternative bids. She, like Betty, wanted to see bids that didn't include unnecessary renovations and improvements. At the bottom line, Smith wanted members to see complete financial statements before any vote on the proposed merger was taken.[25]

In the middle of her concerns about the proposed merger, Betty took note of a memorandum pointing out need for a continuing education center for women in the Bay Area. It asserted that, in order to launch such a center, a cooperative effort across institutions should be considered.[26]

18.4   Year End Subcommittee Follow-Ups

The year 1970 was coming to a close. Betty found herself doing more follow-up to the Policy CSAW report.

Betty and Colson now had the direct attention of the UC president. He informed them he had received a copy of their November letter to the new senate policy committee chair asking for reassurance of action on the Policy

---

[25] 12/14/1970 memorandum from WFC member Josephine E. Smith to the president and members of the board of directors of the WFC re legal aspects of the proposed merger – women's and men's faculty clubs.

[26] 12/18/1970 memorandum from Second Careers for Women Chairman Jane D. Fairbank and Coordinator Susan G. Bell to [unspecified] re continuing education for women in the Bay Area.

CSAW recommendations.[27] He also informed them he had been debriefed about a December meeting Colson had with his VPAA.[28]

The president promised to discuss the Policy CSAW recommendations with the chancellors "in the near future." In fact, he had already mobilized his staff to prepare for such a meeting.[29]

On the same day, the VPA wrote a memorandum to the chancellors regarding the status of women. This was for discussion at the next regular council of chancellor's meeting where the chancellors of all the UC campus would come together to advise the president. It referred to the Policy CSAW report and two of the recommendations in particular: abolishment of the nepotism rule, and provision of paid maternity leaves. Three documents were attached.[30]

One was a white paper titled "Possible Modification of the University's Nepotism Rule" prepared for the upcoming council of chancellor's meeting. It presented the current nepotism rules from the Administrative Manual, Section 113, and from Personnel Rule 7.6. It also presented the options for change: Eliminate the nepotism rules, or else revise them to only retain essential safeguards relating to three situations often described as undesirable. One was favoritism given to candidates for appointment or promotion as a result of family relationships. Another was direct supervisor-subordinate relationship between two closely related employees. A third was certain close working relationships between two closely related employees.

The second attachment was a memorandum from Ervin-Tripp to the VPAA discussing nepotism rules: procedural alternatives, extent of problem, and effects on individuals and the university. She cited a study showing that only about a quarter of American universities had formal nepotism rules. Their real objective was to promote objectivity in personnel assessments.

A nepotism study committee[31] had been assembled by the WFG, which asserted that UC-Berkeley could adopt a more specific set of rules toward this objective than ones that then constituted the nepotism rules. In fact, given that various safeguarding rules for hiring and promotion were already in place, only one specific rule was necessary: "Relatives should not participate in any direct decisions concerning the hiring or promotion of their kin." Persons higher in rank should make such decisions. The committee did not see any other detriments to near relatives that would be above and beyond detriments to friends or collaborators or the like serving in the same depart-

---

[27]12/29/1970 letter from UC President Charles J. Hitch to Betty Scott [separate and identical letter to UC-Berkeley Anthropology Professor Elizabeth Colson].

[28]Note: UC President Charles J. Hitch incorrectly referred to this meeting as if both Betty and Colson had met with VPAA Angus Taylor. Betty wrote in the margin to the letter she received from Hitch: "no [new line] Colson met."

[29]12/29/1970 letter from UC President Charles J. Hitch to Betty Scott [separate and identical letter to UC-Berkeley Anthropology Professor Elizabeth Colson].

[30]12/29/1970 memorandum IN CONFIDENCE from VPA J. Roger Samuelsen to UC chancellors re affirmative action and nondiscrimination in the employment of women (For discussion at the council of chancellor's meeting).

[31]Note: The nepotism study committee was comprised of Ervin-Tripp, UC-Berkeley Computer Science faculty member Laura Gould, and UC-Berkeley Physicist Barbara Sakitt.

ment. Husbands and wives especially had suffered under the nepotism rule. People often marry other PhDs in their own or related fields, and they usually want to live together. These couples have geographical constraints, and one of the spouses – usually the female – often ends up taking a "marginal" position "vastly" below her level of training. Ervin-Tripp warned that the UC might be sued on the basis of nepotism rules, as this had already happened at other universities, and expressed a sense of real urgency about revising them.[32,33]

The third attachment was a copy of an article in *Science*. It described actions the US Department of Health Education and Welfare (HEW) was taking on campuses relative to Executive Order 11246, amended in 1968 by President Lyndon Johnson, prohibiting sex and other discrimination by federal contractors.[34] Research universities were federal contractors and fell under this amended order. Failure to comply would mean losing all federal contracts, which would have been devastating to research universities.

Betty was given a list of Universities and Colleges Charged with Sex Discrimination under Executive Order 11246, as amended; by the end of April 1970, there were over 30 institutions on the list, as well as several professional associations.[35] The article in *Science* specifically described the influence of WEAL relative to the executive order. WEAL had sent letters to campus women across the country informing them of the executive order and its potential to elevate the status of academic women, offering to help them file complaints against their administrations. By the end of 1970, WEAL helped to file over 200 formal complaints, among them complaints against whole university systems, including California. WEAL put considerable pressure on HEW, and HEW pledged to thoroughly investigate all complaints.

The *Science* article also specifically described what was happening to the University of Michigan relative to the executive order. The filed complaint against Michigan echoed complaints of Betty and the women at Berkeley. Michigan was resisting the demands and looking for allies among other universities. Some of the women who filed the complaint were being harassed, as had happened earlier to minorities who participated in the Civil Rights movement. The article proclaimed "the women's liberation movement has a new ally" in the HEW.[36,37]

---

[32] Note: Ervin-Tripp specifically noted Arizona and the University of Michigan had already been sued to force compliance with federal regulations regarding the employment of women.

[33] 12/23/1970 memorandum from UC-Berkeley Speech Professor Susan Ervin-Tripp to VPAA Angus Taylor re replacement of the nepotism regulations.

[34] 12/29/1970 letter from UC President Charles J. Hitch to Betty Scott [separate and identical letter to UC-Berkeley Anthropology Professor Elizabeth Colson].

[35] Women's Equity Action League, "Universities and Colleges Charged with Sex Discrimination Under Executive Order 11246, as Amended," January 31 – April 26, 1970.

[36] 1/5/1971 memorandum from UC-Berkeley Business Administration Professor and Budget Committee Chairman Dow Votaw to VCAA John Henry Raleigh re status of academic women.

[37] Robert J. Bazell, "Sex Discrimination: Campuses Face Contract Loss over HEW Demands," *Science* 170, (1970), 834.

Bernice Sandler began to work for WEAL in 1970. Previously she worked as a staff member for the House Education Committee. If not for her eagle eye while still working for the House, the federal government might not have noticed the universities in relationship to this order.

HEW would soon be taking actions on the Berkeley campus. Betty and Colson didn't liberally use the executive order as a threat. Instead, they were trying to be "very polite" and not antagonize anyone, while trying to get UC-Berkeley to recognize its deficiencies in the treatment of women. Ervin-Tripp reminded Colson that they needed the help of their male colleagues to promote the recommendations in the Policy CSAW report.[38] "At that point, we thought it might be a little bit better to deal with them [their male colleagues] as though they were rational creatures, and appeal to them on the grounds of rationality," Colson recalled many years later.[39]

---

[38] [n.d.?1970] letter from UC-Berkeley Speech Professor Susan Ervin-Tripp to UC-Berkeley Anthropology Professor Elizabeth Colson.

[39] Colson, Elizabeth. "Anthropology and a Lifetime of Observation," an oral history conducted by Suzanne B. Riess, Oral History Center, The Bancroft Library, University of California, Berkeley, 2002.

19

We Intend to Do
(January – March 1971)

Women... hold a general image of a male establishment which
resents their claims to equal opportunity and reward...
-Dow Votaw

After the New Year, the budget committee finally attended to the Policy CSAW's work. They recognized the problems academic women faced, in particular the difficulty women had in obtaining tenure-track appointments and, once obtained, in getting tenured and promoted. The chair wrote: "Women who have worked on campus hold a general image of a male establishment which resents their claims to equal opportunity and reward for good work. The frequency with which they are offered work as lecturers or research associates, rather than academic posts with a future, directly reinforces the view they hold and are now expressing publicly."

The chair described how these problems are rooted at the department and college levels because it is there that recruitments, appointments, and promotions originate: "It is here that... faculty [are] given the intellectual support and other encouragement that provide a good background for the active scholar." Because "... evaluations are in their very nature subject to a great many personal judgments [sic]" the campus can't expect to see much increase in the numbers of women if judgments about whether to hire a woman or the best available man are left entirely to the departments. The chair recommended revising or eliminating the nepotism rule, and relaxing the attitude toward part-time employment. He asserted the university could be proactive to ensure biases don't develop. He recommended women be included, whenever possible, on review committees, but he didn't want to overload them with too many assignments, pledging to use them to the best advantage.[1]

---

[1] 1/5/1971 memorandum from UC-Berkeley Business Administration Professor and Budget Committee Chairman Dow Votaw to VCAA John Henry Raleigh re status of academic women.

19.1   State and System Actions

Betty and the Policy CSAW were also helped when the California Legislature added Section 23185 of the Education Code which stated:

> *It is the intent of the Legislature that the Regents of the University of California shall eliminate all policies which detrimentally and unreasonably affect the employment status of females hired by the university. To accomplish this purpose, the regents shall: (a) Review hiring, wages, job classifications and advancement practices as applied to female employees; (b) Review selection procedures utilized for employment of female employees to determine disparate selection practices; (c) Review opportunity of advancement of qualified female employees to executive positions within departments and divisions.*

This language provided impetus to both systemwide and campus initiatives.

The VPA followed up by sending several reports to the chancellors and affiliated national laboratory directors asking for campus feedback. One was a statistical profile of university employees by sex, and another was personnel practices related to women, both commissioned by the universitywide Office of Equal Employment Opportunity. A third was the Policy CSAW report, already in short supply: The VPA indicated he was sending copies to the chancellors, but there were not enough for the vice presidents, assistant vice presidents, personnel director, or principal officers of the regents.

The VPA included several draft suggestions. One was that chancellors and directors create committees to conduct comprehensive reviews and suggest actions to improve the employment status of women. The committees should use methods (e.g., hear testimony) that lead to accommodating "all constructive differences of perspective and opinion." The president's office should continue to compile and analyze data on the status of women. It should continue to review personnel policy and propose changes that promote the equal employment status of women. Changes had already been proposed to the retirement policy. Under review were policies related to nepotism and maternity leave. Also being considered was new policy for campus childcare centers.[2]

The Council of Chancellors at its regular meeting in January had an agenda item, "Affirmative Action and Non-Discrimination in the Employment of Women." The president's office cited three reasons for including this item. One was the University of Michigan situation, where the HEW had threatened to withhold federal funding unless the university came into compliance with federal equal opportunity employment laws for women. The second was the threat that HEW would take similar actions against other universities,

---

[2]1/5/1971 memorandum from VPA Robert L. Johnson to Chancellors, Los Alamos Scientific Laboratory Director, and Lawrence Radiation Laboratory, Livermore, Director re responsibilities following enaction [sic] of State of California Education Code Section 23185.

including Berkeley. The third was the Policy CSAW report. At the beginning of the new year, Betty and her Policy CSAW colleagues had been frustrated by the fact that it took over six months after the issue of their report for the academic senate to discuss it, but now that all the UC campuses were discussing it, there was reason to hope for some real improvements.

The Policy CSAW had recommended the president take action to replace the nepotism rule with conflict of interest procedures. In contrast, the official pre-meeting notes distributed to the chancellors asserted need for a nepotism rule of some kind to prevent abuses. But the note also asserted the written nepotism policies of the university "may be mild enough" and indicated openness to changes in the application of these "mild" policies: Perhaps the chancellors could make permanent exceptions in certain cases, allowing both spouses to be appointed to tenure track positions. The note stated nepotism was only one aspect of the status of women that needed to be addressed. It suggested the chancellors review their situations and report back. The president wanted to know the problem areas on each campus and what actions had been taken to date to address each area.[3]

Distributed as a special item was a draft statement about childcare centers for children of students and employees. The Policy CSAW had recommended the chancellor take steps to establish such centers on the campus. The statement said there were problems associated with establishing these centers, including funding in a time of tight budgets; estimating the extent of need; determining the types of care needed; and determining the legal aspects of the relationship between the centers and the university. The statement concluded the university system therefore was not in a position to operate such centers; but this did not mean that chancellors could not establish and operate such centers on their campuses following broad university guidelines. The chancellors asked for more data on the issue, wanting to know what industry, the state, and the federal government were doing.[4]

19.2 Faculty Club and Center Proposals

Betty and some of her colleagues were getting more unhappy about proposed plans for the faculty clubs. The proposition was to maintain two separate corporate structures, one for the men's club and one for the women's club, with separate memberships, dues, and boards of directors. Operating expenses and use of facilities would be shared, but the women's club would not assume the men's club's debts. Also the women's club building would be released to

---

[3]1/7/1971 Note for UC President's Council of Chancellors, Agenda Item No. 5: Affirmative Action and Non-Discrimination in the Employment of Women.

[4]1/7/1971 Note for President's Council of Chancellors, Agenda Item No. 5: Affirmative Action and Non-Discrimination in the Employment of Women.

the university for other uses; the men's club building would be renovated, and it would provide adequate space for all needed functions.

Betty and five colleagues were particularly unhappy about the release of their building. There wasn't enough space for visiting faculty members as it was. They poured their dissatisfaction into a memorandum to every voting member of the WFC. They charged that the proposition contained "rather serious mis-information [sic] and mis-statement [sic]." The tone was fervent. Following Betty's style, the women combed the proposition, addressing each contested detail. For example, the proposition called the WFC a "tired old building." Betty and her colleagues took such issue with this statement that they hired the Building Inspector of the City of Berkeley, even using their personal funds to pay for the independent assessment.

Betty and her colleagues also noted the language used in the proposition was all over the map. In one place the proposal was for "combined" clubs, in another it was for "amalgamation" of clubs, and in yet another it was for "merger" of clubs. These three terms – combination, amalgamation, and merger – have different meanings when applied to corporations. "What are we voting for?" they asked. What powers do a board of directors of a corporation have to ask for a vote "in principle," and what powers do they have to "release" a building and capital assets without a formal statement of sale, rental, or gift? The bottom line is Betty and her colleagues had a different vision for the faculty clubs.[5] When interviewed about the proposal by the *The Daily Californian*, Betty once again advocated for one corporate club, saying: "Rules can and must be worked out to allow one Faculty Club."[6]

On February 1, WFC President Mary Lou Norrie delivered her annual report to the 63 voting members present at the 52nd business meeting. She told members the results of the December 14 ballot regarding the future of the two faculty clubs were indeterminate: "This leaves us essentially free to start afresh to consider the future of the WFC." She then received authorization for a "Committee of Five" to further consider the future of the two clubs.[7]

Next Norrie informed members that a women's honor society had submitted a proposal relating to the future of the WFC. Prytanean Alumnae, Inc. is comprised of alumnae of the Prytanean Women's Honor Society, which was formed on the UC-Berkeley campus in 1901 and is the oldest collegiate women's honor society in the US. The proposal was to locate a Center for Renewal Education for Women in the WFC building. This was billed as a "new concept in education" at UC-Berkeley. Prytanean had surveyed its membership in 1967, and such a center ranked first among their interests, as members

---

[5]1/5/1971 memorandum from UC-Berkeley Physical Education faculty member Lucille Czarnowski, May Dorin, Florence Minard, UC-Berkeley Music Library Reference Librarian Harriet Nicewonger, Betty Scott, and Josephine E. Smith to every voting member of the WFC re "Ballot." Note: When not indicated, position is not known.

[6]Ellen Maldonado, "Faculty club merger creates stir," *The Daily Californian*, January 7, 1971, 12.

[7]6/3/1971 letter from WFC Board of Directors President Mary Lou Norrie to Dear Member.

themselves had experienced interruptions in their education because of marriage and family responsibilities. Prytanean raised $1,000 (about $6,457 in 2010 dollars) the year before in order to take action on this important issue.[8] A merger of the two clubs was expected to free up multiuse space for this kind of activity. The proposed Center for Renewal Education would operate a counseling center, make available meeting spaces, and provide a clearinghouse for information. Norrie asked WFC members to please consider the proposal.[9]

Betty strongly supported the idea since it would make the WFC much more relevant to students, especially older returning women students. Betty thought the Policy CSAW's study of women graduate students indicated a great need for such a center. She wrote:

> *[The findings] reveal the striking aloneness that many women students encounter when they come to Berkeley. In many departments, including those with quite a few women students, they feel there is no one to talk to, no one who understands their problems or even no one with whom they can simply converse on a friendly basis. I hope that this isolation can be broken down, that women students will be able to discover that their problems are hardly different from those of other women students, and that, with a little cooperation and perhaps some friendly advice from more experienced students and staff members, they can work out their problems and broaden their horizons... We know that quite a few of the problems are really smaller than they seem and that they can often be solved by someone who understands the complexities of Berkeley. I am often impressed by what a reasonable student with reasonable ideas can work out if only she has some reasonable help. I would regard providing this expert help as one of the more important functions of the proposed center.*

Betty "heartily endorsed" the proposed partnership between the WFC and the, in her opinion, very capable Prytanean Alumnae Inc.[10] Ervin-Tripp also advocated for it in view of the graduate student survey conducted as part of the Policy CSAW report. Ervin-Tripp felt such a center could provide solutions to practical problems many women students faced, and hopefully even incubate solutions to such problems.[11] The Prytanean proposal would have some legs. Within a few months, the proposal would obtain the "enthusiastic support and endorsement" of other individuals and groups.

---

[8]New Concept at Cal. [publisher unknown].

[9]2/1/1971 Minutes of the Annual Business Meeting of the WFC.

[10]4/21/1971 letter from Betty Scott to UC-Berkeley Chancellor Roger W. Heyns.

[11]4/20/1971 letter from UC-Berkeley Speech Professor Susan Ervin-Tripp to Prytanean Alumnae President Mary Lee Jefferds.

19.3   De Facto Discrimination

Ronald Reagan began his second term as governor of California in 1971. The country was in the middle of a national recession, and the California budget situation was in dire straits. A California legislative analyst reported: "For the first time since the 1930s, the state in September 1971 will not have sufficient cash to meet its obligations."[12] In response, Reagan unveiled an austerity budget. Among other cuts, his budget for UC included no increase, even though enrollment was projected to grow by 4,326 students, and no funds whatsoever for capital outlay or salary increases.[13] As a result, the university put a "freeze" on academic appointments.

Along with Ervin-Tripp, May Diaz, and Herma Kay, Betty thought the freeze would "work a special hardship to women." The women wrote to the UC president asking him to take action to protect academic women "from what now seems to be a new form of de facto discrimination." Betty was clearly the leader, as the letter was sent on statistics department stationery. As expected with Betty involved, the argument was based on data. Most of the academic women at UC-Berkeley were in temporary positions: In academic year 1969–70, 29.5% of lecturer, whereas only 3.1% of tenure, positions were held by women; twice as many women were employed in lecturer positions as in tenure-ladder positions, and three times as many if compared to tenured positions. The "tiny proportion" in tenure positions would be protected under the freeze. The "appreciable" group of others would be "in grave jeopardy." Many women in temporary positions were on notice that their positions might not be renewed. Betty and her colleagues urged the president "to take immediate and strong steps to rectify such discrimination against women... If the Lecturers and other temporary ranks are depleted by the freeze," they wrote, "the status of academic women will sink dangerously just at the time when strong efforts should be made to improve women's status."[14]

Betty and Colson soon received a communication from their chancellor. He had received their letter asking what was being done about the Policy CSAW recommendations. He took their questions seriously, but did not deliver the answers Betty was hoping for. He agreed a special pool of money to recruit women would be a good idea; but, because the university was in a budget crisis, he could not set up such a pool at that time. He said every effort should be made to hire qualified women candidates; but there had been so many directives from his office about this and similar matters that another would not be productive. He said the idea of appointing a woman to all promotion

---

[12] "Analyst's Report," *California Journal*, February 1971, 35.

[13] "Reagan's 1971-72 Budget Draws Heavy Attack," *California Journal*, February 1971, 32.

[14] 1/14/1971 letter from UC-Berkeley Anthropology Professor May Diaz, UC-Berkeley Speech Professor Susan Ervin-Tripp, UC-Berkeley Law Professor Herma Kay, and Betty Scott to UC President Charles J. Hitch.

committees where a woman was up for tenure was another good idea; but logistics would make it impossible to do this in every case. He believed all "fair-minded people" would agree that all should work to improve the status of academic women. "This we intend to do," he wrote. He also thought people in each field should make lists of women who should be considered at UC-Berkeley for faculty positions and offered to contribute to the lists.[15]

Knowing that she now had his ear, Betty wrote to the chancellor again the very next day. She expressed concern that the hiring freeze would affect special hardships on academic women. Betty included in her communication to the chancellor a copy of their letter to the UC president, where she and others had expressed the same concerns. She asked for the chancellor's help "in minimizing what appears to be a new form of de facto discrimination."[16]

The chancellor reassured Betty. He told her he was sensitive to her point about de facto discrimination and said they would "do their best to minimize the impact." He wanted her to know he had made his first exception to the hiring freeze, and it was to recruit a woman to the faculty.[17] Betty wanted some of her colleagues to know of the chancellor's reassurance. She sent a memorandum to Babette Barton, May Diaz, Susan Ervin-Tripp, and Herma Kay. She typed the entirety of his message to her into the body of her memorandum so that she would not waste paper on a photocopy.[18]

The VCAA's office had come to count on Betty's opinion. They were investigating the case of a woman lecturer in psychology, whose spring course on the psychology of women was canceled due to the hiring freeze. The VCAA turned to Betty for advice and afterwards reported: "Miss Scott does not think – 'although one can never be sure' – that this is an overt case of discrimination."[19] Note Betty did not assume a priori that the woman was being discriminated against. In general, she could not be expected to accept a position and then look for data to confirm it. To her, such an approach was completely backwards. Betty's approach was to look at all of the evidence and then base her opinion on its overall quality, direction, and strength.

In the meantime, UC-Berkeley wasn't ready to follow the earlier suggestion of the VPA to create a campus committee to conduct comprehensive reviews and suggest actions to improve the status of women. The VCA felt that the creation of such a committee "would not be the most efficient or useful approach" for UC-Berkeley "at this time." What would be most efficient and useful would be for Berkeley to prepare a report on nonacademic women similar to the Policy CSAW report on academic women which they found so

---

[15] 1/18/1971 letter from UC-Berkeley Chancellor Roger W. Heyns to UC-Berkeley Anthropology Professor Elizabeth Colson and Betty Scott.

[16] 1/19/1971 letter from Betty Scott to UC-Berkeley Chancellor Roger W. Heyns.

[17] 1/27/1971 letter from UC-Berkeley Chancellor Roger W. Heyns to Betty Scott.

[18] 1/29/1971 letter from Betty Scott to UC-Berkeley Law Professor Babette Barton, UC-Berkeley Anthropology Professor May Diaz, UC-Berkeley Speech Professor Susan Ervin-Tripp, and UC-Berkeley Law Professor Herma Kay.

[19] 2/3/1971 memorandum from VCAA John Henry Raleigh to VC Robert E. Connick re Dr. Revenna Helson: Research: Psychology (course for the Spring canceled).

valuable. This report would likewise be based on accurate and comprehensive data. The UC-Berkeley administration felt they needed both studies to be able to develop sound recommendations for an all-inclusive affirmative action program for the campus that would meet both state and federal requirements.[20]

The issue of gender equity in employment was now getting broad attention. Promoting or considering the issue at the national level were WEAL and HEW; at the state level were the State of California and UC president's office; and at UC-Berkeley were the campus administration and budget committee. But the full academic senate for whom the Policy CSAW report was originally prepared had not yet taken up the issue. Six months had passed since the report was distributed to senate members. The full senate had not yet put it on the agenda for discussion.

Frustrated by senate inaction, the WFG formed a pressure group, which they named the WFG Subcommittee to Implement the Report on the Status of Women. Nancy Zumwalt chaired the group of four that included Betty, along with Harriett Amster at the Institute for Human Learning and Norma Haan at the Institute for Human Development.[21] The group wrote to the senate policy committee chair, expressing fear that the budget freeze would hamper implementation of Policy CSAW recommendations and perhaps make things even worse for women. Betty and her colleagues asked the chair to do everything he could to get the Policy CSAW report onto the senate agenda.[22] "We believe that implementation of the Policy CSAW's recommendations is a matter of pressing urgency; further delay cannot be tolerated."

The national backdrop to Betty and the UC-Berkeley women's advocacy was intensifying. A January 19 meeting at HEW with higher education officials resulted in an action plan for academic women.[23]

At the beginning of February, the UC-Berkeley chancellor, along with other research university leaders, received a set of preliminary notes from The Association of American Universities (AAU), providing advice about how to best deal with HEW regional offices in matters relating to discrimination against women in employment. Universities generally were not in accord with how HEW was administering the federal rules. The advisory from the AAU also contained suggestions about how to engage with HEW staff members who lacked understanding about aspects of how universities work. There was advice about procedural matters in dealing with HEW. Included was a short list of names of university administrators willing to act as informal consul-

---

[20] 1/20/1971 letter from VCA Robert F. Kerley to VPA Robert L. Johnson.

[21] Note: Nancy Zumwalt's position is not known. Harriett Amster was an assistant research psychologist at UC-Berkeley. In 1971 she was serving as Secretary of the WFG. In 1973 she was appointed to a faculty position in psychology at the University of Texas at Arlington. She served there for 34 years, retiring as a professor in 2007.

[22] 2/5/1971 letter from WFG Subcommittee to Implement the Report on the Status of Women Chairman Nancy Zumwalt to UC-Berkeley Senate Policy Committee Chair and Industrial Relations Professor Lloyd Ulman.

[23] 2/5/1971 letter from AAU Director Charles V. Kidd to Council Members.

tants to other universities in dealing with HEW.[24] The AAU strongly advised universities that it was essential to have competent legal advice in dealing with HEW matters.[25] The AAU notes also contained advice about how to deal with the "30 day" requirement to respond to HEW requests for data and analyses, and about how to deal with HEW regional offices that believed in a so-called "statistical approach." There was discussion about the desire among universities for greater uniformity of procedures and rules among the regional HEW offices; and about HEW ground rules that placed the burden of proof of non-discrimination exclusively on the universities.

Betty was especially interested in issues around the so-called statistical approach. It was a proportional argument: Universities should hire women in the same proportion as represented in the pool of qualified applicants. When adopting such an approach, as did many HEW regional offices, there were many important questions. How should the available pool be defined? How should current numbers in the pool be determined? Who should define the pool and determine the numbers? Are adequate numbers even available?

There were bigger questions. Should the statistical approach be upheld, or should more emphasis be placed on demonstrating good faith? How could actions that take the statistical approach to "absurd extremes" be prevented? How could use of "statistical underrepresentation of women" as "prima facie evidence" be discouraged and instead the principal of "good faith" be encouraged? What kinds of evidence would be needed to demonstrate good faith?

Alongside these national discussions, Ervin-Tripp and the WFG's Nepotism Study Committee stepped up pressure on the UC-Berkeley administration. The women wanted to see hiring and promotion policies and procedures brought into line with national civil rights requirements. The women wanted to know: What progress was being made toward this end?

Ervin-Tripp warned the VPAA: "The clear legal guidelines have been given by Health, Education and Welfare for universities, so that lack of publicly available priorities to aid deans and department chairmen [sic] could be construed as non-compliance, and makes the university vulnerable to enforcement proceedings." She warned: It would not be prudent to use the tightening California budget as an excuse to delay action at UC-Berkeley.[26]

Betty got ready for a meeting with the chancellor by making an outline. She wanted to cover the budget crisis, principal investigator, and nepotism rule issues. She added a fourth category but was interrupted before she finished.

The budget crisis was not just a threat, as California eliminated 119 UC-Berkeley faculty positions before the year's end.[27] Betty had much to say. She

---

[24] 2/5/1971 Notes on Administration by HEWs of Equal Employment Opportunity Programs for Women prepared by AAU.

[25] 1/7/1971 letter from AAU Director Charles V. Kidd to Council members.

[26] 2/8/1971 letter from UC-Berkeley Speech Professor Susan Ervin-Tripp representing the WFG's Nepotism Study Committee to VPAA Angus Taylor.

[27] Kelly St. John, "John H. Raleigh – Berkeley professor," *San Francisco Chronicle*, January 12, 2002.

wanted departments to prioritize recruiting women, both in general and from among existing lecturers; someone designated in each department should oversee the recruitment and promotion of women; when women weren't chosen, departments should justify why. Betty wanted affirmative statements about recruiting women in recruiting materials; a proportion of visitor appointments reserved for women; a pool of money set aside for the recruitment of women; and women given priority in the reappointment process.[28]

The WFG was making the VPAA uncomfortable. He asked a UC-Berkeley VC: Should women be dealing directly and regularly with the UC Office of the President?[29] The ultimate answer from the president's office turned out to be "Yes": The women had every right to such access when it came to matters of policy that affected the entire statewide university system.[30]

19.4   Berkeley Budget Committee

The Policy CSAW had the continuing attention of the UC-Berkeley budget committee and its chair, who expressed concern to the VCAA about the high percentage of women academics in non-tenure-track appointments. The chair's opinion was that the underrepresentation of women in tenure-track appointments was a result of assumptions that women had less drive and originality needed to be successful academics. He acknowledged that, if hiring were gender-neutral, more women would be in tenure-track positions. However, he also said women undoubtedly would have to bear a disproportion of the budget cuts and acknowledged this would aggravate the charges of "massive discrimination against women" against the university.

As a result, the budget committee recommended academic women currently employed in non-tenure-track appointments be given priority in hiring for tenure-track appointments. They also recommended each department prepare a report on the qualifications of non-tenure-track women for tenure-track appointments; the budget committee review these reports and make recommendations to review committees; and women who had strong qualifications be put on priority lists for future appointments to tenure-track positions. The chair felt these procedures would accomplish several things. For the academic women, it would raise their morale. For the departments and deans, it would put them on alert and influence future hiring decisions.[31]

---

[28] 2/10/1971 typed notes by Betty Scott re meeting with UC-Berkeley Chancellor Heyns.

[29] 2/16/1971 letter from VPAA Angus E. Taylor to VC Robert E. Connick.

[30] 2/25/1971 letter from VC Robert E. Connick to VPAA Angus Taylor.

[31] 2/23/1971 memorandum from UC-Berkeley Business Administration Professor and Budget Committee Chairman Dow Votaw to VCAA John Henry Raleigh re Status of Women Academics.

On the same day, the chair wrote a second letter to the VCAA responding to changes being proposed to language in the Administrative Manual on part-time appointments, saying: "We believe that a period when the issue of part-time appointments is being contended so strongly by those who seek more equitable treatment for women scholars and teachers is not an appropriate time to legislate peremptorily on the subject." His committee agreed with the VCAA that there should be further study before confirming the policy on part-time appointments in legislation.[32]

The VPAA assured Ervin-Tripp that he and others in the UC system administration were giving attention to the issues; and the administration was bringing them to the attention of the chancellors and systemwide academic senate. He pointed to referrals the administration had made to committees that worked on issues for all the UC campuses: Women's civil rights requirements were referred to the graduate affairs coordinating committee and the budget committee; maternity leaves were referred to the welfare committee; and nepotism rules were referred to the budget committee.[33]

Toward the end of February, Betty and Colson had a meeting with the UC-Berkeley senate policy committee. Colson did the reporting. They discussed much of what Betty had on her outline of issues for meeting with the chancellor. Betty wrote in her notes that the graduate council "has not done anything because all looks OK." She included the quotation marks and made a note to ask her colleague, statistician David Blackwell, about this.[34]

The meeting was successful from Betty's point of view because there would be broader discussion among faculty leaders. After she and Colson left, the chair sent out a notice of meeting addressed to the senate policy committee and chairmen of the committees of the academic senate. The purpose of the March 4 meeting would be to continue discussions of the Policy CSAW recommendations. Betty and Colson were among those invited to attend.[35]

On the day before the meeting was to take place, the chair of the budget committee wrote a lengthy memorandum to the UC-Berkeley senate policy committee. This was a progress report on the Policy CSAW's recommendations on the status of academic women. It focused on the specific recommendations directed to, or within the purview of, the budget committee. These included the recommendations having to do with nepotism, maternity leaves, recruitment, membership on review committees, review of women faculty, less than full-time appointments, and review of women in official research units.

---

[32] 2/23/1971 memorandum from UC-Berkeley Business Administration Professor and Budget Committee Chairman Dow Votaw to VCAA John Henry Raleigh re Section 101-0-c Administrative Manual.

[33] 2/24/1971 letter from VPAA Angus E. Taylor to UC-Berkeley Speech Professor Susan M. Ervin-Tripp.

[34] 2/25/1971 handwritten notes by Betty Scott re Meeting with UC-Berkeley Senate Policy Committee.

[35] 2/25/1971 notice of meeting from UC-Berkeley Committee on Senate Policy Chairman and Industrial Relations Professor Lloyd Ulman.

Regarding nepotism: The budget committee was in favor of a relaxed policy. A blanket policy wasn't serving the university well. It was depriving the institution of the teaching and scholarship contributions of outstanding women. It had proved to be discriminatory. In fact, the committee had been recommending major changes to the policy over a period of years. They acknowledged the existing policy "prevented objective review of the facts" when considering appointments of women in near relative relationships and was being implemented "exclusively against" women. The committee urged prompt "review, approval and implementation" of a nepotism policy that allowed the chancellor to approve appointments in the best interests of the university.

Regarding maternity leaves and less than full-time appointments: The budget committee had already been urging careful study of the possibility for part-time appointments among tenure-track faculty. The chair reiterated this position in a memorandum to the VC.

Regarding recruitment: The budget committee had already made recommendations regarding recruitment in several recent memoranda to the VC. These aimed to encourage appointment of women in tenure-track positions. The committee's view was that barriers to recruiting and admitting women were rooted in departments and colleges: Once in tenure-track positions, women did pretty much as well as men. It is clear the committee understood the problems faced by women in the recruitment process. They even promoted an appointment process that gave preference to women, at least until the imbalance was ameliorated: A good case could be made that "... a male candidate should have something more than a mere preponderance of the evidence on his side before he will be hired over a qualified female candidate."

Regarding membership on review committees: The chair repeated the position of the budget committee presented in an earlier memorandum to the VC: They would nominate women faculty to review committees whenever possible, especially when the person being reviewed was a woman, "even at the risks of imposing heavy burdens on present tenured women faculty and of stretching a point or two on the proximity of relevant fields of competence..."

Regarding review of women faculty and women in official research units: The budget committee had for several years been conducting annual audits of all review files, for both men and women faculty members, on both tenure-track and research appointments. The audits looked for "oversights, unfairnesses, [sic] or discriminations" in promotions or advancements. When the audits raised questions, departments were asked for explanations or adjustments. Overall in the process, the committee found "little overt unfairness or discrimination" against women. However, it did acknowledge, as it had in the past, that many women were in research rather than tenure-track positions largely because they were women. The chair called attention to his earlier memorandum to the VC where he presented his committee's recommendations for "mitigating the crystallization of these past discriminations."

The chair concluded: Some real progress had been made in response to the Policy CSAW's recommendations. Among the things that could be done quickly, much had been done. Among the things that would take more time, much had been started, but much remained to be done.[36]

Just as the HEW put the burden of proof for non-discrimination of women on the universities, UC-Berkeley proceeded to place the burden of proof on departments. The VCAA acknowledged to deans and department chairs that "both conscious and unconscious factors had been cited and adduced."

The university had affirmed there would be no discrimination against women in employment. But charges of discrimination continued, and as long as that was the case, the burden of proof would lie on the university. The VCAA requested, therefore, that departments assume burden of proof. Departments should demonstrate they had adequately searched for possible women candidates. An adequate search would include consideration of women who were currently employed at UC-Berkeley in non-ladder positions.[37]

19.5 UC Survivor Benefits

Betty continued to be concerned about a discrepancy in survivor benefits, and brought it to the attention of the UC Board of Regents Committee on Finance, asking them to consider changing the benefits so they would be the same for both genders. Betty reported that, at the time, a widow generally needed to have been married for one year AND remain unmarried with a child, OR be disabled, OR be age 50. A widower needed to have been 50% dependent on his deceased wife AND remain dependent on the annuity AND remain unmarried. With such rules, the widow would receive more benefits than the widower. The woman employee would receive fewer benefits for her payments than would the man. Betty explained: This is "inverse discrimination against women resulting in discrimination against men."

How could such differences in survivor benefits be justified? Betty heard several justifications: 1) The man is the "breadwinner." But certainly this doesn't apply in "modern times," Betty asserted, especially when the woman is a university employee. 2) Other retirement systems have similar gender differences. One was the Federal Civil Service, but they had just two months earlier eliminated such discrimination, and so the universities should also, Betty asserted. 3) Paying survivor benefits to "the widowers of all these secretaries" would cost too much, and a consequence might be that payments to widows of professors would be reduced. This was a possible consequence that

---

[36]3/3/1971 memorandum from UC-Berkeley Committee on Budget and Interdepartmental Relations to UC-Berkeley Committee on Senate Policy re Policy CSAW Recommendations on Status of Women.

[37]3/17/1971 memorandum from VCAA John Henry Raleigh to Deans and Department Chairmen re departments assuming burden of proof by conducting searches that demonstrate adequate consideration of women candidates.

Betty lamented, but to her it would be even "more unfortunate if widows were to continue to receive an amount A while widowers receive zero; it would be much fairer if each class receives the amount $\frac{A}{2}$."

In asking the regents to act, Betty proclaimed the differences to be inappropriate, archaic, and illegal. She did her homework and cited the Code of Federal Regulations Title 41, Public Contracts and Property Management, part 60-20.3, Item (c): "There shall be no distinction based on sex in retirement benefits." She pointed out that these regulations were a requirement for affirmative action and therefore applied to moneys from all, not just federal, sources.[38]

Betty felt so strongly about elimination of gender discrimination in survivor benefits that she went before the Retirement Board. Following hers and others' testimony, the board chose to leave the widower benefits unchanged, but changed the widow benefits to start at age 50 rather than 62.[39]

19.6   Berkeley Senate Progress Report

The March 4 meeting of the senate policy committee and senate committee chairs led to a draft progress report, including recommendations, on the status of academic women. The report would be presented to the full senate solely from the policy committee, but it had input from the other committees.

The chair of the senate policy committee called for an April follow-up meeting to discuss the draft report of recommendations. Invited were chairs of the budget committee, graduate council, and welfare committee. Betty and Colson were "especially requested to attend." The report reiterated the positions taken earlier by the budget committee on the subjects of nepotism, recruitment, membership on review committees, review of women faculty, and less than full-time appointments. The report also took up additional subjects.

One was recruitment and support of graduate students. Here they referred to the Policy CSAW recommendations that a woman should be included on the admissions committee, and nontraditional women should be admitted on the basis of their previous academic records. Chairs and deans should give continuing consideration to these Policy CSAW recommendations. This probably wasn't as strong a recommendation as Betty would have liked. Giving consideration seemed to fall short of a full endorsement.

Another additional subject was insurance and retirement. Here they referred to two discriminatory aspects of existing programs. One was that contribution rates were different for men and women at entrance to the programs at the same age. The welfare committee commissioned an actuarial study

---

[38] 3/17/1971 letter from Betty Scott to UC Board of Regents Committee on Finance.
[39] 5/27/1971 letter from Betty Scott to Public Information Officer Jack M. Allard.

that recommended elimination of differing rates. The regents approved this and similar changes to the retirement system at their March meeting. This change actually favored women actuarially as they had longer life expectancy than men. The other discriminatory aspect was that eligibility for survivor benefits was different for widows and widowers. Widows were eligible if they were a specified age or else had minor dependent children. Widowers were eligible no matter what their age, but only if they were dependent in the year before their wife's death. The budget committee proposed that eligibility for survivor benefits be made the same. However, the regents chose instead to enlarge the benefits while keeping a diminished difference. As a result, there would remain some, although lesser, discrimination in survivor benefits.

The report ended with a draft resolution that had three parts. The first called for modification of the nepotism rule. The second called for gender neutral hiring and promotion, while recognizing "the potential instructional advantage of including women among regular faculty members." The third called for establishment of a senate standing committee on the status of women. The charge of this seven-member committee would be to report back to the senate each year on progress toward achieving equal opportunities for women.[40] This committee was established and still exists. However, it now has a broader charge to consider ethnic minorities as well as women, under the name Committee on the Status of Women and Ethnic Minorities (SWEM).

---

[40]3/29/1971 memorandum from UC-Berkeley Committee on Senate Policy Chairman Professor L. Ulman to committees on senate policy, budget and interdepartmental relations, and university welfare; graduate council; and Policy CSAW co-chairs Betty Scott and UC-Berkeley Anthropology Professor Elizabeth Colson re Progress Report on Policy CSAW recommendations.

20

A Little Fire
(April – May 1971)

> *...Nothing is going to happen unless there is a clear indication*
> *that there's going to be a penalty if they don't act.*
> *-Aileen Hernandez*

Women in the UC-Berkeley school of education had formed their own caucus. Their steering committee delivered a call for reform to the school's policy committee, copying everyone in sight. "Very few" women in the school were in positions of teaching, instruction, or lecturing, they wrote. Although there had been more on regular faculty appointments, now there were only two, even though almost half of the students were women. The school's budget was extremely tight, but the committee nevertheless asked that women who were in academic positions not be fired or discharged. More women teachers were needed, not less. The school should solicit female applicants for regular faculty appointments and the open position of dean. In general, it should desist from past policies that supported discrimination against women.[1]

The school of education's problems were by no means unique. Employment of women was a problem across the campus. So was curriculum for women. The year before, Betty had talked with the VCAA about courses needed or desired by women students. While need and interest were rising, the actual number of these courses was declining.

Betty had already pointed out to the VCAA the particular problems in the design department and chairs of the relevant academic senate committees, such as educational policy and courses of instruction. In early April, Betty again went to bat for the design department. Students and others had brought to her attention that massive and swift changes were happening there, both in terms of personnel and curriculum. These were red flags. Betty decided to dig into the evidence. She obtained detailed lists of proposed courses to be dropped, courses to be retained in the new curriculum, and new courses to be added. Here is what she found out: The department was proposing to delete 34 old courses, change the catalog numbers of 24 old courses, and add 44 new courses. The design department was clearly undergoing an overhaul. Betty

---

[1] 4/1/1971 letter from the UC-Berkeley School of Education Woman's Caucus Steering Committee to the UC-Berkeley School of Education Policy Committee.

carefully reviewed the changes and deemed them to be "severe," especially as they affected women students.

The design department prepared the necessary course approval forms, added explanatory notes, and sent the package of change forms forward for approval. The normal approval process would have involved review and approval by the department chair, the college curriculum committee, the college dean, and finally the academic senate committee on courses. What happened was that the college dean "illegally" by-passed the college curriculum committee. Without informing the department, he went directly to the academic senate committee on courses and asked them to approve withdrawal of 29 of the 34 courses slated for withdrawal, without requesting them to approve the addition of any of the 44 courses slated for addition. He justified his action by declaring the "catalogue [must] be at least partially correct." But, as Betty pointed out, his request "felled exactly 50% of the courses" in the department, and these were the ones "likely to be of special use and interest for women." After "felling" the courses, the dean declared the design department did not need as many faculty members. This put three lecturers in jeopardy, one a woman and one a minority.

Betty asked the VCAA: "How can the Dean's behavior be interpreted other than as high handed machinations to destroy the curriculum in Design and thus destroy the faculty in Design?" She asked him to please take action "to undo the Dean's illegal meddling" in the department, declaring the situation urgent and needing his immediate attention. Betty in particular mentioned her concern for the fate of a certain woman faculty member, with a specialty in the history of design and the concept and expression of folk art, whose courses were slated for elimination. The senate educational policy committee was already conducting an investigation of the problems in the design department; the faculty member's two series of courses should be retained, she argued, until the committee completed its investigation.[2]

20.1   Finally on the Senate Agenda

Eight months had now passed since the Policy CSAW report was issued. The academic senate finally put the status of women on the agenda for discussion. The senate met on April 13 and received copies of the senate policy committee's progress report.

The progress report presented was almost identical to the draft discussed two weeks earlier in the policy committee and among the selected committee chairs. In one place the language was relaxed: "primarily against women" replaced "exclusively against women" when talking about the discriminatory

---

[2]4/7/1971 letter from Betty Scott to VCAA John Henry Raleigh.

way the existing nepotism policy was being implemented. In a few other places there was some minor wordsmithing that did not seem to significantly change the meaning. Under recruitment, a statement was added that the administration had accepted the recommendations to encourage appointments of more women in tenure-track positions. It quoted the administration directive "that departments demonstrate that for every new appointment proposed an adequate search has been made for possible women appointees. Such a search should also encompass a review of women currently holding non-ladder appointments on the campus." Under recruitment and support of graduate students, a statement was added that: "the suggestion that women applicants are on the whole better qualified than men applicants should be tested and, if the suggestion is correct, an inquiry should be made as to whether this better qualification is fully expressed in granting admissions and fellowships," i.e., studies of the status of women graduate students should continue.

Betty, it seems, took issue with only one item in the progress report. It was the following statement about the change in retirement benefits: "The Board did not eliminate the distinction [between widows and widowers] but as part of the restructuring and enlargement of survivor benefits substantially diminished its impact where there are surviving minor children or dependent parents." Betty wrote in the margin: "Not true."[3]

At the senate meeting, the chair of the policy committee moved the following resolutions be approved as recommended:

I. Whereas, the overriding objective of faculty recruitment and promotion is to secure the highest quality of scholarship and teaching on the Berkeley campus; and

Whereas, the existence of obstacles to the pursuit of professional careers by women scholars, as discussed in the Report on the Status of Academic Women, and elsewhere, has in fact impeded the full realization of this objective; therefore be it

Resolved, that the President of the University secure modification of the nepotism rule (Section 113 of the Administrative Manual) so that appointment of near relatives to positions in the same department is permitted, subject to the Chancellor's review and his finding that such appointment is justified and in the best interests of the University; and be it further

Resolved, that it is the sense of the Assembly that, in hiring and promoting, departments should not assign preference to a man over a woman on the basis of criteria other than those customarily employed in the evaluation of male candidates, and further that departments should recognize the potential instructional advantage of including women among regular faculty members.

II. Whereas although progress has been made in reducing the obstacles

[3]4/13/1971 Report of the UC-Berkeley Committee on Senate Policy presented to the Representative Assembly of the UC-Berkeley Division of the UC Academic Senate, 4 pages.

> to the pursuit of professional careers by women on the Berkeley Campus,
> the status of women in the University will continue to be a matter of
> concern to the faculty; be it further
>   Resolved, that the Committee on Committees be requested to appoint
> a Standing Committee on the Status of Women, and that the following
> By-Law be adopted:
>   - Status of Women. This committee consists of seven members. This
>   committee shall review the status of women on the Berkeley Campus
>   and report to the Division annually on the progress of the Campus
>   in achieving equality of opportunity for women.

Although several amendments were proposed, the two resolutions were voted
on separately and passed without amendment. Betty was present.[4] Two days
later, *The Daily Californian* reported that the senate had finally discussed the
"pessimistic" and fact-based Policy CSAW report: Although women were being treated more fairly than before, they were still not being treated equally,[5]
but the administration maintained it was "trying."[6]

20.2   Faculty Club Renovations

Betty composed a very thoughtful letter to her chancellor "heartily endorsing"
the Prytanean Alumnae Inc. proposal to house a center for renewed education
for women in the WFC building.[7] It explained in detail why she felt such a
center was needed by women students. Betty sent a copy to the president of
Prytanean, who declared her letter to be "a classic": "If we make it [at] all,
yours is the letter that will have done it... Our Board Members who have seen
it are ecstatic and overwhelmed!"[8]

The chancellor, likely influenced by Betty's endorsement, saw an opportunity. Two donor families under the name Levi Strauss Associates had provided $700,000 (around $3.75 million in 2010 dollars) to the Centennial Fund
in 1967–68. They agreed to use the bulk of the funds to remodel The Faculty
Club building, but there would be some funds remaining. One of two options
was to provide $75,000 (about $400,000) in renovations to the WFC building.[9]

------

[4]4/13/1971 minutes of the UC-Berkeley academic senate, p. iv. Note: The senate attendance record for the 1971 calendar year published in the minutes showed that Betty had
attended all four of the monthly meetings.

[5]Toni Martin, "Faculty Acts on Women's Status," *The Daily Californian*, April 15, 1971.

[6]4/16/1971 letter from VCAA John Henry Raleigh to WFG Secretary Harriett Amster.

[7]4/21/1971 letter from Betty Scott to UC-Berkeley Chancellor Roger W. Heyns.

[8]4/24/1971 handwritten letter from Prytanean Alumnae Inc. President Mary Jefferds to
Betty Scott.

[9]4/12/1971 minutes of the meeting of the WFC board of directors. Note: In the negotiations, the donors reportedly made it clear they favored one faculty club that offered
membership to both men and women.

Less than a week after Betty's advocacy to the chancellor, he proposed the renovations of the WFC with the justification that they would make the building fit to house a much-needed woman's education center:

> *Many women enter or re-enter their careers at a period later in their lives than do most men; consequently, they come to the University with a great deal more maturity, but with less understanding and knowledge about how they can complete their education and use their knowledge in the working world. . . Preliminary explorations of the need for this type of center at Berkeley indicated that there will be an overwhelming acceptance of the concept of such a center and very extensive use of its services.*

The renovations would include a sprinkler system, repair of termite and dry rot damage, other minor structural repairs, and deferred maintenance.[10]

20.3   Berkeley Advisory Committees

Nonacademic Women: One week after the academic senate resolution, the chancellor prepared to announce formation of an ad hoc advisory committee as part of the campus affirmative action program.[11] Since the Policy CSAW report had focused on academic women, this committee would focus on nonacademic women. Such a focus was needed to help improve employment of women on the campus and also in the president's office, and to help complete a picture of employment of women generally. The nine-member advisory committee would conduct a study of the employment of nonacademic women and make recommendations for changes in personnel practices and procedures as indicated by the results.[12]

The WFG was still going strong. The action plan from their April 2 meeting was for their steering committee, with Harriett Amster as secretary, to write a detailed letter to the VCA. The letter strongly supported the formation of an advisory committee on the employment of all women, both academic and nonacademic. It should include academic women, even though there was a strong Policy CSAW report, because there needed to be continuing oversight

---

[10]4/27/1971 memorandum from UC-Berkeley Chancellor Roger W. Heyns to UC President Charles J. Hitch re Project and Funding Approval: Levi Strauss Associates. With attachment: Request for Project and Funding Approval, Renovations of the Women's Faculty Clubhouse to Accommodate a Women's Education Center.

[11]5/12/1971 letter from AAUW Developing Educated Womanpower Committee Chairman Corinne Geeting to UC-Berkeley Chancellor Roger W. Heyns. Note: UC-Irvine, not UC-Berkeley, was reportedly the first campus in the UC system to form a chancellor's advisory committee on the status of women.

[12]4/23/1971 memorandum from International House staff member Jean S. Dobrzensky to VC Robert E. Connick, John Wagner, and Dick Hafner re call for nominations to serve on an advisory committee on the employment of nonacademic women.

of academic women in employment; but this should not overshadow oversight of nonacademic women in employment. Thus, the advisory committee should also include maids, librarians, clerk-typists, administrative assistants, and the like. Furthermore, it should include employees of University Hall in the UC system office, as their problems and practices were thought to be similar to women on the Berkeley campus.

The letter recommended some specific individuals to serve on the committee. The first person mentioned was Betty. It argued that, since Betty was a department chair, she would be in a unique position to contribute.[13]

The UC-Berkeley administration expressed appreciation for the suggestions offered by the WFG, but it wanted to go ahead immediately with nonacademic women, and wait on academic women.[14] The administration announced their program the next day to deans, directors, department chairs, and administrative officers: There would be an ad hoc advisory committee for nonacademic women along the lines previously expressed.[15]

Academic Women: The WFG really wanted the chancellor to form an advisory committee on academic women in addition to the standing committee the academic senate was forming. On April 30, the WFG wasn't aware the chancellor had already put out a call two days earlier for nominations for an advisory committee that was limited to nonacademic women.

The group pressed for a chancellor's committee on academic women. In addition to women in tenure-track appointments, they wanted to include women in non-tenure-track appointments: professional researchers, supervisors, lecturers, specialists, librarians, physicians, continuing education of the Bar specialists, assistant directors, and others. The group hoped to see representation on the committee in proportion to the numbers of women employed in these categories. They would be happy to provide nominations.

WFG Secretary Amster wrote to the chancellor: "It is heartening to see your strong action in the direction of rectifying inequities which exist for women on campus and we can only hope that others in influential positions will be influenced by your lead."[16] However, Amster and others thought legal action would be necessary to propel the actions of Berkeley leaders. Only the day before, she and Hernandez had filed the class action complaint.

---

[13]4/2/1971 letter from WFG Secretary Harriett Amster (and the other members of the steering Ccommittee: Babette Barton, Pres., Beatrice Bain, Barbara Sakitt, and Laura Gould) to VCA Robert F. Kerley.

[14]4/27/1971 memorandum from Janice to Mrs. Jean re getting going on the employment-related problems of women.

[15]4/28/1971 memorandum from VCA Robert F. Kerley to Deans, Directors, Department Chairmen, and Administrative Officers re the UC-Berkeley chancellor's advisory committee on employment of nonacademic women.

[16]4/30/1971 letter from WFG Secretary Harriett Amster to UC-Berkeley Chancellor Roger W. Heyns.

20.4 Class Action Complaint

A serious political pressure group of UC-Berkeley faculty, staff, and community members organized itself as the "League of Academic Women" (LAW). The mission was to "rectify university discriminatory practices." They meant business. LAW pressed for Berkeley to be investigated by HEW right away and federal contract funds to be cut off until Berkeley took affirmative action.[17]

On April 29, LAW and NOW jointly filed a formal class action complaint against UC-Berkeley and the UC system administration. Submitted to the HEW Office of Contract Compliance regional office in San Francisco, it didn't arise out of one specific issue and covered all areas of the university and all women.[18] It charged the university with sex discrimination in the recruitment, hiring, and promotion of women, both academic and nonacademic;[19] and discrimination against women students in admission and fellowships.[20,21] Women had just won a similar suit against the University of Michigan.

A press conference was held at Glide Memorial United Methodist Church in San Francisco on April 29. Aileen Hernandez, President of NOW, explained to news correspondents the nature of the complaint and possible consequences, which were massive:[22] Berkeley would be threatened with annual losses of $29 million in federal funds and with ineligibility from future contracts;[23,24] about a third of UC's annual budget would be threatened. HEW was to decide within 60 days whether the complaint was valid; if valid, HEW would begin work to ensure UC-Berkeley came to compliance.[25] LAW's press release declared funds could be cut off "because of illegal discriminatory policies of the university" and if UC-Berkeley "fails to rectify its existing inequalities in employment."

The legal basis was threefold: (1) Title VII of the Civil Rights Act of 1964 called for equal opportunity in private employment for women and others. (2) Executive Order 11375 called for equal opportunity in federal employment, and in employment by federal contractors, for women and others. (3) Title 41 of the Code of Federal Regulations amended in 1970 included guidelines for

---

[17]n.d. [4/1971] Press Alert from LAW re their class action complaint against UC-Berkeley.

[18]"HEW Allowed to Examine Personnel Records at UC," *The Daily Californian* II(19), February 1, 1972.

[19]n.d. [4/1971] Press Alert from LAW re their class action complaint against UC-Berkeley.

[20]"HEW Investigates UCB," *Gadfly* (Newsletter of the Graduate Assembly), 1(2), June, 1971.

[21]n.d. [?1971] class action complaint: League of Academic Women and National Organization for Women's Class Action Complaint Against The University of California at UC-Berkeley and The Office of the President of the University of California for Violation of Executive Orders 11246 and 11375, Forbidding Discrimination on the Basis of Sex.

[22]n.d. [4/1971] Press Alert from LAW re their class action complaint against UC-Berkeley.

[23]Ruthe Stein, "Complaint Filed Against UC," *San Francisco Chronicle*, April 30, 1971.

[24]Betty Hoffman, "Second Citizens in Academe," *UC News* (draft), [?1971].

[25]Ruthe Stein, "Complaint Filed Against UC," *San Francisco Chronicle*, April 30, 1971.

recruitment, hiring, promotion, seniority, and compensation designed to deter government contractors from discriminating against women.[26]

LAW cited the Wright Institute's report on personnel practices related to women as the basis for their complaint for nonacademic women.

> *[The report] alleges that non-academic women personnel are kept at lower levels even when they have experience and educational qualifications for higher job categories. Additionally, women are systematically overlooked for promotions. Women working with men in equal categories generally receive lower salaries. The report also accuses university affiliated [sic] unions of discriminatory recruitment and training practices. Women are excluded from apprenticeship and training programs for minorities, and the unions are not actively recruiting women from available special sources.*

They cited the Policy CSAW report as documentation of their complaint for academic women.

> *[The Policy CSAW report] asserts that the university has failed to employ women in privileged positions on the faculty, and that the few women who are promoted must wait longer, and receive less pay, than their male colleagues. Less than 3.6% of the professorial positions are held by women, while $\frac{1}{3}$ of the less privileged positions of instructor, lecturer, associate or teaching assistant go to women.*

Thus LAW would use Betty's work, the Policy CSAW report, to support the first comprehensive class action against UC-Berkeley.[27] Unfortunately, no such report was available to cite as the basis for LAW's complaint for students. In the previous fall, the graduate assembly had sent out a questionnaire to graduate women, but they had difficulty getting responses. They did get some responses from departments that had no women applicants or enrollees (e.g., engineering), and from departments that had "no conception of their own discrimination." But departments with critical masses of women students tended not to respond. LAW suspected "that women with glowing recommendations are being passed over in favor of men with poorer academic records. But we have no way to prove it, since these files are not available." HEW on the other hand would have access to all student data and be able to prove discrimination if it existed.[28]

It was Hernandez and Amster who were pressing the class action.[29] There were 85 names on the list, including Betty's. But Ervin-Tripp's strong recol-

---

[26]n.d. [4/1971] Press Release: Further Information by the League of Academic Women re their class action complaint against UC-Berkeley.

[27]n.d. [4/1971] Press Release: U.C. Berkeley Federal Funds Threatened by the League of Academic Women re their class action complaint against UC-Berkeley.

[28]"HEW Investigates UCB," *Gadfly* (Newsletter of the Graduate Assembly), 1(2), June, 1971.

[29]4/29/1971 internal memorandum from Janice to REC re identifying who was pressing the class action suit against UC-Berkeley. Note: Amster was the academic coordinator.

lection some 40 years later was that Betty was not involved in the complaint; in which case Betty's name may have been included because she was co-chair of the Policy CSAW and not because she sought to be included. Ervin-Tripp remembers Betty as one who worked within channels, which makes sense given Betty's military background.[30]

The administration's public response to the class action was to declare they were developing "a formal affirmative action program to provide equal opportunities for women in all job categories," and that they had already begun to implement the recommendations of The Report on Personnel Practices Related to Women and the Policy CSAW report. Specifically, the UC-Berkeley administration announced that, over the past year, departments were required to produce evidence that women were appropriately considered for open faculty positions. More women were hired (8 out of 37 new faculty in the college of letters and science for 1971–72, compared with 4 out of 49 for 1970–71). Women were given new opportunities in administration. Spouses were no longer banned from working together, maternity leave was guaranteed, and discriminatory pension provisions were eliminated.

LAW's public response was that the administration wasn't acting fast enough. They charged, among other things, there were some teaching positions that women never heard about until they were filled, and the university sometimes redefined positions so women would not qualify. According to Hernandez: "We have had a lot of pious statements. Black and chicanos [sic] as well as women have been told (by universities) that they are going to do something, but there has been no action. Nothing is going to happen unless there is a clear indication that there's going to be a penalty if they don't act."

Ervin-Tripp supported the class action. She told the local city newspaper that over the past year the campus had made "a lot of progress... but I don't think they will get anywhere unless there is a little fire under the university."[31,32,33]

20.5   Awakened by Stories and Statistics

The month of April closed with another organization applying pressure to the university administration. This was the American Association of University Women (AAUW). The California State Division of AAUW was large, with some 28,000 members. California leadership of the AAUW wrote a letter to the

---

Karen Nelson-Hurn was the student coordinator. Pauling Fong and Laura Saunders were the nonacademic coordinators.

[30][?7/1971] listing Class Action Complaint Against the U. of C. at Berkeley & Office of Pres.

[31]"Women's Jobs at UC," *San Francisco Examiner*, April 30, 1971.

[32]Ruthe Stein, "Complaint Filed Against UC," *San Francisco Chronicle*, April 30, 1971.

[33]Steve Duscha, "Charge UC 'Unfair' to Women," *Berkeley Daily Gazette*, April 30, 1971.

chancellor. They politely expressed their desire to be positive change agents on behalf of women, enclosing a copy of a booklet, *Alice in Academe*. This booklet was unlike the more clinical, statistical approach used by Betty and the Policy CSAW. The leadership explained this booklet was meant to be lighthearted. The general theme was that women had been stereotyped in their homes and by society, which had led to their underrepresentation in higher education, which was a handicap to a rapidly changing society. This situation could and should be rectified.[34]

Betty continued to statistically monitor the academic women at UC-Berkeley. As the 1970–71 academic year was coming to a close, she asked the administration for a list of women faculty. It was easy to see from the list she received that men had a lion's share of the top academic positions. These were the statistics: As of fall 1970, only 19 (2.3%) of 812 full professors were women. Betty, Elizabeth Colson, Susan Ervin-Tripp, Herma Kay, and Josephine Miles – who had been so active on behalf of academic women – were among the 19. Only 20 (6.6%) of 304 associate professors were women. Only 14 (4.4%) of 322 assistant professors were women. The people in acting or visiting professor, associate professor, or assistant professor positions were also mostly men. It was only when you got to the lower ranks of acting instructor or associate that the percentages of women were larger: 23.9% of acting instructors were women, and 27.7% of associates were women.[35]

In May, the California budget situation had not resolved itself, the academic year was completing, and UC-Berkeley lecturers were closer to unemployment. The situation for women was getting more acute. Betty received a telephone call from a lecturer in Italian who spoke on behalf of herself and three other Italian lecturers. Betty had become a "go-to person" for women who were having employment problems on campus.

Each of the four women had about ten years of teaching experience. All were part time, which suited them. None had a PhD degree. All were teaching language courses; none were teaching literature. All had excellent teaching evaluations. The lecturer reported that at first the four had been told they would be rehired for one more year. They believed that then a single man would be hired to replace all of them, a man who, like them, did not have a PhD. These women had little concrete information about their employment situation. Their department chair maintained they were not faculty and therefore not permitted to attend department meetings. Even the students were permitted to send a representative to observe at these meetings. Why not the

---

[34] 4/24/1971 letter from AAUW President Anita Miller and Special Committee for Developing Educated Womanpower Corinne Geeting to UC-Berkeley Chancellor Roger W. Heyns.

[35] 5/25/1971 memorandum from UC-Berkeley Assistant Chancellor for Budget and Planning Errol Wilson Mauchlans Office to Betty Scott re typed copy of the list of women faculty, titled Women Faculty UC-Berkeley Fall 1970 (Source: PA 604).

lecturers, they wondered? The women in Italian were coming to believe they would not be rehired even for the next year.[36]

The lecturers were not the only ones who needed information. A year had passed since Betty first began looking into gender inequities in the UC retirement system. Changes were proposed that she found objectionable. As far as Betty could tell, all those changes were adopted. But she hadn't been able to get her hands on a copy of the new regulations in order to be sure.[37]

Soon there were more deeply troubling statistics. Fifteen (44%) of 34 departments at UC-Berkeley had not had a tenured woman faculty member for over 51 years, i.e., since the year 1920. Getting specific, in the school of librarianship, women were only 2% of the 13 faculty members, but 77% of the 210 students, and only about 50% of the 35 financial aid awardees. The psychology department had no women and 40 men on their faculty, even though over the most recent four years they had enrolled 57 women graduate students and awarded 23 PhDs to women. Once revealed, statistics such as these indicating the "paucity of academic women at Berkeley" were "surprising even to hard-core male chauvinists."[38] Some simply hadn't thought about the status of women and were being awakened by the statistics.

---

[36] 5/[?]/1971 handwritten notes by Betty Scott after a telephone call from UC-Berkeley Italian Lecturer Elena Burgess regarding problems of women lecturers in Italian.

[37] 5/27/1971 letter from Betty Scott to UC Public Information Officer Jack M. Allard.

[38] "HEW Investigates UCB," *Gadfly* (Newsletter of the Graduate Assembly), 1(2), June, 1971.

Part VI

Affirmative Action

21

Not Easily Erased Overnight
(June – July 1971)

Your office needs to think of ways of making sure that the search
for women will be done in the real spirit of a desire to succeed.
-Susan Ervin-Tripp

HEW investigators were expected at the start of summer. They would investigate the complaints about academic women (month one) and nonacademic women (month two). It was an intense time. If HEW found evidence of sex discrimination, they had authority to force change. The university would need to develop an affirmative action plan for equality in hiring and promotions for both academic and nonacademic women, and in admissions and financial aid for women students. Women were fired up. The LAW flyer proclaimed to women at UC-Berkeley and in the president's office: "You can help make it happen here." LAW "intended to see to it that this [university affirmative action] plan has strong enforcement procedures."

LAW asked academic women to submit stories; LAW would forward these to HEW without identifying information. Women could submit confidential information either to LAW or the Female Liberation Office;[1] they could also volunteer to deliver personal testimony if confidentiality was not an issue;[2] or they could submit formal charges of discrimination directly to HEW.[3] One who chose to submit such a charge was a woman who was dismissed from her faculty position because the French department would not give her tenure.[4]

Summer arrived and with it four HEW investigators. The plan was to interview heads of the 17 departments with academic women and each of these women. The departments were selected after review of departmental data and interviews with representatives of organizations that filed the complaint. HEW would be looking at underrepresentation of women on faculties that

---

[1] n.d. flyer announcing the complaint filed with HEW against UC-Berkeley regarding sex discrimination in the recruitment, hiring, and promotion of women.

[2] "HEW Investigates UCB," *Gadfly* (Newsletter of the Graduate Assembly), 1(2), June, 1971.

[3] 6/2/1971 letter from Equal Employment Opportunity Commission District Director Jules H. Gordon to HEW Office for Civil Rights Regional Director Mr. Floyd L. Pierce.

[4] 5/24/1971 letter from UC-Berkeley French faculty member Anne Prah-Perochon to Office of Equal Employment Opportunity.

had high percentages of women graduate students, concentrations of women academics at lower (non-secure) ranks, gender discrepancies in salary, and various related problems such as nepotism practices.[5] They planned to obtain gender, competence, and experience statistics about the departments. They would request information specific to each individual employee on ethnicity, sex, birth date, job classification, department, hire date, salary, and salary adjustments. These data would be held confidential.[6]

LAW called for specific suggestions from women for an affirmative action plan which they expected HEW to develop by the end of the summer. Based on the experience of HEW with other universities, LAW expected UC-Berkeley to reject the affirmative action plan and enter negotiations with HEW. LAW called for donations of $1.00 to help defray expenses for their work.[7]

21.1   Prizes for Women in Astronomy

Betty was well connected with influential women in astronomy; many knew of her work to support academic women. In early June, distinguished astronomer Margaret Burbidge wrote Betty about the Annie J. Cannon[8] Prize in astronomy that was initiated in 1934 and given every three years.[9] Burbidge had just been offered the prize for 1971; she had great admiration for Cannon and appreciated the honor, but she turned it down.

It was a hard decision. Burbidge wanted Betty to know why she turned it down.[10] This is how Burbidge explained it to the AAS:

> *I [regretfully decline the prize] because I believe that it is high time that discrimination in favor of, as well as against women in professional life*

---

[5]7/12/1971 letter from HEW Contract Compliance Branch Acting Chief Waite H. Madison, Jr. to VC Robert E. Connick. Note: The 17 departments were business administration, dramatic art, education, English, mathematics, music, philosophy, social welfare, anthropology, biochemistry, economics, French, German, history, librarianship, optometry, and psychology.

[6]6/14/1971 letter from HEW Contract Compliance Branch Chief Frank R. Albert, Jr., to UC-Berkeley Chancellor Roger W. Heyns.

[7]?6/1971] notice titled "Berkeley University of California Women re: HEW Complaint." Note: Apparently prepared by LAW.

[8]Note: Annie Jump Cannon (1863–1941) was a great cataloger of stars, having classified 230,000 stellar bodies. Her work helped to create the Harvard Classification Scheme and was influential in developing modern stellar classification systems. She received 25¢ per hour for her work, less than a secretarial wage. Cannon was the first woman to receive an honorary doctorate from Oxford University and to be elected an officer of the AAS. She was the second woman to be awarded the Henry Draper Medal from the NAS. She eventually was given a regular appointment at Harvard.

[9]Note: The Annie J. Cannon Prize is now given annually to a young woman astronomer in North America who is distinguished for exemplary contributions or applications.

[10]6/3/1971 letter from UC-San Diego Physics Professor Margaret Burbidge to Betty Scott.

> be removed, and a prize restricted to women is in this category entirely.
> The Annie J. Cannon Prize is not given for excellence in astronomy – if
> it were, I would be proud and happy to accept it – but for excellence of
> women in astronomy, which is a different thing altogether. Because of the
> small number of women in the field it is not surprising if we all, in our
> turn, are selected for the prize. It would be interesting to know, however,
> how often our names have been excluded from consideration for professor-
> ships, directorships, or chairmanships of major astronomical institutions
> because we are women.

Burbidge added her hope that this prize would be opened up to all astronomers, not just women, and that it be an international prize.[11]

Burbidge took a stand, and Betty backed her. Betty wrote to the AAS:

> I very much admire Dr. Burbidge for the decision she has made. I think
> that her decision is the right decision even though it must have been very
> difficult to arrive at... I hope that Dr. Burbidge's action will give impetus
> to a reversal of the problems women face in the academic community,
> and particularly in astronomy. I join with Dr. Burbidge in urging that
> the American Astronomical Society will open the Annie J. Cannon Prize
> to all astronomers without restrictions.[12]

The prize was given to no one in 1971 after Burbidge turned it down. Betty
would be pleased that the Committee of Presidents of Statistical Societies
(COPSS) prize in her own name, the COPSS Elizabeth L. Scott Award, is
open to both women and men.

21.2   Faculty Club Operations

Two committees met throughout the spring semester to consider the future of
the two faculty clubs. One was the WFC's Committee of Five; the other also
incorporated three members of the men's club. Both groups strove to have
open channels of communications with the two boards.[13]

Betty had attended the April WFC board of directors meeting to discuss
the Proposed Principles for the Joint Operation of the Faculty Clubs. Although she was not yet a board member, Betty was present again at the June

---

[11]6/3/1971 letter from UC-San Diego Physics Professor Margaret Burbidge to AAS Secretary Laurence W. Fredrick. Note: Instead, the AAS transferred it to the AAUW, which
administered it until 2005, after which it was transferred back to the AAS. It remains a
prize only for women.

[12]7/30/1971 letter from Betty Scott to University of Virginia Leander McCormick Observatory staff member Laurence W. Fredrick.

[13]6/3/1971 letter from WFC Board of Directors President Mary Lou Norrie to Dear
Member.

2 meeting when a revised joint operation document was put forward for "urgent" discussion.[14] The WFC board agreed to a trial period for the proposed principles, during which each club would retain its corporate identity, including existing articles of incorporation and bylaws, ownership of buildings, and membership and dues structure. A merger would be considered, including legal details. After a successful trial period, the negotiated merger of the two clubs would be put before the memberships for a vote.[15]

The men's club membership had already agreed to the proposed principles. The WFC board decided to send the joint operations document to their members with a cover letter similar to one the men used. They decided this letter should be revised to not only confirm there were new funds available from an external donor, but also to outline how the funds would be used to implement changes to the women's club building; it should stress that the overall proposal was for joint operation, not a merger, of the two clubs.[16]

The letter was sent out the next day. WFC President Mary Lou Norrie expressed hope that "the proposed combined operation will prove to be a most satisfactory arrangement for the coming two years." She asked for reactions, favorable versus unfavorable, and to have additional comments with identifying information so they could follow up if needed.[17]

The new center for renewal education for women was being established. It would be called the "Center for the Continuing Education of Women." The chancellor asked Betty for help in selecting the director, and she was glad to help. She saw great value in the center for women students at all levels. Betty envisioned a "generation" of "talented women who should now be professors, associate professors, laboratory directors, research directors, deans, commissioners, and so forth, but who actually have been quite ignored for 10 or 15 years." In her words, these women needed "a year or two of quiet work in a scholarly atmosphere with well directed help towards entering the academic mainstream." The new center could provide this, and "In turn, the academic community will benefit from the new resources." The university can then choose among these women to be the professors, directors, and administrators "whom certain individuals claim to be in short supply."[18]

---

[14]6/2/1971 minutes of the meeting of the WFC Board of Directors. Note: The revision of the Proposed Principles for the Joint Operations of the Faculty Clubs was dated 5/25/71.

[15]6/3/1971 letter from WFC Board of Directors President Mary Lou Norrie to Dear Member.

[16]6/2/1971 minutes of the meeting of the WFC Board of Directors.

[17]6/3/1971 letter from WFC Board of Directors President Mary Lou Norrie to Dear Member.

[18]6/5/1972 letter from Betty Scott to UC-Berkeley Chancellor Albert H. Bowker.

21.3   University Doing More

The WFG was continuing to monitor moves, or lack thereof, toward equality for women. The VCAA had taken many constructive steps during the hiring freeze. However, Ervin-Tripp reported the women weren't seeing these implemented on the ground. There was the matter of potential de facto discrimination against women lecturers during the budget crisis. Betty had been particularly concerned about this matter, and with reason. Twice as many women as men lecturers had been given non-renewal notices for the fall. Then there was the matter of departments being required to produce evidence that women had been sought out in faculty searches. This was apparently not happening, at least not the way women had been led to believe it would. Highly qualified lecturers were not always considered for entry level assistant professor positions. Ervin-Tripp wrote to the chancellor: "The irony of this decision [not to consider lecturers for assistant professor positions] is that they are the very women who have most suffered over the years from being overlooked for appointments... your office needs to think of ways of making sure that the search for women will be done in the real spirit of a desire to succeed."[19]

The academic senate had approved a new standing "Committee on the Status of Women" (Senate CSAW). Betty was appointed. Others members were Paul A. Heist (education), Arlie Hochschild (sociology), Frank C. Newman (law), Rosemarie Ostwald (nutritional sciences), Hanna Pitkin (political science), and Frederick S. Sherman (aeronautical sciences). The three men and four women were each given a one-year appointment. Pitkin would be chair. However, she would be in Europe for the summer, so she asked Betty to step in during her absence.[20]

By mid June 1971, the administration was realizing they had to do more toward affirmative action. "Considerable progress" had been made for ethnic minorities. Now it was time for each unit to make and execute formal plans that also included women. The long-term objective was to provide "job opportunities for ethnic minorities and women in proportions related to their representation in the working population [sic]." The written plans were to contain goals, time lines, and identification of an affirmative action coordinator. Preliminary plans were to be submitted to the chancellor's office for review.

The chancellor's office announced the university should establish goals based on qualification, not simply on availability of women applicants. At the same time, "... the University does commit itself to the vigorous recruit-

---

[19] 6/2/1971 letter from UC-Berkeley Rhetoric Professor Susan Ervin-Tripp to UC-Berkeley Chancellor Roger Heyns.

[20] 6/3/1971 memorandum from Hanna Pitkin to Paul A. Heist, Arlie Hochschild, Frank C. Newman, Rosemarie Ostwald, Betty Scott, and Frederick S. Sherman re Betty's handling business of the Senate CSAW during Pitkin's summer absence. Note: Newman also served on the Policy CSAW.

ment of both minority and women candidates having qualifications and potentials comparable to non-minorities and males... Our responsibilities are greater than providing just 'equal opportunity,' they include taking positive measures to insure that ethnic minorities and women are given all possible opportunity to succeed within the levels of their competencies." This announcement came in anticipation of pushback from units about affirmative action for academic positions: faculty, teaching assistant, and research assistant.

The chancellor's office also sent a number of legal and policy documents to the units. One was a Summary of Chapter 60 of Executive Order 11246 – Office of Federal Contract Compliance. This order referred to Affirmative Action Programs. Another was a Summary of Chapter 60-20 of Executive Order 11375 – Office of Federal Contract Compliance. This order summarized the federal SEX Discrimination Guidelines. Yet another was the UC Application of Policy on Nondiscrimination. This explained that sex was essentially eliminated as a qualification for almost all university positions.

Implementation documents were included. "Summary of Responsibilities and Timetables" included responsibilities of deans, vice chancellors, assistant vice chancellors, unit affirmative action coordinators, the campus affirmative action coordinator, the chancellor's office, and the campus personnel office. "Suggestions for Unit Plans for Affirmative Action Programs" presented four steps, with detailed suggestions, for units to follow in creating their affirmative action plans. "Demographic Data" presented data for eight major job categories of workers (officials and managers, professionals, etc.). Two percentages were shown. The first was of individuals comprising that job category, for each of the following groups: Negro, Oriental, American Indian, Spanish Surname, and Women. It was an indicator of "occupational availability." The second was of the total civilian work force in the identified labor market, also broken down by group. It was an indicator of overall market composition. The short-term goal was for the distribution of university workers to look like that according to "occupational availability," and long-term, to look like that according to "overall market composition."[21]

LAW and NOW had already filed complaints against UC-Berkeley and the UC. By mid May, the political science women's caucus and the American Federation of State, County, and Municipal Employees Local 1695 had also filed complaints. HEW was scheduled to commence its investigations on June 7.[22] The UC president's office chose to put on a cooperative face.[23]

HEW asked the university for more data. They wanted lists of all employees: status (full time, part time, or student), name, race, sex, birth date, job

---

[21]6/15/1971 memorandum from the UC-Berkeley Office of the Chancellor to Deans, Vice Chancellors, Assistant Vice Chancellors re establishing formal affirmative action plans for each unit.

[22]5/11/1971 letter from HEW Regional Civil Rights Director Floyd L. Pierce to VPA Robert L. Johnson.

[23]6/25/1971 letter from VPA Dr. Robert L. Johnson to HEW Regional Civil Rights Director Floyd L. Pierce.

classification, department, and current salary. They wanted these sorted in two ways: by department and job classification within department; and by job classification. They wanted lists of all terminations and all new academic hires over the past two years (name, job classification, date of termination, race, sex, and department). They wanted copies of criteria used to select for promotion. They also wanted access to individual personnel files.[24]

An immediate effect of the investigation was that the university began to take a healthy, close look at itself. For example, the administration discovered from most recent data that four departments had a significantly lower graduate student admission rate for women: anthropology, mathematics, public health, and social welfare. The vice chancellor wanted to know if the lower rates were due to discrimination, or to lower qualifications of women applicants. He asked departments to investigate and explain.[25] Public health offered the following explanation: The data were for current applications, which were only 47% complete; the data for the previous year, 1970–71, which were 100% complete, indicated a higher proportion of women applications were accepted (43%) in comparison with men (34%).[26]

It didn't take HEW long to identify a list of issues:

1. Females are underrepresented in faculty positions.

2. Females presently in academic positions are concentrated in non-ladder positions.

3. Recruitment and selection procedures ignore the affirmative action commitment for women made by the University.

4. Females are underrepresented on departmental committees dealing with academic personnel matters; this factor and a discriminatory attitude on the part of some male committee members result in screening out of females in hiring and promotions.

5. Hiring and promotional criteria are applied differently for males and for females.

6. Female graduate students are denied equal opportunity for T.A. positions because they are less likely to be admitted to graduate programs and fairly considered for such positions.

This list would provide the framework for the investigation.[27]

The editor of *UC News* who worked out of the university vice-president for university relations' office wanted to do a story on the status of university women. But the "powers that be at University Hall keep putting me off...," she told Betty. Perhaps the story could be done later, and she pledged to "keep

---

[24] 6/4/1971 letter from HEW Regional Civil Rights Director Floyd L. Pierce to VPA Dr. Robert L. Johnson.

[25] 6/21/1971 letter from VC Robert E. Connick to UC-Berkeley Graduate Division Dean Sanford Elberg.

[26] 7/1/1971 letter from UC-Berkeley School of Public Health Acting Dean Warren Winkelstein, Jr. to UC-Berkeley Graduate Division Dean James F. King.

[27] 7/21/1971 letter from HEW Contract Compliance Branch Chief Frank R. Albert, Jr. to VC Robert E. Connick.

plugging." She produced a draft story, titled "Second Citizens in Academe," which told that it was now official university policy "not to engage in discriminatory practices against any person employed or seeking employment because of race, color, religion, marital status, national origin, sex, citizenship, or age, within the limits imposed by law and University regulations." Her story told the progress UC had made in implementing affirmative action programs across the campuses: Its equal employment opportunity office had made progress in developing an affirmative action plan for women; and the campuses had appointed affirmative action coordinators to provide data, evaluate proposed programs, and coordinate compliance efforts with government agencies.[28] The editor reported that seven UC campuses had appointed special chancellor's advisory committees on the status of nonacademic women, and five had appointed academic senate committees to investigate the status of academic women. The editor called the Policy CSAW report groundbreaking in revealing the "plight of its women faculty members" and "probably the most carefully prepared and complete document of its kind in existence and [it] inspired a rash of similar surveys in colleges and universities across the country."

Pressures resulted in some noticeable positive changes in departments. Psychology had a new woman tenure-track faculty member, the first in 47 years. History had its first woman professor in 13 years. Sociology appointed two spouses as faculty members, ending nepotism rules that had almost always worked against the wife. For the first time, the academic senate appointed a woman to chair the "all-powerful" budget committee.

Recruitment practices seemed like the place to begin to tackle discriminatory practices, which are "not easily erased overnight." The Chancellor's Advisory Council on Non-Academic Women at UC-Berkeley (Chancellor's CSNAW) was working hard formulating these recommendations:

> *1. The Berkeley campus and the Office of the President should have definitive and detailed schedules, goals, and timetables for placing minority group members and women in jobs.*
>
> *2. There should be a strong and active "watchdog" to review all personnel changes in the University. The Personnel Office should be given this authority to meet affirmative action goals, not only in the area of recruitment, but also in the areas of transfers and promotions.*
>
> *3. All recruiting activities for University employment should be consolidated in one office.*
>
> *4. More part-time jobs should be made available to women.*
>
> *5. All applicants, but especially minority and women, should receive professional counseling on job opportunities within the University.*
>
> *6. Sensitivity training programs in the problems and roles of women should be mandatory at all levels of supervision and management.*

---

[28]7/23/1971 letter from UC Office of the Vice-President for University Relations Reporter Betty Hoffman to Betty Scott.

7. Recruiting practices should be changed, so as to eliminate bias in applicant selection.

8. A management study should be immediately undertaken of the UC Berkeley Personnel Office to increase its effectiveness.

They were already planning other recommendations regarding the job classification systems, job training, personnel rules, and promotions and transfers.[29]

21.4   Class Action and Confidentiality

Back in June, the chancellor knew academic women were bringing their grievances to Betty and her colleagues in the WFG. He asked her if she would forward individual cases to his attention so he could investigate them. In Betty's estimation, he was trying to be helpful and improve the status of academic women. But she decided not to comply with his request. If she did, she would have to give him names and specific details. This she was not able to do, because these women had trusted her and come to her in confidence. But Betty was able to tell him the two main reasons women were reluctant to be identified: One was "perhaps she was not so good as she thought and indeed deserved the 'inequity' "; the other was she didn't want to become known as a "troublemaker." Betty lamented: "I am convinced that there are entirely too many problems involving academic women at Berkeley."[30]

In the background to these forward steps, trouble was brewing at the compliance review. The university held that personnel records were confidential and could not be released to HEW for review. HEW held that it had every right to inspect personnel records, and furthermore such review was necessary for the purpose of comparing salaries and ranks across genders.

On July 21, HEW had made another formal request to the VC for immediate access to about 120 faculty personnel files. HEW was already losing patience. HEW wrote to the VC about the agency's attempts to "cooperate with the University as fully as possible," and offered more assurances that there "will be no disclosure of confidential information." HEW talked about the agency's demonstration of good faith, reminding the university that the federal regulations stated it is "the contractor's responsibility to provide access to records as deemed necessary by this office for administration of the Orders." If the university continued to deny access, then the San Francisco office would refer the matter to the Washington office of HEW "for appropriate action." This was indeed a serious matter for the university.[31]

---

[29]Betty Hoffman, "Second Citizens in Academe," *UC News* (draft), [?1971].

[30]6/17/1971 letter from Betty Scott to UC-Berkeley Chancellor Roger Heyns.

[31]7/23/1971 letter from HEW Regional Civil Rights Director Floyd L. Pierce to VC Robert E. Connick.

On July 28, the VC responded that the request "involves a crucial matter of principle in the operation of a scholarly institution." The university had a policy of strict confidentiality when it came to faculty personnel files. There were three basic reasons for this. One was that it needed candid appraisals about intellectual quality in faculty selection and promotion processes. Individuals would not be likely to submit truly honest recommendations without guarantee of confidentiality. Therefore the university had an obligation to not "hand over such appraisals simply because they have been asked for." Another reason was that the university was simultaneously proud of its strong tradition of allowing individuals to freely express themselves and sensitive to political intrusion of any kind. "Confidentiality of personnel files is thus inextricably interwoven with the maintenance of academic freedom," the VC explained. If they granted HEW's request, it would be very hard to turn down future requests by government agencies for personnel records. Yet another reason was to protect the rights to privacy of faculty members. Most of the 120 individuals whose personnel files were requested by HEW had not filed a complaint with HEW and therefore not waived their rights with respect to HEW.

The Vc offered HEW a compromise. He would provide a Curriculum Vitae (CV) of each of the 120 persons whose personnel file was requested. This would include name, sex, date of appointment, rank, salary, complete educational background, complete list of scholarly publications, complete rank and salary history at the university, and justification for appointment. This information, together with selection and promotion policies, procedures, and processes, should "make apparent that any differences in rank or salary between the male and female members of a department are based on objective criteria, not sex discrimination," he wrote. He also sent HEW six CVs randomly selected from English. These, he said, would "vividly illustrate just how unnecessary it is to breach the confidentiality of the one hundred and twenty personnel files." The VC proposed that, should further examination be needed in specific cases among the 120, he would supply all of the evaluations and recommendations involving selections, advances, and promotions of these individuals, as well as the associated official letters of decision and outside appraisal letters, however with identifying information removed.

The VC hoped HEW would accept this compromise. He offered that: "Our concern centers on the procedures utilized to determine the present state of compliance with [HEW's] objectives" rather than HEW's goals of affirmative action per se. The university and HEW fully shared these goals, he said.[32]

---

[32]7/28/1971 letter from Acting VC Robert E. Connick to HEW Regional Civil Rights Director Floyd L. Pierce.

21.5   No Women at the Top

In June 1971, Albert H. Bowker was appointed to be the new chancellor. Betty and Bowker had in common that both were mathematical statisticians.[33]

There had never been a woman president of the UC system, nor a woman chancellor at Berkeley. There was not a single woman vice president, assistant vice president, or vice chancellor at the campus. The class action said this about university top management: "There has never been a woman chancellor or president [at UC], and it is highly unlikely that women were ever even considered as potential candidates for such positions. The absence of women at the highest policy-making levels of University management helps perpetuate the attitudes and practices that keep women in inferior positions."[34]

---

[33]Note: Albert H. Bowker (1919–2008) was a Graduate School Dean at Stanford University and then Chancellor of the City University of New York (1963–1971) before being appointed Chancellor at UC-Berkeley (1971–1980). Bowker was Chair of the Cosmos Club in Washington, DC, and during his tenure was instrumental in opening the Club membership to women. Ingram Olkin, e-mail to author, September 27, 2013.

[34]Betty Hoffman, "Second Citizens in Academe," *UC News* (draft), [?1971]. Note: In 2013, Janet Napolitano would become the 20th president of the UC, the first woman ever to hold this top position. On 3/13/2017, it was announced that Napolitano had selected Carol T. Christ to become the 11th chancellor at UC-Berkeley, the first woman in this position.

22

A Lot of Power
(August – December 1971)

...the investigation...is setting a precedent for the nation, and thus...can ultimately benefit the millions of working women...
-League of Academic Women

Academic women were exchanging information about victories at different universities around the country. For example, by August 1971, Harvard University had established a policy that "non-tenured professors who become pregnant receive one-year extensions of their appointments, not to exceed two years." Betty noted that Stanford University also decided to establish this policy. She hoped the UC would follow suit.[1]

The HEW investigation at UC-Berkeley was continuing. In August, investigators asked for bibliographies. Betty was greatly troubled by this request. She was pretty sure they were unaware of the "unique system" at Berkeley, where lecturers were not allowed to be principal investigators. Because lecturers didn't have research assistants to help write, a consequence was that lecturers, who were more often women, could be expected to have less papers. Because principal investigators often had their names put on papers where they did little work, another consequence was that nominal principal investigators, usually men, could be expected to have more papers. HEW needed to be made aware of this and needed more information in order to evaluate bibliographies in light of these biases. Betty thought maybe each article on a bibliography could be annotated. Not by HEW – that wouldn't be realistic – but by whom?[2]

In late August, academic women got "significantly" more help from the federal government. Order No. 4 had been signed back on January 30, 1970. It provided guidelines to federal contractors and subcontractors in nonconstruction industries for developing affirmative action programs. Now a revised order also required the development of goals, with time lines, "to remedy underutilization of women." In a press release, the US Department of Labor

---

[1] 8/10/1971 letter from Betty Scott to UC-San Francisco Medical Center Adult Development Program Anthropologist Martha S. White.

[2] 8/11/1971 letter from Betty Scott to Institute of Human Learning Researcher Harriett Amster.

expressed confidence that the revisions would have positive impact toward equality for women in America, giving them more opportunities.

Federal non-construction contractors and subcontractors now were required to analyze their workforces for underutilization of women. This included relative to the local labor pool: number of unemployed women; percentage of working women; availability of qualified skilled women; and availability of job-seeking women. It also included availability of qualified skilled women in the potential recruitment area; availability of "promotable and transferable" women within the organization; anticipated "expansion, contraction, and turnover" of the organization's workforce; availability of institutions capable of providing workforce development; and organization-based training that the contractor "is reasonably able to undertake as a means of making all job classes available to women."[3]

The HEW investigation was indicating UC-Berkeley had "a disproportionately high number of women at the lower levels." J. Stanley Pottinger, the "earnest" young Republican lawyer appointed by President Richard Nixon to be director of the US Office of Civil Rights (OCR), pointed out that "most universities" had this same problem. But Berkeley's turn had now come to be under the looking glass, and HEW was getting impatient. HEW ordered Berkeley to open its personnel records to the investigators, but Berkeley so far refused to comply. HEW wanted access to the records to try to determine whether the fact that women were rarely promoted to senior-level positions was the result of sex discrimination. HEW had access to statistics, which could be revealing on some issues but not so much on others. HEW needed to dig into the personnel files on some of those other issues. Pottinger asserted UC-Berkeley was hampering the investigation. He threatened to deprive the campus of any new federal funding unless the impasse could be broken. He had, in fact, taken such action against Columbia University a few weeks earlier. Pottinger told the local press that: "We have a whale of a lot of power and we're prepared to use it if necessary."[4]

The planned center for continuing education for women was being actively supported by the administration, which hoped it would stimulate research activities related to renewed education. The center would definitely be established within the WFC building. According to WFC President Mary Lou Norrie: "the building site is ideally situated to provide the important aspect of high visibility for the various functions and [we] feel that such an operation will strengthen our membership and programs in our service to the entire campus community."[5]

---

[3] 8/26/1971 press release from the US Department of Labor: "Labor Department Acts to Widen Job Opportunities for American Women," 3 pages.

[4] Mitchell Thomas, "Women Employees: UC Warned in Rights Probe" (newspaper article without reference information), n.d. [?1971].

[5] "New Concept at Cal: Center for Renewed Education for Women" (article with unknown publisher), 9/1971.

Plans for joint operation of the faculty clubs were also progressing. Planners hoped to have a formal plan for joint operations by early fall. The women would be given a share ($75,000 – about $400,000 in 2010 dollars) of the Levi Strauss Associates gift given recently to the university. These funds would be used for rehabilitation of the WFC building. Ninety percent of the WFC members indicated they were in favor of joint operations. Their main concern was whether lunch would continue to be served in the women's club building.[6]

22.1   New Advisory Committees in the UC

External pressures on the UC resulted in new efforts toward affirmative action for women. In mid September, UC directed each campus chancellor to appoint two committees on affirmative action for women. One would be for all academic women, including both senate and non-senate members. At UC-Berkeley, this would be a Chancellor's CSAW. The other would be for all nonacademic women. At UC-Berkeley, this would be the already-established Chancellor's CSNAW. [7] These would be in addition to the standing committee set up by the UC-Berkeley academic senate (Senate CSAW).

The Chancellor's CSAW and Chancellor's CSNAW would follow up on recommendations made by the Policy CSAW. These concerned maternity leaves, admissions, hiring, and representation on academic senate committees. The committees would also be concerned with development and implementation of the campus affirmative action plan for women. This would include development of affirmative policies for women who taught but were not members of the senate, a group of temporary employees that comprised 60% of the academic women on the campus. Women lecturers were disadvantaged, and the economic downturn meant funding for them would continue to decrease.

The good news was that the overall percentage of women in tenured or tenure track positions had increased from 5 to 13 percent at UC-Berkeley. Betty told the student newspaper: "Some departments are making a conscious effort to employ women... People are recommending women to the University of California more than before." The bad news was that some departments had gone the other direction and the percentage had actually decreased.[8]

Also, the UC created a systemwide advisory committee on the status of women (System CSW). The nine campuses were represented, as well as the president's office, agricultural extension, and two national laboratories. The committee would provide opportunities for ideas and information exchange,

---

[6] 9/1/1971 The Women's Faculty Club Newsletter.

[7] 9/15/1971 letter from VCAA John Henry Raleigh to UC-Berkeley School of Law Professor and Senate Committee on Committees Chairman Richard Jennings.

[8] Valeria Sopher, "Equality for UC's Academic Females Sought by Women's Status Committee," *Daily Californian*, October 13, 1971.

and for networking, across the campuses. The UC-Berkeley academic senate was represented by Political Science Associate Professor Hanna Pitkin. UC-Berkeley staff women were represented by Business and Finance Assistant to the Vice President Afton Crooks, with whom Betty was impressed. Statistics Professor F.N. David was assigned to represent the Riverside campus.

All of these committees were a signal that the status of women at the UC would be a continuing concern. There were documented inequalities, and it would take time to try to eliminate them.

Pitkin couldn't make the first meeting of the System CSW, held on September 22, and asked Betty to substitute. The meeting was mostly informative. Some campuses hadn't yet, or had only recently, formed committees. Attendees discussed a systemwide maternity leave policy, and each campus gave a brief report on what her campus committee was doing.[9] Betty hand wrote four pages of notes so she could properly report back to Pitkin.

The VCA made some initial remarks. Here are excerpts from Betty's notes:

> ...In addition to particular regulations + emergency programs, how can one change attitudes. HEW will not "save" women... Whole business could be tied up in courts for a long time. Complex fragile social phenomena... UC is sensitive to special purpose legislation encroaching on UC. Therefore UC often opposes legislation as no business of legislature even though is good for women. Want to insist on constitutional independence... [The VPA] is unique in sympathy + attitude in administration. General attitude is "laughing indifference."

Then each campus/extension/lab reported. The group discussed information gathering and sharing; coordination, support, and action; maternity leave; and equal pay for equal work.

The advisory committee decided to meet quarterly. Each campus committee should review their campus personnel policies, especially the leave policies.[10] This was time sensitive, as revised university leave policies were scheduled to be considered on September 24 at the University Staff Personnel Board meeting. The chairman of the board was asked to delay finalization of the revised policies until campus input was heard.[11]

Betty made an important suggestion in her debrief to Pitkin. This was for the Senate CSAW to provide input to the UC-Berkeley affirmative action program plan for women. There wasn't much time. There was a November 3 deadline, just a week away, for the chancellors to prepare plans for their campuses. It might seem odd that Betty made such a suggestion which would seem

---

[9]9/17/1971 memorandum from UC Vice President of the University for Administration Assistant Administrative Analyst Ruth McElhinney to Chairwomen, Advisory Committee on the Status of Women re first meeting of the System CSW.

[10]9/22/1971 handwritten notes by Betty Scott re UC-Berkeley Chancellor's Advisory Committees [sic] on the Status of Women (Chancellor's CSAW).

[11]9/23/1971 letter from UC Vice President for Administration Office of Equal Opportunity Administrative Analyst Rebecca A. Mills to UC Staff Personnel Board Chairman Richard W. Gable.

obvious to some. But the other UC campuses had just one affirmative action advisory committee on academic women, while Berkeley had two (Chancellor's CSAW and Senate CSAW). As of the end of September, the UC-Berkeley chancellor hadn't yet constituted the Chancellor's CSAW for academic women and was only informing and consulting with the Chancellor's CSNAW for nonacademic women. And so it was important that the chancellor also inform and consult with the Senate CSAW.[12] Pitkin made this a formal request, backed by a request from the academic senate chairman.[13,14]

The first meeting of the Senate CSAW was on September 29. The committee included Betty, Education Professor Paul A. Heist, Sociology Professor Arly Hochschild, Law Professor Frank C. Newman, Nutritional Sciences Professor Rosemarie Ostwald, and Aeronautical Sciences Professor Fredrick S. Sherman. Chair Pitkin called her group together at The Faculty Club and reported on her pre-meeting with the chancellor and status of the HEW investigation. The group exchanged ideas on the mission of the committee and action plan for the year.[15] The next meeting would be in October.[16]

22.2 Ideas for AAAS

At the beginning of October, the AAAS was planning a "Symposium on Women in Academia: Evolving Policies Toward Equal Opportunities" to be part of their 138th annual meeting to be held on December 30 in Philadelphia.[17] The AAAS is "the world's largest generic scientific society" and publishes various journals including *Science*.[18] Betty hoped for positive action at the symposium.

It was common practice for symposia to be recorded and for there to be proceedings. Betty thought the AAAS symposia on women should follow this practice. But the last symposium, held in Chicago, was not recorded nor made into proceedings, and it didn't have enough publicity. Betty pressed that the

---

[12] 9/26/1971 letter from [Betty Scott] to UC-Berkeley Political Science Associate Professor Hanna [Pitkin].

[13] 10/3/1971 letter from UC-Berkeley Political Science Associate Professor Hanna Pitkin to UC-Berkeley Chancellor Albert H. Bowker.

[14] 10/6/1971 letter from UC-Berkeley Academic Senate Chairman M.N. Christensen to UC-Berkeley Chancellor A.H. Bowker.

[15] 9/15/1971 memorandum from UC-Berkeley Political Science Professor Hanna Pitkin to the Senate CSAW re plans for the first meeting.

[16] 10/3/1971 memorandum from UC-Berkeley Political Science Professor Hanna Pitkin to the Senate CSAW re announcement of the second meeting on October 13.

[17] 10/13/1971 letter from New York University Graduate School of Business Administration Associate Professor and Social Policy and Urban Affairs Program Coordinator Arie Y. Lewin to Betty Scott.

[18] Note: Information about the AAAS and the journal *Science* may be found at http://www.aaas.org.

upcoming symposium should follow common practice and have more publicity, "especially so that more men will be interested in attending."

In addition to thinking about the symposium, Betty thought the AAAS journal editors should publish all professional academic and research positions for a fixed period of time before the positions could be filled. Betty knew this was already a practice in other countries like England and Australia. She thought it should be done here, too:

> *The reason is that most academic jobs, especially in prestigious univer-*
> *sities, are filled by having the chairman (or his representative) telephone*
> *his friends in other universities and ask "Do you have any bright young*
> *person whom you want to send to my department?" This is what I do*
> *myself at Berkeley. The effect is that not too many people are even aware*
> *that there is an opening. Of course in addition, I receive many solicited*
> *inquiries but these come in sort of a random basis. If the openings were*
> *listed in a regular way then persons seeking a job would all have a chance*
> *to apply; also, chairmen in other universities would have an equal chance*
> *to recommend their students or colleagues whom they consider particu-*
> *larly suitable for the opening. I suspect that even in a small field such*
> *as Statistics the appointing department would have more complete infor-*
> *mation. I hope that it would be easier to control discrimination against*
> *minorities and against women.*

Betty wasn't in favor of forcing academic departments to do the advertising: she thought if they were asked to pay for such advertising "there would be quite a lot of fuss." Instead, she thought the professional societies like AAAS should pass resolutions requiring such advertising. Then the editors of the journals should reserve space for it. That would be progress, Betty thought.

Betty had a leadership position with the AAAS. She was vice president and statistics section chair ("Section U"). She had been a member of their council over the past year and knew channels to use for proposing change in the organization. Betty proposed to the AAAS leadership that her ideas be discussed and, if thought to have merit, then she could draft a resolution. She could also bring her ideas to her other professional societies.[19]

Today it is common practice in this country for academic positions to be nationally advertised. Betty was a major player in making this change happen.

Betty was invited to participate in the AAAS symposium on women. Of the several possible titles she drafted for her talk, the one that the organizer liked the best was "Developing Criteria and Measurements of Equal Opportunities for Women." The organizer also liked her ideas for national advertising of academic positions. Her idea was included in the notes about the symposium that would appear in *Science*. The organizer pursued proceedings and launched a publicity campaign, also as Betty had suggested.

---

[19]10/7/1971 letter from Betty Scott as Vice President of AAAS to New York University Graduate School of Business Administration Professor Arie Y. Lewin.

Speakers with "especially newsworthy topics" were asked to be ready for an "extensive publicity operation." The plan was for the AAAS public information director to ask such speakers to forward 200 copies of their complete text, or else a 2,000-word summary of their text. Betty was told she should expect such a request. The copies would be distributed to newspapers, radio, and TV. Copies would be distributed "in advance to 185 key news outlets all over the world." The other 15 copies would be made available in the meeting press room.[20] This symposium on women was to be noticed by the public.

At the meeting, the AAAS council drafted the following resolution:

> *Whereas the talents and contributions of women in science are not fully recognized and utilized,*
> *Whereas there is no central listing of women in science,*
> *Be it resolved that the Council and Board of the American Association for the Advancement of Science immediately establish an Office for Women's Equality to work toward full representation and opportunity for women in scientific training and employment, affairs of the Association, and in the direction of national science policy.*
> *The tasks of this Office shall include,*
>
> > *1. To develop and undertake programs to improve the status of women scientists*
> >
> > *2. To prepare a directory of women scientists*
> >
> > *3. To write and edit a page on women's equality in <u>Science</u> once a month.*
>
> *The A.A.A.S. Women's caucus furthermore demands that the staff for this Office shall include at least two professional level women who are feminists and an adequate supporting staff.*[21]

Over 50 years later, the Association for Women in Science (AWIS) had this to say about the recognition and utilization of talents of women in science: "While we've made great strides, many invisible barriers to success still exist today."[22]

22.3   Policies and Practices at Berkeley

Insurance inequities were still on Betty's mind. She requested copies of policies from the campus insurance providers (Equitable, Blue Cross, and Kaiser) "so

---

[20] 10/13/1971 letter from New York University Graduate School of Business Administration Associate Professor and Social Policy and Urban Affairs Program Coordinator Arie Y. Lewin to Betty Scott.

[21] 12/30/1971 AAAS Women's Caucus Resolution re creating an office on women's equality.

[22] "Advocacy and Public Engagement," *Association for Women in Science*, accessed June 30, 2015, http://www.awis.org/?197.

that we can see what the fine print looks like in each of them." Even before getting copies, Betty noted the insurance for pregnant women with disabilities was "still unfair." Also, costs for additional persons on the policy were unfair: The third and subsequent persons on the policy were assumed to be children and therefore cost less than the first and second persons, but this ignored the fact that when the second person was a child he/she would be charged the same higher amount as if an adult.[23]

Forces were influencing hiring practices at Berkeley. Internally, there was commitment to equity, and there were study committees. Externally, there were government agency demands for proof of compliance.[24] But, there was evidence that affirmative action hiring practices were not being followed. The VCAA informed the budget committee he had "been told by a reliable source that one Chairwoman during the past academic year has been testing our procedures in regard to the necessity for conducting a search for qualified women for open academic positions. According to her, the reinforcement of the regulation is lax since she sent in recommendations, unaccompanied by evidence of a search for women, and they have gone right through." He declared: "Since we have the policy, we had better be strict in our enforcement." In his letter, the VCAA included "ES" among four sets of initials as blind copy.[25] These most likely referred to Betty: She was one of only a few women department chairs on campus.

The VCAA followed up with a memorandum to academic administrators. Effective immediately, "... certain mandatory requirements must be observed..." First, all proposals for new appointments need to document that qualified women and minorities were actively sought. Second, all departments must document their hiring practices, including reasons why candidates were made offers or not. Third, faculty will continue to be evaluated in the traditional way and without consideration of gender or ethnicity, against criteria for excellence in teaching, scholarship, and service.[26]

Elizabeth Colson was now chair of the budget committee and responded to the VCAA's message about strict enforcement of hiring policy. Colson pledged full support of the budget committee in holding the departments responsible for supplying evidence that qualified women and minorities were sought.[27]

In mid October, the UC-Berkeley chancellor's office circulated for discussion a revised draft affirmative action program plan relating to the employ-

---

[23] 9/30/1971 memorandum from Betty Scott to UC-Berkeley Political Science Professor Hanna Pitkin and UC-Berkeley Anthropology Professor May N. Diaz re insurance inequities.

[24] 10/7/1971 memorandum from VCAA John Henry Raleigh to Deans, Directors, and Department Chairmen re mandatory requirements for hiring practices.

[25] 10/4/1971 letter from VCAA John Henry Raleigh to the UC-Berkeley Academic Senate Budget Committee.

[26] 10/7/1971 memorandum from VCAA John Henry Raleigh to Deans, Directors, and Department Chairmen re mandatory requirements for hiring practices.

[27] 10/8/1971 memorandum from UC-Berkeley Anthropology Professor and Budget Committee Chairman Elizabeth Colson to VCAA John Henry Raleigh re search for qualified women candidates.

ment of women. Prepared by UC-Berkeley Law Professor Melvin A. Eisenberg, the plan addressed unit statistical profiles, communication, employment of near relatives, affirmative action coordinator, review of women in academic non-ladder positions, use of special academic categories, recruitment, maternity leaves, probationary periods, recruitment processes, and graduate student admissions.[28] Pitkin was on the distribution list and brought the draft statement back to Betty and other Senate CSAW members.[29]

On November 1, Pitkin started to get ready for a meeting with the VCAA who was in charge of academic affirmative action, the VCA who was in charge of staff affirmative action, and Eisenberg who had prepared a draft affirmative action plan for women. The Senate CSAW wanted the meeting because they wanted changes to the draft. Pitkin reported to her committee that she had made the notice of meeting as non-threatening to the VCAA as possible "for the moment, though [I] didn't tell any lies."

Pitkin wanted the Senate CSAW to meet two days before she would meet with the administration, at noon in The Faculty Club. There they would orchestrate their response to the draft affirmative action plan. Pitkin wanted the committee to "do homework." Betty should find and bring the data she had gotten from the payroll department the year before which showed employment statistics broken down by department, rank, and gender.[30]

Newman was concerned about the pervasive references to male gender in university policy documents. He wrote specifically to Betty:

> *Do the* Instructions to Appointment and Promotion Committees *still read, "Promotions to tenure positions should be based on consideration of comparable work in A MAN'S own field or in closely related fields"?... [a] revision conceivably might inspire some imaginative consideration of whether all those "he/s," "his/s" and "him/s" truly are necessary in such documents. Might it not reasonably be argued, for example, that by its terms the new Faculty Code of Conduct... applies to men only? But cf. My Fair Lady's "Hymn to Him"!*

Newman wondered if the senate budget committee could make the language of the policy documents more gender neutral.[31]

---

[28] 10/12/1971 letter from UC-Berkeley Law Professor Melvin A. Eisenberg to VCAA John Henry Raleigh.

[29] 10/22/1971 letter from VCAA John Henry Raleigh to UC-Berkeley Political Science Professor Hanna Pitkin. With Attachment: Draft – For Discussion: Outline of Points for Consideration in Berkeley Campus Affirmative Action Programs Relating to the Employment of Women.

[30] 11/1/1971 memorandum from UC-Berkeley Political Science Professor Hanna Pitkin to [Senate CSAW members] Professors Paul A. Heist, Arly Hochschild, Frank C. Newman, Rosemarie Ostwald, Betty Scott, and Frederick S. Sherman re planning to meet with the administration about the affirmative action plans for women.

[31] 11/1/1971 memorandum from UC-Berkeley Law Professor Frank Newman to Betty Scott re gender neutral language in university policy documents.

22.4   Mobilizing a Congressman

In Washington, there were heated debates with some gains on equal rights. In mid October, the US House of Representatives overwhelmingly passed a resolution to let states decide on a constitutional amendment that would extend 14th Amendment guarantees of equal protections to women. But the US Senate still had to approve it, and then 38 states would have to ratify it.[32]

LAW didn't like what it was seeing between HEW and UC-Berkeley and tried to mobilize one of their US House Representatives, California Democrat Jerome Waldie. They told him about the "numerous" "long delay periods" following HEW requests for information, and that HEW had been revising their requests following negotiations with the university. Some results of negotiations were that HEW agreed to submit to the administration a list of names of women and departments they were investigating intensively before proceeding. HEW's latest request was for personnel file information, and the university was so far not complying.

The administration had been permitting HEW to conduct confidential interviews. But LAW now had a report that the administration would begin requiring interviews be in the presence of an administration representative. While LAW had reservations about HEW giving UC-Berkeley the names of women who would be interviewed, they thought this new policy change crossed the line. Many of the women on the interview list were on temporary appointments such as lecturer, secretary, librarian, etc.; they could be retained or let go at the pleasure of the administration. Their jobs would be in jeopardy and their civil rights violated if they were required to interview in the presence of the administration, LAW maintained. Furthermore, they argued, this policy did not follow the intent of the executive orders having to do with contract compliance.

LAW informed Waldie the regional contract compliance office of HEW had sent a letter to the Washington office for official signature. The letter to UC-Berkeley demanded HEW investigators be given access to personnel files, prohibited the administration from being present at interviews, and threatened to withhold grants and contracts. But OCR Director Pottinger was sitting on the letter: He had it for more than three weeks and had not yet signed.

LAW asked Waldie to intervene: "We of the League of Academic Women urge you to take action in order to establish the illegality of discrimination in employment. We believe that the investigation at the University of California at Berkeley is setting a precedent for the nation, and thus, interceding on our behalf can ultimately benefit the millions of working women in the United States." They wanted Waldie to contact Pottinger urging him to sign the letter and not allow the university to "dictate the terms" of the investigation.

---

[32]Washington AP, "Women Gain Equality," *Daily Californian*, October 13, 1971.

They fed specific questions to Waldie for Pottinger and asked him to advise the Secretary of HEW of his communications with Pottinger.[33]

In early December, Waldie responded by contacting the UC liaison in Washington. Waldie shared LAW's letter and asked for the liaison to review and comment.[34] The liaison immediately forwarded the letter to the VCA back at Berkeley. Would the VCA please comment and bring him up to date on the negotiations?[35] The liaison was told UC-Berkeley was still in a holding pattern with HEW. It was agreed that the liaison would call Waldie and indicate it was not the time to rock the boat because "negotiations were in a delicate stage." Furthermore, if Waldie was not satisfied, then the liaison should meet in person with Waldie and provide more detail.[36]

22.5   Chancellor's Advisory Committee Established

Betty was on the UC-Berkeley radar screen. When the VCAA asked the senate's committee on committees to nominate persons to serve on an affirmative action committee on women, Betty was one of 13 nominated.[37]

From the time Pitkin and the academic senate exerted pressure, it took the chancellor's office only three weeks to put out a call for the formation of an advisory committee on the employment of women (Chancellor's CSAW) that included both senate and non-senate women.[38] This was progress, as Amster had raised the question about non-senate academic women already back in April. The women wondered if the delay was due to the turnover in the chancellor position or reluctance of the administration to take advice.[39]

This new ad-hoc committee would be advisory to the chancellor and part of the campus' affirmative action program as required by HEW. The committee's charge was to make evidence-based recommendations of changes to personnel policies and procedures that would improve the employment condition of aca-

---

[33] 10/29/1971 letter from LAW Academic Coordinator Harriett Amster, Administration Coordinator Karen Hurn, Student Coordinator Isabel Welch, and Staff Coordinator Pauline Fong to Member of the House of Representatives The Honorable Jerome Waldie.

[34] 12/8/1971 letter from Fourteenth District US Congressman Jerome R. Waldie to UC liaison in Washington Pete Goldschmidt.

[35] 12/10/1971 letter from UC liaison in Washington Pete Goldschmidt to VCA Robert F. Kerley.

[36] 12/10/1971 note "Jean D. 12.14.71" appended to the 12/10/1971 letter from UC liaison in Washington Pete Goldschmidt to VCA Robert F. Kerley.

[37] 10/5/1971 letter (in confidence) from UC-Berkeley Academic Senate Committee on Committees R.A. Cockrell to UC-Berkeley Chancellor Albert H. Bowker.

[38] 11/2/1971 memorandum from VCAA John Henry Raleigh to Deans, Directors, Department Chairmen, and Administrative Officers re formation of an advisory committee on the employment status of all academic women.

[39] 12/15/1971 Draft of Minutes of First Meeting of Chancellor's CSAW, Wednesday, December 15, 7:45 PM, in Room 5527 Tolman Hall.

demic women. They would hear testimony and proposals, but not grievances. Members would be selected from the Senate CSAW and from among individuals who represented the diversity of academic non-senate women; they would be appointed by the chancellor for a one-year period.[40]

The WFC board of directors met on November 9. Members discussed progress made by the operating committee, agreeing that problems should be ironed out by year's end at the latest. Some of the problems included menus: The women's club had stopped serving dinners, and the women were having trouble maintaining their diets at the men's club. Other problems included atmosphere: The men's club didn't seem to have the community feel that had existed previously at the women's club. Betty was now vice president of the WFC.[41] She wasn't able to attend the November 9 meeting, but her name came up when members discussed the call for nominations for the new Chancellor's CSAW. The club decided to nominate Betty and five other women.[42,43]

Twelve individuals were appointed out of a large number of excellent nominations.[44] Among those recommended by the WFC board, Betty, Margaret Uridge, and Mary Murai were appointed. Others included some familiar names: Harriet Amster who was appointed chairman of the committee, Hanna Pitkin, and Lucy Sells.[45] Betty and Pitkin were now members of both the Senate CSAW and the Chancellor's CSAW.

Amster didn't waste time, calling the group together one week after membership was announced.[46] Sells wrote the draft minutes.[47] Members agreed to regularly send information to the chancellor. They would draft an operations proposal, including a budget. For one thing, they needed to send Amster to the statewide meeting of CSAW chairs from the nine campuses, and neither

---

[40]11/2/1971 memorandum from VCAA John Henry Raleigh to Deans, Directors, Department Chairmen, and Administrative Officers re formation of an advisory committee on the employment status of all academic women.

[41]Elizabeth L. Scott, "Biography," n.d.

[42]Note: The four other women were UC-Berkeley English Professor Ann Middleton, UC-Berkeley Industrial Research Specialist Margaret Thal-Larsen, [position not known] Margaret Uridge, UC-Berkeley Physical Education Supervisor Doris White, and [position not known] Mary Murai.

[43]11/9/1971 Minutes of the Meeting of the Board of Directors of the WFC.

[44]11/29/1971 letter from UC-Berkeley Chancellor Albert H. Bowker to UC-Berkeley School of Social Welfare employee Mildred Alexander.

[45]12/7/1971 memorandum from VCAA John Henry Raleigh to Deans, Directors, Department Chairmen, and Administrative Officers re announcing members of the Chancellor's CSAW.

[46]12/10/1971 memorandum from UC-Berkeley Education Professor Harriett Amster to Betty Scott re first meeting of the Chancellor's CSAW.

[47]12/16/1971 letter from UC-Berkeley Sociologist Lucy W. Sells to UC-Berkeley Education Professor Harriet Amster. Note: Two acronyms were being used for the group at this time: CACSAW and CCSAW.

the system vice president's office nor the UC-Berkeley chancellor's office was willing to fund travel to these meetings.[48]

The Chancellor's CSAW was sensitive to the different categories of non-academic women in their recommendations: They would make separate recommendations for graduate students, research positions, and "lecturers et al." Recommendations would include revisions of policies and procedures for hiring and financial support within these categories. For example, private family resources including spouse's income should not be considered in criteria for hiring, promotion, tenure, retention, or financial support. The plan was to interview individuals from each job category to get input on relevant policies and procedures and the most recent draft affirmative action plan.[49]

22.6 Class Action Impasse

On November 18, Betty went to hear Pottinger talk at the UC-Berkeley law school. Someone took careful notes and shared them with Betty. HEW was still trying to get complete access to UC-Berkeley personnel files. HEW and the administration were still negotiating. HEW wanted to get to the bottom of why there were so many women lecturers and so few women professors. Pottinger reported the university was "cooperating to a significant extent, but when we get to the waters edge, they contend that we have no right to the files." The stumbling block was a gap in what HEW and UC-Berkeley thought was legitimate information. UC-Berkeley's concern was about how HEW would use the information in the personnel files.

It was reported at the talk that the president had told Herma Kay he believed affirmative action was or may be discrimination against white males. Pottinger responded off the record. "A big crock." "Balderdash." "The kind of argument one expects to here [sic] from a backwoods cracker farmer in Mississippi."

Pottinger addressed the question of whether only Berkeley or all UC campuses would suffer suspension of contracts. He was looking for a way to suspend UC-Berkeley contracts only. He worried about the reaction of congress to an across-the-system suspension and wanted a hearing before UC-Berkeley contracts would be suspended.

Berkeley was probably not the most uncooperative campus. Although Pottinger said UC-San Diego and UC-Irvine had been more cooperative, he implied Berkeley was being more cooperative than Columbia.[50]

---

[48] 12/15/1971 Draft of Minutes of First Meeting of Chancellor's CSAW, Wednesday, December 15, 7:45 PM, in Room 5527 Tolman Hall.

[49] 12/15/1971 Selective Summary of Minutes of Meeting of UC-Berkeley Chancellor's CSAW.

[50] 11/18/[1971] "Some brief notes on Pottinger's talk at Boalt yesterday," Signed "hafner."

Then the VCA got a Mailgram from Washington. It was not the kind of telegram anyone would want to receive. The OCR had reviewed the impasse between UC-Berkeley and HEW. The verdict was that "further progress in the present situation was not possible at the regional level." The recommendation was for "formal enforcement action against U.C., Berkeley"; further communication from the Washington office would follow.[51]

22.7   Berkeley Follow-Ups

The WFG steering committee continued its vigilance on behalf of campus women. They reviewed the VCAA's October 7 memorandum to deans, directors, and department chairs on hiring practices. They generally found the memo to be "helpful and necessary," but also found a major omission: It didn't address women holding non-tenure-track appointments. The VCAA should amend his memo to make it clear these women should be reviewed as part of departmental searches.[52]

The UC-Berkeley academic senate was on board for women. Their assembly passed the following resolution on December 2:

> *Whereas, the overriding objective of faculty recruitment and promotion is to secure the highest quality of scholarship and teaching in the University of California; and Whereas, the existence of obstacles to the pursuit of professional careers by women scholars, as discussed in the reports from the several campuses on the Status of Academic Women, and elsewhere, has in fact impeded the full realization of this objective; therefore be it*
>
> > *1. Resolved that it is the sense of the Assembly of the Academic Senate that, in hiring and promoting, departments should not assign preference to a man over a woman on the basis of criteria other than those customarily employed in the evaluation of male candidates, and further that departments should recognize the potential enrichment of scholarship and teaching resulting from inclusion of more women among regular faculty members; and be it further*
> >
> > *2. Resolved that the Assembly of the Academic Senate endorses and encourages the development of University procedures that will assure effective recruitment of qualified women and members of ethnic minorities; and be it further*
> >
> > *3. Resolved that the University should evaluate all its faculty members according to traditionally professed standards. Presumed excellence in*

---

[51]11/26/1971 Mailgram from HEW OCR Contract Compliance Division Director Kiely to VCA Robert F. Kerley re announcement of formal HEW action against UC-Berkeley.

[52]12/2/1971 letter from Anthropology Professor and WFG President May Diaz to VCAA John Henry Raleigh.

> *teaching, scholarship and service shall be expected for entry, and certified excellence in these categories shall be required for tenure, irrespective of sex or ethnic origin.*

Traditional standards, applied equally across all people, won the day.[53]

On December 14, Eisenberg distributed a revised affirmative action plan. The Senate CSAW was given less than two weeks to comment on this, "the most important document this campus has had on affirmative action."[54,55] Eisenberg pledged to incorporate as many comments as possible.[56]

The Senate CSAW continued to be concerned about women in temporary academic positions. A "disproportionately high percentage of the academic women on the campus" were in this group, and the ongoing budget crisis would pose "differential hardship" on them. The year before the administration had set up a buffer fund to help retain some of these positions, but not this year: All temporary positions were eliminated. Pitkin urged the chancellor to set up such a fund to protect against a drastic fall in the numbers of women: "Moreover, the women here involved have been uncertain of their status for two years running now, and simple human concern suggests that some definite action be taken soon."

Pitkin then offered two additional suggestions to the chancellor. One was that his office set up a "data bureau" containing information on all UC-Berkeley academic women: names, skills, interests, etc. Campus leaders could use the bureau as a resource when looking for qualified people to fill positions. "It seems to us that initiating such a bureau, at least for women in temporary positions would be a direct action you could take that would genuinely help them in a practical way, in addition to having considerable symbolic value as a demonstration of concern," she argued. The second suggestion was that the Senate CSAW be informed of the numbers of women affected by the budget crisis. Pitkin asked that each campus unit report the number of positions lost, and how many were filled by women. This would give everyone a "more accurate sense of the magnitude of the problem."[57]

After the first Chancellor's CSAW meeting, Sells recommended they solicit from all directors and chairs the name of a person who would serve as a liaison to their committee. This person would serve as a two-way conduit for information, and there would be a "two-way form of consciousness raising." She proposed to meet with women faculty and students in each of the 48

---

[53] 12/2/1971 Resolution passed by UC-Berkeley Assembly of the Academic Senate.

[54] 12/14/1971 memorandum from UC-Berkeley Law Professor Melvin A. Eisenberg to the Senate CSAW re request to review and comment on the most recent draft of the UC-Berkeley affirmative action program.

[55] [?1/3/1972] draft letter from the Senate CSAW to UC-Berkeley Chancellor Albert H. Bowker.

[56] 12/14/1971 memorandum from UC-Berkeley Law Professor Melvin A. Eisenberg to the Senate CSAW re request to review and comment on the most recent draft of the campus affirmative action program.

[57] 12/17/1971 letter from Chancellor's CSAW Chairman Hanna Pitkin to UC-Berkeley Chancellor Albert H. Bowker.

departments in the college of letters and sciences and probe what it was like as a woman to work in that department. She thought there would be "plenty of grist here for both [her] dissertation and the needs of the Committee."[58]

Two weeks after the first meeting of the Chancellor's CSAW, Chair Ameter met with the chancellor. She had requested a copy of the current draft affirmative action plan. She knew there had already been three drafts and a fourth was expected shortly. Amster wanted to make sure the Chancellor's CSAW would have the opportunity to advise the affirmative action coordinators and chancellor about the most current draft. Amster asked that time be built into the development timeline for the Chancellors CSAW to review the draft plans and provide suggestions. The chancellor asked Amster to let him know how much time they would need.

Amster also wanted to make sure people in the different employment categories had a chance to review the most current draft of the plan and make recommendations. This should be done before the plan is finalized in order to "PREVENT EMBARASSMENT."[59] The chancellor told Amster he believed "that the main problems are with these categories, and that the problem is not mainly one of discrimination." Amster did not agree and she told him so. In fact, she thought she "persuaded him to reconsider."

The Chancellor's CSAW wanted specific data. Amster asked the chancellor for a list of department affirmative action coordinators. She asked for a printout of academic salary titles, broken down by race and gender, listing salary and seniority. She asked for a list of individuals within units, broken down by salary code and sex. She also asked for student financial aid information.

The chancellor welcomed the Chancellor's CSAW's suggestions about developing anonymous grievance procedures. He seemed agreeable to developing protections for lecturers who were being threatened with release or salary reductions and asked the committee to "spell out" for him their recommendations. He was in favor of changing the current system where one person, usually the department chair, could decide the fate of a non-senate academic person, and he asked the committee for suggestions. The chancellor would call a meeting of all of the deans and inform them of the problems in hiring non-senate academics and "what discriminatory biases should be avoided." He thought long-time lecturers should be considered for tenure-track positions if they wanted to be, even if they hadn't had much research activity. He thought collective bargaining could bring better conditions for this group of employees. He welcomed the Chancellor's CSAW to submit a budget proposal to support legitimate and relevant activities, and he expected to fund it.

---

[58] 12/16/1971 letter from UC-Berkeley Sociologist Lucy W. Sells to UC-Berkeley Education Professor Harriet Amster.

[59] 12/29/1971 Agenda of UC-Berkeley Education Professor Harriet Amster's Meeting with UC-Berkeley Chancellor Albert H. Bowker.

22.8   Still Negotiating Confidentiality

On December 27, Eisenberg distributed what he called the final draft of the affirmative action program. He told the Senate CSAW that this draft reflected the comments of the committee.[60]

UC-Berkeley had given HEW many documents, including vitae, biographies, salaries, and de-identified letters of recommendation. But they still hadn't given HEW direct access to their personnel files. The chancellor's position was that the personnel files contained much information that was irrelevant to a class action investigation. He gave as an example "letters which make extravagantly glowing claims about UC faculty which cannot be substantiated, i.e., 'lies.'" UC-Berkeley was waiting for HEW to justify their demands. The two sides were still negotiating.[61]

---

[60] 12/27/1971 memorandum from UC-Berkeley Law Professor Melvin A. Eisenberg to the Senate CSAW re distribution of the final draft affirmative action program.

[61] 12/29/1971 UC-Berkeley Education Professor Harriet Amster's Summary of Meeting with UC-Berkeley Chancellor Albert H. Bowker, December 29, 1971, indicating more recent information in parentheses.

23

Weak, Grudging, Incomplete (January – February 1972)

Affirmative action need not be a grudging defensive reaction to criticism.
-Senate CSAW

At the beginning of 1972, UC-Berkeley was working against a January 15 deadline to get their letter of response and affirmative action plan into the HEW office. Then HEW would prepare a letter of findings.

The Senate CSAW set their next meeting for January 3. Betty and the other members were asked to review documents relating to the current Eisenberg draft affirmative action plan. They had concerns about the current draft: "Our advice on drafting has been completely ignored when offered, or unsought." They also had concerns about not having access to official documents: The university administration's response was that it would endanger the court case to make these public. The chair of the Senate CSAW decided to ask the administration for clarification of the difference between making the documents public and sharing them with the committee.

The chair drafted a letter to the chancellor. She asked her committee to improve it, and to identify policy changes to promote the status of women that might have a chance of getting passed by the policy committee and then the academic senate.[1] The draft letter said the Senate CSAW was "shocked" to find the hundreds of hours spent by three committees (Senate CSAW, Chancellor's CSAW, and Chancellor's CSNAW) on the draft affirmative action plan had been "useless." The Senate CSAW and Chancellor's CSAW had previously in August reported eight criticisms, but only two were incorporated into the most recent draft of the plan; the others disappeared "without explanation."

In fact, the Senate CSAW thought the whole Eisenberg draft affirmative action plan that everyone worked on had essentially been abandoned. Instead, the new plan consisted of a series of memos stitched together; it was "weak, grudging in tone, incomplete, not even meeting minimal requirements of the HEW guidelines... we are frankly appalled that such a piece of work could have come out of your office," they told the administration. The Senate

---

[1] 1/3/1972 memorandum from the Senate CSAW Chair to the Senate CSAW re the current draft UC-Berkeley affirmative action plan.

CSAW had hoped for a model piece of work and instead was looking at "an embarrassment." The committee's draft letter argued:

> *Affirmative action need not be a grudging defensive reaction to criticism. This is an occasion for the university which proclaims constantly its allegiance to excellence to produce an excellent product. Affirmative action is not even viewed in this document as an expansion of the university's opportunities, but as mere equality of employment, the narrowest possible perspective. This program can be seen more broadly, as a chance for the university to draw into its staff, faculty and management those gifted persons who by virtue of their sex or ethnicity have not had access to those positions which their aptitudes could enrich. More than half the population is excluded from or seriously disadvantaged in the competition for intellectual, scholastic, and professional recognition of merit. Thus the university could theoretically be twice as good as it is...*

They thought the chancellor should advance a plan that reflected the many hours of "creative effort" put to the task by campus committees and staff.[2]

In the end, the chancellor did not throw out the Eisenberg plan. The Senate CSAW resumed their work to try to find ways to improve it.

23.1   Carnegie Assignment

Betty's presentation at the AAAS symposium on women in academia had gone well. The organizer was anxious to publish the papers in a proceedings and asked Betty to send a copy of hers as soon as possible, even if it wasn't yet in final form, so that publication planning could proceed.

Sometime before 1972, Betty was given an assignment for the CCHE that came from former UC President Clark Kerr, who had created and was leading the commission. Back in 1967, Kerr was given substantial funding and near carte blanch to carry out an expansive mission of "promoting research and reflection on higher education and its role in society" and to "provide recommendations on the most vital issues facing American higher education."

Kerr used the CCHE to engage leading scholars and practitioners to think outside the box in carrying out research that targeted six broad policy areas: social justice; provision of high skills and new knowledge; effectiveness, quality, and integrity of academic programs; adequacy of governance; human and financial resources available to higher education; and purposes and per-

---

[2][early 1972] draft letter from the Senate CSAW to UC-Berkeley Chancellor Albert Bowker.

formance of higher education institutions.[3] The status of women in higher education fit squarely within these areas, and Betty was one of the leading scholars asked to contribute, likely because of her solid research-based approach. She had been given a three-page handwritten outline from Kerr the month before. The report was to be about 25 pages of text with a few tables, a brief annotated bibliography, policy guidelines, and appended documents.[4] Beyond Kerr's outline, Betty was apparently given considerable latitude to be creative.

The result of individual and group efforts across the nation was that there was now a wealth of information on the status of women that Betty could reference for her CCHE report. One of her AAAS colleagues suggested she consider a comprehensive review of hiring, tenure, and salary procedures at colleges and universities across the US and how these affect the status of women academics.[5] But Betty also planned to prepare new analyses for the report using data from UC-Berkeley. Unfortunately, there would be selection bias: The data would only include people who were already on the faculty, or who had already achieved graduate student status, i.e., it would only include people already successful. One way to deal with this selection bias was to match women with men on selected characteristics already known to be significant in determining status or changes in status: date of degree, department, prestige of institution, age, etc. Once matched, the samples of men and women could be compared across other characteristics not used for matching.

Betty continued to feel strongly that the simple additive linear regression approaches used by Helen Astin and Alan Bayer could be improved. Betty approached Astin about the matter of interactions, and Astin agreed with Betty that they were probably important. Also, Betty had strong opinions that reporting only correlations and corresponding significance probabilities was not sufficient. The magnitude of differences in the variables (e.g., salary between men and women) also needed to be reported.[6] Betty immediately began to think mathematically: $i=satisfaction$ with quality of education; $j=type\ of\ college$; and $k=sex$. She began writing equations for probabilities under different hypotheses and formulating statistical tests.[7]

[3] John Aubrey Douglass, "The Carnegie Commission and Council on Higher Education: A Retrospective," *Center for Studies in Higher Education*, accessed March 4, 2013, http://www.cshe.berkeley.edu/publications/carnegie-commission-and-council-higher-education-retrospective. Note: In the end, there would be 37 policy reports and 137 sponsored research/technical reports produced under this agenda.

[4] [about 12/1971] handwritten three-page outline from CCHE Chief Clark Kerr titled "Higher Education and the Role of Women in American Society: Ten Years Too Late."

[5] 1/4/1972 letter from New York University Graduate School of Business Administration Associate Professor and Social Policy & Urban Affairs Program Coordinator Arie Y. Lewin.

[6] [n.d.] brief report from [?Betty Scott] titled "Suggestions on Study of Status of Women."

[7] [?1/12/1972] handwritten page of equations by Betty Scott.

23.2   Angela Davis and UCLA

Betty was a member of the AAUP which has the mission of "advancing academic freedom and shared governance, defining fundamental professional values and standards for higher education, and ensuring higher education's contribution to the common good."[8] She had enough trust in the organization to vote her conscience on their initiatives: "I know that the California political climate is difficult. But I think that this is all the more reason to be sure to vote what we believe. I hope that this sentiment applies most especially to the AAUP. This is what I have always believed."[9]

The AAUP National Committee was expected to cast an important vote. Censure is a considerable stigma that can affect such important factors as hiring, fundraising, etc. It is a "profound punishment." This particular censure issue involved the young, temporary, black philosophy professor Angela Davis, a radical feminist and activist who had a relationship with the Black Panthers. The UC regents, upon the urging of California Governor Ronald Reagan, failed to renew Davis' contract in spring 1970 after she let it be known she was a member of the Communist party.[10] The regents were accused of firing Davis without due process. The AAUP conducted an investigation. One of the regents accused the AAUP report as displaying "evident pervasive bias." The AAUP counteracted by charging the regents with "dubious effort to avoid coming to terms with the calamity their actions have brought to the University of California." Some maintained the whole matter was a display of the contentiousness between the regents and faculty.[11]

Later, in April, Betty as an executive committee member of the AAUP Berkeley chapter would be asked to cast a vote on whether the chapter would support the censure of UCLA. Betty wouldn't be able to attend the meeting where censure would be discussed. She wouldn't feel the chapter had all the information they needed about the case, but the regents and UC president were reportedly not being totally forthcoming with the AAUP, and so Betty somewhat hesitantly would decide to send a message to the executive com-

---

[8] *American Association for University Professors*, accessed March 5, 2013, www.aaup.org.

[9] 4/9/1972 letter from Betty Scott to UC-Berkeley Physiology-Anatomy Associate Professor and Associate Dean Marian C. Diamond.

[10] Note: Angela Davis (1944–) is a retired professor of philosophy at UC-Santa Cruz. In a situation unrelated to her spring 1970 dismissal from UCLA, she supported three prison inmates accused of murder. During the trial in August 1970, there was an escape attempt in the courtroom when several people were killed. Davis was brought up on charges that included murder based on evidence that the guns used were registered in her name. She spent 18 months in jail and then was acquitted in June 1972, right before the censure vote. See "University Censured for Dismissing Angela Davis," *Jet Magazine*, May 25, 1972, 8.

[11] Carl Irving, "Angela Davis Firing May Blacklist UC," *San Francisco Examiner*, April 25, 1972.

mittee that she would vote for "any reasonable censure." Later the AAUP National Committee proceeded to censure UCLA over the Davis matter.

23.3 Advisory Committee Concerns

Betty continued to send out copies of the Policy CSAW report. One went to the chair of the Arizona State University Board on Equal Opportunity for Women, who said: "This report makes truly fascinating reading. UC-Berkeley's dearth of women with professorial ranks, the continued restriction on husband and wife employment before actual proof of nepotism, and the comprehensiveness and statistical soundness of your report all impressed me." The chair reported to Betty that the Arizona Board of Regents had rescinded their nepotism rules a few years before and had used the Policy CSAW report to develop several suggestions for her own campus' affirmative action plan for women. She thanked Betty again "for extending help to us."[12]

The Chancellor's CSAW continued to work on the Eisenberg draft affirmative action plan. They wanted to make sure it was optimal in terms of protection and benefits for women in both non-tenure and tenure-ladder positions. Where they thought the draft fell short was in the area of protections. The remedy for this, they reasoned, would be a board of appeals. Such a board could have a critical role in enforcement of affirmative action for women.[13]

The Chancellor's CSAW was also concerned about who the academic senate would appoint to the powerful budget committee that "represents the [academic senate] in academic appointment and promotional matters and in the allocation of resources."[14] The budget committee recently had been given two additional responsibilities related to women: to ensure the recruitment of women for professional positions; and to ensure that campus women who were not in tenure-ladder positions would be reviewed for such positions. The Chancellor's CSAW thought the leadership of the academic senate should accordingly "consider the feminist attitudes of those who are appointed to serve" on the committee in addition to the normal qualifications.[15] The reference to feminist attitudes was removed from the final draft of the letter, but the lead-

---

[12] 1/25/1972 letter from Arizona State University Board of Equal Opportunity for Women Chairman Susanne M. Shafer to Betty Scott.

[13] 1/24/1972 draft letter from Chancellor's CSAW Chairman Harriett Amster to UC-Berkeley Chancellor Albert Bowker.

[14] "UC-Berkeley Budget and Interdepartmental Relations," *Academic Senate*, accessed March 6, 2013, http://academic-senate.berkeley.edu/committees/bir.

[15] 1/24/1972 draft letter from Chancellors CSAW Chairman Harriett Amster to UC-Berkeley Law Professor and Chair Richard W. Jennings.

ership was asked to "consider the attitudes toward women of those who are appointed to serve, and in addition, include women on the committee."[16]

The Senate CSAW and Chancellor's CSAW were working collaboratively. At the end of January, the chairwomen of the two committees, Hanna Pitkin and Harriett Amster, jointly wrote to the chancellor. Women were clustered in temporary positions and were more affected by budget cuts. Departments should be prevented from not renewing or reducing the percentage of time of women's contracts: "... this de facto discrimination will become aggravated unless administrative action is taken immediately... Departments should not be permitted to appoint or re-appoint white males in these categories until assurance is given that the women are being re-appointed."[17]

On January 28, the Senate CSAW sent the VCAA a lengthy reaction to the "final" Eisenberg draft affirmative action plan for academic women and women graduate students. Eisenberg had been responsive to the committee's suggestions; they were "pleased with many aspects of the draft"; and there was agreement on "many issues, perhaps most." Yet, they found the draft to be incomplete in some respects: Certain topics and groups (e.g., librarians, undergraduates) weren't addressed, and parts were unsatisfactory. They offered suggestions to improve the text, including operationalizations, clarifications, standards, and the like. They furthermore thought the plan fell short by providing only for "punitive enforcement of requirements": It missed an opportunity by not providing for positive incentives, encouragements, and inducements. Also, they regretted the plan did not contain provisions for women who had complaints to seek redress: The current plan should specify who would hear formal complaints of sex discrimination. Also and importantly, the Senate CSAW found the draft to be deficient and inconsistent in terms of goal-setting and enforcement: Their letter insisted on specification of both immediate, short, and long-term goal standards for gender equity, along with timelines. They wrote that these are important even in a university shared-governance setting: "In an important sense, these matters are the heart of any affirmative action plan, the point to which one looks in determining whether declarations of principle are seriously meant and will in fact be carried out."

But the Senate CSAW was cognizant that this was in the middle of a budget crisis, suggesting the strong possibility of "differential hardships" for women. They appreciated the urgency of having a plan to protect and improve the status of women. Thus, while it should be modified before its final adoption, there should be immediate adoption of agreed-upon proposals. Waiting until the affirmative action plan was finalized "would seem unjustified."

Also, the Senate CSAW advocated for appointment of an acting affirmative action coordinator for women. This person should be a woman and her responsibilities should be limited to women (i.e., not shared with minority

---

[16]1/28/1972 letter from Chancellor's CSAW Chairman Harriett Amster to UC-Berkeley Department of Law Chairman and Professor Richard W. Jennings.

[17]1/28/1972 letter from Chancellor's CSAW Chair Harriett Amster and Senate CSAW Chair Hanna Pitkin to UC-Berkeley Chancellor Albert Bowker.

groups). Other suggestions in the Senate CSAW's letter to the VCAA specifically included many of the topics addressed in the Policy CSAW report such as proportional representation on the faculty and administration as well as graduate admissions and awards, child care facilities, maternity leaves of absence, nepotism rules, principal investigator rules, and career interruptions.

The letter had Betty's fingerprints all over it. The following is an example:

> *We suggest that the Affirmative Action Plan itself must explicitly establish a realistic, Campus-wide goal of long-range expectations against which particular units can measure themselves and by which they can set their course. Such a goal is needed in addition to the specification of particular requirements to be enforced immediately. In the abstract, one might expect that in the absence of all discrimination, the proportion of women on the faculty would show only random variation from the proportion of women in the population as a whole. In the meanwhile, it is worth noting that for Berkeley simply to match the percentage of women holding regular faculty appointments in all colleges and universities in the country in 1966 (15 percent) would require quadrupling the number of women in the Academic Senate (as of 1969–70). To reach a proportion of women in parity with the proportion of doctoral degrees we ourselves awarded to women in the period 1966–67 to 1968–69 would require tripling the number of women in the Academic Senate. (See Senate Policy Committee Report, May 19, 1970; and Academic Employment of Women at Stanford, a report to President Richard W. Lyman from Anne S. Milner, consultant to the President on affirmative action for women, submitted October 13, 1971).*

This statement included a mixture of measurement, probability, evidence, analysis, and precision. It was pure Betty. Insistence that a data bureau be established "now," and the stress on the need for good data and what constituted good data, showed Betty's continuing influence.[18]

Changes to the nepotism policy were in process; the necessary approvals had been obtained on the Berkeley campus but had yet to be approved by the UC president. Betty still had time to object to the proposed language.

While considering the language, Betty received a call from a UC-Berkeley art history professor. It was about a special art history student, Margaret, who had been employed by a medical physics professor as a clerk on a grant-funded drug project. Margaret was married to Tom, who was also employed by the medical physics professor and preparing to do research in the drug area. Betty was informed that Margaret was told she could not continue in her position because a husband and wife could not both be employed at UC; Margaret had to drop out and take an office clerk position in San Francisco.

Betty spoke with the professor's assistant, a lab administrator, a university policy officer, a campus personnel officer, an equal employment officer,

---

[18]1/28/1972 letter from Senate CSAW Chairman Hanna Pitkin to VCAA John Henry Raleigh.

someone on the staff personnel board, several other individuals, and the medical physics professor. She learned that Tom and Margaret had definitely been told they both could not be employed. Betty kept meticulous notes. In the end, she had telephoned at least eight offices in an effort to provide help to a married student couple who she did not know personally and was not even totally sure needed help, but who had apparently been affected by university nepotism rules.[19]

23.4   Class Action Confidentiality Agreement

On January 31, after four weeks of discussions, an agreement was finally reached where UC-Berkeley would allow HEW investigators into the personnel files. Administrators expressed they felt lucky they got an agreement at all, as HEW had the power to go into the personnel files without university permission.[20] HEW credited the agreement to a new interpretation of the executive order that would allow them to withhold funds from the Berkeley campus alone, rather than penalizing the whole UC system.[21] HEW would be able to examine individual personnel records relating to "recruitment, selection, employment, placement, termination, rate of pay, and other conditions and benefits available to employees, for both academic and non-academic employees." They would be able to interview employees or students without a university official being present.[22] The formal agreement included "safeguards" to ensure confidentiality. For example, while the university would reveal the position titles and sexes of persons who wrote letters of recommendation, and the races/ethnicities if known, actual identifying information (names of individuals, committees, departments, and outside institutions) would be deleted from the letters. HEW upheld the possibility that they could breach confidentiality if the records "constitute a necessary part of the evidence to be presented in any enforcement proceeding."

Pitkin, the chairwoman of the Senate CSAW, was generally in favor of opening the personnel files: "HEW cannot be in a position to protect individual rights and liberties unless it can get information on how those rights and liberties are being damaged." Hochschild, a member of the Senate CSAW, was also in favor: "the substance of discrimination is really found in the subtleties

---

[19] 2/2/1972 (and various later dates) handwritten notes by Betty Scott re art history student's employment at UC-Berkeley in relationship to the nepotism rule.

[20] Toni Martin, "Open Files Raise Confidentiality Issue," *The Daily Californian* II(21), February 3, 1972.

[21] Joyce Wolfe, "Complaint Filed Against Campus: HEW, Women and Berkeley Officials Tangle in an Investigation of Sex Bias," *Journal of Educational Change, University of California, Berkeley* 3, no. 6 (1972).

[22] "HEW Allowed to Examine Personnel Records at UC," *The Daily Californian* II(19), February 2, 1972.

of wording in letters." She thought the issue was important enough to risk any dangers of opening the personnel files.[23]

An editorial in the student newspaper explained the situation this way:

> *At first glance, the agreement appears to be a victory for the women's movement – an indication that the HEW investigation is progressing. However, that any government agency should have access to university personnel files is in itself unacceptable, and the end result, which we hope will be the elimination of a discriminatory hiring policy, is partially negated by the erosion of privacy. The University, in its standard refusal to go along with any kind of progressive change, has created this situation. Because the administration was insensitive to the early demands of campus women to end discriminatory hiring practices, the League of Academic Women, under the auspices of the National Organization for Women (NOW) was forced to file a suit with HEW.*

The editorial offered that implementation of an acceptable affirmative action plan for women would be the answer to problems with HEW.[24] In fact, the outcome of the HEW investigation would be such a plan, informed by the university's proposed affirmative action plan as well as the evidence that HEW uncovered in its investigation.[25]

23.5 Employment and Benefits

The Senate CSAW met again on February 8. The question was raised: What should they tackle next? There were still tasks from the Policy CSAW report that needed to be completed. Those should be tackled. What should be the timing? Action items should be proposed and submitted to the academic senate office in March. The Senate CSAW agreed to proceed based on the current situation for women, not based on the HEW investigation.

Betty volunteered to work on the lecturer problem. The Senate CSAW wanted a pool of money to be set aside to protect women lecturers: Most women employed were lecturers, and the budget problems were putting them at risk. Betty also volunteered to work on nepotism, and to review the Policy CSAW report for uncompleted tasks.[26]

Two days later, Chair Pitkin reported two things to the Senate CSAW. One was a request from the VCAA and Eisenberg to meet and discuss the

---

[23]Toni Martin, "Open Files Raise Confidentiality Issue," *The Daily Californian* II(21), February 3, 1972.

[24]"Open Files" [editorial], *The Daily Californian* II(19), February 1, 1972, 3.

[25]"HEW Investigates Discrimination," *The Daily Californian* [unknown issue], ?February 2, 1972, 7.

[26]2/8/1972 handwritten notes by Betty Scott re Senate CSAW meeting.

most recent comments on the draft affirmative action plan. The Senate CSAW had planned to meet again soon anyway, and so Pitkin aligned the purpose of this next meeting. The other thing was the VCAA's answer to Chancellor's CSAW Chair Amster's letter charging de facto discrimination and suggesting protections for women lecturers. The VCAA asked for evidence of such discrimination: He wanted to "know the facts of the matter."[27]

The committees on the status of women were getting more proactive and their voices were getting louder. For example, the Chancellor's CSAW heard that the VCAA might be stepping down from his position and returning to the faculty. Amster conferred with Betty and the other members of her committee, and everyone was in agreement to nominate Associate Dean Marian Diamond for the position. Amster wrote to the chancellor: If he would consider a nomination, then they all wanted to nominate Diamond.[28]

Betty faithfully attended Chancellor's CSAW meetings. At the February 11 meeting, good news was announced that a woman lecturer was now a professor. A subcommittee of the Chancellor's CSAW was formed (Washburn, Amster, Walker) to draft the agenda and formulate goals for the upcoming meeting with the chancellor on May 7. There were announcements about other upcoming meetings.

Betty and the Chancellor's CSAW continued to be concerned about the numbers of women on the faculty. Betty wrote in her notes from the February 11 meeting that there were 15 women among the 788 professors, and 49 among the 1,351 in ladder positions. She wrote a note to remind herself to request a copy of the academic job category list so the Chancellor's CSAW could determine the numbers in individual categories. She made another note to remind herself that these would be head counts: If a person was in more than one department, then the person might be counted more than once. She also made a note that the catch-all category needed to be "looked at carefully." Then Betty noted that the agenda for the next meeting should include a request for a factual investigation of current employment and uses/misuses of the job classification system.[29]

The UC had its own retirement plan. Sex inequities in benefits had already been tackled and resolved by the beginning of 1972: Women and men were paying the same premiums into the plan.

Previously even though they received the same benefits, women had to pay higher premiums than men under the argument that women were expected to live longer. Betty argued: "The data for the last 50 years at Berkeley casts some doubt on this assumption for university faculty. For Berkeley faculty, there is no evidence that women live longer than men. It would be interesting

---

[27] 2/10/1972 memorandum from Senate CSAW Chair Hanna Pitkin to Senate CSAW re IMPORTANT CHANGE OF PLANS.

[28] 2/11/1972 letter from Chancellor's CSAW Chair Harriett Amster to UC-Berkeley Chancellor Albert Bowker.

[29] 2/11/1972 handwritten notes by Betty Scott from the meeting of the Chancellor's CSAW.

to see data from other universities since, undoubtedly, the number of women faculty at Berkeley is too small to form a large enough sample to entirely settle the problem." But even if the data were inconclusive, there was a strong case to be made for spreading risks across all employees and having all pay the same premiums no matter their age or sex. This is the argument that led to equal premiums for men and women at the UC.

Not all retirement plans were as enlightened as the UC plan. The large national TIAA-CREF plan,[30] for example, was still charging women higher premiums. Betty heard complaints from faculty members who moved to UC and had retirement money in TIAA-CREF. She wrote a letter of protest on their behalf to the chairman of TIAA-CREF urging the removal of discrimination against women by spreading the risk across all employees and having all pay the same premiums, just as was being done at UC.[31]

TIAA-CREF wasn't convinced. Their chairman forwarded Betty's letter to their executive vice president and actuary, who acknowledged that most other retirement plans by then had established equal premiums for women and men. But, TIAA-CREF had data that "has repeatedly shown that female death rates are lower than male rates at all ages. The amounts vary from time to time, with the current female expectation of life at age 65 exceeding the male by about four years, a truly significant actuarial difference." The actuary challenged Betty to show him details of her UC-Berkeley mortality studies.[32]

23.6 Feeling the Way

The hearing for the class action lawsuit was scheduled for March 13 at 7:30 am in the San Francisco Federal Building, 17th floor, Courtroom 4, with Judge Renfrew presiding. Also, there was good news for Chancellor's CSAW chair Amster: The school of education recommended her appointment as associate professor.

Pitkin, Betty, and other members of the Senate CSAW continued to be vigilant about high profile administrative appointments in the UC system. Now they learned from a UCSF academic senate committee member that this medical campus was searching for a new chancellor. The process was that the campus committee on committees would nominate individuals to serve on a five-person search committee to advise the president on the chancellor

---

[30]Note: TIAA-CREF began through the Carnegie Corporation and Carnegie Foundation and has been offering retirement services to teachers for over 100 years. It now offers full financial services to individuals in academic, medical, cultural, governmental, and research fields.

[31]2/11/1972 letter from Betty Scott to TIAA-CREF Chairman William C. Greenough asking for removal of sex discrimination against women in the TIAA-CREF retirement plan.

[32]2/22/1972 letter from TIAA-CREF Executive Vice President and Actuary Robert M. Duncan to Betty Scott.

selection.[33] Were any women selected to be among the five? The Senate CSAW didn't know. Pitkin drafted a letter to the UC-Berkeley chancellor asking him to urge the UC president to include a woman on the committee:

> *It would be a great symbolic (and therefore also real) boon to women in the entire University, and therefore also on this campus, if the faculty panel that President Hitch selects to advise him on choice of a new Chancellor were to include at least one woman. It would also improve the University as an institute of higher learning... I would also urge you to urge him to consider selecting a woman... to be the new Chancellor over there. That would do even more for the University, its intellectual climate, and the situation of women and minorities in it.[34]*

But Pitkin wasn't sure this was any of the Senate CSAW's business. She asked her members to please let her know their opinions. She would then decide whether to send the letter to the UC-Berkeley chancellor.[35]

The Senate CSAW decided that searches for high level administrators at other UC campuses was its business. Pitkin then sent a letter to the UC president urging him to include a woman on the search committee, and to consider a woman for the UCSF chancellor position. Pitkin copied the UC-Berkeley chancellor, asking him to add his voice of support to these recommendations. She also asked the chancellor about child care. His predecessor had committed to providing funds for a center that would serve the entire campus. Pitkin believed this one "single action" could have the greatest practical benefit to women "toward making their intellectual capacity available to the University." Five months had passed. Where was the child care center, she asked?[36]

The *Journal of Educational Change* was preparing to publish a lengthy article about UC-Berkeley versus HEW. The leaders of both the Senate CSAW and the Chancellor's CSAW spoke out about the university's efforts toward improving the status of women. Pitkin expressed the view that: "The University's first affirmative action draft was one of minimal, grudging compliance with a threatening and unwelcome external force. After months of consultation it has improved, though, in fact, it is presumptuous of the University to appoint a man to draft it."[37] Amster expressed the view that "the University is trying to remedy the problem at least cost and with least reorganization of

---

[33]2/23/1972 memorandum from Senate CSAW Chair Hanna Pitkin to the Senate CSAW re the Chancellor's CSAW matters.

[34][?2/23/1972] confidential draft letter from [the Senate CSAW] to UC-Berkeley Chancellor Albert Bowker.

[35]2/23/1972 memorandum from Senate CSAW Chair Hanna Pitkin to the Senate CSAW re Amster's proposed resignation from the Chancellor's CSAW and other matters.

[36]2/28/1972 letter from Senate CSAW Chair Hanna Pitkin to UC-Berkeley Chancellor Bowker.

[37]Joyce Wolfe, "Complaint Filed Against Campus: HEW, Women and Berkeley Officials Tangle in an Investigation of Sex Bias," *Journal of Educational Change*, University of California, Berkeley 3, no. 6 (1972).

the rest of the system. Our emphasis, on the other hand, is to get the kind of remedy that will benefit the most women."

Just before the article was published and on February 29, Amster submitted her resignation from the Chancellor's CSAW upon the advice of the committee and having carefully considered the conflict of interest situation where she was also active in the class action lawsuit against UC-Berkeley.[38] The Chancellor's CSAW asked Vice-chair Fraenkel-Conrat to act as interim chair until the chancellor could appoint a new chair.[39] Betty marked her calendar. The next meeting of the Senate CSAW was planned for March 3. They would meet with the VCAA and Eisenberg on March 8.[40] The next meeting of the Chancellor's CSAW was planned for March 7. The committee planned both a pre-meeting to set the agenda and a postmortem meeting.[41] A page had been turned. There was still much to do.

---

[38] 2/29/1972 letter from UC-Berkeley Education Associate Professor Harriett Amster to UC-Berkeley Chancellor Albert Bowker.

[39] 3/1/1972 letter from Chancellor's CSAW Vice-Chair Beatrice Fraenkel-Conrat to UC-Berkeley Chancellor Bowker.

[40] 2/28/1972 memorandum from Senate CSAW Chair Hanna Pitkin to Senate CSAW re upcoming meetings.

[41] 2/24/1972 memorandum from Senate CSAW Chair Harriett Amster to Members of the Chancellor's CSAW re her proposed resignation from the committee.

24

Time for Action
(March – June 1972)

. . . so it's going to take a long time to bring in women.
-Albert Bowker

In 1972, where were the women who worked on the Berkeley campus? They were visible as receptionists, secretaries, and typists. They were hard to find among the faculty. They were invisible among the administrators.

An article in *Journal of Educational Change* reported many women were given lesser positions than men with similar backgrounds and experience and were afraid to speak out for fear of retaliation. One mustered the courage to say: "I think the real reason for my nonadvancement was male insecurity. Men would open the door to my office and say things you wouldn't believe. Like: 'You're trying to be better than we are.' 'You should be home with your kids.' 'You say yes to all the men.' 'You shouldn't be wearing a white coat.'..." Another said: "I looked around and saw that most women stayed at the lectureship level for 20 years, that the few women scientists who had achieved professorships wore big black shoes and were bitter toward men."

24.1 Affirmative Action Delays

It had been over a year since the "anticipated yet nonetheless startling" results of the Berkeley campus study on women appeared in the Policy CSAW report. Even though the university did not dispute the results and acknowledged the discrimination, not much had changed. Only 7% of tenured faculty members were women. Still a third of departments had not hired a woman into a tenured position within the most recent 20-year period.

Also, 64% of academic women remained in non-tenure track positions without the voting privileges needed to try to improve their situations. This group was shrinking due to the budget crisis: The campus was losing about 30 academic positions each year, and this was working against goals of increasing proportions of academics who were women. The chancellor said, "...so it's going to take a long time to bring in women." He rationalized:

367

> *Part of the reason that many women are not at Berkeley already is a lack of sensitivity to the problem. . . Berkeley is not a natural place for single academic women to live anyway. Berkeley is not much fun for single, professional women. Practically all the social life here is momma-poppa, couples, so in the past it did not attract many single women.*

New York was a much more fun place for single women to live, and so there were more distinguished academic women to be found there, he surmised.

The *Journal of Educational Change* summed up the HEW threat:

> *On the surface it would seem that HEW has become a powerful ally of the women's movement. However, HEW can only be as effective as it is willing and able to employ its enforcement powers, and it is this willingness which is presently being tested at Berkeley. . . Administrators here seriously doubt that HEW would jeopardize both the financial structure of the University and with it one of the nation's most vital research banks in order to protect the rights of women to become professors.*

Although HEW's chief declared they were "ready to take a stand," HEW and the university had for six months been staging "a political pas de deaux":

> *For its part, the University finds itself caught in a difficult position. Officials do not want to maintain male chauvinism on campus, nor do they wish to challenge HEW's credibility. Yet they are inadvertently doing both in the effort to protect what they call "Berkeley's interests."*

These interests included protecting confidential personnel information. The administration reportedly may have feared that if such information was released to HEW, then it could also end up in the hands of the FBI or CIA.

Another problem from the university standpoint is that HEW did not yet have a set of national standards or guidelines for campus affirmative action programs. Frustration on campuses across the nation over the uncertainties of the situation was the order of the day. J. Stanley Pottinger understood his office at OCR was caught in the middle: "The University thinks we are wild-eyed zealots who are totally tearing down the university system, and the women's groups think we have done much too little."

If HEW withheld federal funds, then it could expect several things to happen: a lawsuit from UC-Berkeley; pressure from congressmen mobilized by their constituents who lost jobs; pressure from government agencies that depended on canceled research; pressure from businessmen who feared they might be subject to similar government actions; and a "flood" of individual lawsuits from women. Pottinger declared: "And any politician who discounts the impact of the educated middle-class white female vote would be committing political suicide." Also, he understood that there was only so much HEW could do. They had 150 compliance officers to monitor 2,000 universities. If HEW mandated back pay or sex quotas, then it could be expected there would be red tape to tie up progress for years. This meant HEW would have to rely a good deal on the good faith of the universities.[1]

On March 31, Judge Renfrew ordered a delay of ruling on the class action lawsuit. The suit would now be heard at the end of July. This would be 40 days after the HEW issued its findings.[2]

24.2 Progress on Subcommittee Recommendations

Maternity Leaves: Other universities continued to work on academic women's issues and seek out Betty. In mid March, an inquiry came from the chair of a CSAW at The University of Connecticut. They had noticed the UC-Berkeley Policy CSAW's recommendation that there be paid maternity leave, wanted to know if it had been adopted, and if so, how it had been operationalized in terms of length of leave and budget allocation.[3]

Betty's response was vague: Everything was in a "state of flux." UC-Berkeley still didn't even have unpaid maternity leave. In fact, she pointed out, academic personnel didn't even have sick leave, and the administration said, "Have your babies during the summer." But Betty was hopeful, because now there was a campus affirmative action coordinator, and soon there was a draft affirmative action plan that included maternity leave.

However, the statement about maternity leave wasn't yet "firmly tied down." Betty didn't think the final draft would include paid maternity leave. She thought the length of leave should be up to the mother and her doctor. The matter of whether the leave should include both mother and father still needed to be settled. There could even be part-time rather than full-time leave. Or there could be a substitute instructor for a few weeks, rather than actual leave, just as was the current practice in cases of illness.[4]

Non-Senate Principal Investigators: Later in March, Betty and Pitkin went to see the VCR about the issue of non-senate women being principal investigators. They learned a lot. Pitkin immediately wrote down her own ideas and sent them to Eisenberg to see if he wanted to incorporate any of them into the current draft affirmative action plan. She suggested including clauses that urged principal investigators to pursue affirmative action goals

---

[1] Joyce Wolfe, "Complaint Filed Against Campus: HEW, Women and Berkeley Officials Tangle in an Investigation of Sex Bias," *Journal of Educational Change* 3, no. 6 (1972).

[2] P.A. Heist, A.J. Hochschild, F.C. Newman, R. Ostwald, E.L. Scott, F.S. Sherman, and H. Pitkin, "Report of the Committee on the Status of Women," in *Meeting of the Berkeley Division of the University of California Academic Senate* (Berkeley: University of California at Berkeley, 1972), 6.

[3] 3/15/1972 letter from University of Connecticut History Professor Emiliana P. Noether to Betty Scott.

[4] 4/4/1972 letter from Betty Scott to University of University of Connecticut History Professor Emiliana P. Noether.

when they searched for and hired junior researchers, and that required research unit directors to seek permanent principal investigator status for women who met the qualifications. The "exceptions" section of the current rules could be used to accomplish this.[5] It is not clear what Betty thought about Pitkin's primary suggestions: "Urge" is not as strong as "require," and women would still be exceptions.

Nepotism: At the end of March, Betty was still fired up about the nepotism rules. The VCA announced a recent rule change (a so-called rule 7.6 on February 25) that applied to non-academic employees including students. Betty thought the new rule was "quite ambiguous." The cover letter had a negative tone. In Betty's opinion, it was not only worse than the rule itself, it was even "harsher" than the previous rule. Betty felt the new rule and cover letter would make it "difficult to approve the appointment of two relatives." She thought the mere existence of the cover letter would inhibit managers from asking for exceptions. Betty wanted the nepotism rules to be eliminated, period, and she wanted the cover letter to be eliminated for sure.

Betty thought it was time to coordinate efforts toward joint action across the Senate CSAW, the Chancellor's CSAW, and the Chancellor's CSNAW. Together, the three committees could take on all of the nepotism rules, but this one in particular, as well as a number of other issues that could affect all women on the campus.[6] Betty immediately followed up with Fraenkel-Conrat,[7] Pitkin, and other members of the committees.[8]

Report: At the same time, Betty and the Senate CSAW were taking stock of what progress had and had not been made for women on the campus in 1971–72. Their draft report included a summary of the class action lawsuit activities and formation of multiple advisory committees on women. It mentioned that the Senate CSAW had welcomed two representatives of the graduate assembly to its regular meetings as a way of obtaining more input on women graduate student issues. It also mentioned that the chancellor gave his new affirmative action coordinator responsibility for both minorities and women. This was contrary to the women's recommendation that there be separate coordinators for each group. The draft report furthermore outlined the areas of disagreement with the current draft affirmative action plan. These included inadequate articulation of goals, timetables, and enforcement mechanisms; and lack of articulation of positive incentives to encourage or reward affirmative action. The Senate CSAW was either dissatisfied with, or dissatisfied with the clarity of, the administration's positions on nepotism, child care,

---

[5] 3/21/1972 letter from Senate CSAW Chairman Hanna Pitkin to UC-Berkeley Law Professor Mel Eisenberg.

[6] 3/30/1972 letter from Betty Scott to UC Office of the Vice-President for Business and Finance Staff Member Joyce E. Davis.

[7] 3/31/1972 letter from Betty Scott to UC-Berkeley Space Sciences Laboratory Scientist and Chancellor's CSAW Interim Chair Beatrice A. Fraenkel-Conrat.

[8] 3/31/1972 letter from Betty Scott to UC-Berkeley Department of Political Science Professor and Senate CSAW Chair Hanna Pitkin.

maternity or family leave, and part-time study. They continued to think the nepotism rule should be replaced by rules for conflict of interest.

The Senate CSAW understood its responsibility to be that of "reporter and policy advisor" to the academic senate on the status of academic women. At the same time, it was repeatedly being asked to deal with complaints of individual women, or specific issues about the treatment of women. The committee tried to help even though these complaints and issues were outside of its charge. Women needed somewhere they could turn, especially women whose issues couldn't appropriately (e.g., because of their status) be addressed within the committee on privilege and tenure or the grievance processes.

It was so difficult to get anything done! Patience was becoming more difficult. It was taking so long to get things done! The bureaucracy, even in the presence of good will, was becoming discouraging. There was study, restudy, and commentary. Where was the action? Where was the campus child care center; center for continuing education for women; resolution of the class action lawsuit; special fund to protect academic women on temporary positions in this time of budget crisis; data bureau to help such women find alternative positions within the university; and survey data the administration had obtained from units to determine the magnitude of the problems for women?[9]

Lucy Sells was still working with Betty and the Senate CSAW. Early in April, Sells wanted to get together with Betty to work on more department data, including the statistics department where Betty was chair. The admissions rates for the statistics department were 59% for men and 42% for women. Sells asked Betty: "Are women applicants that much less qualified than men?"

Score Card: A member of the Senate CSAW provided an evaluation of progress on the Policy CSAW recommendations, based on the 1970–71 budget committee report and what she knew of discussions within the Senate CSAW. The score card didn't look very good: Of the 14 recommendations, 2 were done, 5 resulted in some changes, 3 were being studied, and 4 had apparently been ignored.

24.3   Affidavit, Dissent, Conferences, AAAS

On March 27, Betty gave an affidavit before a notary public in relationship to the HEW investigation. She declared she was a member of the Policy CSAW and an author of their original report; the report took 11 months to prepare; and the recommendations were now being acted upon by the senate.

Regarding the data, Betty declared the Policy CSAW had used all available information, including that provided by the administration: "The administra-

---

[9] 4/5/1972 Draft – For Committee Use Only: Report on Developments in the Status of Women on the Berkeley Campus and the Activities of the Committee on the Status of Women in 1971–72.

tion customarily gathers information from departments and research units for use in academic planning, building plans, graduate student enrollment; and makes plans based on statistical analysis of the identical type utilized in the Report of the Subcommittee." Regarding analysis of the data, she declared that: "The best possible methods of analysis were applied when dealing with the factual data analyzed in the Subcommittee Report. When conclusions were merely tentative, they were identified as such."

Betty's affidavit was supportive of the relationship between her and the Policy CSAW on the one hand, and the UC and UC-Berkeley administrations on the other. She declared the Policy CSAW had the cooperation of the administration. She also declared that: "To the best of my knowledge, there has been no contradiction of any of the facts relied upon in the Report by the administration, including the President of the University, the Chancellor of the Berkeley Campus, or any other responsible official of the administration."[10]

Not everyone agreed with all of the recommendations of the original report. For example, Betty owned some stock in the Westinghouse Electric Corporation, and it came time to cast a vote on a ballot. Noticing all members of the board of directors were men, she wrote to the chair and suggested they appoint a woman. His response was: "They had no discriminatory policy about who could serve on the board... It is just that no woman was available who had the 'background, experience, ability and the judgment' needed to deal with the 'complex situations of today's business world.'"[11]

Also early in April, Betty made plans to attend a women's conference at Radcliffe College. She indicated her preferences for session attendance: (Monday) Work and the Family, then Biological and Psychological Bases of Sex Roles; (Tuesday) Gender Roles and Development of Person, then Life Styles of Women. She planned to attend the final dinner. Betty indicated her fields were mathematical statistics and applications of statistics; her particular research areas in women's studies were women in higher education and women in science. She indicated she was interested in trying to change UC-Berkeley and improve educational and job opportunities for women.[12]

Emory University Department of Statistics and Biometry Professor Donna Brogan[13] was organizing a session on women's issues for the upcoming Montreal ASA meeting. She invited Betty to deliver a paper. Betty politely declined: She didn't have funding to attend the meeting or anything new to present. Betty then suggested a possible topic for another speaker. She was worrying about efforts to increase the number of women on the faculty. There

---

[10]3/27/1972 Affidavit of Betty Scott, Marked Exhibit AH.

[11]3/22/1972 letter from Westinghouse Electric Corporation Chairman D.C. Burnham to Betty Scott.

[12]4/6/1972 form TO: Women's Conference – Radcliffe College, completed by Betty Scott.

[13]Note: Donna R. Brogan (unkn –) is a statistician specializing in the design and analysis of complex sample surveys. She had many 'firsts': First woman faculty member in statistics/biometry at Emory University, first woman full professor and first woman chair of the biostatistics department in Emory's Rollins School of Public Health, and first president of the caucus for women in statistics. She has received the COPSS Elizabeth L. Scott Award.

were expected to be almost no opportunities in higher education over the next decade: First there was "backlash that is already developing" due to budget crises; then there were declining birthrates. One thing the conference session could do, Betty suggested, was to plant seeds among young women and eventually grow them into faculty members. For example, how can they ignite the interest of high school girls in mathematical sciences? Betty told Brogan she was glad there was going to be a meeting on women's issues because it was important to maintain interest and continue to apply pressure.[14]

In May, Betty gave another talk on the status of academic women, this time at the sub-regional meeting of the College Entrance Examination Board (CEEB) in Salt Lake City. It was well received, and the luncheon conversation about equal opportunity was stimulating. The director of the CEEB Western Regional Office wrote this: "The immediacy of your data and the practicality of your recommendations were irresistible." Both Betty and he had worried that her talk wouldn't fall on sympathetic ears, but they were wrong: "Perhaps we did not predict [the sympathetic ears] but it seems evident that our Mormon friends are reevaluating their current posture on these matters," he wrote.[15]

Betty already had a relationship with the AAAS by serving as one of their vice presidents and participating at one of their national meetings on women in science. Back in April, the AAAS board of directors had decided to create a ad hoc advisory committee on women.[16] The leadership asked Betty to serve and summoned her to a meeting in Washington. This new committee would work with the AAAS executive officer to review programs of the AAAS and other science-related professional organizations "as they pertain to women." It would work with the AAAS committee on minorities in science to mobilize participation of women in these organizations and advance participation of women in science in general. The committee would be chaired by American University Mathematics and Statistics Professor Mary Gray.[17]

Betty's first reaction to the invitation wasn't positive. She thought, rather than form another committee, AAAS staff should take on the project. Already becoming weary of committee work that led to little or no action, she wrote: "I do believe that there has been enough discussion already and now it is time for action." But Betty's second reaction was to cooperate. If the AAAS

---

[14]5/11/1972 letter from Betty Scott to Emory University Statistics and Biometry Professor Donna R. Brogan.

[15]6/1/1972 letter from College Entrance Examination Board Western Regional Office Director Robert G. Cameron to Betty Scott.

[16]5/11/1972 letter from Betty Scott to Emory University Department of Statistics and Biometry Professor Donna R. Brogan.

[17]Note: Mary Lee Wheat Gray (1939–) is a mathematician, attorney, and statistician. She is former Chair of Statistics at American University. Her early work was on ring theory, but she progressively became involved in more human rights issues which brought her into law and statistics. Gray was presented the Presidential Award for Excellence in Science, Engineering, and Mathematics Mentoring by President George W. Bush. She is a fellow of the American Mathematical Society and the American Statistical Association. In 2012 Gray won the Committee of Presidents of Statistical Societies' Elizabeth L. Scott Award.

firmly believed another committee would help improve the status of women, then she would gladly serve on it but hoped it would be short lived.[18] She also hoped the committee could work on previous recommendations for the AAAS, including a data base of women scientists, a dedicated monthly page in *Science*, and a staff member to handle questions regarding employment of women.[19]

The AAAS responded to Betty that it was understaffed and needed volunteers to help take on the issues of women. It wanted to assemble a small number of women leaders in science to help "perceive more clearly what roles it may take effectively in the interest of women in science." The AAAS had assembled a similar group of minority leaders and the effort was illuminating: They didn't feel the needs of women and minorities could be met with a single group and firmly believed another committee would help improve the status of women.[20]

Betty learned the idea of a AAAS internal office for women was still on the table. She waited anxiously to receive the various proposals, including plans and budgets, for the office. She hoped to have them in plenty of time before the next AAAS committee on women meeting. Betty wanted to go to the meeting "prepared," and she hoped something could get done.[21]

The AAAS ad hoc committee on women would meet in Washington at the end of May. It would decide to recommend that AAAS set up an office on women and sponsor a women's program at the December AAAS meeting. The program, "Fact and Fiction With Regard to Sex Differences," would be approached as science and arranged by Betty.[22] It would be approached as science.[23] They would also discuss other matters toward mobilizing the participation of women in AAAS in particular and science in general.[24]

Gray began organizing women mathematicians across the nation and beyond. Betty took note. There had been a small gathering of women mathematicians in Atlantic City. They decided to found the Association of Women Mathematicians (AWM). The group would work toward equality for women in hiring and advancement. Gray reported: "Various proposals were advanced; the thought running through all of them was that in unity there should be some strength." Gray served as the first president of the AWM, from 1971–1973, and was highly influential. Here is an account:

---

[18]4/17/1972 letter from Betty Scott to AAAS Executive Officer Dr. William Bevan.

[19]5/11/1972 letter from Betty Scott to Emory University Statistics and Biometry Professor Donna R. Brogan.

[20]4/20/1972 letter from AAAS Executive Officer William Bevan to Betty Scott.

[21]5/12/1972 letter from Betty Scott to Yale University Biologist Mary E. Clutter.

[22]12/18/1972 report from The American University Mathematics and Statistics Professor and AAAS Ad Hoc Committee on Women Chair Mary Gray to AAAS Council re Report of the AAAS Ad Hoc Committee on Women.

[23]12/27/1972 press release from The American University Mathematics and Statistics Professor and AAAS Ad Hoc Committee on Women Chair Mary Gray regarding activities and complaints of the AAAS ad hoc Committee on women.

[24]12/6/1972 letter from The American University Mathematics and Statistics Professor and AAAS Ad Hoc Committee on Women Chair Mary Gray to AAAS Board of Directors.

Mary Gray decided that women needed to be more integral to the central decision-making process of the [AMS]. Shortly after she became president of the AWM, she carefully read the AMS bylaws and discovered that the AMS Council meetings were officially open to all members. One of her first acts as first AWM president was to show up in the room where the Council was about to meet and sit down. She was asked to leave. She replied that according to the bylaws, the AMS Council meetings were open to all members, and she was a duly paid member. The response was that there was a gentleman's agreement that only board members would be present during board meetings. "I'm not a gentleman," was her now-famous reply. "I'm staying."[25]

One of the first steps the women mathematicians took was to try to "get more women in positions of influence so that the oft-expressed concerns get more attention." They immediately petitioned the AMS council to elect another woman member. Gray was willing to be nominated. She solicited support from other members of the AMS, men as well as women. A few years later, Gray would be elected the second female vice president of the AMS.

Gray also urged people to form local AWM chapters.[26] She had participated in a panel on women in math at the 1971 summer meeting of the Mathematical Association of America and in May 1972 published a follow-up article in an expository journal on math and the profession. Her article articulated the issues, and also her own opinions and impressions.[27] Forty years later, in 2012, Gray acknowledged an "enduring mystery": AWM has many members, while the Caucus for Women in Statistics has few.[28]

24.4   New Senate Recommendations

The UC-Berkeley academic senate met at the end of May. The Senate CSAW presented an excellent eight-page summary of what had been done so far. The summary included data, accounts of events, summaries of policy changes, recommendations, and a resolution. They wrote: "In the year since the creation of this Committee the status of women on the Campus has improved significantly in some respects but declined on others; there have been important achievements, but progress is slow and not enough has been accomplished." It was taking too long to get things done, they declared.

---

[25]Patricia Clark Kenschaft, "Change is Possible: Stories of Women and Minorities in Mathematics," *American Mathematical Society*, (2005), 134.

[26][?1972] memorandum from American University Mathematics Professor Mary Gray to Women Mathematicians re forming a new organization, the AWM.

[27]Mary Gray, "Women in Mathematics," *American Mathematical Monthly*, (1972), 475.

[28]Mary Gray, e-mail to author, September 16, 2009.

The philosophy of the Senate CSAW was hard to argue with:

> *Despite the necessities imposed on the University by federal prohibitions against sex discrimination, the Committee has consistently urged that affirmative action at Berkeley be undertaken affirmatively, not defensively in reluctant compliance with externally imposed demands. We share the view expressed by the Senate in 1970, that existing "obstacles to the pursuit of professional careers by women scholars" impede the "overriding objective" of securing for Berkeley "the highest quality of scholarship and teaching." And we have urged that a great university must take a leadership role in combating prejudice and discrimination, promoting rationality, genuine merit, and social justice.*

The committee described its consultations with the chancellor to be both extensive and cooperative, and the graduate division to be "most cooperative."

The Policy CSAW summary called for more data, because little or no data had been collected or published since the report. They needed more time and funds to conduct needed studies. Perhaps some of the offices on campus – institutional research, graduate division, etc. – could undertake some of these studies. Data they did have for the past few years showed "no dramatic change" in proportions of women in various academic categories. There was no change in number of women administrators; also "no striking change" in proportions of graduate students admitted, funded, or completing who were women.

There had been some progress on the hiring of near relatives. The new rules allowing the chancellor to approve such "appointments if in the best interests of the university" resulted in 14 appointments between September 1971 and February 1972. Nevertheless, the Senate CSAW saw the new rules as being "unnecessarily restrictive and an obstacle." The policy should be liberalized, they opined, and replaced entirely by rules for conflict of interest.

The Senate CSAW also made a strong argument for more liberalized rules for principal investigator status, which was restricted by a rule that "projects must be appropriately related to the teaching responsibilities of the University." They argued: "While we share the concern for the importance of teaching, we also see how the University is pedagogically impoverished by obstacles to the professional careers of women... teaching might in fact be improved by a somewhat more generous interpretation of this requirement."

The Senate CSAW noted that many more women had been appointed to academic senate committees since their original report was issued. Numbers of women on senate committees went from 9 in 1969–70, to 17 1970–71, to 28 in the current year. Over the same time period, numbers of women in academic senate committee chair positions went from zero to one to three, with one of the three being chair of the powerful budget committee. While these were positive trends, there were so few women in positions that would qualify them for senate committee membership, that these few women almost immediately became overburdened with committee work.

The report contained 11 new recommendations:

To the President:

1. The President of the University be requested, in revising current policy on maternity leave, to make such leave available not just in connection with childbirth, but also in connection with other usually burdensome family responsibilities, not just to women but also to men.

2. The President of the University be requested, in current reconsideration of policy on part-time study, to make such study available to men and women who are precluded by family responsibilities from studying full time, and to make family leave available to students of either sex as leave for military service is now available to men students.

To the Chancellor:

3. The Chancellor be requested to form a pool of F.T.E. positions available to departments for the recruitment of outstanding women to the faculty.

4. The Chancellor be requested to proceed as quickly as possible with establishment of a program of child care on the Campus, for children between the ages of six months and six years, the program to be available to the children of faculty, staff, and students at reasonable cost.

5. The Chancellor be requested to appoint an Ombudswoman to hear complaints of individuals concerning inequitable treatment allegedly caused by sex discrimination. The Ombudswoman should refer complaints to the Committee on Privilege and Tenure or to the regular grievance procedure whenever these are appropriate remedies; she should not function as a substitute for these established channels.

6. The Chancellor be requested to notify deans, directors, and department chairmen that regular part-time positions now are available to both men and women, instructing them that such positions can be used to promote affirmative action.

7. The Chancellor be requested to liberalize the Berkeley policy on approval of requests for employment of near relatives in non-academic positions (currently articulated in a memorandum dated February 25, 1972) to permit such appointments unless they are unjustified or conflict with the best interests of the University, and to encourage them where they will in fact promote affirmative action.

8. The Chancellor be requested to instruct the Office of Institutional Research to collect and periodically to publish data on faculty ranks by sex in each department, earned degrees by sex in each department, undergraduate majors by sex in each department, and the promotion practices of organized research units with respect to women (salaries, rates of advancement, etc.).

To the Budget Committee:

9. The Budget Committee be requested to reconsider the present in-

terpretation of the requirement that proposals for extramural funding "be appropriately related to the teaching responsibilities of the University," in light of budgetary restrictions making it difficult for able women to obtain ladder positions that would automatically qualify them as principal investigators.

10. The Budget Committee be requested to conduct a study of possible reasons for observed differences in the rates of promotion of men and women faculty.

To the Graduate Division:

11. The Graduate Division be requested to sponsor research on the relative qualifications of women and men applicants to graduate school, and for fellowships, teaching assistantships, and research appointments, by department and level of student.

These recommendations were followed by a resolution to abolish the nepotism rule and replace it with rules to prevent conflict of interest; and to instruct the Senate CSAW to draft such rules and put them forward.[29] The resolution passed overwhelmingly (102 in favor, 22 opposed, 7 abstentions).[30]

24.5   Carnegie Deadlines, Topics, Connections

David Blackwell wrote a note to Betty asking her, as department chair, to please excuse him from the May 29 statistics department meeting. He needed to meet with Clark Kerr's group at that time. He wanted Betty to know that "WOMEN" was an item on the Kerr agenda. "Are you sure CK isn't expecting you at the meeting?" he asked Betty. Betty hadn't been asked, and so she asked Blackwell's advice about whether she should request to be at the meeting. "I am very much concerned but no one said anything to me," Betty wrote. Blackwell immediately called Kerr's office and verified that women were on the agenda, and Betty would definitely be asked.[31]

Also in May, Betty received a telephone call from CCHE Associate Director Virginia Smith.[32] Betty was given a time line for her report on the status of women. May 5: Have an outline to the Carnegie Commission. May 8: The out-

---

[29]P.A. Heist, A.J. Hochschild, F.C. Newman, R. Ostwald, E.L. Scott, F.S. Sherman, and H. Pitkin, "Report of the Committee on the Status of Women," in *Meeting of the Berkeley Division of the University of California Academic Senate* (Berkeley: University of California at Berkeley, 1972), 2.

[30]5/30/1972 University of California Minutes of the Berkeley Division, Academic Senate, 4.

[31][?5/29/1972] handwritten notes between UC-Berkeley Statistics Professor David Blackwell and Betty Scott re Betty's attendance at the 5/29/1972 meeting of Clark Kerr's group.

[32]Note: Virginia Smith (1923–2010) was a lawyer and economist. She was the first woman to serve as an assistant vice president at UC and was the eighth president of Vassar College. She also served as an associate director for the CCHE and director of the Fund for the Improvement of Postsecondary Education.

line will be reviewed by the technical advisory committee and Carnegie staff.[33] June 12: Have a draft report to Carnegie. June 23: Meet with the Carnegie board in Minneapolis. October: Have the draft ready for the publisher.[34]

Betty was working on the CCHE assignment and staying close to Kerr's outline. She put flesh on each of the following topics:[35]

Women have the same ability as men (academic ability)...

Women have same rights, should have same opportunities...

Society needs women of high ability; should be trained and educated in accordance with ability...

Women have some handicaps...

Women have some extra (uncounted) contributions...

Factual situation in various ways (use tables and plots)...

Present problems in trying to increase percentage of women...

Policies and recommendations...

Special...

What we should not do...

Where should we be in 1980? in 2000?

She moved a few topics around, combined a few, and put more flesh on some before sending the new detailed outline to Kerr.[36]

Betty was frustrated by the assignment and was very direct with Kerr:[37]

This outline is not satisfactory, I am sorry to say. It does not seem to have any focus, it is bogged down in details. But it is the best I can write now. Although I have learned a lot about this subject in the last three years, I am not sure that I know enough to write a proper report for the purpose at hand. Throw it out if you wish, please.

To say Betty was having trouble with the outline was an understatement.

Betty's network was growing as a result of her involvement with Kerr and the CCHE report. She learned from a former University of Illinois president that the Educational Testing Service (ETS) was conducting a research study, "Women Doctoral Recipients and Their Current Status." She wanted to include some of the results in her report, even if they were still tentative.[38]

Betty even got close up with HEW as a result of her CCHE connection. Kerr suggested she meet with Pottinger in relationship to her report. On June

---

[33] 5/6/1972 handwritten notes by Betty Scott re her telephone conversation with CCHE Associate Director Virginia Smith about the CCHE report on the status of women. Note: The CCHE Technical Advisory Committee had 14 members, one of whom was Betty's statistics colleague David Blackwell.

[34] 5/6/1972 handwritten notes by Betty Scott re her telephone conversation with Virginia Smith about the CCHE report on the status of women.

[35] [? between 12/1971 and 6/10/1972] Higher Education and the Role of Women in American Society, Betty Scott's "Outline Following Clark Kerr," 3 pages.

[36] 6/10/1972 Higher Education and the Role of Women in American Society, Betty Scott's "Draft Outline," 12 pages.

[37] 6/11/1972 letter from Betty Scott to CCHE Chair Clark Kerr.

[38] 7/12/1972 letter from Betty Scott to ETS Research Psychologist John Centra.

13, she met with him and one of his aides in his office. At first Pottinger didn't seem to know who Betty was or why she was there. Once he understood, he was very cooperative, and they had a good, relaxed, informal, and serious discussion. His confidence in Betty increased as their conversation continued.

Betty networked well. Pottinger told her HEW had finished their investigation of UC-Berkeley and expected to present their findings in early July. He also told her the HEW guidelines for affirmative action were drafted but not ready for distribution. Pottinger asked Betty if she would be willing to provide input to the current draft. Of course she would!

Betty asked Pottinger his opinions about emphases and recommendations for her CCHE report. His list included abolishing nepotism rules, launching child care centers, instituting family leaves, establishing grievance procedures, and including women on search committees. He also mentioned the need to continuously analyze employment patterns, and to establish goals and timetables. Pottinger felt particularly strongly that the report should "go well beyond the implications of the law," that it should "emphasize that it is to the university's advantage to admit and to employ women and minorities."

Betty also asked Pottinger about differing campus experiences with affirmative action programs. Pottinger responded that some were reacting to requirements out of fear. He thought they instead should react on the basis of basic fairness. They should take affirmative action seriously. Having faculty involvement was critical to the success of university affirmative action efforts. Having ongoing compliance oversight by committees or authorities was also critical. One of the many things these committees could do is watch out for cover letters that effectively change policy, just as the VPA's cover letter restricting the nepotism policy for non-academic employees.

The two discussed how difficult it was in tight budget times to increase the percentages of women. They discussed "the consequent deflation of women's hopes, [and] the tendency of white males to blame their employment problems on the employment of women."[39]

---

[39]6/13/1972 CONFIDENTIAL typed notes by Betty Scott re her meeting with HEW OCR Director Stanley Pottinger about the CCHE report on women in higher education.

Part VII

Salary Equity Studies

25

Facts of the Matter
(July – December 1972)

> ... *discrepancies of $3,000 or of $6,000 or even $10,000 are*
> *common, especially in the more elitist institutions...*
> *-M.G. Darland et al.*

Betty was becoming an information magnet as a result of all of her involvement with women's issues in higher education. For example, CCHE Associate Director Virginia Smith sent her excerpts of a report that examined proportions of civilian employees who were women in different countries around the world, as well as proportions of higher education students who were women. Smith thought Betty would find the information to be interesting.[1]

For another example, an important report on "Sex Discrimination in Academe," by Helen Astin and Alan Bayer, came to Betty from two directions. One was from Bayer himself, who Betty had met at a meeting in March. The other was in June from the UC-Berkeley graduate dean with whom Betty had established a relationship based on her local advocacy work.

Betty had already taken note in 1969 of Astin's pioneering work with linear regression analyses of the effects of sex on salary. She saw some glaring gaps in Astin's work. One was the exclusion of people not teaching full time, i.e., teaching less than nine hours. This created a "very strong bias," Betty thought, because it excluded both lower paid instructors and highly paid researchers; it excluded a third of the women and over half of the men. Then there was how Astin did her regressions. She used simple additive models. Betty expressed "reservations" about this approach. Then there was the fact that Astin only reported estimated overall differences in salary between men and women. Betty wanted to see estimates for differences within different job classifications.

Betty now had the data from the extensive CCHE-ACE national questionnaire conducted in spring 1969 and used by Astin. There were 60,028 college and university teaching faculty respondents. Betty set out to do regressions on these faculty data, and also on data from undergraduates, graduate students, and researchers. She asked to speak with the graduate dean to get his input

---

[1] 7/31/1972 memorandum from CCHE Associate Director Virginia Smith to Betty Scott re international comparisons of women in the labor force.

to her CCHE report and see if there were specific questions he would like her to try to answer.[2,3]

The university was putting "special efforts" into increasing the proportions of women in higher rank positions. Betty continued to receive requests from other campuses for information on other efforts at UC-Berkeley.[4]

As promised, J. Stanley Pottinger put Betty on the distribution list for review and comment of the draft HEW affirmative action guidelines. This put Betty in a position of potential significant influence. The HEW OCR sent the formal request to Betty on July 27: "I believe that your unique perspective will enable you to provide constructive suggestions as to how we can make this an effective and useful tool for securing compliance with the requirements of the Executive Order. Please bear in mind that your suggestions must be legally supportable by existing rules and regulations." Betty was asked not to copy or otherwise distribute the memo, as the guidelines were still considered a preliminary draft. She was asked to send her comments by April 15.[5]

Betty's work on academic women became noticed by the UC-Berkeley institute of governmental relations. The institute had a publication series, *Public Affairs Report.* They invited Betty to contribute an article on "The Women's Civil Rights Movement." Betty was glad the institute was interested in women's civil rights. But she didn't feel she had the time to write the requested article, as she was "immersed" in her CCHE report. Also, she didn't feel qualified to write on a subject that was broader than academic women. She suggested the institute narrow the subject and approach Susan Ervin-Tripp to write the article. Ervin-Tripp was Pitkin's successor as chairwoman of the Senate CSAW. With her strong public advocacy orientation, Betty was quite interested in the *Public Affairs Report* and asked to be put on their mailing list. She also connected the dots between the mission of the reports and hers and Neyman's joint work on weather modification experimentation.

The associate dean in the graduate division was asked by two staff women to assess whether there was sex bias in graduate admissions. Having a novice's interest in statistical thinking, he took the most recent data, which were for fall 1971, and estimated the following: The chance of admission for applicants was 0.486 for men and 0.415 for women; the chance of enrolling for admitted applicants was 0.503 for men and 0.619 for women; therefore the chance of enrolling for applicants was 0.244 for men and 0.257 for women.

Many would have stopped at the first comparison: Men applicants had a higher chance of admission than women applicants. This could be taken as evidence for discrimination. However, the associate dean argued that, when

---

[2]M.G. Darland, S.M. Dawkins, J.L. Lovasich, E.L. Scott, M.E. Sherman, and J.L. Whipple, "Women in Higher Education: The Facts of the Matter," in *Proceedings of the Twelfth Annual Meeting of the Council of Graduate Schools*, 55-65, New Orleans, LA, 1972.

[3]"UC-Berkeley Study of Faculty Salaries," *University Bulletin* 22, no. 24 (1972), 117.

[4]9/22/1972 letter from Betty Scott to University of Southern California Assistant Education Librarian Suzanne Harkins.

[5]7/27/1972 letter from HEW OCR Higher Education Division Acting Director Robert E. Smith to Betty Scott.

looking for bias in graduate admissions, one must also take into account the chance that an admitted applicant will actually enroll. Here, women had higher chances than men. His conclusion was: "... men and women have approximately equal chances of becoming graduate students at Berkeley, women having a slight edge over men (probably just sampling error)."

The associate dean copied Betty on his results and conclusion. She would have taken note. By his conclusion, he was incorporating bias in the graduate admissions process, which could be attributed to the university as an institution, with bias in the enrollment decision process, which could be attributed to the individual applicant and not the university. It could be argued that affirmative action was intended to mitigate university institutional biases, not biases of individual applicants.[6]

25.1 "Facts of the Matter" Manuscript

Betty collaborated with five others to prepare a report, on "Women in Higher Education – The Facts of the Matter," with partial support from the CCHE and NIH. It presented facts on women faculty, graduate students, and undergraduates. Betty and her collaborators presented the report to the council of graduate schools at the end of November.

The report began with stories about outstanding famous women who were treated unequally in comparison with men in areas of salary, employment, admissions, using the biggest telescopes, and treatment in faculty clubs. Nepotism rules were specifically mentioned. "The same or similar patterns persist today," and the thinking of women as well as men needs to be changed.

The facts on faculty centered on results from the very large CCHE-ACE survey. Results were similar to what smaller studies had shown at individual universities. Women were especially underrepresented within research universities, clustered at lower ranks, and less represented in higher paying departments like engineering, business, and physical sciences. Women tended to be older, unmarried, have less children, and do more undergraduate teaching. The status of women was "declining on almost every measure."

In the report, Betty and her co-authors discussed a recent study. Using linear regression methods, Astin and Bayer estimated that full-time women faculty, on average, were underpaid by $1,040; and discriminated in their rank assignment by one-fifth of a step. They explained that these discriminations in rank and salary were lower bounds in not taking into account discriminations in the 32 predictor variables (e.g., type of employing institution, field, research opportunities, etc.).

---

[6]8/21/1972 report from UC-Berkeley Graduate Division Associate Dean Eugene A. Hammel to Ms. Dobrzensky and Ms. Seiple titled "Admission to Graduate School, by Sex."

Betty and her colleagues extended the work of Astin and Bayer by including all of the surveyed women and a 0.25 random sample of surveyed men. They asked new research questions. Were the salary discriminations similar at different types of institutions? Were they similar in different fields? Could better estimates be obtained when higher order interactions were taken into account? Were all of these results similar when faculty teaching less than nine hours was included in the analyses, i.e., when all surveyed women were included? The "Facts of the Matter" report included preliminary results for these questions. Betty and her colleagues produced regression results for the 1,183 men and 312 women at Research I universities who were in the biological and physical sciences. They produced coefficients for men only and women only. This allowed them to look at more specific differences between men and women in each of the predictor variables. They reported on the 21 predictor variables, including four interactions, that significantly improved the precision of the salary estimates. Betty and her colleagues compared the actual salaries with the predicted salaries.

Betty thought the primary equity question was: Does a person of one sex earn more than what was predicted for the salary of the other sex? The answer, she and her colleagues found, was a resounding "yes," where "...discrepancies of $3,000 or of $6,000 or even $10,000 are common, especially in the more elitist institutions and in the fields where women are scarce." Women were being underpaid, or on the flip side of the coin, men were overpaid. It appeared the salaries of women should be "markedly increased."

The facts on graduate women centered on trend data from the US Office of Education and Bureau of the Census. These data showed a leaky pipeline. The percentage of women among undergraduates was larger than among graduate students, which in turn was larger than among faculty in high prestige universities. The latter percentage was "often zero." The facts on graduate women were this: "...women tend to be *more* likely to attain the degree, women finish *faster* on the average, women are *less* likely to receive financial aid, and those women who do obtain the doctorate are *as* likely to be employed and to publish papers and are *more* able than the corresponding men."

The facts on undergraduate women centered on data from the CCHE-ACE survey of undergraduates, and the ETS. These data showed women were higher achievers than men in both grade point average and class standing. The authors concluded: "Selection in the field of study and guidance in the selection of women students, as well as career choice counseling, are areas that need drastic change. As now taught in universities, and as now practiced, they strongly discriminate against women academically and economically."[7]

One of the effects of this work was that women on campus and elsewhere started paying more attention to salaries. According to Ervin-Tripp, "One of my friends in anthropology told me in the seventies that when salaries were

---

[7]M.G. Darland, S.M. Dawkins, J.L. Lovasich, E.L. Scott, M.E. Sherman, and J.L. Whipple, "Women in Higher Education – The Facts of the Matter," in *Proceedings of the Twelfth Annual Meeting of the Council of Graduate Schools*, 55-65, New Orleans, LA, 1972.

leaked she found her male contemporaries made \$20K a year more than she did. It turns out to be still true. She also complains still that many of the highest paid have low productivity. They butter up the administration, she says." It was important that Betty used productivity controls in her regressions, such as number of articles and number of books.

25.2 Faculty Club Accepts Women

At the end of July, The Faculty Club made a move independent of the ad hoc merger study committee. The board of directors at their June meeting voted to oppose any plan that would involve use of the women's club building as part of a merger. The board chair had written to the chancellor informing him of this vote and the five reasons behind it. (1) An ad hoc committee report to the previous chancellor back in 1970 recommended only one building be used after a merger, and the circumstances had not changed. (2) An analysis indicated it would be difficult to break even with continued use of the women's club building. (3) It would be difficult to service and maintain two buildings. (4) Use of part of the women's building for the women's center for continuing education would reduce use of the building for club purposes. (5) Continued use of two buildings would make the financial position of the merged clubs even more precarious. The men told the chancellor they could easily amend their constitution to include women as regular faculty and board members: "We have hoped to be able to contribute to solving the problem of the women's building, but our experiment to date suggests that the attempt to do so would only jeopardize the success of the merged clubs' future operation."[8]

In October, merger talks were continuing. The two clubs were using different accounting and record-keeping methods, and the women noted some resulting difficulties. There was considerable discussion at their October 11 meeting about how finances in a joint operation should work. Betty thought the WFC should collect its own rents and pay all bills pertaining to the operation of their clubhouse; this passed unanimously.[9]

Tensions were growing between the two clubs. As the men understood the accounting problems, these resulted from trying to operate the women's club as an accounting subunit of the men's club, and from trying to operate two separate offices. They didn't see these problems as being serious enough to warrant a major change of procedures, especially as the procedures were intended to be only temporary. However, they expressed impatience: Not much progress had been made "toward achieving a universal membership club."

---

[8]7/26/1972 letter from UC-Berkeley Institute of Business and Economic Research Professor J.W. Garbarino for The Faculty Club Board to UC-Berkeley Chancellor Albert H. Bowker.

[9]10/11/1972 minutes of the WFC meeting.

They suggested the women turn the complete operation of their club "back to the Club's Board as soon as practicable": This would solve the accounting problems. The men's club would immediately change their constitution and by-laws to allow the enrollment of women on an equal basis to men: ". . . in our opinion, this is needed to show good faith in the matter of the [Haas] gift." The men suggested that, in the meantime, the merger discussions should continue. Their board chair wrote: "As you know, I have never believed that the approach to a merger should be through creation of a new corporation. As a legal matter, that seems unnecessarily cumbersome to me. We think that amending the constitution and by-laws of the present corporation would be simpler and easier."[10] To the women, this proposal seemed less like a merger and more like a takeover.

The women called the attorney they had already retained to assist them in merger negotiations. On October 30, the attorney wrote to the chancellor: ". . . I have become a little concerned over the apparent lack of mutuality which I have observed developing in the course of the past few months." It was this same attorney who had suggested a two-year trial period before a consolidation of the faculty clubs. It was intended that adjustment problems be ironed out during this period. A formal contract had been prepared for this trial period, and the attorney expressed shock to learn that it had not been signed by the two organizations.

Now there were accounting problems between the clubs. The attorney wrote: "The suggestion is then made by the [men's] Faculty Club, which has everything to gain and nothing to lose by absorbing the Women's Faculty Club, that membership be open to women in the Faculty Club." He continued:

> In my view, and in the view of those ladies with whom I have spoken, this does not constitute a consolidation of two entities with some assurance of adequate and reasonable representation but rather an attempt to absorb or engulf the smaller of the two clubs by the larger. The Women's Faculty Club has functioned through the years with a membership which was made up of a large number of people who could not qualify for membership in the men's Faculty Club. It has always managed to keep its head above water and one of the reasons why there was not an immediate merger of the two clubs was because of the large outstanding indebtedness and the continual losses of the Faculty Club whereas the Women's Faculty Club was operating at least even with the board most of the time and generally managing to make small improvements from slight profits as they accrued.

---

[10] 10/17/1972 letter from UC-Berkeley Institute of Business and Economic Research's J.W. Garbarino "For the Board" to UC-Berkeley Physical Education Professor Mary Lou Norrie.

The attorney was in favor of one club for both sexes. But:

> *Should there be no merger, then it would appear that the very generous [Levi Strauss] gift which has been partially expended would be in some jeopardy. On the other hand, in order to save this gift, no one has the right to ask the Women's Faculty Club to give up a substantial percentage of its existing members, to give up its Club House, to be swallowed up by an organization which is so deeply in debt that without generous help it will never be able to come out even; and finally, to do all this without consultation with some sort of business-like committee which can create a true and fair merger rather than an arbitrary attempt to deplete the ranks of the Women's Faculty Club by draining off its membership through a change in the Faculty club by-laws.*

The attorney was generally annoyed about the manners of faculty: "It seems ridiculous that grown men and women should have to resort to attorneys, or Chancellors as far as that goes, to settle such trivial matters of social intercourse between members of the Faculty." Faculty manners are sometimes difficult to understand both from the outside looking in and the inside looking around.

In December, the *San Francisco Examiner* and *Chronicle* called The Faculty Club the "former stronghold of male supremacy." They reported that the 89-year-old club had not only voted to accept women members, but they had also gotten a full liquor license. Notably, they reported details of the liquor license at the same time as the November 15 vote to strike the word "men" from the constitution of The Faculty Club, which had passed by a 3:1 margin.

Were all of the women happy at the news? Not even. They felt the men's vote had pretty much negated seven years of work on a merger between the two clubs. The women had been bypassed. WFC President Mary Lou Norrie put it this way: "I think the move, although well intended, was a little simplistic. I don't think changing a word or two in the constitution is the way to make a viable merger." She didn't think the vote would cause women to immediately switch over to The Faculty Club. She thought women would wait until they saw evidence of a viable merger. Others thought women were tired of waiting for a viable merger and would, in fact, switch their memberships before too long. In any case, for 70 years women were forbidden from attending The Faculty Club annual Christmas feast. It was always a "stag affair." Now women could attend for the first time. Four women would even be in the skits this year, one of which was "Deck the Halls with [Chancellor] Bowker's Folly," and another of which was "HEW to the Line."[11]

---

[11] Nancy Dooley, "Traditions Tumble at UC Faculty Club," *San Francisco Examiner* and *San Francisco Chronicle*, December 3, 1972.

25.3   AAAS and Women in Science

In the fall of 1972, there were only five women full professors in the sciences at UC-Berkeley. Two had been promoted to full professor just the year before. Another was educated in Europe, where the system was more open to women. Betty was one of the other two. For change toward equality, there needed to be more women in the sciences. One way was to encourage high school girls to study, and work toward careers in, the sciences. Women in science were so rare, that when one went to talk to a group of high school girls, the girls were reportedly "amazed" to hear a talk by a woman.

Women in science began to assemble to attack problems of discrimination. Thirty UC-Berkeley women attended a first meeting on October 30 in the student union lounge. Betty was among them. The women brainstormed about ways to promote their cause. Based on her experience advocating for academic women in general, and having learned that there could be strong roots of discrimination at the department level, Betty suggested that there be remedial action for the sciences at the department level. There were many other suggestions, even that there be an auto and bike repair training program for women, because they thought many women were being overcharged by "unscrupulous male auto mechanics."[12]

Few people knew about the internal office for women the AAAS was setting up. Betty thought the entire science community needed to know more: They would not be satisfied just knowing that another committee had been set up. Betty thought the ad hoc committee for women and the new office should be promoted and clarified, and she expressed her opinion to the executive officer. She thought some notices should be published in *Science*.[13]

Betty immediately received an update on activities related to the new office. It had been named the "Office of Opportunities in Science." A new director was expected by December 1. The office would be concerned with "a full range of problems that affect the welfare of the individual scientist in his role as a scientist," i.e., with "career development in science." Also, there would be small panels to advise the office director. One would be on women. There would be a committee on opportunities in science, comprised of chairs of each of the advisory panels plus several additional members. The AAAS would announce these developments early next year, in its leadership report published in *Science*.[14]

The AAAS plan to update its membership about the new office wasn't much and wasn't what Betty wanted. She wanted something visible. Some "old-fashioned publicity." Something to change the image of AAAS. What

---

[12]Richard Colman, "Scientists: Women Meet – Plan New Hiring Tactics," *The Daily Californian* (Berkeley, CA), November 1, 1972.

[13]11/13/1972 letter from Betty Scott to AAAS Executive Officer William Bevan.

[14]11/17/1972 letter from AAAS Executive Officer William Bevan to Betty Scott.

she wanted was a quarter page in *Science*. Betty complained to her AAAS committee colleague Mary Gray and asked: What can we do to follow up?[15]

On November 27, the AAAS leadership finally sent out a formal notice about the new office to Betty and other members of their committee on women in science. Betty had already been told about the new office, the search for its director, the panels, and the committee on opportunities in science. But there was one new thing in the letter: The new office, director, panels, and committee would take over the work of the committee on women in science. The letter indicated the committee had served its function: It was no longer needed and the AAAS board had terminated it.[16]

Gray was not happy that the AAAS had abolished their ad hoc committee on women. She decided to fight. In December, she wrote to the board of directors protesting the "inept handling of this matter." She complained that the committee had received "very little cooperation from the AAAS office." According to Gray, the committee hadn't been consulted on plans for the office of opportunities in science. Furthermore, these plans treated women as one of several minority groups, even though women, unlike other minority groups, constituted 51 percent of the population, and even though both the committee on women and the committee on minorities opposed such treatment. Also according to Gray, then the executive officer communicated these plans to only a few individuals on the committee on women, not to the whole group nor its chairman (Gray), and then only in response to Betty's inquiry. Then the executive director scheduled a meeting of the committee on women with the new director of the office of opportunities in science without mentioning the board's vote to abolish the committee on women six weeks earlier.

Gray protested on behalf of her committee. She told the board the executive director had conducted "this whole exercise. . . in bad faith. . .":

> *Indeed, the whole affair has destroyed the credibility of the AAAS and dissipated the good will acquired by the passage of the resolution on women by the 1971 AAAS Council meeting. It was my impression that the AAAS wished to change its image, to reach out to those whom it had previously ignored; however, I see the same elitism, the same autocratic behavior, the same lack of human concern which have characterized the science establishment in the past. Since the communication has been so poor, I feel that I should be invited to address the Board at its December 9th meeting to present an alternative to its actions or proposed actions. I suggest that the Ad Hoc Committee on Women should be continued, that it should be involved in the activities of the Office of Opportunities in Science and serve as liaison to the Women's Caucus of the AAAS, and that there be a concerted public relations effort to change the image [of]*

---

[15]11/21/1972 letter from Betty Scott to American University Mathematics Professor and AAAS Ad hoc Committee on Women Chair Mary W. Gray.

[16]11/27/1972 letter from AAAS Executive Officer William Bevan to Betty Scott.

science, and in particular the AAAS, presents to women and to the public in general.[17]

Gray sent her letter to members of the, now former, committee on women, expressing her views: "I basically feel that the AAAS – in the person of [the executive director]. . . – is attempting to ignore the legitimate concerns of women and in particular to phase out 'troublemakers.' The new Panel is likely to have an establishment-oriented, non-activist character. . ." Gray was adamant that the AAAS leadership be "sensitive to the concerns of women."[18]

Since the protesters charged him with, in his words, "ineptitude and bad faith and suggested the same of the Board," the executive director presented a lengthy response. He argued there was confusion about the board's intentions and protesters' expectations. He outlined all the steps he had taken to get the office on women started and funded, and to recruit a director. It sounded like he had taken the initial recommendation from the committee on women to form the office and had run with it, without looking back and conferring with the committee along the way except to solicit their individual recommendations for director.

The executive director provided an operational and budget rationale. He wrote a justification for the communications delay, explaining the need for board action and subsequent paperwork that took several months. He explained his workload in detail, and also some serious operational problems confronting him. The AAAS executive director then pointed out that Gray was never actually officially appointed by the board or elected to chair the committee. At the bottom line he also expressed regret that it was taking so long to move women's issues forward within the AAAS. He closed his letter by apologizing profusely to the committee.[19] The next day, the director officially informed the former committee on women of the appointment of a director of the office of opportunities in science.[20]

Gray didn't let up. A few days later, she prepared a report to the AAAS council outlining the activities and grievances of the committee on women. They had little cooperation from the AAAS office. They had no success in getting another committee meeting set up, getting publicity in *Science*, or selecting the director of the new office of opportunities in science. The AAAS office had proceeded to plan the activities of the new office without consulting the committee on women. The AAAS had set up an oversight committee on both women and minorities, contrary to the recommendations of both

---

[17] 12/6/1972 letter from American University Mathematics Professor and AAAS Ad hoc Committee on Women Chair Mary W. Gray to AAAS board of directors.

[18] 12/12/1972 letter from American University Mathematics Professor and AAAS Ad hoc Committee on Women Chair Mary W. Gray to members of the former AAAS ad hoc committee on women.

[19] 12/14/1972 letter from AAAS Executive Officer William Bevan to [the former AAAS committee on women] Rhoda Baruch, Mary Gray, Fann Harding, Claire Nader, Mina Rees, Theresa Tellez, Virginia Walbot, and Betty Scott.

[20] 12/14/1972 letter from AAAS Executive Officer William Bevan to the former AAAS committee on women.

the committee on women and the committee on minorities. The oversight committee would have only three women on it. As Gray put it:

> *There are a variety of women in science – in universities, government and industry – in teaching, research and administration – young women and old women, minority group women. Three people cannot adequately represent these women, cannot represent the diversity of the Women's Caucus. Moreover, the influence of even these women will be so diluted by being fed through its chairman into a larger committee, whose primary concerns are not those of women, as to render it ineffective. However, even if the panel on women were expanded and allowed to function separately with direct access to the Board of Directors, if its advice is to be ignored or indeed if it is not to be consulted, it will serve no useful purpose.*

Where was the cooperation? Where was the consultation?[21]

With the resignation of Harriett Amster, Beatrice Fraenkel-Conrat moved from interim chair to chair of the Chancellor's CSAW. Fraenkel-Conrat led the committee to a review of the most recent draft affirmative action plan. At the end of August, she sent the committee's review to the chancellor. The issues identified were not new: needs more provision for enforcement; more provision for grievance procedures; better methods to determine goals and timetables; and specific policies on child care and maternity/family leaves. She asked the administration for annual data reports to be disseminated to the whole university community. These should focus on academic hiring and promotion, and graduate student admissions and financial support (fellowships, teaching assistantships, research assistantships), broken down by sex and year. Fraenkel-Conrat made a pitch to have the Chancellor's CSAW made into a permanent committee. Such a committee would be able to continue to make important contributions to campus affirmative action. It could review appointment procedures and employment policies, toward ensuring fairness for women. It could also provide advice to the campus affirmative action coordinator.[22]

In October, HEW at last issued its affirmative action guidelines for higher education. These directed each campus and the UC president's office to develop affirmative action programs. In all, there would be ten programs within the UC system. The university began to review all of its personnel policies to bring them into compliance with the guidelines. At the top of the list were the maternity/family leave and retirement policies.[23]

Rather than solving every problem for academic women, the appointment of an affirmative action coordinator for the Berkeley campus brought new complaints. In mid December, Ervin-Tripp expressed concern that the Senate

---

[21] 12/18/1972 report from American University Mathematics Professor and AAAS Ad hoc Committee on Women Chair Mary W. Gray to AAAS council re report of the AAAS ad hoc committee on women.

[22] 8/25/1972 letter from Chancellor's CSAW Chairman Beatrice A. Fraenkel-Conrat to UC-Berkeley Chancellor Albert H. Bowker.

[23] [?10/1972] loose page from UC-Berkeley Chancellor's records titled "VII. AFFIRMATIVE ACTION PROGRAMS."

CSAW, which she was now chairing, and the coordinator did not have a close relationship. The coordinator's reaction was as defensive as the AAAS executive director's had been. She outlined the attempts she had made to work with the women's committees on campus. But, she told Ervin-Tripp, she had to balance the needs of women and minorities.[24]

Thus, at the end of 1972, with the institutionalization of permanent administrators to handle their needs, academic women had something to cheer about. But the fact that the roles of these administrators were broader than the needs of just women meant the academic women didn't get all of what they wanted. The committees that had worked so hard for academic women were being asked to settle for having their needs balanced with needs of other groups.

At year's end, Gray remained unhappy with the AAAS's treatment of women. In a press release, she announced the many programs about women's issues held at the December AAAS meeting, but also aired the former committee on women's grievances about consultations, responses, and actions of the AAAS. She wrote: "The AAAS must become less establishment oriented and more people – oriented on the issue of women as on many other issues."[25]

25.4   Inadequate Input at Berkeley

At the end of December, Betty was still fired up about the lack of faculty and staff input into the UC's statewide affirmative action plan. The statewide academic senate committees, she noted, had not been consulted about the statewide plan. Betty wrote to the UC president: "... as a woman faculty member, I feel very deeply concerned." She urged the president to "consult now."[26] On the same day, Betty wrote a similar letter to the UC-Berkeley chancellor about the lack of recent input into the campus plan. She urged the chancellor to "consult now." She then wished him the best in the new year.[27]

It took only one day for Betty's letter to get across campus. The chancellor replied immediately. There were explanations about communications that did happen and the time it took to prepare a new draft plan. The chancellor cited the tight time line given by HEW. He informed Betty that the campus plan was completed just "yesterday," and copies were immediately sent to Betty and other members of the three women's committees, as well as to relevant

---

[24] 12/19/1972 letter from UC-Berkeley Affirmative Action Coordinator Colette M. Seiple to UC-Berkeley Rhetoric Professor and Senate CSAW Chair Susan Ervin-Tripp re relationship between the affirmative action coordinator and the Senate CSAW.

[25] 12/27/1972 press release from American University Mathematics Professor and AAAS Ad Hoc Committee on Women Chair Mary W. Gray re activities and complaints of the AAAS ad hoc committee on women.

[26] 12/26/1972 letter from Betty Scott to UC President Charles J. Hitch.

[27] 12/26/1972 letter from Betty Scott to UC-Berkeley Chancellor Albert H. Bowker.

committees of the academic senate. To satisfy the time line, he told Betty he needed their input "on or before January 15," so that a final revision could be in the hands of HEW "on January 15." This was not a lot of time – less than two weeks – to make any serious revisions. The chancellor told Betty he was "hopeful" she would review and comment "by January 5."[28]

---

[28] 12/27/1972 letter from UC-Berkeley Chancellor Albert H. Bowker to Betty Scott.

26

Focusing on Salary Data
(January – July 1973)

May I urge you again to take positive action to build a better
University.
-Elizabeth L. Scott

The Senate CSAW and Chancellor's CSAW were appalled at the draft campus affirmative action plan they had been given less than two weeks to review. The two committees had put hundreds of hours into making recommendations for the plan. Only two of their eight recommendations had any impact on the plan. "We are dismayed to find that all of this work has been useless," the chairs of the committees told the chancellor. The committees saw the draft as being patchwork in nature, "grudging in tone, incomplete, and not even meeting the intent of the HEW guidelines." Goal specifications, incentives, and details of monitoring were nowhere to be found, they charged.

Committee Chairs Susan Ervin-Tripp and Beatrice Fraenkel-Conrat told the chancellor, "We are frankly appalled that such a piece of work could have come out of your office." They reiterated that they hoped the campus would produce a plan that was innovative, "a model," "forward-looking," and "something to be admired." Instead what they saw didn't even resemble the Eisenberg draft. It was an "embarrassment," "ineffective," and "impoverished." "Affirmative Action is not even viewed in this document as an expansion of the university's opportunities, but as merely equality of employment, the narrowest possible perspective... Surely we are all determined that the university be at the forefront in identifying and developing human excellence," they wrote. They implored the chancellor to reconsider the plan.

Betty pointed out the women had also not had "any connection whatsoever" with the systemwide plan. It was important to the academic senate and affected women that plans be written with "effective cooperation."

Betty was comfortable working with the academic senate policy committee, as the Policy CSAW she had co-chaired a few years earlier was its subcommittee. Now that she thought the senate wasn't paying enough attention to the problems of minorities, she went back to the committee for help. Betty was willing to see issues of minorities given equal treatment with women. This would be fine, she reasoned, if it was needed to help minorities. Betty accordingly asked the committee to recommend to the senate that either an affirmative action committee be set up to cover both minorities and women, perhaps with a subcommittee on women, or else a committee on minorities that would parallel the Senate CSAW. Betty didn't have a strong opinion about which of these options should be adopted, but felt strongly that one should be, in order to ensure problems of minorities were addressed.[1]

Betty continued to put her statistical expertise to work. Her latest project was to estimate and project the employee turnover rate. She confessed to her provost: "The striking conclusion is that the situation is dismal for women and for minorities. I am sure that you realize this and I probably did too, but I did not have it down in front of me in black and white."[2]

Betty was getting increasingly worked up about what she and other women saw as lack of reasonable affirmative action efforts. She didn't like the UC president's January 5 draft plan for the system, and now there was a March 5 "tentative" plan for the Berkeley campus that she didn't like either. She thought the campus had written a minimalist document that didn't meet all the federal requirements and would be hurtful to women and minorities. Her vision was that UC-Berkeley should have the best possible plan, one that included innovative new programs and make Berkeley a leader in the nation.

Betty was so angry she didn't trust herself to speak out. She had the opportunity to speak out to her chancellor the week before at the Silver Jubilee dinner for the Berkeley Symposium on Mathematical Statistics and Probability. She wanted to point out to him how bad the situation was; but then again, she knew that he knew very well how bad it was. She asked herself: What could she say to change the fact that he seemed to be doing nothing?

Betty let off steam to the VC whose assignments included preparing the campus faculty affirmative action plan. She declared she was "disturbed beyond belief," felt "obliged to protest once more," and wanted to "repeat my plea." She asked the VC to please "listen and take action": Berkeley may be "only asking for trouble" with the current document.

The VC had asked the women to comment on a set of tables appearing in the March 21 draft five-year UC-Berkeley affirmative action plan. Betty strongly objected to the groupings of large, diverse units, calling these "utter nonsense": The groupings would only serve to allow departments who weren't taking affirmative action to hide within their grouping. She couldn't verify the figures on the page and thought the text was unclear. Betty repeated:

> ...*it is nonsense to use the doctorates granted during the last five years only. There are many women who obtained the doctorate 10 or even 15 years ago who are underemployed. These are the women that we need to seek out. It seems to me that* <u>*underutilization*</u> *is what Berkeley is supposed to correct. I wonder if the Chancellor's Office knows what this means? I can assure you that women do.*

---

[1] 4/3/1973 letter from Betty Scott to the Berkeley Division – Academic Senate Committee on Senate Policy Chairman Professor G.F. Chew.

[2] 1/26/1973 letter from Betty Scott to UC-Berkeley Provost G. J. Maslach.

Betty urged him again "to take positive action to build a better University."[3]

Betty continued to monitor and analyze the status of academic women. She was working on a new report on tenure and promotion. The Policy CSAW report was now somewhat out of date. However, Betty declared to a colleague that the "problems have not really changed."[4]

26.1 More Information Exchange

Women continued to exchange reports. Back in January, Betty received a copy of a report on the status of women at UC-Davis, and she arranged to have a copy of a CCHE report on privilege and tenure sent in exchange.[5]

Bernice Sandler had high regard for the University of Wisconsin's affirmative action plan. Betty wrote to UW asking for a copy: "We would like to use it as ammunition in Berkeley, where we are having quite a few problems."[6]

Word had gotten out that Betty had worked on discrimination in retirement benefits. In mid February, the Oregon Women's Equal Employment Opportunity Director asked Betty for her results,[7] but she unfortunately didn't have any more to send. She had looked at the 1920–1970 UC-Berkeley faculty records and found no difference in life expectancy between male and female retired emeritus professors, but her sample size was too small for definitive results. Betty had heard about other studies of retirement benefits, but she hadn't seen anything published. When Berkeley changed policies to charge and pay equally for men and women, Betty didn't have any reason to continue this work. "I am sorry I can't be more helpful," she said.[8]

By this time Betty, who was well trained and published in astronomy, had some 25 years of experience in a statistics department. She had plenty of time to reflect on her own career and had some insights on differences and similarities between statistics and astronomy, but she had not lived a career in an astronomy department. When someone at the University of Virginia's McCormick Observatory asked her about discrimination against women and minorities, Betty was suitably tentative and focused on her studies of women in higher education and science. She reported looking at two measures of discrimination against women: percentage losses of salary and employment

---

[3]4/9/1973 letter from Betty Scott to UC-Berkeley Vice Chancellor Mark N. Christensen.

[4]4/16/1973 letter from Betty Scott to Smith College Mathematics Professor Marjorie Senechal.

[5]1/8/1973 letter from Betty Scott to UC-Davis Genetics Professor Kathleen M. Fisher.

[6]1/29/1973 letter from Betty Scott to University of Wisconsin Special Assistant to the Chancellor Cyrena Pondrom.

[7]Note: Mary Gray and Ingram Olkin assisted the women in this suit. Ingram Olkin, e-mail to author, September 27, 2013.

[8]2/17/1973 letter from Betty Scott to State of Oregon Bureau of Labor, Civil Rights Division, Women's Equal Employment Opportunity Director Elanor M. Meyers.

opportunities. She found no evidence of more discrimination against women and minorities in astronomy than other fields: "Yet Astronomy [sic] is one field where one can pinpoint (and prove) very general and specific discrimination against women." She felt lucky she had been as successful as she had been. Looking at data from the most highly regarded research universities, she felt that "special efforts are needed to help women." Betty made four suggestions: (1) abolish anti-nepotism rules; (2) provide adequate fee-for-service child-care facilities; (3) provide equal opportunities to students in admission, training, financial support, recommendations, career services, and role models; and (4) provide equal access to instruments and equipment. Betty specifically mentioned the observatories should provide equal access to large telescopes.

The elaboration Betty provided with her third suggestion was brand new:

> *You may not be aware that men and women are given <u>different</u> vocational aptitude tests. If I take the man's test, it advises me to be a physicist or a mathematician or an artist. If I take the woman's test, it advises me to be a programmer or an X-ray technician. In junior high school and high school, girls are <u>forbidden</u> to take courses in mechanical drawing, in shop, in electronics; girls are strongly advised not to take second-year algebra, trigonometry, physics.*

Betty did take second-year algebra, trigonometry, and physics. She was in a unique position to offer that these "practices must be turned around."[9]

At the end of April, Betty received some information about the American Astronomical Association. It had 3,000 members: 209 were women, 84 with PhDs; and 18 of the 84 published in the *Astrophysical Journal* in 1971. She noted a UC-Berkeley astronomy department search committee summary alleging that her own "research interests at present are entirely non-astronomical."[10] Betty surely would have taken issue with this statement for, while her most recent publications had been on weather modification, she was working hard on a study of field and cluster galaxies that she would publish the next year.[11]

Betty also reviewed information on a search in the Italian department. She noted that women were thrown out of the applicant pool for various reasons, but different standards were used for a man in the pool.[12]

---

[9]3/8/1973 letter from Betty Scott to University of Virginia McCormick Observatory's C.R. Tolbert.

[10]End April 1973 handwritten notes by Betty Scott on the status of women in astronomy and Italian.

[11]J. Neyman and E.L. Scott, "Field galaxies and cluster galaxies: Abundances of morphological types and corresponding luminosity functions," in *Confrontation of Cosmological Theories with Observational Data (Longair MS, ed.)*, (1974), 129.

[12]End April 1973 handwritten notes by Betty Scott on the status of women in astronomy and Italian.

26.2 More AAAS Activity

The session Betty had organized at the AAAS meeting on "Fact and Fiction" had been well attended and received.[13] She began the new year 1973 by thanking Mary Gray, University of Maryland Economics Professor Barbara Bergmann,[14] and others for their excellent presentations. Betty hoped tapes of all of the sessions would be widely distributed and noted the AAAS would soon advertise purchase of their session tape. A national group of women in science was still a relatively new idea, and the women enjoyed all the benefits one has from being part of a group. Betty looked forward to seeing her women colleagues again at future meetings. She noted the next AAAS meeting didn't yet have a session on women and hoped this would soon be remedied.[15]

The AAAS committee on women may have been officially dissolved, but they continued to function as a group. On January 3, Betty sent a memorandum updating them about the AAAS council meeting. The committee reports were mailed to her and other council members in advance of the meeting; but Gray's committee on women report was not included and had to be distributed at the meeting; and Betty had to leave the meeting early to catch her flight. "There was not mention of problems of women before that time," she reported. Nonetheless, the new AAAS office of opportunities in science director, Janet Brown,[16] left a favorable impression on Betty. She was very able and active, Betty thought, and could do much if supported. Betty urged her colleagues to support Brown and influence the new office.[17]

As a member of the AAAS board, Betty carried resolutions from the AAAS women's caucus to the council.[18] The women's caucus wanted separate offices of women in science and opportunities in science. They wanted members who gave gifts to the AAAS to be able to earmark their donations. They also wanted to be able to distribute literature at the AAAS meetings relating to "problems of science and its role in society."[19] Betty also suggested to the

---

[13]1/3/1973 memorandum from Betty Scott to members of the former AAAS ad hoc committee on women re the December AAAS council meeting.

[14]Note: Barbara Bergmann (1927–2015) was a "trailblazer for the study of gender in economics" who spent most of her career on the faculty of the University of Maryland and then American University. Accessed 7/22/2015, http://www.nytimes.com/2015/04/12/business/barbara-bergmann-trailblazer-for-study-of-gender-in-economics-is-dead-at-87.html?_r=0.

[15]1/3/1973 letter from Betty Scott to Barbara Bergmann, Mary Gray, and four others.

[16]Note: Janet Welsh Brown (unkn – unkn) served as director of the AAAS office of opportunities in science for seven years. She eventually left to become executive director of the Environmental Defense Fund.

[17]1/3/1973 memorandum from Betty Scott to members of the former AAAS ad hoc committee on women re the December AAAS council meeting.

[18]1/10/1973 letter from AAAS Executive Director William Bevan to Betty Scott.

[19]5/1/1973 letter from AAAS Executive Director William Bevan to Betty Scott.

council that meetings of the women's caucus be listed on the AAAS program; in this way, all AAAS members would be made aware of the caucus meetings.[20]

The AAAS committee on opportunities in science had its first meeting in March. Their board agreed to several requests: make a policy statement about equal opportunities; and not hold any more functions at the Cosmos Club in Washington, DC because it would not accept women members. The UC-Berkeley chancellor was president of the Cosmos Club at the time; founded in 1878, the private social club for scientists and intellectuals would not begin admitting women members until 1988. The AAAS committee on opportunities wanted to take action immediately to coordinate existing registries of women professionals. It wanted to elect women to AAAS leadership positions, which meant women should be placed on nominating committees.[21]

Brown continued to make a favorable impression on Betty. They were forging a relationship. Brown sent Betty reports, and Betty sent back encouragement: "I am very glad that you are taking action and making suggestions. I am particularly impressed by your observation that the proportion of women and minority speakers increases markedly when there are women on the organizing committee." Betty also sent suggestions: Brown could publish an article in *Science* that described her office. Betty offered to help Brown in any way.[22]

Brown acted on Betty's suggestions over the next year. Brown would send out a letter aiming to increase participation of women in the 1975 annual AAAS meeting in New York. She emphasized that when the session organizer was a woman or minority, then there tended to be more women or minorities on the panel; and because such sessions tended to draw more women or minorities to the audience, it would be good to have such sessions across the program and not just in special topics. She asked for help conceptualizing and organizing such sessions, and asked that her office be copied so they could monitor the process.[23] Brown also acted on Betty's suggestion that there be an article about her office in *Science*. It appeared in the "Association Affairs" section of the journal and described the office's charge and mission. It reported that the number of women participating in annual meetings had increased 50% between 1972 and 1974; but that the percentage of participants who were women remained less than 9%. "Also discouraging is the fact that the largest concentration [of women and minorities] was found in symposia dealing with issues of special interest to minorities and women," the article reported. It also reported there were more women or minorities on AAAS committees (although six sections still had none) and an affirmative action policy had resulted in the

---

[20] 1/3/1973 memorandum from Betty Scott to members of the board of the AAAS re resolution by women's caucus of AAAS.

[21] 5/8/1973 letter from University of Georgia Biochemist Virginia Walbot to members of the women's caucus of the AAAS.

[22] 4/4/1973 letter from Betty Scott to AAAS Director of the Office of Opportunities in Science Janet W. Brown.

[23] 4/15/1974 letter from AAAS Office of Opportunities in Science Director Janet W. Brown to colleagues.

hiring of two women into senior department head positions. There would be a major push by the AAAS to educate more women and minority scientists.[24]

The AAAS board met and acted on the women's caucus resolution that Betty delivered in January. At the beginning of May, Betty received a mixed verdict. The board decided not to create an independent office of women in science, thinking they should give the office of opportunities in science a chance to function for women. But they did decide to allow earmarking of gifts and distribution of literature.[25] They also acknowledged that the title of a session, "Science and Man in the Americas," at their next national meeting was probably ill chosen; however, it was too late to do anything about it. Also, they agreed to put notices of the women's caucus meetings on the program.[26]

Betty sent the results of the board action to Gray.[27] "It seems to me that we should take advantage of whatever we can," she told Gray. This meant scheduling a meeting of the women's caucus at the next AAAS meeting, sending resolutions to the board in a more timely manner, and asking permission to distribute some literature at the AAAS meetings. Betty asked Gray for suggestions on the topic of women in science, and on sessions that could be organized by women.[28]

In May, the office of opportunities in science sent out a fact sheet about changes in the AAAS bylaws. These had to do with the way section chairs and members-at-large would be selected. Under previous bylaws, they were selected by committees. Under the new bylaws, they would be elected by the whole section membership. This change was meant to "democratize the decision-making," but women and minorities weren't so sure. They thought the changes might make it more difficult to achieve more women and minority leaders by having to "work within a complex electoral system" rather than "influence a willing Board and Executive Officer." It was deemed important to not split the women's vote, and so the fact sheet stated: "It will be very important to have women or minority candidates running in every one of these slots, but it is important that they not run against each other."[29]

---

[24] "Office of Opportunities in Science (AAAS News)," *Science*, (1974), 919.

[25] 5/1/1973 letter from AAAS Executive Director William Bevan to Betty Scott.

[26] 5/8/1973 letter from University of Georgia Biochemist Virginia Walbot to members of the AAAS women's caucus.

[27] Note: The AAAS ad hoc committee on women had 12 members, including two AAAS office representatives. The document was not dated. Betty was no longer listed as a member.

[28] 5/7/1973 letter from Betty Scott to American University Mathematics and Statistics Professor Mary Gray.

[29] 5/1973 fact sheet by the AAAS office of opportunities in science re new AAAS bylaws: Changes in section organization and elections.

26.3   Faculty Club Relationships

Relationships were strained when the boards of directors of the two Berkeley faculty clubs came together for a joint meeting back on January 22. WFC President Mary Lou Norrie presided. The different cultures of the two clubs were evident, even in the lists and attendance of board members: There were seven members of The Faculty Club board, two of whom were absent; the WFC board was much larger with eleven members, all of whom attended, including Betty. The women controlled the agenda. Norrie explained the purpose of the meeting and presented a brief history of the six years of actions taken toward a merger. WFC Secretary Margaret Uridge presented a report of the WFC review committee she was chairing. Time was set aside for discussion.[30]

The report recapped previous actions taken on the proposed merger. In 1967 the chancellor had appointed an exploratory joint ad hoc committee. In 1970 this committee was replaced with a proposal-development committee appointed by the boards of the two clubs. In January 1971 a proposal from this committee that involved releasing the women's club building was rejected by the women. In February 1971 a "Committee of Five" of the WFC prepared a joint operations plan that was rejected by representatives of the men's club. In June 1971 a modified joint operations plan was approved by memberships of both clubs. Next the report outlined the "unsatisfactory situation" of current joint operations. In July 1972 the WFC took back management of their building, citing frequent complaints about lack of maintenance and upkeep. In November 1972 the women's club took back the billing and collection of dues of its members at the suggestion of the president of the men's club. Over time there were many complaints about the type of food and quality of food service in the men's club. Also over time the principle of joint operations that members be appointed by their board of directors was followed by the women but not the men. The manager under the joint operations reported to the men's club but largely ignored requests and suggestions from the women's club. Without prior consent of the women's club, billing practices under the joint operations followed those of the men's club.

Finally, the report outlined recommendations and called for discussion. These oozed feelings of justice, pride, and practicality. Form a new corporation, to be called The Faculty Center of the UC-Berkeley, to include both men and women on an equal basis; include an agreement on how to pay the debt that had been accrued by the men's club; retain the WFC building, and connect it structurally to The Faculty Club; ensure adequate space for parking;

---

[30]1/22/1973 agenda of the joint meeting of the boards of directors of The Faculty Club and the WFC.

draft bylaws and separate house rules; and hire an outside vendor to conduct a management survey, that should include the topic of food.[31]

The Faculty Club leadership had their own perspectives about relationships with the WFC. The women's attorney suggested he meet with the men's club leadership; and the men were eager to meet with him "very soon" for a number of reasons, citing specific accounting difficulties. In their view the women were not sending payments promptly and were "in effect borrowing large amounts of money from us without paying interest or a service charge."

Women were being allowed to use the men's club facilities without paying dues, but this arrangement was not reciprocal. The men did not want this arrangement to extend beyond July 1 since they did not view it as being fair to their members. They wanted terms of a merger to be settled by then. The men's club secretary-treasurer asked the women's attorney to convey the July 1 deadline to the WFC board. This would be their notice of the men's intention to terminate the current arrangement.[32]

The WFC board of directors called a special meeting for July 13. Betty was there. The boards needed to agree on an "intent to work toward consolidation of operations" resolution so that the UC regents could accept a bid to renovate and add a dining room addition to the men's club building. The men's club drafted a resolution and asked the women to consider it.

At the meeting, the women approved the draft resolution with amendments. In their version, the WFC restated "its agreement with the principle of combined management operations" but mentioned the unresolved problems relating to billing, accounting, facilities management and planning, and food services. They made the resolution contingent on "the continued existence of the two club buildings and upon the construction of an enclosed connecting passage between the WFC building and the new dining pavilion."[33]

A result of these actions was the drafting of a formal memorandum of agreement. This MOA referred to the "planned development of a new Faculty Center" that incorporated the women's contingencies. It gave each party the right to exit from the MOA with 60 days notice if the plan for the center wasn't realized. It called for joint operations with a single food service; and a management analysis to help resolve the billing, accounting, facilities management and planning, and food services problems. It stated the intent to effect complete combined operations during the Fall Quarter of 1973. The presidents of the WFC and The Faculty Club signed the MOA on July 20.[34]

[31] 1/19/1973 report by the WFC review committee, "REPORT to the Joint Boards of Directors Meeting of the Faculty Club and the Women's Faculty Club Monday, January 22, 1973."

[32] 4/12/1973 letter from The Faculty Club Secretary-Treasurer Phillip E. Johnson to Attorney Henry Poppic.

[33] 7/13/1973 minutes of the meeting of the UC-Berkeley WFC board of directors with attached resolution by the board of directors of the WFC.

[34] 7/20/1973 MOA for joint operations of The Faculty Club and the WFC, signed by the presidents of the two clubs.

26.4   Time for Affirmative Action at UC

On May 8, Betty and the Chancellor's CSAW met with the chancellor and his cabinet. Betty's notes indicated a focus on searches. The chancellor asked what more could be done to ensure fairness, and what kind of advisory committee should be employed. Discussion included inadequacies of percentages of women applicants and hires, as well as of the search processes themselves.[35]

Prior to 1973, the UC system had never done a detailed analysis of the composition of its workforce by job category, looking specifically at women and minorities. Now it had, and results showed that, at the nine UC campuses, only 11.3% of assistant professors, 6.8% of associate professors, 3.7% of professors, and 6.9% of deans, directors, and provosts were women. In these and other academic and staff job categories, it was reported there had been too few women; more women had been placed over the past few years; but there were still too few women; and the women were being paid less than men. It was unclear whether the analysis was motivated by the HEW report that apparently – the report was prepared the year before but never made public – alleged sexual and racial discrimination in employment at UC. In any case, the UC administration asserted the results of the analysis would serve as a benchmark to monitor future progress for women and minorities.[36]

Meanwhile, the US Office of Education had conducted its own national study. The results were similar. Women were not receiving "equal pay for equal work": Males had a 21% pay advantage over women across institutions. Women were underrepresented among instructional faculty in general and full professors in particular, being clustered at the bottom ranks. The report concluded that "in no case do women come up to parity in pay with men," even at community colleges where the differential between men and women was smallest; and that over a ten-year period, the picture for women had been fairly stable, the biggest gains being at the instructor level.[37]

On April 25, Betty had sent a letter to the UC retirement system board raising the issue of equity for men and women in survivorship benefits. On May 7 she received the tentative and unofficial reply from the board chair indicating that the board would probably equalize the benefits; but additional funds would be needed to do the equalization, funds that would probably have to be raised by decreasing benefits to all new employees, both male and female.[38]

On May 14, Betty as statistics department chair received a request from the ASUC for information about student organizations in her department. Betty

---

[35] 5/8/1973 handwritten notes by Betty Scott at a meeting of the Chancellor's CSAW with UC-Berkeley Chancellor Albert H. Bowker and his cabinet re hires and hiring processes.

[36] "Women, Minorities: Report on UC's Top Jobs," *San Francisco Chronicle*, May 10, 1973; "UC's Sex and Ethnic Employee Data Complete," *UC News*, May 15, 1973.

[37] William Hines, "Women Professors Earn Far Less Than Men, Study Shows," *Los Angeles Times*, May 11, 1973.

[38] 5/7/1973 letter from UC-Berkeley Law Professor David E. Feller to Betty Scott.

reported that the statistics student organizations helped and guided both new and continuing students, organized picnics, and so forth. Betty offered: "The importance of student activities and opinion goes much farther." Betty chose to be blunt with the ASUC senator who wrote the request. She admonished him: "Kindly refrain from addressing department Chairs as 'Sir.' "[39]

By May 21, the UC president had approved an affirmative action plan for each UC campus. There was also a separate plan for the president's office. It was now time to get to work on the comprehensive work force "utilization" analysis that was called for by each plan. This analysis would compare data on availability pools to data on the actual workforce for women and minorities. If women and minorities came up short, then a plan would be made with goals and time lines to remedy the underutilization. It was also time to intensify work on equality in recruitment, training, and promotion. Plans were made to report efforts in all of these areas to the California legislature by year end.[40]

26.5   Lack of Senate Quorum

On May 14, Betty received a request from the academic senate to serve on the Senate CSAW again for 1973–74.[41] She readily agreed.

At the end of May, Ervin-Tripp, Betty, and the Senate CSAW were preparing an annual report to the senate. Their work over the past year focused on reviewing and commenting on the draft campus and system affirmative action plans. The report summarized efforts to improve the status of women in the areas of hiring and promotions. Evidence was presented that small improvements had been made in these areas over the past year; however: "The present rate of new appointments is not sufficient to make more than a very slow change in the sex ratio of the faculty." Evidence was presented that women were being hired at lower salaries, and that there was maintenance of a "maldistribution [sic] of women by rank as well as step," with women receiving less than men. Also, women continued to be scarce in leadership positions. The draft report discussed the hiring policy in some detail. There reportedly were "weaknesses in the framing and implementation" of the policy which led to some continuing actions that were decidedly not affirmative.

The committee wanted to see proactive measures taken, such as the appointment of a campus coordinator to work with academic search committees. Proactive measures were preferred over defensive ones, such as retrospectively

---

[39]5/14/1973 letter from Betty Scott to ASUC Senator Eric Jaeger.

[40][?5/1973] loose page titled "III. Affirmative Action Personnel Programs."

[41]5/14/1973 memorandum from UC-Berkeley Academic Senate Committee on Committees Secretary R.A. Cockrell to S.M. Ervin-Tripp (Rhet), Chair, B.B. Barton (Law), J. Cason (Chem), L.A. Henkin (Math), A.J. Hochschild (Sociol), D.H. Pyle (Bus Ad), and Betty Scott (Stat) re appointment to the Senate CSAW for 1973–74.

reviewing dossiers of candidates that had already been selected by search committees and sent forward to the administration for hiring approval.

The draft Senate CSAW annual report, which had Betty's stamp all over it, pointed out that goals and timetables were absent from the campus affirmative action plan submitted to HEW. The projections were pessimistic: Going from the then-current 5% or so of women on the faculty to a goal of 14% within 20 years would require a consistent 30% hiring rate for women over this period. But the discussion was optimistic: "If searches are very energetic and ingenious, and quality judgments are fair, these dramatic changes in hiring rates should happen as a routine consequence."

The Senate CSAW hoped to give more attention to student issues, in spite of the fact that the plan failed to mention students. The committee was concerned with student attrition, starting at the lower levels, and particularly in math and the sciences. A recent study by Lucy Sells showed that women in high school did not take as much math as men due to lack of social support: 57% of men had four years of high school math, compared with only 8% of women. This situation seriously constrained women's choices of majors in college. Programs were needed to provide support for women to take more math and science in high school, and for women to continue in all fields, including math and the sciences, through the doctorate.

The committee reported spending the most recent semester "trying to recover lost ground" on their work on the campus and system affirmative action plans. There were some improvements in the status of women to report, especially in the areas of nepotism, maternity and family leave, Ombudspersons, part-time positions, and the center for continuation of women. But there were also some remaining gaps which they articulated in some detail, and there were some areas such as child care that reportedly remained "untouched."

The draft report concluded that "the central weakness of the current affirmative action implementation lies in the lack of an affirmative action coordinator for academic affairs." The committee thought that overseeing and coordinating action was a full-time responsibility most effectively carried out by a respected academic with talents to work effectively with both internal and external constituencies. The appointment of such a person, they offered, could provide momentum for progress on many fronts.[42]

The senate policy committee was working hard on minority and women's hiring issues. They thought the proportion of women among appointees should mirror the proportion in the pool of qualified candidates. They accordingly prepared an affirmative action proposal for the senate that called for 25% of all future faculty appointments to be women. The policy committee brought the proposal forward for discussion and action at the May senate meeting.

The senate assembly was still comprised of all 1,700 eligible faculty members. It was not yet an assembly of elected representatives. Herma Kay had

---

[42]5/21/1973 draft report, "Report of the Committee on the Status of Women, To the Representative Assembly."

assumed the position of chair. The vote on the policy committee proposal was blocked by senate apathy: 75 members needed to be present; only 50 or so showed up. This was the second consecutive meeting that had less than a quorum. Action would have to wait until a future meeting.[43]

26.6   Carnegie Report Published

In July, Betty was still struggling with her work for the CCHE, now located on the tenth floor of a building on the west perimeter of the campus.[44] She was working with Jeanne Lovasich[45] on undergraduate women, and with others on graduates and faculty women. Betty passed along some comments about her report work to CCHE staff member Margaret S. "Peg" Gordon.[46] Betty also passed along to Gordon a subsection she had written that had given her quite a bit of trouble and that she had worked hard to condense.[47]

The CCHE report on women was shaping up, now under Gordon's direction. By the end of July there was a "final version" of the "second final version," with appendices. The report was written (or as Betty put it, "maybe rewritten") by Gordon, who still planned to make some changes based on reviews by commission members and "quite a few women" like Bernice Sandler and Kay. The report would go to the printer by the end of the week, and Gordon wanted Betty's last minute comments and suggestions.[48]

Gordon included Betty's salary studies in the report. She also considered including some of Betty's recent work on comparisons of ability between men and women. It was of "great interest," but the CCHE report was at the eleventh hour, and Betty's report on ability comparisons was too long. Gordon suggested Betty instead publish the work on ability comparisons as a journal article.[49] This suggestion didn't have legs because, as Betty recognized, much of the research done by others on intellectual differences was of poor quality.

---

[43]Maria Lenhart, "Apathetic Senate May be Doomed," *The Daily Californian*, May 25, 1973.

[44]8/1/1973 letter from Betty Scott to UC-Berkeley Rhetoric Professor Susan Ervin-Tripp.

[45]Dennis Pearl, telephone interview with author, February 17, 2014; Brian S. Yandell, telephone interview with author, June 24, 2014. Note: Jeanne Lovasich was a prominent figure in the Stat Lab, the main research staff member. Her title was Supervising Statistician, and she was head of computing. She performed calculations for Betty and other statistics researchers, first on a mechanical calculator, and later using the mainframe computer. Lovasich was sometimes a coauthor on Betty's publications.

[46]Note: Margaret S. "Peg" Gordon (1911–1994) was a professor of economics at UC-Berkeley, where she also was associate director of the Institute of Industrial Relations. She wrote a number of books. She also was an associate director at the CCHE where she was responsible for over a dozen reports.

[47]7/12/1973 letter from Betty Scott to CCHE's Peg [Margaret S. Gordon].

[48]8/1/1973 letter from Betty Scott to UC-Berkeley Rhetoric Professor Susan Ervin-Tripp.

[49]7/24/1973 letter from CCHE Associate Director Margaret S. "Peg" Gordon to Betty Scott.

Betty had been "bogged down" by her CCHE work. She thought she should have written the report faster and finished it sooner. She took responsibility for the delay and was glad the report was finally going to the printer.[50]

The CCHE report, "Opportunities for Women in Higher Education: Their Current Participation, Prospects for the Future, and Recommendations for Action," was published in September. It had chapters on the many roles of college-educated women: women entering higher education; women as undergraduates, graduate and professional school students, faculty members, and academic administrators; affirmative action; and needed campus facilities. Within 21 major themes, it contained many recommendations for action.

CCHE did not give Betty an authorship on the report. The author was listed as "Carnegie Commission on Higher Education." Betty instead received a significant acknowledgment: "A particularly valuable contribution was made by Elizabeth Scott, Chairman, Department of Statistics, University of California, Berkeley, who prepared early outlines and drafts of the report, and who is responsible for some of the special statistical analyses included in this report especially in Appendix C." Immediately after the acknowledgment, the report said: "We also want to acknowledge the helpful suggestions of Roger W. Heyns, president of the American Council on Education." Perhaps Betty was gratified to be mentioned before the ACE president.

Betty's Appendix C was titled, "Statistical Analyses Based on Carnegie Commission Survey of Faculty and Student Opinion, 1969." It was almost 50 pages of tables and graphs. Topics included, for example, women as a percentage of individuals at successive levels of higher education, and women as a percentage of individuals on the faculty. But most of the appendix was Betty's multilinear regression analyses of faculty salary differences (women versus men) for different subpopulations (type of university; field).

Among Betty's 29 variables used in the regression were a number of productivity measures: number of articles; number of books; association with a research institute; number of sources of research support; number of sources of paid consulting; research versus teaching inclination; administrative activity; consulting; outside professional practice; and hours taught per week. There were a number of interaction variables: date of birth and number of articles; sex and number of children; date of birth and number of children; and sex, marital status, and age. When the regression results were not what she expected, Betty dug deeper for possible contributing factors.[51]

It is not clear what Betty thought about the final report. One reviewer thought it was a good reference, especially praising the facts and figures

---

[50]7/31/1973 letter from Betty Scott to Carnegie Council on Children Administrator Elga Wasserman. Note: Prior to joining the Carnegie Council on Children, Elga Wasserman served for four years as the highest ranking woman administrator at Yale University; she was instrumental in the transition of Yale University to a coeducational institution.

[51]The CCHE, "Opportunities for women in higher education: Their current participation, prospects for the future, and recommendations for action" (New York: McGraw-Hill Book Company, 1973).

(Betty's domain); described it overall as "reasoned and reasonable"; but criticized the "overdone mood of moderation":

> ... *the book's tone and perspective is that of the Carnegie Commission's quasi-reverence for the basic goodness and soundness of American colleges and universities as they stand. Thus, while a multitude of worthwhile proposals and recommendations are put forth, the basic sentiment of the Commission is stated speedily (p. 8) and succinctly: "Generally we should seek the maximum gains for women but at the minimum cost to academic institutions and to society" (p. 8). The fundamental policy proposed, therefore, is one of accommodation on the part of the aggrieved minority gender and acceptance of the slowness needed for gradual change.*[52]

It seems Betty would have agreed with this criticism. Perhaps the acknowledgment and lack of authorship suited Betty just fine.

With all the politics and struggles, even Betty found herself momentarily leaning into the "lib." Some thought her most recent manuscript for the Carnegie Council on Children volume wasn't convincing. Her original draft had been more evidence based, but her most recent draft had been influenced by tapes of a meeting that had more of a "lib" flavor. Betty knew she needed to return to the flavor of the original draft and her own foundations.[53]

---

[52]Michelle Patterson, "Book Reviews: Opportunities for women in higher education: Their current participation, prospects for the future and recommendations for action," *Higher Education* 4, (1975), 251.

[53]7/31/1973 letter from Betty Scott to Carnegie Council on Children Elga Wasserman.

27

Society's Problem
(August – December 1973)

Seems unlikely to be completed during my lifetime. We now
propose an ongoing projection of research and, hopefully, action.
-Elizabeth L. Scott

Betty had been working on salary comparisons between men and women using UC-Berkeley payroll data. On August 1 she sent to Susan Ervin-Tripp a "complete but condensed" copy of comparisons based on the November 1972 payroll data. Betty wanted to use these to track whether the women were rehired the next year and, if so, whether their salaries increased or decreased. Tracking using payroll data was going to be a challenge, because women often changed their names. Betty decided it would be best to copy the women's data onto punched cards and then use the computer to help with the tracking. Creating a "decent file" was slow work, and Betty wished she had more time to devote to it. She was convinced there needed to be a "big separate study" that used a comparison group of men. Because sensitive materials would need to be reviewed for such a study, Betty thought perhaps the affirmative action office should conduct it, maybe in connection with the Senate CSAW and the senate budget committee. In the meantime, on August 3, Betty sent her set of completed salary comparisons to Beatrice Fraenkel-Conrat. For most employee classifications, the difference in weighted salary between women and men was negative. It was disappointing, but expected.[1,2]

In Betty's view, her research on academic women was action-oriented. In early August, she completed a form from the center for continuing education of women. It asked for information about her "current research on, by, and about women," to be included in a directory that would be disseminated campuswide so that people interested in similar research questions could find each other. Under the "Project Title," Betty wrote:

I do not have such regimented research as this form implies. Several of us
are concerned with Women in Higher Education – what might be called
the facts of the matter: salary comparisons, employmer [sic] rates, faculty

---

[1] 8/1/1973 letter from Betty Scott to UC-Berkeley Rhetoric Professor Susan Ervin-Tripp.
[2] 8/3/1973 letter from Betty Scott to UC-Berkeley MBVL Biology Research Scientist Beatrice Fraenkel-Conrat.

projections; possible reasons for small proportions of women in sciences; comparative abilities of men and women and reasons for differences, if any; and so forth.

Under the "Expected Completion Date" part of the form, Betty wrote: "Seems unlikely to be completed during my lifetime. We now propose an ongoing projection of research and, hopefully, action."[3]

27.1   Vision for a Faculty Center

The merger deal between the WFC and The Faculty Club still wasn't coming together. The regents decided the documents signed by the two boards "to agree to merge" were insufficient to permit release of the Strauss funds. So the regents' general council drew up a contract for joint operations that would be sufficient to release funds. Signed by the two club presidents, this contract would have allowed renovation and building of the new wing to begin – except that the Senior Men's Hall (now Senior Hall) was standing in the way, right between the two faculty clubs where the proposed new wing of the proposed merged faculty club was to be built.

Completed in 1906 with private financing, this is a large, rustic redwood log cabin designed by renowned architect John Galen Howard to represent classic California structures. *The Berkeleyan* gave this perspective: "Referred to it [then] as 'the heart of the university' (with the adjacent Faculty Club being the mind), it was the first official campus meeting place for senior men, who ruled student life in those days."[4] The hall had its ardent supporters who threatened to get a restraining order to keep it, and who began a fund-raising campaign to place it on the National Register of Historic Places.

The chancellor backed down. Plans for the new faculty club wing were scrapped. The regents' joint operation contract was rescinded. A new study of the needs of the two clubs was commissioned. The forces of change were diminishing. Betty listened as her fellow board members began talking about reopening the dining room in the WFC building.[5]

In fall 1973, the chancellor asked the faculty clubs for their needs, and he received a lengthy memo from the women that articulated their vision:

The Board of the Women's Faculty Club is thinking in terms of a Faculty Center, coordinating the two faculty club buildings structurally, by a connecting "breeze-way," and under joint management, with possible merger

---

[3]8/6/1973 letter and form from UC-Berkeley Center for Continuing Education of Women Director May N. Diaz and UC-Berkeley Sociologist Lucy Sells to Betty Scott.

[4]Julia Sommer, "Log Cabin Secrets in the Heart of Campus," *The Berkeleyan* (Berkeley, CA), October 14, 1998.

[5]8/2/1973 minutes of the noon meeting of the UC-Berkeley WFC board of directors.

at a later date into one incorporated Faculty Center. The two buildings complement each other in structure, atmosphere and by offering different types of food, furnishings, etc. they would provide the Faculty Center with a variety of services to accommodate the differences in taste of the faculty regardless of sex.

The women had come to envision "a less segregated organization." Yet, it was clear from the memo that they were proud of the establishment, and fiscally responsible maintenance, of their own club and building. It was important to them that they not be swallowed up by the men.

The women outlined in detail the needed renovations and structural improvements to achieve their vision of a Faculty Center to serve both sexes. They addressed needs for a structure to connect the two faculty club buildings, the porch, the plumbing and fixtures, heating and wiring, kitchen and kitchenettes and associated equipment, structural partitioning, furnishings, and the like. They concluded by saying: "It is our strong conviction that neither faculty club should be 'absorbed' by the other – but by combining the two clubs into a Faculty Center, each would contribute its unique attributes and better serve the needs of the academic community."[6] The women hadn't yet seen the men's statement of needs, and so they decided to be proactive and let the chancellor know what they thought should minimally be done to the men's club building: Convert the card room into a cocktail room that everyone will be comfortable in; convert the Howard Room into a lounge that everyone will be comfortable in; do something about the rough spots on chairs and tables, "thus reducing the hazards of snagging knit stockings, trousers, skirts, etc."; and make a restroom for women in the west end.[7]

27.2 California Assembly Committee Hearing

One of the people who participated with Betty at one of the statewide workshops for the Chancellors' Advisory Committees on the Status of Women (System CCSW) was a sociology PhD candidate. After participating in the candidate's study of successful academic "role model" women, Betty received a thank you letter. The candidate told Betty she found the life experiences of Betty and other interviewees to be "exciting and inspiring.": "Each woman's stoary [sic] would make an extremely interesting autobiography, and I hope that someday you will have the opportunity to write one."[8] The candidate was correct that Betty's story was interesting.

---

[6] 10/1/1973 memorandum from WFC President Margaret D. Uridge to UC-Berkeley Chancellor Albert H. Bowker re the faculty clubs improvement project.

[7] 10/5/1973 letter from WFC President Margaret D. Uridge to UC-Berkeley Chancellor Albert H. Bowker.

[8] 10/23/1973 letter from UC-Berkeley PhD candidate to Betty Scott.

The candidate worked for March Fong who was a Democrat and member of the California Legislature.[9] Fong represented a portion of Alameda County that included Berkeley. Among her primary interests were education and training. In 1973, Fong was chair of the California Assembly Committee on Employment and Public Employees. This committee was concerned with equal opportunity in state employment and was getting ready to put UC on the hot seat.

The assembly committee scheduled a one-day public hearing in Berkeley on November 8 in the auditorium of the Boalt Hall law school building on "Equal Opportunity in State Employment for Women – An Examination of Hiring and Advancement practices of the University of California." Basically, the committee wanted to know more about the HEW investigation[10] and evaluate the issues about, and proposed solutions to, sex discrimination at the university: "It is our intention to fully and methodically examine the University's policies of recruitment, hiring, promotion, advancement and retention as they may relate to possible discrimination on the basis of sex." According to Fong, "The issues and problems facing the University reflect a larger problem all society must solve relating to sex discrimination, and it is with this in mind that the Committee will hold the Berkeley hearing."[11]

The candidate assisted Fong by sending out an agenda that included presentations from HEW on their findings and from UC on their affirmative action plans, and a panel discussion by UC women. Betty readily agreed to testify and be on the panel. She was identified on the agenda with the CCHE report on the status of women. Senate CSAW Chair Ervin-Tripp and six other women would also be on the panel. The panel discussion would be followed in the afternoon by brief presentations on various topics. PhD Candidate Lucy Sells, for example, would present student issues. To facilitate preparation for the hearing, participants were invited to a pre-hearing briefing with Fong.[12]

Materials for the hearing included a white paper that explained the background, problem, and university action taken. The background in a nutshell was that federal legislation required the university to practice equal opportunity for women in employment, otherwise federal funds could be withheld; and there was a class action lawsuit. The problem, which was getting worse, was that women were hired at ranks below their qualifications and being passed over for promotions. The university response in a nutshell was the formation of

---

[9]Note: March K. Fong (1922–) was first elected in 1966 to represent the 15th Assembly District. She was elected as a representative three more times and then was elected to the position of Secretary of State.

[10]10/5/1973 letter from California Legislature Assembly Committee on Employment and Public Employees Chairman March K. Fong to UC-Berkeley Genetics Associate Professor Patricia St. Lawrence.

[11]10/30/1973 press release from California Assembly Committee on Employment and Public Employees Chairman March K. Fong "For Immediate Release" re hearing on equal opportunity in state employment for women.

[12]10/24/1973 memorandum from UC-Berkeley PhD candidate to Chuck Coles re legislative hearings, with attached handwritten notes.

an affirmative action plan that many womens' groups found objectionable; and funding was prohibiting advancement of affirmative action goals. Distributed materials also included a list of 11 questions, with many subquestions, that the university was asked to be prepared to answer. Questions pertained to the HEW report, university response to HEW, available data, relationship between goals and practices, budget constraints, salary setting methods, monitoring committees, promotion opportunities, training opportunities, child care centers, and student financial support.[13]

The Fong hearing took place as scheduled. Two sides of the story were told. University administrators defended system and campus affirmative action efforts: Recruitment and hiring practices were changed, and undergraduate women were encouraged to aspire to typically male-dominated professions. The academic senate chair said numbers of faculty women remained small because they only took "the most qualified people," and women in this group were scarce. The senate chair's home department of physics had 60 faculty members, none of whom were women. "Hopefully we will be able to find a woman within the next four years," he told the committee. Next a women graduate student at UC-Berkeley "told the story of a woman who was denied a position in Berkeley's physics department because her husband was hired there and was 'making enough money for both of them.'" The woman worked for no salary and won a Nobel prize. "What kind of qualifications could they be talking about?" the student asked the committee.

LAW presented a critique of the university's "current weak Affirmative Action efforts" and advocated for further legislation. A UC-Irvine anthropology professor, then the chair of the statewide UC-AFT (UC branch of the American Federation of Teachers union) affirmative action committee, concurred that pressure of additional legislation was needed. Her testimony gave specific examples as to why.

First, most campuses did not hire full-time affirmative action coordinators. Those that did made the hire at a low level. Those that did not added the responsibilities to an administrator who already had many high-level responsibilities and may not have had a commitment to affirmative action; one was quoted as saying he was quite willing to assume the challenge of "pimping for academic women." Second, there were problems associated with various campus committees created to address affirmative action issues. Almost invariably they were asked to function without resources (no budgets or staff support). Committees or individuals producing much needed data and reports were invariably asked to do so as volunteers and at the expense of their other work. Third, there was continuing use of the old boy system for faculty recruitment. The university had been slow to make directives in affirmative action hiring. Invariably additional funding was not provided to advertise positions and increase women and minority applicants.

---

[13]10/30/1973 press release from California Assembly Committee on Employment and Public Employees Chairman March K. Fong "For Immediate Release" re hearing on equal opportunity in state employment for women.

The statewide UC-AFT chair talked about lack of gains in proportions of women on the faculty. She confessed how it felt to be a faculty woman. In person, male faculty members, librarians, and staff members often perceived her to be a student. On the telephone, she was usually perceived to be a secretary. "This is ridiculous and appalling, for the problem is not, as with minorities, one of availability," she declared. The bottom line, said the statewide UC-AFT chair, was that "women are losing faith in UC assurances that it is implementing Affirmative Action or will do so soon." She told the legislative committee that women were accordingly looking for other ways to mobilize: "We hope you will investigate the situation very thoroughly and we intend to help you do so..."[14]

Betty, who continued to conduct payroll studies for the SWEM, testified on salary inequities. She informed the committee that women at prestigious universities were making on average $2,000 to $5,000 less than men in similar positions.[15] Ervin-Tripp testified that affirmative action coordinators should report to the chancellor. The administration, defending the current setup, countered that this would be impossible: "...he or she would be caught between the constraints of running the program and satisfying complaining parties." The committee was told that HEW had not yet taken action on the UC affirmative action plan, and so no federal funds had been held up to date. HEW representatives declined to attend the hearing, saying that negotiations over the UC affirmative action plan were "at a critical point."

Fong took a very strong position at the hearing. She declared: "The University will only meet the minimum requirements of HEW which are totally inadequate to meet all of the needs of University women." She intended to write a proposal that would include a number of things missing from the current UC affirmative action plan: a dedicated affirmative action officer, grievance procedures, incentive and monitoring systems, and salary equity.[16]

Betty read the research paper written by the PhD candidate who assisted Fong. Betty had plenty to say about the paper, "Success Correlates for Academic Women," and she said it. While she found it interesting and well-written, and she was glad for the research, there were a number of points on which she had criticisms.

The candidate wrote that UC-Berkeley refused fellowships to women. Betty pointed to her own experience to show this was not true. Betty had a Lick Observatory Fellowship. It not only was a big honor, but paid well in comparison with other fellowships. Of the observatories with which Betty was familiar, only Mount Wilson "openly said that women need not apply":

---

[14]11/8/1973 testimony of UC-Irvine Anthropology Professor and UC-AFT Affirmative Action Committee Chair, "Berkeley Federal Funding in Jeopardy: Assembly Sex Discrimination Hearings," [source unknown].

[15]"Women Assail UC Job Policy," *San Francisco Chronicle*, November 9, 1973, 4.

[16]Laura Thomas and Rick Feldman, "Campus Women Attack UC Affirmative Action," *The Daily Californian*, November 9, 1973.

It is unfair and really untrue to state "She denies that the facility (fac-ulty???) directly discriminated, since they never openly asked her to leave;..." Contrary to what you state, the faculty encouraged me to stay and to continue graduate work. They made arrangements [sic] for me to spend summers at Lick Observatory (paying only a small amount for food), to spend a summer at the Mt. Wilson Observatory working with the most famous astronomer of our time (Hubble), to obtain special ob-servations, etc., etc. When I was also doing war work, they made special arrangements so that I could continue writing papers in astronomy. I think that what you need to realize is that many people do things because they like to do them. This is probably true of all astronomers. I like to do astronomy, the Berkeley faculty likes to do astronomy and they like to see students who want to do astronomy. The same theme runs through many fields.

Betty also pointed out there was faulty statistical reasoning in the paper and made an offer to help the candidate understand what was wrong.

Betty presented blunt criticisms to the general reasoning in the paper:

I am distressed by what you write in the latter part of the paper – in fact, I would say that what you write indicates shocking socialization. Why do you not expect every one who is a professor at Berkeley (except for the one or two mistakes who are still with us) to be "active, inde-pendent..." If anyone thinks they are not, I would say that there is a lack of understanding. Why should "working long, hard hours because of being so personally involved, committed to, and happy with their work" be a "masculine" concept? It sounds as if you have been brain washed to believe that women cannot be involved, committed or happy (I am not going to entertain the possibility that women cannot work long years [sic] since they almost all do). Why should you expect faculty in the same in-stitution to have sex differences in attitude? You forget that most of the women in the ACE-Carnegie study are not in research-oriented institu-tions. The values, attitudes, etc., that you describe are those of successful-in-research-oriented-combined-with-good-teaching universities.

Betty went on to reveal more of her views on women and the professorate:

Relating well to colleagues and being a professor do not follow from each other at all. For men or for women. Here again there is no justifica-tion for distinguishing between men and women. The fact of the matter is that not all professor [sic] relate, no matter where, no matter when – they are human beings. My limited observation suggests that the percent-age of nice, easy to get along with, interesting to talk to, etc., humans among professors is just as large and perhaps larger among professors than among people in other occupations. It makes me angry when you suggest that women have to do some plain and fancy cheating, even get married, to become professors. There is no good evidence that intelligence, ability, getting the job done, working long hours, being interested in talking about ideas and writing about ideas, etc. – all the qualities to become a professor at Berkeley – are in any way sex linked. There is good evidence that women who are professors do have these characteristics in plenty, perhaps more than their male colleagues.

Betty pointed to some statements that she regarded as "utter hogwash" and "nonsense": "...I feel very strongly that it is time for women to think through what they are saying and not fall into traps, especially ten-year old traps."[17]

27.3 Three Salary Manuscripts

In August, the Chancellor's CSAW asked the chancellor to address advisory committee structures, budget matters, and reporting structures for the affirmative action coordinator. Betty noted he expected the campus affirmative action plan would be ready to send to HEW by August 27.

The committee particularly wanted the chancellor to commit part of his discretionary funds for the coming year to recruiting academic women, supporting the women's educational assistance program, expanding the graduate minority science program to other fields, and monitoring students. They did not include in their list such things as accelerated faculty hiring, employee development, and special recruitment, as they thought funds for these initiatives could be found from other sources.[18]

Women Generally Receive Less: In late November, Betty was working with the UC salary "1972 PER221" computer report. Something seemed to be wrong with one of the listings, "Personnel in Title Code Sequence – by Campus." Either Betty didn't understand the definition of "percent of full time," or else there was a programming error. She asked the chancellor's office to track down answers. Betty was not shy to express frustration about lack of quality control: "I suppose that the lesson is to be sure that the listing received is clear and without blanks and holes! It is very hard to fill in later."[19,20]

Ervin-Tripp followed up with the chancellor's office about the PER reports. Now there was a problem: They wanted the reports back, but Betty wasn't done with them because they had not yet provided answers.

Also, the Senate CSAW was interested in looking at advancement patterns over time by sex and ethnicity. Could the latest compilations be provided for

[17] 11/26/1973 letter from Betty Scott to UC-Berkeley Sociology PhD candidate.

[18] 8/13/1973 agenda for the Chancellor's CSAW meeting with UC-Berkeley Chancellor Albert H. Bowker.

[19] 11/25/1973 letter from Betty Scott to UC-Berkeley Chancellor's Office's Jose Rodriguez.

[20] 1/30/1974 letter from UC-Berkeley Chancellor's Office's Jose Rodriguez to Personnel Division Salary Administration Coordinator Lewis B. Perry.

Betty to analyze? "This analysis would be done by Professor Scott with the same conditions of extreme care which she has employed in the past with the payroll," Ervin-Tripp promised.[21]

The "1972 PER221" listings resulted in a white paper, "Women Generally Receive Less: Inequities in Employment, in Salary, and in Other Benefits at the University of California, Berkeley." The white paper does not list an author or date, but is presumed to be Betty's and written in late 1973 or early 1974. The section titles tell the story of the paper:

I   *The Percentage of Women Employed in Academic Positions, Especially in High Academic Positions, Is Low*

II *The Percentage of Women among New Appointments to the Professorial Series is Increasing but Women are Coming in at Lower Ranks*

III *The Average Salary Rate for Women is Lower than for Men in Almost Every Academic Title Grouping*

IV *Out of the 100 Highest Academic Salaries, None Goes to a Woman*

V   *Faculty Women Receive Lower Salaries than Faculty Men of the Same Ability and Performance*

VI *Women Employees of the University of California Receive Smaller Fringe Benefits than Men of the Same Salary*

It is well documented and packed with statistical reasoning and thinking.[22]

Betty's salary work was getting noticed. Individual women on the Berkeley campus began preparing cases for salary adjustments. Betty helped them compile data pertinent to their cases and made sure they didn't misstate their cases to their department chairs. In turn, the women were grateful for the effort Betty was directing to the issue of pay equity.[23]

Comparison of the Status: Betty produced a draft manuscript with Jeanne Lovasich, a "Comparison of the Status of Men and Women Faculty, Especially as Regards Salary." The draft aimed to compare the status of women and men. Betty had a page of outstanding questions in her mind and hand wrote them onto the manuscript:

What about other types of universities and colleges?
What about more recent years?
What about minority persons?
What about analyses at the department level?

---

[21][n.d.] letter from UC-Berkeley Rhetoric Professor Susan Ervin-Tripp to UC-Berkeley VC Mark Christensen.

[22][?1973 or 1974] white paper presumably authored by Betty Scott, titled: "Women Generally Receive Less: Inequities in Employment, in Salary, and in Other Benefits at the University of California, Berkeley," 14 pages.

[23][n.d.] letter from UC-Berkeley Department of City and Regional Planning Assistant Professor Ann R. Markusen to Betty Scott.

What about non-teaching (or non-ladder) academic appointments?
All above refers to universities and colleges. What about PhD holders with
other types of employment? Management? "successful"; stay-out for an
interval
What about rank, promotions, tenure?
Do not use tenure as a predictor variable, or rank
Entry level salaries and salaries of new appointees through the years
Promotion rates through the years (see figures)

These questions may have been for herself as well as Lovasich.[24]

Application of Multiple Regression: Leading a group of four women researchers that also included Lovasich,[25] Betty presented her salary results at the Allied Social Sciences meeting in New York at the end of December, and at the 1973 annual meeting of the ASA. George Washington University Statistics and Economics Professor Joseph Gastwirth recalled the session, which was organized by Emory University Biostatistics Professor Donna Brogan and federal statistician Marie Wan on behalf of the Caucus for Women in Statistics, and which included papers by Betty, himself, and others:

> *It was the first session on discrimination. Betty had a paper on UCB salaries. It was a thorough analysis. She did a very nice job delivering the paper. She had very detailed data on productivity measures. She got a R^2 of .86 which is very high. The hardest problem with academics is getting the universities to give you data – they don't want to do a thorough job with equity studies – they want to do a fairly routine job. Betty used approaches used previously... It was a lively session. The audience was reasonable, not huge, and it was basically sympathetic. The session was relatively well received. Her regression study at UCB was one of the most detailed and careful/thorough. She was quite good. She put a lot of time into it. Betty used a very acceptable method. She was a pioneer in the area of using regression in discrimination. In 1973 the courts had not yet accepted formal statistical hypothesis testing, but they did accept some statistical reasoning.[26]*

Betty and her colleagues published their paper, "Application of Multivariate Regression to Studies of Salary Differences between Men and Women Faculty" in the *Proceedings of the American Statistical Association*. They also had aspirations to publish the study in a "regular journal such as *Science*."[27]

---

[24][?n.d. 1973 or 1974] draft manuscript, "Comparison of the Status of Men and Women Faculty, Especially as Regards Salary," by Jeanne L. Lovasich and Elizabeth L. Scott, 10 pages.

[25]"Women Paid Less Than Men, UC Study Shows," *The Daily Californian*, February 20, 1974, 16.

[26]Joseph L. Gastwirth, e-mail to author, August 11, 2014.

[27]2/12/1974 letter from Betty Scott to Association of American Colleges Project on Status and Education of Women Bernice Sandler.

In the study, Betty and her co-authors used regression methods with 29 predictor variables from the CCHE questionnaires to examine the extent to which salary differences were due to "relatively objective factors" like differences in performance versus discrimination. They estimated faculty salaries in various categories of institutions and fields. With six pages of regression coefficients appearing in the proceedings version of the study, they concluded that: "The apparent discrimination in faculty salary due to sex is strong. . ."[28]

---

[28] M.G. Darland, S.M. Dawkins, J.L. Lovasich, E.L. Scott, M.E. Sherman, and J.L. Whipple, "Application of Multivariate Regression to Studies of Salary Differences Between Men and Women Faculty," in *Proceedings of the Social Statistics Section* (Washington, DC: American Statistical Association, 1973), 120.

28

Women Generally Receive Less (January – April 1974)

*. . . the horrendous statistics it contained actually radicalized one
of our faculty members. . .*
-Tina Frost

The year 1974 was busy in terms of women's issues. So many things were happening at the same time. It was a mosaic of activity, with Betty at the nexus. The amount of pressure for social change was tremendous. Things were going to change; the question was how and how much. Betty felt the pressure. She was guiding and funneling change, in certain directions, as best she could.

28.1 Disclosing Berkeley Salary Inequities

At the turn of the year, Betty was immersed in salary equity analyses. She wanted to make sure her work influenced affirmative action efforts on the Berkeley campus. But she didn't see any evidence of it in the official affirmative action plans that the chancellor's office was preparing. Betty wrote to the VC: "I have noticed that the plans prepared in your office do not refer to details of inequities in salary between men and women (indeed, this subject is often omitted)." She enclosed the white paper, "Women Generally Receive Less,"[1] which contained estimates of how much women faculty at Berkeley were being underpaid. It also contained her proposed methods for bringing payments to women in line with the payments to men. Betty requested permission to send copies of the white paper to others on the campus.[2]

The academic senate elected Betty to the committee on committees. Now she would be instrumental in appointing members and chairs to all senate committees, and in influencing membership on important committees (e.g., the budget committee) to the benefit of women. Betty would be busy.

---

[1] [?1973 or 1974] white paper presumably authored by Betty Scott, titled: "Women Generally Receive Less: Inequities in Employment, in Salary, and in Other Benefits at the University of California, Berkeley," 14 pages.

[2] 1/6/1974 letter from Betty Scott to VC Mark N. Christensen.

The VC responded to Betty's suggestions about the campus affirmative action plans. He would keep her comments in mind in future discussions with HEW. He promised to add material about salary equity, specifically referencing Betty's work. Referring to Betty's new assignment to the committee on committees, the VC declared: "I hope we will both see the day when there are sufficient women on the faculty that all of the women on campus do not have the overburden of committee assignments which seems quite common."[3]

Word needed to get out about salaries at UC-Berkeley. The March Fong hearings had taken place on November 8, and Betty had testified about shortages of women on the faculty, especially at the higher levels. Now Betty wanted Fong to be aware of the salary differences between men and women faculty. She sent Fong a copy of the white paper, "Women Generally Receive Less."

Betty pointed out to Fong that the average salary for women faculty at UC-Berkeley was "much lower" than for men. On average it was $306 per month less within the same job title. For directors it was the largest, at $741 per month less. "Month after month, year after year, women are receiving less, much less than men with similar positions," Betty told Fong. She also pointed out to Fong that the salary differentials weren't explained by objective attributes or performance factors. It could be verbalized either way: Men were "overpaid compared to women," or women were "underpaid compared to men." "Very much underpaid," Betty told Fong.[4]

28.2 Invited Speakerships

Betty accepted speaking engagements at all levels. At one end of the spectrum, she spoke to national and international audiences of professionals. Her most recent invitation was to speak at another AAAS symposium, this one on "Science Manpower in the Seventies – Will Supply Meet Demand?" Betty wanted to take up the topic of participation of women and minorities in the labor force, especially in science. She would draw upon her CCHE results.[5]

At the other end of the spectrum, she spoke to girls at elementary schools. One invitation was from a school in El Cerrito, California. Thank you letters indicated she made quite an impact there. One girl wrote:

> *Thank you for coming to our class the other day. I was a little surprised when I heard you were a women [sic] because most scientist [sic] are men. You taught us some things about science. Also you answered some questions we asked.*

---

[3]1/17/1974 letter from VC Mark N. Christensen to Betty Scott.

[4]1/26/1974 letter from Betty Scott to California State Legislature Assembly Committee on Employment and Public Employees Assemblywoman March Fong.

[5]1/24/1974 letter from Betty Scott to Stanford University Statistics Professor and Dean of Graduate Studies Lincoln E. Moses.

Another girl wrote:

> *Thank you very much for coming to our class. I enjoyed meeting you!!*
> *you are a very SMART!!! [sic] I wish the [sic] was some other way to*
> *thank you but: Thanks Again!!*

Yet another girl wrote:

> *Thank you very much for coming to show us some of your work. It was*
> *exciting to [sic] the water vapor, helium, hydrogen, and neon gas. I liked*
> *water the form of the prisoms [sic] change shape. Thank you for letting*
> *us keep the slides. Thanks to you I learned more about prisoms [sic] and*
> *the stars. I bet it is hard to get a Doctor's degree. Thank you for showing*
> *the things you showed us.*[6]

These young girls found a new role model in Betty.

Even the small liberal arts colleges began to take up the cause for academic women. For example, Reed College in Portland, Oregon held a public panel discussion on "Women in Academia" in January. Some 100 people showed up. This was huge for a college as small as Reed. Betty sent a copy of the CCHE report on academic women to one of the student leaders of the panel, Tina Frost. The report was a "terrific help" to the panel and was "making waves." According to Frost:

> *In fact, the horrendous statistics it contained actually radicalized one of*
> *our faculty members: she's shifted her stance from the it-is-enough-that-*
> *I-succeed-so-as-to-prove-it-can-be-done position to the I've-got-to-get-on-*
> *policy-making-committees-and-start-doing-something-constructive-about-*
> *this-awful-situation position, which is a great step forward.*

Frost told Betty that panelist Mary Gray "in particular was great, especially at handling the 'you're trying to make us hire incompetent women instead of all those brilliant men with families to support' objections." The students declared Gray a heroine; they needed all the heroines they could get.[7]

COPSS had a visiting lecturer program that began in 1963 and would continue into the 1990s.[8] In 1974, the ASA committee on women wanted to use this program to influence educational and career decisions of women. The field of statistics was dominated by men, women statistics graduate students were not sufficiently encouraged nor were they given equal access to financial aid, and women statisticians were not given equal opportunity for employment, salary, and advancement. This "situation" needed to be "corrected," the committee thought. The committee wrote to Betty and others signed up

---

[6] 6/7/1974 handwritten letters from three girls (Sherri Pleasants, Anna C. Stenberg, and Diana Lee) in Ms. Newlund's class at Castro Elementary School in El Cerrito, CA to Betty Scott.

[7] 1/22/1974 letter from Reed College mathematics student Christina (Tina) Frost to Betty Scott.

[8] Ingram Olkin, "A brief history of the Committee of Presidents of Statistical Societies (COPSS)," in *Past, Present and Future of Statistical Science*, ed. X. Lin et al., (2014), 3.

to participate in the visiting lecturer program asking them to intentionally encourage "budding" women statisticians: They should inform about career opportunities for women in statistics, and about women statisticians who have been success stories; and let the committee know if the institutions they visited seemed to be encouraging women and helping them get financial aid.[9]

28.3   AAUP Joint Committee Venture

Founded in 1915, the AAUP "has helped to shape American higher education by developing the standards and procedures that maintain quality in education and academic freedom in this country's colleges and universities." The AAUP has a number of topical standing committees: Academic freedom and tenure, academic professionals, accreditation, institutional governance, professional ethnics, and more.[10]

Committee W is the committee on women in the academic profession. It has worked on a broad range of issues, such as life-work integration, sexual harassment, affirmative action, and equity in retirement benefits. Back in 1973, Gray was appointed chair of Committee W. She would provide strong leadership to the committee for many years.

Committee Z is the committee on the economic status of the profession. It conducts an annual faculty compensation survey and prepares an "Annual Report on the Economic Status of the Profession" that "describes national trends in faculty compensation, as well as issues in the financial organization and condition of higher education..."[11] Back in 1973, Harvard University Political Economy Professor Robert Dorfman[12] chaired Committee Z. Dorfman had a background that overlapped with Betty's. He earned his bachelor's and master's degrees in mathematical statistics from Columbia University, then participated in World War II as an operations analyst for the US Army Air Forces, after which he earned his PhD in economics at UC-Berkeley, one of his academic specialties being linear models of production and allocation.

In 1974, Committees W and Z were planning a joint venture.[13] A backdrop was that, in 1963, the US had legislated fair labor standards in the

---

[9]1/21/1974 letter from New York Life Insurance Company Assistant Vice President Mary Virginia Bowden for the ASA committee on women in statistics to Betty Scott.

[10]"Mission and Description," *AAUP*, accessed March 11, 2014, www.aaup.org/about/mission-description.

[11]"Standing Committees," *AAUP*, accessed March 11, 2014, http://www.aaup.org/about/committees/standing-committees.

[12]Note: Robert Dorfman (1916–2002) had strong statistics training at Columbia. See: P.A. Samuelson, "Harold Hotelling as a Mathematical Economist," *The American Statistician* 14, no. 3 (1960), 21.

[13]1/24/1974 memorandum from AAUP Staff Member Margaret L. Rumbarger to AAUP Committee W re agenda for Committee W meeting February 1–2, 1974.

form of an Equal Pay Act: These were later revised to include professional, administrative, and executive workers and thereby faculty and other educational employees. In mid 1973, national higher education associations such as the National Association of State Universities and Land Grant Colleges (NASULGC) notified their constituents of this change. Even though the academic context was still somewhat unfamiliar territory to the US Department of Labor, many in the academic community began to see the act as the best legal means of remedying gender inequities in salary in their institutions. AAUP and labor department staff both advised that institutions should take action on their own initiative, rather than waiting to be acted upon in the courts, as the courts had shown sympathy for institutions that had been proactive at providing salary relief.[14] At their 1973 annual meeting, the AAUP had passed a motion that Committees W and Z would take on a collaborative, "systematic study of compensation differentials between male and female faculty members."[15] In December 1973, Gray had prepared a discussion paper on equal pay for equal work.[16] It would be a catalyst for the joint work of the two committees.

Representatives of Committees W and Z met on January 23, including Gray, Dorfman, Barbara Bergmann, and three AAUP staff members. Gray in October of 1973 had hand-selected Bergmann to be a member of Committee W. Bergmann had already published in the area of discrimination in employment and pay of minorities and women, and she had helped with a discrimination case at the University of Maryland when a woman didn't get tenure there and pay equity came up as part of the case. With academic specializations in sex roles in the economy and labor economics, and having served as a senior staff member to President John F. Kennedy's Council of Economic Advisors and on other national government advisory boards, Bergmann was a perfect member for Committee W as the pay equity issue came to the fore. The joint meeting resulted in three recommendations:

1. In addition to the data collected in the regular compensation survey questionnaire (i.e., number of full-time faculty members, salaries and fringe benefits which are not broken down by sex), the number of full-time faculty members broken down by rank and sex within rank should be requested and published by institution. This change would be introduced in the 1974–75 regular questionnaire. The institution would continue to report salary and fringe benefits by rank but not broken down by male/female.

2. Committee Z had at its last meeting in December, [sic] decided to request institutions to supply information on tenure-nontenured faculty ratio starting with the 1974–75 questionnaire. This decision has now been

---

[14] 1/23/1974 memorandum from AAUP Staff Member Margaret L. Rumbarger to Members of AAUP Committee W and file re Equal Pay Act.

[15] AAUP Annual Meeting Report, Summer 1973, 142.

[16] Mary Gray, "Equal Pay for Equal Work? A Consideration of Academic Salaries" (draft white paper, 1973).

> *amended and institutions will now be asked to report tenure/nontenured*
> *faculty ratio by rank and sex.*
> *3. The development of the "do it yourself" kit... will begin in the sum-*
> *mer.*[17]

The do-it-yourself kit was a tool that would be developed to assist institutions with self-evaluations of gender inequities in salary.

These were landmark recommendations. At the AAUP annual meeting, a resolution would be passed: "RESOLVED, That Committee Z in cooperation with Committee W undertake a systematic study of compensation differentials between male and female faculty members."[18]

The AAUP had conducted annual surveys on the economic status of the profession since 1958.[19] The original purpose was to raise faculty salaries by publishing salaries and compensation ratings by institution.[20] The decision in 1974 to begin collecting data and reporting the distribution of faculty by sex, but not salary and benefits by sex, as part of the annual survey was a difficult one to arrive at. It was thought there could be as many as 80 institutions that would drop out of the survey if asked to report salary and benefits data by sex. Bergmann thought this wouldn't "be the end of the world." Dorfman argued that the quality of the results depended greatly on having a virtually complete level of response and worried about what nonresponse would do to the reputation of this "unique survey that has come to have very high prestige, authority, and value... probably the highest quality and most complete annual survey that any industry is fortunate enough to have." He wanted to proceed with caution. A plan was developed to have AAUP send out a questionnaire asking institutions how they would react to adding questions to the survey that would allow publication of salary information by sex. Dorfman asked Bergmann if she would please help to draft the questions with the aim of avoiding bias to the responses, adding:

> *I should be very glad to take other steps also, but always with the real-*
> *ization that we are treading on treacherous grounds. I feel, like you, that*
> *the AAUP has a responsibility to plan a role in diminishing and perhaps*
> *ending discrimination against women on college and University faculties.*
> *I also feel that the annual survey is to be used, and not merely preserved,*
> *so that timorousness that prevents us from using it where needed defeats*
> *its real objectives. We have to thread our way cautiously between these*
> *objectives, as I think I said before.*

---

[17] 1/28/1974 memorandum from AAUP Staff Member Margaret L. Rumbarger to members of Committees Z and W re decisions made at the meeting of January 23 between Z and W representatives.

[18] 6/24/1974 letter from AAUP Staff Member Maryse Eymonerie to Betty Scott.

[19] 11/14/1974 letter from AAUP Staff Member Maryse Eymonerie to Exxon Educational Foundation's Frederick Bolman.

[20] [?1974] "Request for a grant to support the development of a 'Higher Education Salary Evaluation Kit' to assist administrators and faculty groups to evaluate and rectify possible salary inequities."

Dorfman pledged to Bergmann that they would "push it forward as quickly as seems reasonably safe."[21] Bergmann was "very glad" to read Dorfman's letter to Committee W at their next meeting, but in spite of his objections Committee W decided to recommend to Committee Z that data on salary be collected and published by rank and sex, starting with the 1974–75 survey.[22,23,24]

Committee Z did conduct an institutional survey to determine whether the committee could expect institutions to provide sex differentiated data. In spite of results that weren't totally encouraging, the AAUP would go on to mandate the collection of such data beginning in 1974–75.[25]

Around this same time, Betty serendipitously sent a copy of her paper, "Application of Multivariate Regression to Salary Differences Between Men and Women Faculty," to Maryse Eymonerie who was AAUP Association Secretary, Annual Survey of Faculty Compensation Director, and Committee Z Staff Member. Betty said: "I hope you can use this study, for example, in the work of Committee Z and Committee W."[26] Betty also sent a copy of her Application paper to AAUP Staff Member Margaret Rumbarger, again expressing hope that it would be of use to Committees W and Z.[27]

Gray thanked Betty for the Application paper. In exchange, Gray offered some information to Betty. AAUP's Committee W on women had begun working on salary analysis software for campus use. The committee also made a request to the AAUP that their salary survey summary statistics be reported by sex. It was time for AAUP to select a new general secretary. Did Betty have any suggestions? It should be "someone whose first reaction to everything won't be "No," Gray offered.[28]

---

[21]1/30/1974 memorandum from Harvard University Economics Professor and AAUP Committee Z Chair Robert Dorfman to University of Maryland Economics Professor and AAUP Committee W Member Barbara Bergman re collecting data and reporting distributions of salary and benefits by sex.

[22]2/5/1974 letter from AAUP Staff Member Maryse Eymonerie to Harvard University Economics Professor and AAUP Committee Z Chair Robert Dorfman.

[23]2/11/1974 memorandum from AAUP Staff Member Margaret L. Rumbarger to members of Committee W re draft minutes of Committee W meeting, February 1–2, 1974.

[24]3/14/1974 letter from AAUP Staff Member Margaret L. Rumbarger to Harvard University Economics Professor and AAUP Committee Z Chair Robert Dorfman.

[25]6/24/1974 letter from AAUP Staff Member Maryse Eymonerie to Betty Scott.

[26]2/9/2014 handwritten memo from Betty Scott to AAUP Staff Member Maryse Eymonerie re Betty's Application paper.

[27]2/14/1974 letter from Betty Scott to AAUP Staff Member Margaret L. Rumbarger.

[28]2/26/1974 letter from American University Mathematics and Statistics Professor and AAUP Committee W Chair Mary Gray to Betty Scott.

28.4 "Underutilization" Methods at Berkeley

Betty had been critical of the figures for underutilization of women and minorities being used in the most recent draft affirmative action plan for UC-Berkeley. They were produced using some sort of "square-root method," and Betty didn't like it. Her strong opinion was that the statistical method resulted in badly biased estimates that were too low. She called it "malicious nonsense." She asked that the figures be removed immediately, before further distribution to department chairs or "leaks" to newspapers.

HEW guidelines were explicit on how to define underutilization and did not refer to any kind of "fit." Yet the Berkeley plan referred to a "method of calculating a 'reasonable' fit." Any references to a "fit" should be removed, Betty told the VC. "Underutilization is a matter of facts," not of punishments or quotas, and should not be confused with such. In Betty's view, references to "fit" were "statistical machinations," and Berkeley should not stoop to these. She sternly wrote to the VC: "I trust that you will delete them immediately."

The Senate CSAW and Chancellor's CSAW had advised that the affirmative action coordinator should report directly to the chancellor, or otherwise to the VC. Instead, the coordinator was hired at a low level, and it had been frustrating for everyone. By the end of January, the coordinator resigned. Betty urged the VC to reconsider the level of this position. "What about accepting the advice this time," Betty asked him.[29]

Erich Lehmann had now succeeded Betty as chair of the statistics department. The VC's way of dealing with Betty's severe criticism of the statistical methods for underutilization was to ask Lehmann to get his statistics faculty to discuss and critique a four-page rationale for the methods prepared by the administration.[30] They were to come to an "independent consensus" on how to determine "correct estimates" for underutilization. The VC hoped Betty would be part of that independent consensus.[31]

Betty was glad the VC was seeking the advice of the statistics faculty, but only some faculty members received the written rationale and were asked to participate in discussions about it, and she was not among them. Betty eventually received a copy after her colleagues' comments had been submitted to the administration. It being the case that Betty had no input whatsoever to the review submitted by her department, she submitted her own criticisms of the methods to the VC. She had two major points.

The first point had to do with the use of the Binomial probability distribution. At this point in her reasoning, Betty asserted it was not correct to

---

[29] 1/28/1974 letter from Betty Scott to VC Mark N. Christensen.

[30] 2/7/1974 handwritten white paper (4 pages) to UC-Berkeley Statistics Professor and Chair Erich Lehmann from Moore [for the UC-Berkeley administration] re methods for utilization analyses.

[31] 2/7/1974 letter from VC Mark N. Christensen to Betty Scott. Note: Betty's objection was to the estimates on p. 45, columns 4 and 8.

assume a Binomial situation for underutilization (although she did not specify an alternative). Furthermore, the Binomial variance was approximated, which she pointed out was unnecessary given that the Binomial distribution and its confidence intervals could easily be read from existing published tables. Betty also pointed out the formulas were not valid when the expected number of women employed in a hiring unit in an occupational category (determined from the total number of employees in the unit in the category and the proportion of women qualified for employment in the category) was small.

The second point had to do with the research question to be answered by the utilization analyses. Betty didn't see any point in "flagging" units that were doing well at hiring women. She thought, and verified with the provost, that the question should be: Which departments are not employing women when they should be? In Betty's opinion, as described above, the proposed methods would never flag departments that had small expected numbers; since this was about one-third of the departments at Berkeley, the research question could not be answered with the proposed methods. "I regard this as an important error," Betty explained to the VC, "just when affirmative action is most needed to present to students and staff a wider view and to provide role models, this method will never flag."

Betty was also concerned that the administration was using two columns of estimates she thought were just wrong. One was the availability of women. In Betty's view, these estimates were too low. They didn't take into account the historical underutilization of women. The other was the current underutilization of women. In Betty's view, these estimates were way too low. They were calculated using a low range of availability. "This is a flagrant error... [with this method] minorities will almost without exception be listed as having zero underutilization," Betty complained. What would Betty have done differently? She would have calculated a range of underutilization: She was confident department chairs could draw conclusions that were not "too optimistic" from a table containing upper in addition to the lower bounds. She would have paid attention to units with high underutilization or low actual numbers, especially units with no women whatsoever.[32]

Betty hand wrote some notes regarding a draft manuscript and put them into her file. "Change over from men overpaid to women underpaid." This was more politically correct. "Since many of the predictor variables are presumably themselves colinear, the individual significance probabilities do not necessarily give a full picture of the importance of particular predictors. Nevertheless, they are of interest."[33] This was being statistically careful.

---

[32] 2/12/1974 letter from Betty Scott to VC Mark N. Christensen.

[33] [?1974] handwritten notes by Betty Scott re review of an unidentified manuscript.

28.5   Communicating Methods; Explosive Results

In mid February, Betty decided to share her Application paper with Helen Astin and Alan Bayer, urging them to continue their regression work on gender equity in salaries using more recent data. She suggested they improve their work by publishing regression coefficients in addition to correlation coefficients; unlike correlation coefficients, regression coefficients contained a lot of information that individual readers could use to assess individual salaries. Betty recommended they determine regression coefficients for different categories of schools and different fields. If she were to have done her own study over again, Betty would have used a complete sample and stratifed it by type of school. Her reasoning was that different types of schools used different pay systems. Betty told this to Astin and Bayer. Betty also told them that she and her colleagues had developed some statistical software for these types of regressions. She offered to help, saying: "I hope that you can do the study soon because now is the time when it will be most used."[34]

Betty and Elizabeth Colson had co-chaired the Policy CSAW, but several years had passed and Colson had fallen out of touch with women's issues. Now Betty sent Colson a copy of her Application paper. Colson read the paper and found the salary differences between women and men to be "startling" – this in spite of many assurances by the administration over the past few years that the gap would be closed. Colson had been told by women colleagues in the unionized New York City higher education system that there the qualifications for the faculty ranks and salaries had been fully spelled out. Colson wondered to Betty whether things were better in New York than at Berkeley.[35]

Betty also sent a copy of her paper to the budget committee. She urged the subject of salary equity receive more attention. She was confident:

> *Our study is statistical so there may be some reason(s) to explain the large differences other than by discrimination. However, since we have such excellent prediction (R = about 0.8), I think this is very unlikely. Clearly, it is quite safe to bet that no appreciable explanation beyond discrimination will be found. So I hope that the Senate Committees will look into the differences. Also, I hope that the Budget Committee will note that there are two factors, at least, making up the differences: one is that women are appointed at lower salary than similar men and the other is that women are advanced more slowly than men of equal attributes and performance. The reports of the Budget Committee over the last years indicate that women are still entering at lower rank and steps than men.*

---

[34]2/12/1974 letter from Betty Scott to UCLA Education Professor Helen Astin and Florida State University Sociology Professor Alan Bayer.

[35]2/13/1974 letter from UC-Berkeley Anthropology Professor Elizabeth Colson to Betty Scott.

Once again, Betty was statistically careful and concluded her letter to the budget committee chairman with this caveat: "From the report one has no clue whether this [conclusion of discrimination] is justified, but it does seem surprising that it happens year after year."[36]

Next, Betty sent a copy of the Application paper to Louise Taylor who was a staff member in the chancellor's office. The methods could be used, she explained to Taylor, to compare Berkeley to other institutions, or to compare Berkeley men and women with the same profiles (abilities and performance). Betty hoped the office would find the report to be useful and they would give the subject of salary equity more attention. Betty took issue with a comment made earlier by Taylor that faculty flow charts could not be made for departments or other small units. "I was surprised to hear you say this and perhaps I misunderstood you," Betty wrote. Betty then explained that statisticians worked on small samples and populations all the time and had done so for over 50 years. She told Taylor she saw no reason why the methods of hypothesis testing and estimation developed by statisticians for small samples could not be used in faculty flow charts.[37]

Then Betty sent a copy of the Application paper to George Washington University Medical Center Physiology Professor M. Elizabeth Tidball.[38] A few years earlier, Tidball had published studies showing graduates of women-only colleges were much more likely to be recognized for career accomplishments; and the availability of women faculty role models made a significant difference in the success of women students. Tidball was "terribly impressed" with Betty's salary equity study and called it "elegant" and "extremely useful."[39]

William Kruskal,[40] who was "interested in everything," was also interested in Betty's Application paper and hoped it would be published soon. He suggested all kinds of people who might also be interested in receiving a copy of the study. He scrutinized the report and commented to Betty on various details. Kruskal recognized and admitted to Betty that he had some male prejudices, and he explained his observations in light of consistency with these prejudices.[41]

Salary equity study results were important for the overall cause of women, and Betty accordingly shared her results with Bernice Sandler. Betty also

[36]2/14/1974 letter from Betty Scott to UC-Berkeley Physics Professor and Budget Committee Chair William Fretter.

[37]2/14/1974 letter from Betty to UC-Berkeley Chancellor's Office Staff Member Louise Taylor.

[38]Note: M. Elizabeth Tidball (1920–2014) was a researcher and advocate for higher education for women. She was the first woman professor of physiology at George Washington University School of Medicine.

[39][?1974] handwritten letter from George Washington University Medical Center Physiology Professor "Lee" M. Elizabeth Tidball to Betty Scott.

[40]"William Henry Kruskal," accessed July 26, 2013, http://www-history.mcs.st-andrews.ac.uk/Biographies/Kruskal_William.html. Note: Kruskal was an expert in nonparametric analysis of variance.

[41]3/11/1974 letter from University of Chicago Statistics Professor William Kruskal to Betty Scott.

updated Sandler on what was happening at Berkeley. The administration was holding to their so-called square-root method of determining underutilization. HEW wasn't accepting it, and there were many meetings. HEW put federal moneys to Berkeley on hold. "Of course, that is not the only thing wrong but it does seem to be a major [sic] block," Betty reported to Sandler.[42]

Sandler was now a fan. She said of Betty's Application paper: "It's damn good, and I hope you can get it reprinted in *Science*," Sandler wrote. "It needs to be read by a great many people..." Sandler wanted to know how people could most easily get copies of the report.[43]

Betty's Application paper had already been published in a statistics proceedings, but she wanted to see it appear in a more visible journal, one with readership beyond the field of statistics. Many females and minorities in higher education were looking for help.[44] She considered submitting the paper to the *AAUP Bulletin*, but thought they might not be interested in another paper on salary estimation. She considered *Science*, but thought they might not be interested in a technical paper. She leaned toward submitting the paper to *Science*. The editors might not want to publish it, but Betty wanted to give it a try.[45] In fact, "many persons" including Sandler encouraged Betty to submit the paper to *Science*, thinking the paper was not too technical.

Betty proceeded to submit the paper to *Science*. "We feel that essentially any reader of Science [sic] could apply the appropriate parts of our results to his or her abilities and performance and thus compute his (her) expected salary," Betty told the editor, expressing hope "that many people and institutions will do this and make some progress towards reducing salary discrimination."[46] This expression of how her results could be used would eventually see the light of day in the form of a "Salary Equity Kit" that would be developed and published under the auspices of the AAUP's Committee W and Committee Z.

Science sent Betty's paper out for review and forwarded the two anonymous reviews to her. One focused on the fact that many of the variables Betty had used were ordinal, i.e., their measurement was into categories that had a natural order. Some of the variables were naturally ordinal, e.g., "highest degree," which was coded $0 = BA$ or less, $1 = MA$, and $2 = doctorate$. Other variables were reduced to ordinal levels, e.g., "hours taught per week," was coded into nine categories that ranged from $1 = none$ to $9 = 21$ *or more*, rather than measured as actual hours. The reviewer wanted to know the implications of using ordinal variables in the regression analyses.

---

[42] 2/12/1974 letter from Betty Scott to Association of American Colleges Project on Status and Education of Women Executive Associate Bernice Sandler.

[43] 3/15/1974 letter from Association of American Colleges Project on the Status and Education of Women Executive Associate Bernice Sandler to Betty Scott.

[44] 2/25/1974 letter from AAUP Staff Member Margaret L. Rumbarger to Betty Scott.

[45] 2/14/1974 letter from Betty Scott to AAUP Staff Member Margaret L. Rumbarger.

[46] 2/14/1974 letter from Betty Scott to Science Editor D. Philip H. Abelson.

The second of the two reviews from *Science* was glowing. It read:

> *This is the most scholarly report of employment data within the university structure that I have seen. The group presenting this material is a sophisticated segment of the statistics establishment and has analyzed the data with unusual precision. The material presented will come as no surprise to anyone reading the avalanche of reports on discrimination against women in academia but this paper does more than an anecdotal survey. It subjects the data to tough minded analysis and interpretation. Furthermore, it has asked some novel questions that have been overlooked before. The answers to these questions are powerful tools for the destruction of some complacent assumptions that things have really changed significantly for academic women. Any scientist ought to be able to understand the paper as written even though standard statistical shorthand is often used.*[47]

But the paper never did see the light of day in *Science*.

The Daily Californian published an article about Betty's salary equity work. It reported the primary finding of her Application paper: Across the nation and among those with regular faculty appointments, women who had equal qualifications to men were being paid on average $1,500 less per year (over $6,500 in 2010 dollars), and often they were underpaid much more; this difference was not explained by women interrupting their careers to raise a family. Explanations that involved less qualifications and more career interruptions among women were simply not supported by the data. The data also showed that, as the number of years since the PhD increased, the salary differences increased; the differences were wider in research universities and the biological and physical sciences; and about 20% of women were underpaid by $4,000 per year or more (almost $17,500 in 2010 dollars).[48]

These results also made their way into the *University Bulletin*. Now all faculty and staff at all UC campuses would be aware that equally qualified women were being paid an average of $1,500 a year less than men, and that 20% of equally qualified women were being paid $4,000 a year less. The data in the study were collected in 1969. When interviewed for the *University Bulletin* article in 1974, Betty said: "One might hope that salary differentials due to sex are less pronounced now. But preliminary information from a follow-up study taken during the 1972–73 academic year shows that there has been no appreciable change in salary differences between men and women."[49]

Betty had sent a copy of her Application paper to the co-director of the Student Lobby of the Associated Students of the University of California, who used it in her conversations with the administration. The legislature was expecting to get an inequity report in May, and Betty's work would be included.

---

[47][?1974] two anonymous reviews from *Science* re Betty's Application paper.

[48] "Women Paid Less Than Men, UC Study Shows," *The Daily Californian*, February 20, 1974, 16.

[49] "UC Berkeley Study of Faculty Salaries," *University Bulletin* 22, no. 24 (1994), 117.

The co-director had one word for the academic affairs data that were being prepared for the report: "Explosive."[50]

28.6   Responding to "Blasts" and Supporting Individuals

Betty was bracing herself. She had heard that one of the UC-Berkeley political science professors was "preparing another blast" about women and minorities, this time reportedly to be published in the Sunday edition of the *Washington Post*. The professor had been putting out "blasts" since 1972 and had become known for his protests against HEW's requirement that Berkeley add more women and minorities to its faculty. "It would be good to counteract [the professor] some," Betty wrote to one of her associates. Her evaluation of the professor was realistic and practical: "We can regard him as beneath our dignity but he does seem to have access to the press."

Betty's associate had written a paper and asked her to review it. She thought the paper's bitter tone was fully justified and wouldn't change it. Some aspects of Betty's writing style and personal philosophy were revealed in her correspondence, even if the nature of his paper was not:

> *The beginning part, actually the first half, is better than the latter part. It is tighter and less sweeping. There are a few places that are rather purple prose. This is a matter of style. Personally, I feel that too many adjectives are not effective. I marked a couple of these passages by a vertical line in the margin... Not everyone is bad and even the bad ones occasionally do good (with some exceptions, I presume). This much I can document: for some persons the probability of doing the right thing is reasonably high. I have seen these persons in action, and I have seen them opposed by others and even defeated by others. I think they need support. Your paper does not seem to include them and you may wish to include such an example.*

Betty ruminated that she didn't know what it would take to motivate change at Berkeley. She admitted to recently having lost her temper: "I am really having backward effect," she confided to her associate.[51]

Some academic women across the country who decided to challenge negative employment decisions by taking their universities to court found Betty after seeing her salary equity studies described in an issue of *US News and World Report*. One woman was preparing to go to federal court after the EEOC

---

[50] 3/29/1974 letter from ASUC Student Lobby Co-Director Linda Bond to Betty Scott.

[51] 2/23/1974 letter from Betty Scott to James C. Goodwin. Note: Betty wrote to James C. Goodwin at a Berkeley address; she did not include his position. Myrna Oliver, "Paul Seabury, 67; UC Professor, Expert on U.S. Foreign Policy," *Los Angeles Times*, October 20, 1990.

found her to be "unlawfully terminated, underpaid, and retaliated against for filing the EEOC complaint." She knew that her career in higher education was over, unless she was able to get a positive ruling in the court.

The complainant sent Betty statistics, details, and computations galore, and she hoped Betty would send some data in return. The complainant was already being supported by the WEAL. She was told she already had a very strong case, but she thought Betty's help could make the case even stronger.[52]

28.7   Class Action Conciliation Agreement

The VC represented the chancellor at the February 26 meeting of the UC-Berkeley academic senate. He expressed caution about the HEW guidelines for affirmative action: These "had been formulated in Washington but there had been little experience to establish whether these guidelines were practical for institutions such as Berkeley." Betty took note of the minutes of the meeting. While he acknowledged progress at Berkeley, the VC also thought much more progress still needed to be made. He pledged continuing commitment to affirmative action on the part of the administration.[53]

It wasn't often that the chancellor made appearances on his campus, but at the end of February he held a town hall meeting at The Faculty Club. Eighty people showed up, including representatives of LAW. More may have shown up had there not been a requirement that people either buy their own lunch or pay an entrance fee. The chancellor reportedly took sharp criticism on affirmative action. A representative of the American Federation of State, County, and Municipal Employees asked him why women and minorities weren't included in the meetings with HEW investigators. He replied that these were formal negotiations, "not a glorified encounter group." According to *The Daily Californian*, he said he couldn't see another way to carry out the formal negotiations. Some of the women and minorities had difficulty understanding this, as they were the ones who were affected by the proceedings.[54]

HEW negotiators and UC-Berkeley officials had begun holding daily meetings on February 4 to try to reach an agreement.[55] On March 4, it was finally announced that the "extensive negotiations" had concluded. There was a conciliation agreement that covered equity in all areas of employment, both aca-

[52]2/26/1974 letter from Oklahoma State University Music Assistant Professor Linda Greer Morales to Betty Scott. Note: Linda Greer Morales (1939–1997), a pianist, had a graduate degree from the Juilliard School for the Performing Arts in New York City and was a music professor at Oklahoma State University.

[53]2/26/1974 UC Academic Senate, Minutes of the Representative Assembly, Berkeley Division IV(3): i.

[54]Laura Thomas, "Bowker Criticized," *The Daily Californian*, March 1, 1974.

[55]William Bates, "HEW Lifts Federal Contract Boycott," *The Daily Californian*, [publication date unknown].

demic and nonacademic,[56] and that included detailed steps Berkeley needed
to take toward affirmative action, including specification of analytic and statistical procedures.[57] According to the agreement, Berkeley would put a plan
into effect by September 30 that would be a model for the other UC campuses.

Federal money would start to flow again to the university. About $2.9
million, almost all in contracts from the National Aeronautics and Space Administration (NASA) to the Berkeley Space Sciences Lab, had been held up
for over three months – since November 16, 1973 – pending the agreement.[58]
Now Berkeley and NASA could collaborate on two key experiments for the
Apollo-Soyuz space mission: The Extreme Ultraviolet Telescope, and the In-
terstellar Helium Glow Experiment.[59] Manned spacecraft launched by both
the US and Soviet Union would dock over the Atlantic Ocean and conduct
almost 20 hours of goodwill joint activities.[60] There was considerable pressure
for HEW officials to reach an agreement on the affirmative action plan because
of the importance of Berkeley's involvement in this mission.[61]

By unfreezing the federal moneys before the Berkeley affirmative action
plan was put into place, HEW had essentially given UC more time.[62] LAW
called this agreement "irresponsible on the part of HEW."[63]

Betty had established a connection with the HEW's OCR in 1972 through
J. Stanley Pottinger, and she decided to send a copy of her Application paper
to his office. They immediately thought Betty's report would be useful in their
contract compliance investigations; and they saw a use for it in their attempts
to develop standards for university affirmative action programs and policies.
They told her they "felt that the meeting which we had with you in Berkeley
was of help... [in] administering the Executive Order program in colleges and
universities," and, "I know that we can count on your continued support."[64]

---

[56] "UC-HEW Agreement on Affirmative Action," *The Daily Californian*, March 4, 1974.

[57] William Bates, "HEW Lifts Federal Contract Boycott," *The Daily Californian*, [publi-
cation date unknown].

[58] "UC-HEW Agreement on Affirmative Action," *The Daily Californian*, March 4, 1974.

[59] Martin A. Barstow and Jay B. Holberg, *Extreme Ultraviolet Astronomy* (Cambridge:
Cambridge University Press, 2003), 9.

[60] "The Apollo-Soyuz Mission," *NASA*, accessed July 26, 2013, http://www.nasa.gov/miss
ion_pages/apollo-soyuz/astp_mission.html.

[61] William Bates, "HEW Lifts Federal Contract Boycott," *The Daily Californian*, [publi-
cation date unknown].

[62] "UC-Berkeley, U.S. Agree on Hiring Plan," *San Francisco Chronicle*, March 2, 1974.

[63] "NASA Contracts With Berkeley Space Laboratory," *The Daily Californian*, March 4,
1974.

[64] 3/12/1974 letter from HEW OCR Higher Education Division Director Mary M. Lepper
to Betty Scott.

28.8   Time to Improve Pay Reporting at Berkeley

Betty had been working with the UC-Berkeley payroll data for the past few years. All along she had been concerned about the quality of the affirmative action reports produced by the chancellor's office using these data.

A year earlier Betty was given the raw payroll data for the entire university and used it to perform random checks on the affirmative action reports for the departments. She found the reports were not reliable. This year she didn't have the raw data to perform the random checks, but she did have the data for her own department of statistics. Betty saw many differences between the reports she was able to produce for the statistics department and the official reports produced by the administration.[65] One of the problems was an apparent lack of consistency in the way head count was defined in the current versus previous year's tables. Another problem was the rounding down of FTE to whole numbers. Betty thought this was "nonsense, and biased," and there was no excuse for not reporting FTE to two decimal places.

The administration should have produced documentation to explain how their summary data were computed, but Betty couldn't get her hands on either a write-up or copies of actual computer programs used. She also couldn't find anyone in the administration – programmers or others – who could answer her specific questions about the reports. The people who did the programming were apparently no longer employed at the university.

Maybe the many problems Betty saw in the official reports were not real, she thought. She was, after all, very tired. She had just given an evening seminar, and it was now 1:00 am. So Betty circled the many apparent errors in red and sent them to the VC. She proceeded to compare the official reports across departments. There she also found apparent errors in the reports for other departments. Why weren't the instructions, computer programs, or output checked before the reports were distributed? These latest official affirmative action reports were simply not satisfactory. They were "worse than useless," Betty told the VC. They were "sloppy, inaccurate, even misleading." They were in many respects worse than last year's reports, she criticized.

Betty suggested it was "time for action": It was time to get an affirmative action coordinator who was able to work effectively with programmers to produce accurate reports. Betty accordingly asked the VC for the raw payroll data "without further delay." She would have her team, as they had done the last year, compute some departmental tables by hand and follow some individuals over time. "This just has to be done, I am sorry to say," Betty told him.

Betty pressed again to see the instructions to the programmers. This time she was successful. She saw there were errors in the instructions. But it was

---

[65]Note: Here Betty was referring to the affirmative action reports PER 1123, PER 1125, PER 1223, and PER 1225.

worse than that. There were also errors in the data. That sealed it. It was essential that Betty have access to the actual payroll data. She was on a mission to offer suggestions to the administration on how to produce a reliable set of official affirmative action reports.[66]

Betty was now the last word at UC on academic affirmative action data. The System CCSW invited Betty to give a 30-minute talk on her academic salary studies at their April 1 universitywide meeting.[67] A presentation about staff salary studies was also on the agenda.[68] Staff salaries were so relevant to what Betty was doing that she took over a page of notes about their objectives, peculiarities, barriers, etc.[69]

On April 1, the VC sent Betty one of the data listings she had requested. He was clearly nervous about loaning out the listing. He asked Betty to return it within six weeks, and to take "extreme care" when handling it.[70]

Betty had still not received any response to her earlier letter of complaints about errors. She turned to Susan Ervin-Tripp for help. Ervin-Tripp was now chair of the SWEM. The committee would surely need some results soon to present to the senate. Could Ervin-Tripp please "get some action?"

Betty thought Ervin-Tripp would be interested in a calculation she had published in the CCHE study. It answered the basic question: Suppose 6% of faculty are women in a certain year, and suppose further that 24% of the total employment are women each year, then how will the percentage of women on the faculty change through the years? Betty showed this calculation for the years 1973, 1978, 1983, and 1988. The percentages of the faculty who were women were 6.0, 9.1, 10.6, and 10.1, respectively. Betty thought these calculations could be very useful to the campus if done for subgroups of the faculty (by rank, ethnicity, or some other category). For example, if done for latter rank faculty, the calculations could be very useful to the committee on committees in making senate committee appointments.

28.9   Big Telescopes Story Challenged

Now on the national lecture circuit, Betty was talking openly about women not being allowed to use the big telescopes at Mount Palomar and Mount Wilson observatories. One of her lectures was to freshmen women at Purdue

---

[66] 3/13/1974 letter from Betty Scott to VC Mark N. Christensen.

[67] 3/13/1974 letter from System CSW Coordinator Betty Yu to Betty Scott.

[68] 4/1/1974 agenda for the System CSW.

[69] 4/1/1974 handwritten notes by Betty Scott re Office of Personnel Career Opportunities Development Program Analyst Jack Martin's talk on a staff salary inequities study.

[70] 4/1/1974 memorandum from VC Mark N. Christensen to Betty Scott re loan of a copy of PER 221, Academic Personnel Listing in Title Code Sequence.

University in connection with their experimental program to try and retain more women in science.[71]

On March 27, Purdue University issued a press release quoting Betty as saying that no woman had ever been permitted to use the 200-inch telescope at Mount Palomar. The release made its way to the Business and Professional Women's Foundation in Washington, DC. One of the functions of the foundation was to be a reference center on women; they often collected press releases and statements. The foundation librarian Jeanne Spiegel wanted their archives to be accurate, and so she proceeded to contact the Hale Observatories at Mount Palomar asking if Betty's statement was true.[72]

Hale Observatories' Director HW Babcock personally responded to Betty's allegation, saying: "There is no such discrimination." Babcock quoted the institution's guest investigator policy, which stated that "Approval of observing programs and allotment of time on the telescopes will be made on the basis of scientific merit of the proposal and suitability of the available instruments." He reported that the policy had been in place for seven years (since 1967) and was widely disseminated, including in the *Publications of the Astronomical Society of the Pacific* and at least two other professional publications. Then he quoted from their records regarding applications for guest investigator privileges. For 1973, there were 54 applications including 3 women. Forty of these, including the 3 women, were accepted. For 1974 up to April 12, there were 48 applications including 3 women. Twenty-nine of these were accepted, including the 3 women. Babcock declared he was glad to set the record straight. He stated that "well known" women astronomers had all used the telescopes at Hale Observatories.

Spiegel was thorough. She decided not to let the matter drop. "Did Purdue misquote you?" she asked Betty.[73] Betty hadn't seen the press release, but she did remember what she said at her talk:

> ... *I described my experiences as a statistician: what statistics is, anyway; the kinds of problems I study as a statistician, including statistical problems in astronomy; my training in science; and the prospects for the future for women in science, especially statistics. I tried to stress that the situation for women in science is improving rapidly. Among the evidence for this, I used my own experiences. In particular, I said that when I was a student in Astronomy, I had been encouraged to spend a summer at the Lick Observatory and at the Mt. Wilson Observatory. There, and*

---

[71] 4/29/1974 letter from Betty Scott to Business and Professional Women's Association Librarian Jeanne Spiegel.

[72] 4/1/1974 letter from The Business and Professional Women's Association Librarian Jeanne Spiegel to Mount Palomar Observatory. Note: The Business and Professional Women's Association began in 1919. Their vision is: "Working Women Helping Women Work." See http://bpwfoundation.org/.

[73] 4/16/1974 letter from Business and Professional Women's Foundation Librarian Jeanne Spiegel to Betty Scott. Note: Several of these women astronomers were young at the time, c.f., Judith Cohen (unkn –), Sandra Faber (1944 –), Eleanor Helin (1932–2009), Susan Kleinmann (unkn –), Vera Rubin (1928 –), and Joan Vorpahl (unkn –).

elsewhere, I was advised many times not to get a Ph.D. in Astronomy.
I think this advice was well intentioned and merely reflected the times. I
knew that women then were not allowed to use the 60" and 100" telescopes
in their own names (The 200" was under construction at that time.) [sic]
I assumed that when a woman PH.D. [sic] from Berkeley was employed
to carry on the observing programs of astronomers away on war research
during World War II, the chain would be broken. But it was not. You may
wish to look up the article about Dr. Margaret Burbidge in Time March
20, 1972, page 38, see especially top of column 2.[74]

Burbidge had come to the US from Britain with her physicist husband. The
Burbidges were "the most famous couple in astronomy." Both had hoped to
get Carnegie Fellowships in astronomy at Mount Wilson Observatory. He got
one, but she wasn't eligible because she was a woman. Instead she settled for a
job as a researcher at CalTech, which is less than an hour by car from Mount
Wilson. The *Time* article explained it this way: "There was also a more serious
problem. As a woman, Burbidge found that she could get precious observing
time at Mount Wilson Observatory only if her husband applied for it and
she pretended to act as his assistant." This was hard for Burbidge because
she liked big telescopes: "As big as I can get my hands on," she would say.
Furthermore, while working on the mountain, the Burbidges could not stay
at The Monastery dormitory, because only men were allowed, "...so they
had to camp out, sometimes in the middle of winter, in a primitive summer
cottage heated only by a wood-burning stove. After an all-night session at the
telescope they would often trudge through waist-deep snow and try to kindle
damp logs into a blaze to boil the kettle for tea." Burbidge recalled: "It was
my first exposure to the discouragement women scientists encounter in the
U.S." At first the Burbidges accepted this arrangement. Then one day Mrs.
Burbidge was refused use of an observatory truck to haul her scientific gear up
the mountain. That did it: The Burbidges formally protested the antiwoman
rule..."[75],[76] They won.

Betty went on to explain to Spiegel:

> *There has never been a problem (that I know) making difficulties for*
> *women to use the big telescopes of the Lick Observatory of the University*
> *of California. But there certainly have been problems at the Mt. Wilson*
> *and Palomar Observatories (now Hale Observatories). Dr. Vera Rubin*
> *was the first woman to use the 100" telescope, in her name. This was*
> *only a few years ago. I am informed that she received a special invitation*
> *from the Director inviting her to use any telescope except the 200". A few*
> *years later, she was the first women [sic] to use the 40" Schmidt at Palo-*

---

[74]4/29/1974 letter from Betty Scott to Business and Professional Women's Foundation
Librarian Jeanne Spiegel.

[75]Timothy Green, "A great woman astronomer leaves England – again," *Smithsonian* 4,
no. 10 (1974), 34.

[76]"Science: The Stargazer," *Time*, March 1972.

mar. Both of these occasions were so remarkable that they were noted in the Vassar (I recall Vassar – a women's college) Alumni Magazine. Dr. Rubin has not yet used the 200" telescope; I confirmed this by telephone on Friday. I believe that no woman has yet used the 200" telescope but I also believe that it is now only a matter of time. This is what I told the students at Purdue. About two years ago, I again checked the observing schedule for the 200" and still did not find a woman listed. I just do not have the time to do this tonight, but I really think that I would have heard.

Betty told Spiegel that Babcock's letter wasn't clear. There were many telescopes at the Hale Observatories. Babcock wasn't specific about which telescopes were used by women. Betty was almost positive the women had been using the smaller telescopes, but not the 200".

Betty then sent a copy of her letter directly to Babcock. Betty really hoped Babcock would tell her she was wrong and send her some names of women who had used the 200", but she wasn't holding her breath.[77]

Spiegel was persistent and wanted to make sure of the facts. She responded to Betty's "long, informative letter," asking to be kept informed of anything heard from Babcock. Spiegel reread Babcock's letter and realized he didn't refer to the size of any of the telescopes when he said women had used them. She admitted assuming that Babcock was referring to the 200" telescope in his response, since her letter to him had specifically asked about it. She asked Betty if it was possible that no woman had ever asked to use the 200" telescope, including Rubin. "Perhaps we will stimulate and hasten the time when women can use the 200 inch telescope by our correspondence and questions, if they have not already used it," she wrote to Betty.[78]

---

[77] 4/29/1974 letter from Betty Scott to Business and Professional Women's Foundation Librarian Jeanne Spiegel.

[78] 5/3/1974 letter from Business and Professional Women's Foundation Librarian Jeanne Spiegel.

Part VIII

AAUP Higher Education Salary Evaluation Kit

29

Persuasive Analysis
(April – December 1974)

*If we could entice her it will be a real coup. She is a fine woman
and a first-class statistician, just what we need.*
-Robert Dorfman

In April 1974, the VC sent Betty three copies of the "Academic Personnel Listing in Title Code Sequence" (PER 221) for October 31, 1973. She
was away in Warsaw, Poland. Susan Ervin-Tripp reacted in her absence: How
could Betty possibly complete the new analysis in six weeks, when the administration still hadn't given her the information needed to complete the
analysis of the previous year's payroll? They had asked the administration
repeatedly over the past year for the clarification information, and there had
only been mis-communications and troubles tracking down what was needed.
This year there were likely to be more problems, such as accuracy of ethnicity
coding. Could the administration find a way to make the working relationship
smoother, so she and Betty didn't have to keep prodding them?[1]

Betty was learning that, because the university's data systems were designed for administration, they had limitations for producing research reports.
However, the university understood that, in order to meet affirmative action
needs, the systems would need to be adapted to produce the detail and accuracy that Betty was looking for, and so there was hope for the future.[2]

When Betty got back from Poland, she found the PER 221 waiting for
her. She computed head count (people at least 50% time) and FTE (full-
time-equivalent people based on portion of salary received). Again the current
listing appeared to have errors. Betty hoped that SWEM would "...come
down hard on slow, sloppy, inaccurate data collection and reporting. There is
no possible excuse for what we get nor for when we get it."[3]

Betty found many "apparent," "serious" errors in the listing. She immediately began to seek clarification on them but knew she would not be able to
complete her work in time to meet the VC's six-week deadline. Betty echoed

---

[1] 4/3/1974 letter from UC-Berkeley Rhetoric Professor and SWEM Chair Susan Ervin-
Tripp to VC Mark N. Christensen.
[2] 4/18/1974 letter from UC Office of the President's Virginia Leimbach to Betty Scott.
[3] 4/23/1974 letter from Betty Scott to UC-Berkeley Rhetoric Professor and SWEM Chair
Susan Ervin-Tripp.

Ervin-Tripp: Many of their questions about the previous year's listing had not been answered. "I think it is fair to say that only the Lord knows when we will receive answers to the problems with this year's payroll. It seems to be in worse shape than last year's. How can this be?" She asked to keep a copy of the PER 221 until all issues were resolved and clean hand counts could be made.

Betty's frustration overflowed:

> *I request that the University Administration make a correct payroll listing and correct summary counts from the tapes. It is absurd and tedious and slow to count by hand as Mrs. Lovasich and I are now doing when tapes are right here. Only Mrs. Lovasich and I are doing the work, in order to preserve the confidential nature of the PER 221. I point out that this is not a proper job for a Professor of Statistics nor for a Senior Statistician, certainly not year after year after year.*

Betty then requested a budget of $1,000 ($4,371 in 2010 dollars) from the VC. She didn't care how the moneys came to her – they could come either through the SWEM or through emergency allocation or however. The moneys would be used to employ someone to do the work that Lovasich "should have been doing" while she was working on the salary studies.[4] Betty was soon given what she asked for, from affirmative action committee funds.[5]

Betty continued to fight with the data reports. She didn't have major problems with the data file per se; 50 years of trend summaries for many categories could in theory be produced from these data. But the PER 221 report produced from these data simply wasn't accurate. She could easily verify this by looking at the data for her own department. When she compared for the statistics department the PER 221 data with the official computer printouts that came from the accounting office, she found what, in her judgment, were many errors. For example, the PER 221 listing for the statistics department showed that David Blackwell was being paid by the statistics department when at the time he was actually being paid by UC-Santa Barbara (he was on an education abroad program). For another example, the PER 221 listing did not show the administrative stipends paid to Erich Lehmann for being department chair and to Neyman for being Stat Lab director. For yet another example, the PER 221 listing for the statistics department did not include the department's share of the FTE for Paul Lévy's student Michel Loève specializing in probability theory who had a 50:50 split appointment with the mathematics department, and Blackwell's student Aram Thomasian specializing in information theory who had a 50:50 appointment with electrical engineering. There were others who were not included in the FTE count who should have been, or else they were included but not at the correct level. Because of the way soft money was handled, and because women and minorities were more often on soft money, their salaries and FTEs were more

---

[4]4/23/1974 letter from Betty Scott to VC Mark N. Christensen.
[5]5/13/1974 letter from Mary to Mildred [last names not given].

often miscounted. Betty asked that the PER 221 listing be redone to agree exactly with the payroll accounting tapes, to include all payments made, and to include accurate head counts and minority classifications. "Please help us as soon as possible... This is a big job!" she implored the administration.[6]

The UC system office was sympathetic to Betty's pleas. They learned the Berkeley chancellor's office was trying to provide reports that would come closer to meeting Betty's needs. They also "engaged in an intensive effort to correct the October 1973 summary reports," declaring that it was "incredibly difficult to obtain detailed and accurate information from a computerized system designed to produce pay checks." Problems had to do with incorrect title codes, late or incorrect processing of forms, and retroactive pay checks. "These kinds of inputs," Betty was told, "account for some of the differences between the PER 221 and the Department of Statistics accounting records."

The office went on to explain why the PER 221 reports would not meet Betty's needs. First, reports never included name with ethnicity, unless specifically requested by the chancellor; this was to preserve confidentiality. Second, reports were to give accounts of salary rates, not total compensation; this is why Neyman's director stipend and Lehmann's chair stipend were not included. Third, reports never included retroactive salary payments because this would result in delays of a month or two, which would not serve the needs of reporting. Fourth, reports were to show head counts by title sequence; this is why Loève and Thomasian were included in the departmental head counts, even though they had split appointments. The solution, the system office offered, was not to change the purpose of the PER 221 reports, but to produce different reports that would better meet the needs of affirmative action.

The system office had recently made special efforts to improve their reports. But, they explained, "data collection is a huge problem when it involves, as it does in the University, the coordination of records from departments through administrative offices to the data processing centers for input into information system records." They acknowledged more improvements were needed and assured Betty that "we are working as hard as we can to get good information within the limits of our computer system and the problems connected with keeping track of some 70,000 employees at nine campuses."[7]

Betty implored the VC to provide ethnicity information: "This is an urgent request to supply it to us. We can match by hand. On the [academic senate] Committee on Committees we saw again the barrel of errors being turned out. I cannot suggest any excuse worth putting on paper for such sloppy, unchecked listings." She needed the ethnicity data, and the ethnic summaries were, in her words, "known to be complete nonsense." She was frustrated.[8]

The provost had a "Listing of Ethnic Identification." Betty asked to borrow it so she could prepare her academic senate report. She also needed it over the longer term to study the progress of faculty minorities. Betty made it

---

[6]5/3/1974 letter from Betty Scott to UC Staff Member Virginia Leimbach.

[7]6/28/1974 letter from UC Staff Member Virginia Leimbach to Betty Scott.

[8]4/23/1974 letter from Betty Scott to VC Mark N. Christensen.

clear she wanted his listing and not the "nonesense [sic] listing" that had been distributed to the senate. She promised she would keep the listing in a locked file cabinet in a locked office; only she and Lovasich would have access; and they would use and return the listing expeditiously. Then – she was so frustrated she couldn't stop herself – she told the provost how "inefficient and tedious" the process had been, given the many, messy errors in the data tapes and reports.[9]

The provost responded almost immediately. He wanted to be helpful but had some issues: The ethnicity listing Betty was asking for was being heavily used by the administration to prepare the next HEW report due July 31 and the next segments due September 30. It wasn't possible to quickly get a new listing; in fact, it would have taken about a month and a half because the new listing would have to be generated by "University-wide," which involved getting in line to use the computing staff and equipment at the system-level. But the provost wanted to find a way to help, and so he proposed that Lovasich be sent to the administration offices in California Hall to make use of the ethnicity report when it was not being used by his staff for the HEW reports. "I hate by the above proposal to reveal a priority of the HEW reporting over cooperation with a Senate committee," he wrote. "However I have no other resource available to me in order to get this reporting done, for which we are legally liable at this time." He hoped his proposal would be satisfactory to Betty.[10] A month later, the system office reported to Betty that they had been making special efforts to improve ethnicity information. The reports that Betty wanted to use would reflect these improvements.[11]

29.1   Carnegie Study Backlash

The CCHE report on opportunities for women in academe was being widely read across the nation. Betty's part of the study was the most extensive research on salary inequities to date. It was based on 1969 data, and faculty survey data from 1972–73 sadly showed no improvement.[12]

It was inevitable that the CCHE report would be both celebrated and challenged. One major challenge was in the form of a 167-page report, "Antibias Regulation of Universities: Faculty Problems and Their Solutions," that ironically was also commissioned by the CCHE.[13] It was authored by Richard

---

[9]5/16/1974 letter from Betty Scott to UC-Berkeley Provost George Maslach.

[10]5/20/1974 letter from Provost George Maslach to Betty Scott.

[11]6/28/1974 letter from UC Staff Member Virginia Leimbach to Betty Scott.

[12]5/20/1974 "Report to the Berkeley Division from UC-Berkeley Rhetoric Professor and SWEM Chair Susan Ervin-Tripp," pp. 13-17.

[13]Richard A. Lester, "Antibias regulation of universities: Faculty problems and their solutions" (New York City: McGraw-Hill, 1974), 167.

A. Lester, a Princeton University labor economist who had many bona fides in the areas of wage determination, unemployment, and affirmative action.[14]

A pre-publication copy of the Lester report came into Ervin-Tripp's possession, noting there was an introduction written by Clark Kerr that supported Lester's findings. Ervin-Tripp found the most offensive part of the report to be the chapter on availability of women for high rank positions in universities: "Essentially [Lester] argues there is a danger in enforcement of goals and timetables at this level, because women are unable to compete for the highly competitive levels where working 90 hours a week rather than 85 may make the difference." This part of the report also included a direct attack on Scott's work in the earlier CCHE report on women: "[Lester] assumes there are real qualitative differences between the men and women compared there," according to Ervin-Tripp. But Betty needn't have felt singled out. Lester also attacked certain aspects of the CCHE's own proposals.

According to Ervin-Tripp, Lester showed bias. One example was his interpretation of "the post-war drop [in decade-by-decade PhD ratios] as due to Feminine Mystique rather than to the public policy represented by the G.I. Bill." Ervin-Tripp wanted the biases and offensive arguments to be rebutted "based on a better interpretation of available data," and soon: "I am afraid the effects will be considerable; the material feeds popular belief just as nicely as Jensen's intelligence studies do, and will be popular in university circles and court testimony for that reason." Ervin-Tripp thought there was time to prepare a formal reply, because it would be a few months before the report would be published. She called on Betty, Helen Astin and Bernice Sandler to write the reply, because they were researchers in the field.[15]

Astin asked Ervin-Tripp to send her a copy of the Lester report. Astin had already noticed stories about the report on the radio and in the newspaper, stories that had "concerned and distressed" her. She immediately agreed that the women had a "responsibility to prepare a critique" and suggested they involve other researchers, especially her colleague Alan Bayer.[16]

Betty was serving with Clark Kerr on the UC-Berkeley academic senate committee on educational policy and had just seen him over the summer. Kerr had gone over to Betty, and they talked. Betty's conclusion from the conversation was that Ervin-Tripp's group should go ahead and try to get the *New York Times* to publish a rejoinder, but it did not look hopeful.[17]

---

[14]Justin Harmon, "Richard A. Lester Dies at 89; Influential Economist and Dean of the Faculty at Princeton University," *Princeton University*, accessed October 4, 2013, https://www.princeton.edu/pr/news/97/q4/1231-lester.html. Note: Lester served as chairman of President Kennedy's Commission on the Status of Women; he also presided as dean of the faculty over Princeton's first affirmative action program where he developed a program that allocated special funds to attract women and minorities to the faculty.

[15]7/9/1974 memorandum from UC-Berkeley Rhetoric Professor and SWEM Chair Susan Ervin-Tripp to Betty Scott, Helen Astin and Bernice Sandler re writing a rebuttal to Lester's report for the CCHE.

[16]7/17/1974 letter from UCLA Higher Education Professor Helen S. Astin to UC-Berkeley Rhetoric Professor and SWEM Chair Susan Ervin-Tripp.

[17]8/9/1974 letter from Betty Scott to UCLA Higher Education Professor Helen S. Astin.

29.2   Top 100 Salaries at Berkeley

The May SWEM report to the academic senate was all about salary equity. Betty's methodology was outlined: A woman's salary discrepancy due to sex could be estimated using the regression equation for males for the woman's field and institution type; while holding constant many explanatory and interaction variables, one would ask what the woman's salary would be if she were a man. Next Betty's national results were reported. The average academic woman was underpaid about $1,500 in 1969. In biological and physical sciences, the underpayment was about $2,320. In comparison with like men, about 80% of women were underpaid, with one in five being underpaid by more than $4,000. Salary differences were greatest for older women. Then Betty's UC-Berkeley results were reported. The average academic woman was underpaid about $3,672 in 1972. Men earned more than women in almost every job title. Looking at the highest 100 academic salaries, none went to women.

SWEM wanted to find remedies for existing underpayments, but they also wanted to prevent future underpayments. Conditions that gave rise to the underpayments had been carefully delineated in the 1970 Policy CSAW report, yet four years later little had been done to correct them. Women were still being hired at lower steps, and the probability of promotion was still higher for men. SWEM suggested that both the law and principles of equity called for correcting existing, and preventing future, discrepancies.

SWEM put a resolution in front of the academic senate. An important part was a call to establish a faculty committee on salary equity that would "design and implement a procedure for identifying cases with an a priori likelihood of an inequity." The idea was that the committee would screen salaries and produce a list of individuals whose steps, ranks, or salaries were found to be inequitable, including where they should be placed to be equitable. The budget committee would then consider changes of status for these individuals in light of additional qualitative information.[18]

The academic senate decided immediately to establish the "Special Committee on Salary Equity" (SEC). Genetics Associate Professor Patricia St. Lawrence[19] was appointed chair. Betty's methods for estimating underpayment would be used. There would be a pilot study at Berkeley to analyze

---

[18]4/29/1974 report from UC-Berkeley Rhetoric Professor and SWEM Chair Susan Ervin-Tripp to the UC-Berkeley academic senate.

[19]Note: Patricia St. Lawrence (1922–1996) was an associate professor of genetics at UC-Berkeley and the first woman faculty member in the department, eventually serving as chair from 1987–1989. Throughout her career she was a fierce advocate for women and minorities. Among her many accomplishments for equal rights, she helped to found the Berkeley Faculty Union in 1967, later serving as president, and she helped to found WAGE (We Advocate Gender Equity).

salary data from all departments to identify individuals with salary inequities, estimate the amounts of the inequities, and refer the cases for special review.[20]

Betty's analysis of the previous year's UC-Berkeley payroll data (October 1972) had indicated academic women overall were being paid on average (weighted by FTE) $306 per month less than academic men. Her analysis of the data for the most recent year (October 1973) indicated women were paid $324 per month less. It was troubling to Betty that the new deficit for women was larger, and she thought a study of the causes was sorely needed.[21]

By the beginning of June, Betty found a particularly startling result. All of the top 100 academic salaries in October 1972 went to "MEN" [Betty's emphasis]. There was not a single woman among them. Betty sent the details to Ervin-Tripp, with a copy to the VC. She had the hand tabulations done twice as a double-check: Among the top 100, the highest salary was $3,233 per month ($16,658 in 2010 dollars); 13 had a salary of $3,000 or more; and the lowest salary among the top 100 was $2,683. "If the payroll is correct, there are NO women in the top 100 academic salaries," Betty wrote. "We know that the payroll is not completely shown – for example, departmental chairmen get additional stipends that are not shown – but I do not know of any woman who would be moved yet enough. I am pretty sure that no women are in the top 100." Lovasich did the calculations for October 1973: The highest salary was $3,500 per month; 28 had a salary of $3,000 or more; and the lowest salary among the top 100 was $2,808. In this year, there was one woman in the top 100 academic salaries; she had been given a position-title change at some point during the year. Betty was so careful with her data handling that she even sent her work tables to Ervin-Tripp along with her letter.[22]

29.3 Statistical Scrutiny of Berkeley Affirmative Action

One of the quality control procedures that a statistician will perform on a data set is to compare data from that set to like data from another set. Betty had performed this procedure when she compared data on the statistics department from two different sources. There she found discrepancies that opened a conversation about how to get the information needed for affirmative action reporting, and this moved the university forward toward systems that would produce such information. Betty performed this procedure on other data sets. For example, she compared data from the statistics department with graduate student employment data tables produced by the graduate division.

---

[20]n.d. letter from UC-Berkeley Genetics Professor and SEC Chair Patricia St. Lawrence to department chairs.

[21]5/6/1974 letter from Betty Scott to UC-Berkeley Rhetoric Professor and SWEM Chair Susan Ervin-Tripp.

[22]6/1/1974 memorandum from Betty Scott to the SWEM.

When she found some errors in the graduate division data, where they had counted TAs and RAs twice, the graduate division redid the counts and asked her to check them again, which she was more than willing to do.[23]

Betty also had problems with the departmental reviews in the affirmative action reports. She wanted to know why so few departments were reviewed. She also had problems with the complexity of affirmative action forms for applicants, suggesting they be required only for, say, the top 20 candidates: "If you make the red tape performance of affirmative action so unpleasant and so time consuming, you will defeat the purpose. If you insist on having such forms filled in for more than 20 candidates, you will find that some secretary is doing the job, with more or less random entries."[24]

Betty scrutinized the affirmative action plans submitted to HEW in April and distributed to the campus in May. Her first criticism was statistical: What was the cover letter referring to when it talked about "appropriate statistical tests" that were "under consideration and/or agreed on among the campuses"? She wrote: "The correct statistical procedures should be used; this is important. I do not find any statistical tests on some points; in other cases I find incorrect procedure [sic] and also incorrect data."

In all, Betty had four single-spaced typed pages of criticisms. She criticized the statement about appointment of near relatives because it included the phrase, "also applies to student employees." Women students were the near relatives who invariably would lose their jobs under this policy when they got married or when the husband was appointed as a research assistant in the same department. She criticized the occupation codes that were used. They were "too gross and the classification is applied peculiarly." Why weren't the subcategories being used that were adopted for the university three years prior? There were references to statistical tests. Which statistical tests?

Betty also raised a number of questions about the affirmative action program for academic employees. Who is the faculty assistant to the VC, and how can interested individuals provide input into the program? Who is the administrative analyst, and how can individuals who are knowledgeable in either data analysis or affirmative action (or both, as in Betty's case) provide their expertise toward insuring accurate and informative data? "We are agreed, I believe, that present data are wrong (badly wrong) and incomplete. Action is required. How can this be accomplished?" Betty wanted to know.

---

[23] 7/16/?1974 handwritten note marked "URGENT" from Betty Scott to Jeanne [Lovasch] re checking the counts of TAs and RAs for the statistics department against those produced by the graduate division. Note: Much of this work was apparently done by hand rather than by computer. UC-Santa Cruz Applied Mathematics and Statistics Professor David Draper remembers that Lovasch's title was Computer and she used the old fashioned Marshant calculators to do calculations for Betty and Neyman; the year after Neyman died, the university "quietly closed down" the Computer job category (David Draper, e-mail to author, October 31, 2014).

[24] 5/19/1974 letter from Betty Scott to UC-Berkeley Rhetoric Professor and SWEM Chair Susan Ervin-Tripp.

Betty also had some problems with the section on utilization analyses in the affirmative action report. First, she thought there should be more clarity in how lecturers were included. Also, the two demographic classes, "women" and "minorities," were cross-tabulated with the two graduate student funding classes, "with" and "without" support: The report suggested an "interim statistical method" be used on these tables that amounted to a chi-square test with correction for continuity. Betty explained that this method was

> . . . *not valid for small number [sic] of students. The uneven proportions in a b versus c d [entries in the 2x2 table] mean the exact hypergeometric distribution of probabilities is not symmetrical. The Yates correction does not help here (it helps wash out discreteness) and sometimes harms. Since the hypergeometric distribution is well tabled (and not hard to compute either), it would be much better to use the exact probabilities rather than any sort of chi square approximation.*

Then Betty pointed out that "the data in the tables are wrong anyway!" She was completely on top of the statistics department faculty and graduate student personnel data. She knew exactly how many students, non-minority TA/RAs, women TA/RAs, etc. there were. The numbers given in the reports were wrong. "What is going on here?" Betty wanted to know.

Actually, Betty thought women and minorities should be separated into four classes, rather than two: male minority, female minority, male non-minority, and female non-minority. She thought graduate student training should be reported in five categories, instead of two: research assistant (RA), teaching assistant (TA) but not also RA, fellowships but not also either RA or TA, other award paying $200 per month or more, and none of the above. Then 4x5 tables should be created. These would be informative. They could be analyzed using a chi-square approximation if there were roughly 40 students in a department (this, she said, would depend "somewhat on the proportion who are women and minority"); otherwise generalized hypergeometric probabilities should be used. If the deviations between observed and expected values were too large to be attributed to chance, then one could, she pointed out, analyze which categories were creating the discrepancies.[25]

29.4   Campus Attitudes toward Affirmative Action

Affirmative action had become an overarching issue at UC. Campuses had been directed to maintain academic personnel records that could be used for reporting. At the beginning of May, the system distributed existing documents to the campus. These included the December 17, 1973 Affirmative Action Plan;

---

[25]5/19/1974 memorandum from Betty Scott to UC-Berkeley Rhetoric Professor and SWEM Chair Susan Ervin-Tripp re comments on April 30, 1974 submission to HEW.

Staff Personnel Policy on Employment; February 26, 1974 Affirmative Action Planning Document; and March 7, 1974 Conciliation Agreement that clarified and summarized the February 26 agreement. These would be assembled into a continuous narrative that would comprise the system's affirmative action plan and be submitted on September 30 to HEW. The system wanted to circulate the documents as widely as possible before the narrative was constructed so that questions or comments could be used to improve the plan.[26]

Betty now considered herself both a statistical and an affirmative action expert. She expressed herself to the committee on educational policy, of which she was a member: Because "records will have to be kept and data provided," the new regulations "will be a bother" to people on the campus, there may be federal "interference," or there may be lawsuits; but the attitude of the campus should be "constructive rather than confrontation":

> *The present HEW regulations have been both good and bad for Berkeley and for the women here, in particular. Certainly, the opportunities for admission, progress, and employment are much fairer and, presumably, as equal as any sort of imposed regulations could provide. However, in my opinion, the confrontation attitude of several years has consumed a lot of energy, time and money that has been largely wasted producing mostly frustration. There now appears to be a more positive attitude again and I hope that this can be reinforced. HEW is not about to "go away"; let us build better educational policy out of these regulations.*

Further regulations were needed and this shouldn't be a surprise, Betty opined, because the California constitution had for "many decades" forbidden sex discrimination in admissions, yet such discrimination existed and persisted, especially at the graduate and professional student levels.

Then Betty addressed details of the regulations and their potential influence on educational policy. The campus should "take positive action" and "take the initiative" to undertake annual evaluations that are "simple to carry out, clear, and accurate." The campus should ensure that remedial action be taken in all areas when needed, not just in athletics; provide stronger affirmative action in recruitment and admissions; support gender equity in financial assistance; and, even though not covered by HEW regulations, eliminate sexism in academic programs, curricula, and textbooks. While Betty knew that her statements by themselves provided a productive vision for the future, she challenged Ervin-Tripp's committee to write an even stronger statement.[27,28]

---

[26] 5/3/1974 memorandum from VCA Robert F. Kerley to Deans, Directors, Department Chairs, Administrative Officers, Academic Senate Chairs, and Interested Members of the Berkeley Campus Community re UC Affirmative Action Plan.

[27] 8/14/1974 memorandum from Betty Scott to UC-Berkeley Assistant to the Chancellor and Political Science Lecturer Jack H. Schuster re proposed HEW regulations concerning sex discrimination.

[28] 8/17/1974 memorandum from Betty Scott to UC-Berkeley Rhetoric Professor and SWEM Chair Susan Ervin-Tripp re HEW Title IX at Berkeley.

Now Betty was firmly connected with the Berkeley salary data systems. In November, the provost ordered a new (October 31, 1974) PER 221; this time he ordered both a printout and a tape disc, both for Betty to use.[29]

29.5   Expertise Needed by AAUP

Back in early February, AAUP Committee W had passed a resolution on equal pay for equal work, stating "that institutions be urged to move rapidly to identify and eliminate existing pay differentials based on sex, in accordance with considerations of equity and the requirements of the law."[30]

In April, Robert Dorfman invited Betty to become a member of Committee Z. He had received a copy of her Application paper and said this:

> *I found it to be an excellent and utterly persuasive analysis. On the other hand, I was discouraged by the large number of equations, the plethora of variables, and the indications (in Table 1) of base ingratitude that makes universities penalize long years of service (variable 12) and heavy teaching loads (variable 23). I suppose that there is a lot more work that ought to be done... One of the important tasks before the committee is to devise a simple and practical statistical procedure by which the existence of male-female discrimination on a campus can be documented and its extent measured... Seeing this paper made me wonder whether you have a continuing interest in the problem...*

Dorfman wondered: If Betty did have a continuing interest, would she join Committee Z?[31] He hand wrote a note to Maryse Eymonerie: "E.L. Scott is much better statistician than [another person]."[32] He later wrote: "If we can entice her on to Committee Z it will be a real coup. She is a fine woman and first-class statistician, just what we need."[33]

When the invitation came from Dorfman, Betty was characteristically overextended, "especially by the problems of women in higher education" and because she expected to be on sabbatical. She told him: "I feel that I should return to astronomy." It took Betty a few months to decide, but then she did

---

[29]11/4/1974 letter from Provost George J. Maslach to UC Director Lundborg.

[30]2/26/1974 memorandum from MLR (presumably AAUP Staff Member Margaret L. Rumbarger) to WBW [unknown] re AAUP Committee W proposals for the resolutions committee.

[31]4/29/1974 letter from Harvard University Economics Professor and AAUP Committee Z Chairman Robert Dorfman to Betty Scott.

[32]4/29/1974 handwritten note from Harvard University Economics Professor and AAUP Committee Z Chairman Robert Dorfman to AAUP Staff Member Maryse Eymonerie re Betty's competence as a statistician.

[33]5/2/1974 letter from Harvard University Economics Professor and AAUP Committee Z Chairman Robert Dorfman to AAUP Staff Member Maryse Eymonerie.

agree to serve on Committee Z, because she thought "the economic problems of women faculty need more work and are of importance."[34,35]

Dorfman had earned his PhD in economics at UC-Berkeley and served on the faculty there for five years. Betty and Dorfman clearly knew each other. He wrote to Betty: "Give my best regards to Dr. Neyman."[36] Betty wrote back: "It is nice to hear from you. I hope that you will be continuing as Chairman [of Committee Z]. Then I'll have a chance to see you." She told him she had been in his neck of the woods a number of times in the spring, "but to worry about problems in the statistics of skin cancer."[37]

Bergmann had offered the idea to Committee W that there should be a "Higher Education Salary Evaluation Kit" (Kit) to "assist administrators and faculty groups to evaluate and rectify possible salary inequities." The idea was twofold. (1) With the added sex variable, the AAUP annual survey of faculty compensation would show that differences existed between men and women; and (2) the Kit would demonstrate and measure discrimination by relating, on individual campuses, differences in salary to potential explanatory variables such as length of service and experience, including sex. The Kit would contribute to impartial salary determinations and be an important part of the AAUP effort to eliminate sex discrimination in academe.[38]

In the official invitation, Eymonerie told Betty:

> *I feel very strongly that collecting data by sex is not sufficient to determine inequalities. I believe the evaluation of inequities must be done at the local level (i.e., on an institutional basis). I recommend the development of a "do it yourself kit" which would help administrators and faculty groups to evaluate and correct differentials in salary between male and female faculty. With the cooperation and inputs from both Committees, this becomes a special and important project which needs your expertise.*

Because Betty was already involved in the evaluation of salary differentials at UC-Berkeley, she would be a perfect addition to Committee Z. In fact, She and Bergmann would be expected to play key roles in the development of this "do it yourself kit." Dorfman and Gray would provide oversight.

Eymonerie went on to say:

> *The kit, in my opinion, should probably contain the following:*

---

[34]6/12/1974 letter from Betty Scott to Harvard University Economics Professor and AAUP Committee Z Chairman Robert Dorfman.

[35]5/27/1975 memorandum from AAUP Staff Member Maryse Eymonerie to Committee Z members re status report on Committee Z projects. Note: Betty would serve on Committee Z with some big names, including University of Pennsylvania Economics Professor Robert Summers, the father of Lawrence Summers.

[36]4/29/1974 letter from Harvard University Economics Professor and AAUP Committee Z Chairman Robert Dorfman to Betty Scott.

[37]6/12/1974 letter from Betty Scott to Harvard University Economics Professor and AAUP Committee Z Chairman Robert Dorfman.

[38]6/20/1974 letter from University of Maryland Economics Professor Barbara Bergmann to University of Maryland Director of Institutional Research Richard Good.

1. Description of input needed (e.g., for each individual full-time faculty: sex, highest degree and year when awarded, contracted salary and length of contract, field, etc.).

2. A statistical package or computer program [using regression methods] with certain options which would make it adaptable to particular situations.

3. Interpretation and evaluation techniques of outputs.

4. Criteria and formulas to adjust salary.

A service should be provided by AAUP or another entity to conduct the salary equity analysis for institutions not equipped to conduct it themselves.

Bergmann offered to develop a statistical package for the Kit. She proposed to secure the approval of her administration at the University of Maryland to provide necessary data for a AAUP pilot study and to test regression methods that would be used.[39] There was much discussion back and forth among Betty, Bergmann, Dorfman, Gray, Eymonerie, and others about the variables that should be used in the regression. Eymonerie thought it would be good to remember that their goal was to help institutions and faculty to evaluate gender differences in faculty salaries, and "not to get into possible discrimination because of race, religion or other reasons ..."[40] Dorfman explored the use of Harvard University data for a AAUP pilot study, but found he would need to conduct a special survey to obtain data for all of the 16 variables he wanted to include in the regression; not all were available from administrative records.[41] He wanted to convince Fred Mosteller to supervise the Harvard project.[42] But Dorfman's efforts were becoming stalled due to lack of funding, and Dorfman and Bergman were not coordinating their efforts in the sense that they were planning to use different variables in their regressions.[43]

Because the AAUP budget was tight, they got a $10,000 grant to revise the 1974–75 questionnaire for the annual economic survey to include sex and tenure status. Committee W included the Kit in its 1974–75 Committee W budget, dedicating as much of the budget as it could to the Kit.

By the end of the year, the Kit was Committee W's top priority.[44]

In September, Eymonerie sent a message to Betty and the other members of Committee Z. The AAUP had been invited to participate in President Ford's meeting of health, education, and welfare leaders and to attend a con-

---

[39] 6/24/1974 letter from AAUP Staff Member Maryse Eymonerie to Betty Scott.

[40] 7/12/1974 letter from AAUP Staff Member Maryse Eymonerie to Harvard University Economics Professor and AAUP Committee Z Chair Robert Dorfman.

[41] 9/19/1974 letter from AAUP Staff Member Maryse Eymonerie to Betty Scott.

[42] 8/30/1974 memorandum from Harvard University Faculty of Arts and Sciences Associate Dean for Academic Affairs Phyllis Keller to Harvard University Faculty of Arts and Sciences Dean Henry Rosovsky re Dorfman's proposal of a study of salary differences at Harvard.

[43] 9/19/1974 letter from AAUP Staff Member Maryse Eymonerie to Betty Scott.

[44] 11/13/1974 memorandum from DIP [unknown] to JD [unknown] re Committee W activities.

ference on inflation. The AAUP Secretary General wanted expert input to take to these assemblies.[45] Betty responded:

> ...*I want to urge the importance of increasing employment. I see the terrible situation in this area where people are not finding employment month after month so that they eventually give up. I expect them to be a public charge forever – and crime is encouraged thereby. They feel the system is cheating them and they are right. It is frightening to feel the doubling price of milk, sugar, coffee, even beans during the last year (during the last three months for simple cleansers used around the home), but it is terrifying to see the huge increase in the number of persons using food stamps, persons standing in line at soup kitchens, persons scrounging in clothes barrels, etc. Some sort of employment that is reasonable, even public projects, must be found for these people.*

And then Betty related the problem to her own environment:

> *The problem is very acute for young research workers, for young students seeking part-time work, for older women whose job has been wiped out. I think that some educational training programs should be set up, and more low interest loans for students (which are not restrictive on sex or age) should be set up. Teaching and research do require continuing support; it is not possible to turn them on and off.*

She also offered that she was not an expert on inflation, and she was "very depressed about Ford as a person."[46]

Bergmann did develop a statistical package and conduct a pilot study. Her results were reported in the campus newspaper and published the next year in the *AAUP Bulletin* with a graduate assistant co-author.

The study concluded that women on the University of Maryland-College Park campus were being underpaid. Bergmann used regression methods to build a model for male salaries; she included as independent variables sex, doctorate (yes/no), date of last degree, 10- or 12-month appointment, and department. Then she used the same model to predict women's salaries. The results were that 73% of women had salaries lower than what would be predicted if they were men. The university president told Bergmann "this wasn't the time" to discuss a settlement. He further criticized the regression model as providing a "very crude kind of measure of the problem... It is very approximate and subject to all kinds of statistical errors."[47,48]

---

[45]9/4/1974 memorandum from AAUP Staff Member Maryse Eymonerie to Betty Scott re soliciting Betty's comments on inflation.

[46]9/10/1974 letter from Betty Scott to AAUP Staff Member Maryse Eymonerie.

[47]Marcia Kass, "Women paid less, study reveals: Dorsey debates AAUP report," *The Diamondback* (University of Maryland-College Park), November 15, 1974.

[48]Barbara R. Bergmann and Myles Maxfield Jr. "How to analyze the fairness of faculty women's salaries on your own campus," *AAUP Bulletin*, Autumn 1975.

29.6 UC Committee of Statisticians

It became increasingly clear that both UC and HEW needed to be educated about the use of statistics in affirmative action. They needed the basics: why statistical tests should be applied; what needs to be considered and is involved in applying tests; how to choose an appropriate test; how to decide whether it is appropriate to apply a test at all; and "the fact that 'statistical test' is not a magical incantation."[49] They also needed to be educated about the relationship between the choice of test and corrective action.[50]

The UCLA executive vice chancellor proposed to the VP that all campuses in the UC system use uniform regression methodology to study discrimination on their campuses. The results should be used to carry out a case-by-case review of the salaries and statuses of individuals who had lower than expected salaries, and a review of departments where statistical evidence pointed to possible discriminatory practices. He proposed that UCLA's approach be used, which was based on the model Betty had developed for UC-Berkeley.[51]

At the end of July, Betty, Ervin-Tripp, F.N. David (FND), and another professor from UC-Davis met with the UC assistant vice president for academic and staff personnel to discuss the system's approach to salary inequity review and affirmative action.[52] As an apparent outcome of that meeting, two weeks later, the assistant VP sent invitations to Betty, FND, UC-Berkeley Biostatistics Professor Chin Long Chiang, and UCLA Biostatistics Professor Raymond J. Jessen. The invitation was to become a member of a UC-wide statistical advisory committee (SAC) to the VPAA. The charge was to determine statistical methods – either statistical tests or different types of procedures – that would detect underutilization of women and minorities.

Betty took notes at the first SAC meeting, attended by all members except Jessen. The administration hired a statistical consultant to write a draft committee report.[53] The group was debriefed about what the administration had done so far to determine underutilization and discussed a basic statistical approach to estimate it.[54] The group decided: An appropriate statistical test

[49]7/22/1974 memorandum from LAJ [presumably UCLA Mathematics Professor Louis A. Jaeckel] to GJM [presumably UC-Berkeley Provost George Maslach] re statistical tests.

[50]n.d. "Use of Statistical Tests for Underutilization" (outline), 2 pages.

[51]7/19/1974 letter from UCLA Executive Vice Chancellor and Physics Professor David S. Saxon to VPAA Angus E. Taylor re a uniform salary inequity model for the UC system.

[52]7/23/1974 letter from UC Assistant VPAA to UC-Riverside Statistics Professor and Chair F.N. David, UC-Berkeley Rhetoric Professor Susan Ervin-Tripp, UC-Davis Political Science Professor Dale Rogers Marshall, and Betty Scott.

[53]8/6/1974 letter from UC Assistant VPAA to UC-Berkeley Biostatistics Professor Chin Long Chiang, UC-Riverside Statistics Professor and Chair F.N. David, UCLA Biostatistics Professor Raymond J. Jessen, and Betty Scott. Note: Raymond J. Jessen (1910–2003) had expertise in statistical sampling. Louis A. Jaeckel graduated from UC-Berkeley with a PhD in Statistics under Erich Lehmann in 1969.

[54]8/8/1974 handwritten notes by Betty Scott re first meeting of the SAC.

was a one-sided exact test of a proportion; the definition of the proportion in the availability pool was problematic; and the administration would determine the significance levels and corresponding actions. The group heard a proposal by the UC assistant counsel to have categories of action that corresponded to different significance levels as shown below:

> *10% significance level: subpar utilization, warranting full affirmative action efforts, but not to be considered underutilization. Full scrutiny of all personnel actions.*

> *5% significance level: underutilization, warranting above efforts, plus setting of goals+time tables.*

> *1% significance level: extreme underutilization, warranting above, plus direct control of hiring by the Chancellor.*

For example, if the p-value for a department was less than or equal to the 5% significance level, then the department would be judged to have underutilization, warranting full affirmative action efforts, full scrutiny of all personnel actions, plus setting of goals and timetables. Smaller p-values would be stronger indications of underutilization. The proposal was apparently an attempt to avoid lawsuits waged by women and minorities. The statistical consultant drafted a report summarizing these ideas and included an example.[55],[56]

Betty had her own ideas. She thought the statistical consultant's draft "does not go far enough" and found the suggestions and example to be misleading. She pointed out that with small sample sizes it may not be concluded that there is no underutilization, just that underutilization is not detected. Especially, she thought that a 0.10 significance level was too low of a starting point in determining discrimination. She thought they needed to start with a 0.50 significance level, which corresponded to less than expected utilization, warranting increased affirmative action. She instructed that it was important to look at a statistical test's performance characteristic because, in this case, the test at the proposed significance levels will never identify underutilization when the number of people in the population is small and the proportion in the availability pool is small. A simple solution, she pointed out, would be to include the significance probability of 0.50. Finally, she felt strongly that the tests should be applied at the level where the employment takes place, which is the unit (i.e., department) level. But because the power of these tests will be low, combinations of units (e.g., divisions, colleges, campuses) should be examined using Fisher's test criterion for combining independent tests.[57]

---

[55]8/12/1974 letter from Louis A. Jaeckel to SAC. Note: Exact tests were proposed because many department sizes were small.

[56]C.L. Chiang, F.N. David, and E.L. Scott, "Letters: Statistically speaking," *New York Times*, June 15, 1975.

[57]8/14/1974 memorandum from Betty Scott to the SAC re suggestions on Jaeckel's draft report from the first meeting.

All did not go well after that. The SAC met only once and decided more study was needed. The statistical consultant's draft report was reviewed and commented on by the committee, but it was never revised, even though the committee was told it would be. Then the UC-Berkeley administration went ahead and prepared a revision of their affirmative action plan, dated September 30. It contained an abbreviated (three-page) version of the first draft report from the SAC, attributing recommendations to the committee. The full text of the report and comments made by each of the four SAC members were reportedly included in a Supplement, Section A. The administration submitted the report to HEW before the SAC saw it.[58]

"Newspaper people" got hold of the revision and asked Betty to comment. Betty finally got a copy of the three-page "operative part," read it, and summarized her feelings: "I am distressed about the whole thing so far as any committee of statisticians is concerned. They ignored what any one of us said and wrote what [the assistant council] wanted."[59,60]

LAW met and discussed the situation. They decided it was grave. One of the LAW leaders wrote to Betty a few days later, advising her to "dissocate [sic] yourself from the submission as rapidly as possible... you might stress... the lack of validity of the method and its gross simplification and distortion... it is important that [the assistant council] be caught at his own game, as I am sure you realize." The LAW leader went further to suggest to Betty that she might take more drastic measures than just writing to the VC, the regional HEW office, and the Washington HEW office, that she might also want to see the chancellor "and demanding that he issue a disclaimer re the submission's statistical methods lest you sue the University for maligning your academic reputation... I feel that the stronger the action the better. I was very touched by your feeling of helplessness in the situation which you expressed at the L.A.W. meeting and I do think such a threat from you might force some kind of recantation... time is of the essence..."[61]

The issue found its way into the press. LAW planned to send a letter to HEW to urge rejection of the September 30 affirmative action plan. The plan, they said, was "not only totally inadequate, but also misleading, evasive, incomplete, inaccurate and an outrageous insult to the groups affected." LAW charged UC with manipulating the statistical data to claim no underrepresentation. The UC-Berkeley provost shot back, saying to the press that "the statistical data used by Berkeley were not 'the University Administration's unique creation,' but were determined by a University-wide committee of statisticians, which included women and minority members."[62]

---

[58] 10/18/1974 letter from VPAA Angus E. Taylor to HEW San Francisco OCR Director Floyd L. Pierce, with attached 10/11/1974 letter from SAC members to VC Ira Michael Heyman.

[59] 10/12/1974 letter from Betty Scott to UCLA Biostatistics Professor Raymond J. Jessen.

[60] 10/11/1974 letter from SAC members C.L. Chiang, F.N. David, and Betty Scott to VC Ira Michael Heyman.

[61] 10/12/1974 letter from LAW leader Isabel Pritchard to Betty Scott.

[62] Cindy Kadonaga, "Group Urges Rejection of UC Affirmative Action Plan," *publisher unknown*, October 1974. Note: The publisher was not noted on the clipping, but it is known that Kadonga was a staff writer for the *LA Times* around 1985.

Betty, FND, and Chiang prepared an official protest. They wrote to the VC, thinking the affirmative action recommendations were inaccurate and incomplete: "The statements presented there do not reflect the committee's discussion. . ." They objected "strenuously" to having their names on the "misrepresentation" and insisted the university send a revised statement to HEW "without delay so that we will not be misleading them. . ."

The three statisticians made three points. One was that statisticians could not define either underutilization or limits. What they could do was suggest procedures to identify discriminating units, procedures that would have to be thoroughly studied before applied. The labels (e.g., subpar utilization, underutilization, extreme underutilization) attached to the significance probabilities were the work of the attorney and should not be attributed to the statisticians. The second point was that the procedure discussed by the committee would not have much chance of detecting discrimination in units where the number of faculty was small, as was the case with many of the units: "It is not appropriate – indeed, it is foolish – for the University of California to set out a procedure that obviously is faulty, that obviously will never point out discrimination against minorities," they wrote. The solution was to revise the procedure to look at divisions and colleges, which were larger, in addition to the units. The third point was that attribute sampling methods be investigated. The chancellor was a statistician and expert in these methods. But application to the problem of women and minorities would require that tables and charts be computed for small proportions. In the meantime, they suggested "every unit that has less than parity should be asked to review its search and hiring procedures that will increase the number of women and minorities employed."[63] Betty invited the fourth statistician on the statistical advisory committee, Jessen, to join in the protest. After all, his name was also attached to the methods in the September 30 revised plan.[64]

SWEM lined up behind Betty, FND, and Chiang. They drafted a very strong memo to the VC: "The misuse of statistical data and the misrepresentation of the work of the university-wide committee of statisticians who were asked to study utilization analysis has come to our attention. . . there is no legitimate basis for attaching utilization categories to specific probabilities (P-values). . . The attribution to these statisticians of the method. . . is an abuse of their professional reputations and of their good faith. . . this exploitation of the reputations of distinguished colleagues goes beyond the typical benign neglect." The draft memo called on the VC to rescind the September 30 revised plan and support the approach recommended by the universitywide

---

[63] 10/11/1974 letter from SAC members C.L. Chiang, F.N. David, and Betty Scott to VC Ira Michael Heyman.

[64] 10/12/1974 letter from Betty Scott to UCLA Biostatistics Professor Raymond J. Jessen.

statistics committee. The chairs of SWEM called on Betty to revise their draft memo to the VC so it could be sent by Friday.[65]

The ASUC also lined up behind the three statisticians. They issued a statement calling the revised plan "a failure" and asking HEW to reject it.[66]

Betty made an official statement to the academic senate that included four points. One was that it was improper to send a revision of the affirmative action plan to HEW without consulting any committees of the academic senate: "The fact that the Chancellor's Office was rushed is not a good excuse. Berkeley has been preparing its Affirmative Action Plan for many years – starting when Heyns was Chancellor," she said. The second point was that the revised plan implied that the SAC prepared part of it, which was "inaccurate and incomplete. The Statisticians' objections, as stated by [the VC], are strong. It is not proper to so mistreat a faculty committee," Betty said. Betty's third point was that the revised plan was not proper:

> *It will never, ever, detect discrimination against minorities or against women in fields where women are not very common members (such as English) and also the department is large. There are two kinds of errors that can be made when stating that a department is underutilizing women or minorities:*
> *(i) State that a department is underutilizing when it is not,*
> *(ii) Do not state that a department is underutilizing when in fact it is.*
> *The September 30 submission ignores the second kind of error; the error that I consider the more important to avoid. In any effort to detect discrimination the Plan is a farce.*

Betty's fourth point was that it was "absurd to say that goals and timetables imply anything hard and fast," and every department should have them.[67]

The VC responded promptly and wrote to the local HEW office in San Francisco. He forwarded the statisticians' letter, saying the VPAA would respond to HEW about the issues raised, as the VPAA had formed the SAC.[68] The VPAA responded to HEW and was conciliatory. He saw the present procedures as dynamic, respected Betty and the other members of the SAC, wanted the committee to be a continuing committee, and would confer with them about the details in their letter of protest. He agreed with the committee that the administration has the responsibility to define underutilization; it was not a statistical question. He took full responsibility for the decisions

---

[65][?10/1974] draft memorandum from SWEM Co-Chairs UC-Berkeley History Professor Diane Clemens and UC-Berkeley Banking and Finance Professor David Pyle to VC Ira Michael Heyman re revision of the affirmative action personnel program, UC-Berkeley, September 30, 1974.

[66]Larry Spears, "U.C. asks $589.2 million," *Oakland Tribune* (Oakland, CA), October 19, 1974.

[67]11/26/1974 letter from Betty Scott to UC-Berkeley Academic Senate Staff Member Pat Hatfield.

[68]10/21/1974 letter from VC Ira Michael Heyman to HEW San Francisco OCR Director Floyd L. Pierce.

on how to define underutilization. He would do his best to clear up misunderstandings between the UC attorney and the SAC. He would organize further discussions on what tests were appropriate for analyzing underutilization.[69]

The VPAA didn't waste any time in organizing a next meeting of the SAC. He wanted to discuss the recent UC-Berkeley experience with statistical testing for underutilization; and respond to the California legislature's request for a review and report on systemwide salary inequity. UCLA had already drafted a proposal, and the VPAA asked the committee to review it.[70]

The next (second) meeting of the SAC took place at the end of November. The VPAA told the committee he wanted statistical tests, with instructions on how to use them, and wondered how to respond to changes that might be imposed by HEW in the future. Chiang responded that the committee should be informed before any work is attributed to them. FND asked whether the VPAA was prepared to have different cutoff levels for different departments. Betty asked if he was prepared to have different actions that corresponded to each cutoff level. Chiang offered that perhaps the rule needed to be more complicated, involving appointments over the last several years, rather than acting like all could be accomplished in one year. It was agreed that Betty, FND, and Chiang would discuss these matters further among themselves.

The HEW OCR director wrote to Betty. His office had received a copy of her October 11 memorandum to the VC. He and his staff in both Washington, DC and San Francisco found her comments to be "very useful in our review of the University's affirmative action program."[71]

29.7   Nominated for Affirmative Action

It is important to determine availability pools for faculty hiring, because they are used for comparison with actual faculty hiring. Betty strenuously objected to using an availability pool comprised of only Americans who earned their degrees within the last five years. She knew that women of her generation had been treated unfairly and as a result had ended up working in research positions or not working at all. She insisted that a pool of people who graduated in the last five years was biased. She also knew the US Supreme Court had ruled that foreign students must be included in the pool.[72]

---

[69] 10/18/1974 letter from VPAA Angus E. Taylor to HEW San Francisco OCR Director Floyd L. Pierce.

[70] 11/18/1974 letter from VPAA Angus E. Taylor to SAC members Chin Long Chiang, F.N. David, Raymond J. Jessen, and Betty Scott.

[71] 12/13/1974 letter from HEW OCR Director Peter E. Holmes to Betty Scott. Note: Jesson also wasn't present at the second meeting of the SAC.

[72] 7/20/1974 letter from Betty Scott to Princeton University Statistics Professor and Chair G.S. Watson.

Betty had become so knowledgeable about issues of diversity involving women and minorities that the VC asked if she would consider accepting the position of faculty assistant for affirmative action.[73] He wanted a tenured faculty member on a 50% assignment to provide guidance and assistance internally to all campus administrators, committees, and groups in implementing the affirmative action policy, as well as to be a liaison with groups, committees, organizations, government agencies, the UC president's office, and other higher education institutions. In addition, the person would establish a local availability pool for temporary appointments and an enlarged availability pool for all positions; and assist him in investigations of complaints and appeals, administer annual statistical analyses, and draft reports.[74]

In many ways, Betty would have been perfect for this position. But she did not accept it.

At the end of the year, Betty was nominated for reelection to the WFC board of directors. The club's president, Margaret D. Uridge, had approached Betty about this, and Betty "demured," saying she would be on sabbatical. But Betty "didn't flatly turn it down. On the basis of that 'reluctant acquiesance,'" [sic] the nominating committee listed Betty on the ballot.[75]

---

[73] 9/4/1974 letter from VC Ira Michael Heyman to Betty Scott.

[74] 8/26/1974 memorandum from VC Ira Michael Heyman to all faculty members re call for applicants for the position of Affirmative Action Officer for Academic Personnel.

[75] 12/30/1974 letter from WFC President Margaret D. Uridge to Betty Scott.

30

High Stakes
(1975)

*There are no special circumstances in higher education that
absolve it from affirmative action.*
-Elizabeth L. Scott

In 1975, AAUP Committee W was becoming more active and planning to
develop the Kit. Even with proposed additions to the annual budget, AAUP
needed a grant to support the Kit's development. The grant came from the
Exxon Education Foundation in New York, established in 1955, which had
a program to support the development and testing of widely applicable new
ideas in education. The grant to AAUP was to address the "chronic complaint"
among academic women of salary inequity.

The grant was for two purposes. One was to subsidize making changes to
the AAUP annual economic survey: adding sex and making related changes
to allow for assessment of salary by sex within rank.[1] Betty was on AAUP
Committee Z that would begin to collect the data by sex in the summer.
She was practical and told Eymonerie that "adding more questions will some
day bring too many questions. And these question seem hard to answer."[2]
By the time of Committee Z's annual report later in the year, almost 90%
of surveyed institutions would provide the requested sex data.[3] Later, when
Betty found out the collection of sex data would not necessarily be permanent,
she objected strongly to Eymonerie: "I want to request that definite plans be
made for the collection of sex data. I think that two more years is not enough.
If there are plans to stop, the schedule should be discussed."[4] In response,
Eymonerie made plans to continue collecting the data indefinitely.[5]

The other purpose of the grant was to develop the Kit. The AAUP explained to Exxon: "The modifications made to the regular annual survey are,

---

[1] Note: AAUP Committee Z began to collect annual data on faculty salaries and fringe
benefits in 1957–58. In 1976, while Betty was still a member, the AAUP transferred the
responsibility for collecting these data to the National Center for Education Statistics,
hoping they would have time for more research-oriented projects.

[2] 8/30/1975 letter from Betty Scott to AAUP Staff Member Maryse Eymonerie.

[3] 10/22/1975 memorandum from AAUP Staff Member Maryse Eymonerie to members
of the [AAUP] council re status report on the activities of Committee Z.

[4] 8/30/1975 letter from Betty Scott to AAUP Staff Member Maryse Eymonerie.

[5] 9/9/1975 letter from AAUP Staff Member Maryse Eymonerie to Betty Scott.

however, not sufficient. It is incumbent upon us under our professional responsibility to assist administrators and faculty groups to identify, correct, and eliminate any and all forms of pay inequalities on the basis of sex... the stakes are high... a comprehensive study must be made by recognized experts in the field..." More specifically, it wasn't enough to modify the survey to be able to show the differences between average salaries of men and women within rank; steps must also be taken to "demonstrate and measure discrimination." The AAUP argued the Kit could accomplish this by relating salary to causal factors including sex: "If sex should turn out to be a significant causal factor, as is widely suspected, the existence and severity of sex discrimination will be established." Operationally, a statistical package needed to be developed by a respected member of the academic statistical community that could be used at different academic institutions to help diagnose salary inequities. It is logical that this eventually turned out to be Betty, as she had conducted two of the six "relevant studies" listed in the Exxon proposal.

Exxon granted $50,600 to the AAUP to develop the Kit, including $15,000 for a 0.75 FTE statistician to be responsible for the research and technical components. The AAUP general secretary appointed a special advisory committee to assist, including Robert Dorfman, Barbara Bergman, Mary Gray, and Betty. Betty readily accepted the appointment, as she was already working on the salary equity problem.[6,7,8,9,10] The AAUP put out a press release. The purpose of the Kit, it said, was to "provide technical assistance to faculty groups and administrations for studies on salary inequities between men and women."[11] The AAUP said that "for the first time in higher education both faculty and administrators will have the tools necessary to make determinations as to where salary inequities exist."[12]

The committee had its first meeting in mid April. The agenda focused on delineating the activities and a time line for completing a Kit. Betty hand wrote the questions: What steps should be taken? What methods should be used; what variables? What will happen with kit; how to be used? How to choose pilot institutions; via Committee W's; via Univ admin? Possible methods were discussed at the meeting. The committee proposed pilot studies be conducted at different kinds of institutions. They determined that deliverables

[6][?1974] "Request for a grant to support the development of a 'Higher Education Salary Evaluation Kit' to assist administrators and faculty groups to evaluate and rectify possible salary inequities."

[7]11/14/1974 letter from AAUP Staff Member Maryse Eymonerie to Exxon Educational Foundation Executive Director Frederick Bolman.

[8]1/31/1975 letter from Exxon Education Foundation Executive Director Frederick Bolman to AAUP General Secretary Joseph Duffey.

[9]3/7/1975 letter from Betty Scott to AAUP Staff Member Maryse Eymonerie.

[10]6/16/1975 press release from the Exxon Education Foundation: "Exxon Education Foundation Grants $50,600 for College Faculty Salary Evaluation Kit."

[11]4/21/1975 press release from the AAUP Office of Public Information re call for planned salary equity studies that might be used as models for the Kit.

[12]"The Association receives EXXON grant," unidentified AAUP newspaper clipping, 1975.

would include standardized and consistent procedures, including a description of needed inputs, a statistical package that included regressions, a set of guidelines for interpreting and evaluating the computer output, and criteria and formula for adjusting salaries. Announcements would be made in professional newsletters and journals, and results would be published.[13]

Betty was motivated. She returned to campus and made inquiries about doing a pilot project at UC-Berkeley. She was optimistic the chancellor's office would agree, because they had to do it for their affirmative action program anyway. Betty also advocated that the project include librarians, research faculty, and non-ladder teaching faculty such as lecturers.[14]

30.1 Faculty Club Relationship Pains

Both faculty clubs at Berkeley used membership dues to subsidize the cost of lunches in order to keep prices competitive. Recognizing that women were eating more at the men's club than vice versa, The Faculty Club decided "unilaterally" to impose a 20% surcharge on the meal charges of WFC members. To Betty and the rest of the WFC board, the surcharge seemed peculiar in light of the continuing plans to merge the two clubs.[15,16] The result of the hard feelings surrounding the surcharge was that communications between the clubs were becoming more legalistic.[17] The letter that went from the president of the WFC to the president of The Faculty Club did not mince words: The move to add a surcharge was "a discriminatory effort at exclusiveness, and, certainly in regard to the members of the Women's Faculty Club, as a sexual discrimination in attitude and its effect – just the opposite to efforts at good faith towards development of one Faculty Center."[18]

At the May meeting of the WFC board, Betty reminded the board that The Faculty Club dues were quite a bit higher than the WFC dues. At what level would the dues for The Faculty Center be set? Women's salaries were lower, and Betty worried that women would be paying a larger proportion of their salaries as dues. The board agreed that the dues were an important question, in addition to The Faculty Club's debt.[19] Around this same time,

---

[13]4/16/1974 handwritten notes by Betty Scott at the first meeting of the AAUP advisory committee to plan the development of the Kit.

[14]4/21/1975 letter from Betty Scott to AAUP Staff Member Maryse Eymonerie.

[15]1/1975 Newsletter, WFC, UC-Berkeley, p. 1.

[16]3/3/1975 Minutes of the Meeting, Board of Directors, WFC, UC-Berkeley.

[17]4/7/1975 Minutes of the Meeting, Board of Directors, WFC, UC-Berkeley.

[18]4/16/1975 letter from WFC President Margaret D. Uridge to The Faculty Club President and Chemical Engineering Professor Alex T. Bell.

[19]5/5/1975 Minutes of the Meeting, Board of Directors, WFC, UC-Berkeley.

a joint committee was formed, with members from both clubs, to develop operating principles, a charge, and bylaws for The Faculty Center.[20]

By the July meeting of the WFC board, Betty wanted to resign. At the meeting, possible replacements for Betty were discussed.[21]

When October came, there was a draft merger proposal. The merged clubs would be called The Berkeley Faculty Club. The Faculty Club building would be called the West building, and the WFC building would be called the East building. All debts and assets of both clubs would be assumed by the new club, except for the restricted savings accounts of the WFC, which would be used only for the East building. Members of both clubs would become members of the new club. The draft merger proposal had an air of tentativeness about it; at the end of the document, two alternatives to a merger were outlined.[22]

30.2   UC Progress and Problems

The UC affirmative action report described progress between 1973 and 1974 in decreasing the underutilization of women and minorities. Specifically, the growth in positions for women and minorities was greater than the overall growth in positions, and the proportions of new appointments to women and minorities exceeded their availability. But the most significant news was that the OCR's two-year review of Berkeley's affirmative action plan was completed. The campus was found to be in compliance with federal executive orders for nondiscriminatory hiring and promotion on February 18, 1975.[23]

LAW, which filed the original complaint with HEW in 1971, wasn't happy with the HEW acceptance, charging that the availability pool data used in the plan were "either so old or so limited that they were in fact based on existing discrimination practiced at other campuses." UC administrators pledged to use better data when they became available. The newspaper reported that, at the end of the day, the UC was exhausted rather than triumphant.

HEW described the plan as "sound and viable," one that could be a model for the nation. UC was required to tighten up its "utilization analysis." Thirty-four of UC-Berkeley's 75 departments were given goals to hire more women or minority faculty, the numbers and timetables being different for each department, depending on the numbers of women in the availability pools. If the goals were not met, HEW would further examine UC hiring.[24]

---

[20]"May – June, 1975," *The Faculty Club: A Monthly Newsletter for The Faculty Club of the University of California, Berkeley*, no. 4 (1975).

[21]7/7/1975 Minutes of the Meeting, Board of Directors, WFC, UC-Berkeley.

[22]10/15/1975 Merger proposal [The Faculty Club and the WFC], Joint Merger Committee.

[23]UC Report, "III. Affirmative Action Personnel Programs," [?1975].

[24]Larry Spears, "Berkeley's action affirmed by HEW; Labor remains," *California Monthly*, March 1975.

In mid April, Betty wrote a memo to the VC critiquing the amendment of the affirmative action personnel program. She declared the amended definition of underutilization was "not a satisfactory procedure. I hope that it will be revised so as to be fair to minorities, to women in small departments, and to women in fields where the proportion of women is not large... Furthermore, I hope that the availability will be revised so as to be realistic. Betty thought the result of the unsatisfactory procedures would be that departments wouldn't be encouraged to search any harder for women and minorities, and small departments or departments with small proportions of women or minorities that had been discriminating would continue to be able to do so. Betty emphasized, as she had done before, that procedures needed to be carefully examined "using reasonable availability pools which will be larger than the pools shown" to see how they would work in practice. Betty articulated the questions in statistical terms, trying to get the administration to understand what was needed:

> *(a) How frequently will the procedure ignore cases where minorities and women should be employed but are not considered (an error of the first kind)? (b) How frequently will the procedure point to a Department as underutilizing minorities or women when in fact it is not with the apparent disparities due only to chance decisions (an error of the second kind)? Both of these errors will be possible under the present amendment. As the paragraph following the definition of Underutilization notes, with the present amendment, minorities will almost never be found to be underutilized and women rarely in small departments or fields where women are few. In order to be able to find the cases where women and minorities should be employed but are not, the procedure should be adapted to the size of the unit and the proportion of women and/or minorities (not knock off 1.0 across the board), the availability pool should be realistic, and the units should be looked at individually and also grouped into larger units in a natural way (in the way appointments are made and approved). On the other hand, under the present amendment, there is a nonzero probability that a department may be listed as underutilizing women when due to chance alone it has been unlucky in its efforts (trying to employ a woman who goes to Harvard instead, for example). The frequency of these second kind of errors should be small under the present amendment, very small. Since these are goals only and the Department is in no way harmed thereby (except perhaps by paperwork), I believe that this frequency could be increased markedly without hardship. Certainly, a more reasonable cutoff than 1.0, use of more accurate pools, and the combining of units as well as individual attention would be proper.*

Betty had a strong statistical opinion, and she spoke truth to power.[25]

On November 4, UC would issue a new set of affirmative action guide-

---

[25] 4/18/1975 memorandum from Betty Scott to VC Ira Michael Heyman re "Amendment of the Affirmative Action Personnel Program: 1. The Definition of Underutilization and Reformulated Goals and Timetables."

lines prepared in response to its meetings earlier in the year with OCR. The campuses would have to revise their affirmative action programs accordingly. Some consequences of this process were that a firm employee data base was established and standards for reporting were set so that annual data could be compared over time. UC made it clear to the campuses that it was tired of just gathering data, working on definitions, and making reports on affirmative action. It wanted to see a shift to "action-oriented programs," for example in the areas of recruitment, faculty development, and staff development; and more state support for its programs so they could be continued and expanded.[26]

In the previous year, Betty, Chin Long Chiang, and F.N. David as members of the SAC had submitted a proposal to the UC administration to conduct a salary equity study. The administration found the proposal useful toward setting up such a program on all campuses. Now they asked to convene the SAC again to update them and ask for additional suggestions.[27] The VC and VP acknowledged to HEW that they agreed with the SAC's point that the statisticians' role was not to define underutilization. But an article that appeared in the *New York Times* ran Berkeley through the mud over the issue, without interviewing either Betty, Chiang, or FND. The article sufficiently outraged the three statisticians that they wrote a letter of response, calling the article "abrasive," "not an accurate or fair account of Berkeley's efforts at affirmative action," and "wrong and inappropriate." Betty even wrote onto the draft: "She [the author of the article] just ladles out the muck. Such yellow journalism is not worthy of the *New York Times*. Let Berkeley now go forward in peace!" This is what Betty really thought, but she struck the sentences before sending the letter to the *Times*.[28,29,30]

30.3   Berkeley Pilot Study

Next Betty wrote a detailed proposal to the VC to start a "flagging" study at the UC campuses. The proposal, which Eymonerie described as "superb,"[31] was related to the AAUP Kit efforts: It was independent of the SEC, but endorsed by its chair. Specifically, Betty proposed to estimate the salary of

---

[26]UC Report, "III. Affirmative Action Personnel Programs," [?1975].

[27]3/17/1975 memorandum from VPAA Angus E. Taylor to SAC members C.L. Chiang, Betty Scott, and F.N. David re proposal for academic salary inequity study.

[28]C.L. Chiang, F.N. David, and E.L. Scott, "Draft of letters: Statistically speaking," [May or June] 1975.

[29]C.L. Chiang, F.N. David, and E.L. Scott, "Letters: Statistically speaking," *The New York Times*, June 1975.

[30]Cindy Kadonaga, "N.Y. Times report on bias biased?" *The Daily Californian*, May 1975.

[31]5/27/1975 memorandum from AAUP Staff Member Maryse Eymonerie to AAUP Committee Z members re status report.

a woman if she had been a white male with the same experience and ability, "flag" individuals who had an apparent salary inequity, and estimate the amount of "apparent inequity"; these individuals would be referred to the budget committee or administration for additional review. Accurate flagging would be easier than trying to get accurate estimates of salary. Other universities had already conducted salary equity studies using a variety of methods that had different costs and reliability. Betty wanted to compare these methods using UC data. As she explained to the VC, "It is convenient to employ a method that is easy to use on data already available (such as field, age, highest degree with its data, etc. as predictors in a linear equation). On the other hand, it is important to have a method that will work well in the sense that persons who have an appreciable salary inequity will be flagged with high probability, that persons who have no inequity will not be flagged too often because this would waste the time of the Budget Committee and the Administration in useless review, and that will be regarded as fair and reasonable." Betty had already discussed the problem with a number of UC administrators, the chair of the SEC, and FND. She proposed to start with faculty at the rank of lecturer and above at Berkeley.

There were basically three categories of "flagging" methods, and Betty described them in detail. (1) A "pairing" method, used for example at the University of Toronto, involved pairing a woman with several peer men; when her salary was less than the average of the men, the case was referred for review. (2) A "regression at the institution" method, used for example by Betty in her CCHE study, involved creating a linear model to estimate the salary of white males and use them to estimate the salary of a woman; when her salary was less than predicted, the case was referred for review. (3) A "peer-group regression analysis" method, a variation of the second method, involved assuming the UC campus salaries were approximately equal to those of the universities in the same Carnegie category (e.g., Research I), and then using the regression coefficients from Betty's CCHE study while adjusting the constant term to account for changes in UC salaries since the survey was conducted. Betty asked the VC for some help in comparing the three methods. She needed the administration office to collect the data from the individual personnel files, and to provide funding for computing and programming.

Betty expected the university would benefit by gaining estimates of the reliability, accuracy, and costs of the different methods. She thought she should do this work because: "(i) flagging is required already but almost no work has been done to find efficient reliable [sic] methods for flagging; (ii) some flagging is now being done by methods that are suspect – flaggings without statements of reliability and accuracy are pretty worthless, indeed flagging without sufficient reliability is harmful; (iii) I have some experience with such studies as well as the necessary skills; (iv) I want to cooperate with the various committees and authorities who are already required to do the flagging." Betty proposed to prepare a summary statistical report to the VC and relevant committees and authorities, and to make the report available to other institutions, the AAUP, and other interested parties. She then wrote: "I must apologize for the length of this letter. I have tried to make everything clear." [32]

One of the people copied on Betty's letter was William Kruskal. He was glad, but not surprised, that Betty was continuing this line of research, declaring that her work would "mightily help to keep it [the inquiry] from becoming the construction of a mechanical device":

> *I had fears about [the Kit becoming a mechanical device], especially when I read the language of the AAUP announcement that spoke of a "kit" – a word that connotes to me a carpenter's kit or a surveyor's kit. Still, no matter how wisely a flagging procedure is set up, my experience is that it tends to become less flagging in nature and more flogging or mandatory, as its use becomes routinized and it spreads to the hands of individuals who do not know its background. (A good analog in the development of statistics is the rigid, superficial, and distorted exposition of Neyman-Pearson testing theory in some contemporary textbooks.) One consequence of establishing flagging procedures like those you describe is, I predict, that some others – non-minority males – who feel their salaries unfairly low will ask to be fed into the flagging machinery. Some will allege discrimination of various kinds, e.g., against physically handicapped people, or easterners, or whatever. Since about half the non-minority males will be below the regression estimates, that could amount to lots of flagging and lots of review. Presumably you will consider, at least as a start, variant methods in which the flag goes up only if the difference between regression and actual salary is larger than (say) $1500, or maybe 10%. Choice would doubtless depend on intrinsic residual scatter, and on review costs.*

The amount of thought that Kruskal put into the flagging was noteworthy. He wondered about the rhetoric that would result when the procedures were used to highlight "remarkably high salaries." He thought it would be important to examine the robustness of the methods to changes in details. He worried that data errors could cause troubles, especially when the regression methods were used by novices. He reminded Betty that there could be tricky differences in fringe benefits that would affect comparisons across institutions. The bottom line, Kruskal told her, was that "your penultimate paragraph is great, and I send special huzzahs for [your] point. . . on needing information about reliability and accuracy." He told Betty he would send copies of her letter to Chiang, FND, and other "interested friends." He closed by saying: "I gather that Fred Mosteller has not persuaded you of the virtues of the title 'Chair.'" [33]

---

[32] 5/15/1975 letter from Betty Scott to VC Ira Michael Heyman.

[33] 5/27/1975 letter from University of Chicago Statistics Professor William Kruskal to Betty Scott.

30.4 Initial AAUP Kit Progress

The VC gave Betty $500 to compare the methods. He thought he could get her the needed data from the individual personnel files. But he was less clear about providing her with funding: "The first priority of the campus must be in recruitment and appointment...I am aware, of course, of the value of an effective mechanism for highlighting equity needs, but it does not rank at the top of the list." They should talk further about funding after Betty got a better sense of the real costs.[34] A few weeks later, the VC authorized $1,500 through the university affirmative action office for her project.[35]

In July, Betty and her team completed a draft report based on trial runs of CCHE data and UC-Berkeley statistics department data. Next they planned to do anthropology department data, noting there were a few women on that faculty, unlike most departments where there was only one or none. Betty noted that, with fewer predictors, more women are flagged, but that:

> ... the R is 0.8 and does not go down much with fewer predictors. What if add [sic] some our [sic] University claims to use, such as university service, professional standing (probably do not have data on this), etc.? I now doubt that we will ever get the R big enough that will flag well even if the sex bias is as much as $3000 per year.. [sic] We are now also thinking of looking at progress up the salary ladder through the years. Probably can only get data for last 10 years so what if already biased then [sic]. However, some of our predictors are biased in the regression method.

Betty complained about the quality/accessibility of the Berkeley data.[36]

In mid July, Betty reported to Pat St. Lawrence, who was still heading the SEC, that there had been some hoops to jump through to get the anthropology data, but Betty's team had succeeded. They had tried adding actual, rather than grouped, number of papers as in the CCHE study, and this had "worked wonders" for prediction: There was "excellent" prediction with this variable, together with only age or year of PhD. They also had tried adding number of PhD committees chaired, but this did not help the prediction.

By the end of August, Betty's team had updated their report to include the anthropology data. They used two methods. (i) Regression methods involved first computing regression formulas for white male salaries in a department (statistics or anthropology). A number of regressions were computed, with different predictor variables. The data collection form included many variables that could be used for prediction: department, faculty rank, annual salary, year of birth, highest degree (BA, MA, PhD), year of highest degree, date of hire

---

[34] 5/22/1975 letter from VC Ira Michael Heyman to Betty Scott.

[35] 6/13/1975 letter from UC-Berkeley Faculty Assistant for Affirmative Action Professor Olly Wilson to Betty Scott.

[36] 7/9/1975 letter from Betty Scott to [Kit consulting statistician].

at UC, number of papers published, number of books published or edited, percent of administrative activity, number of hours teaching per week, salary base (9/10 or 11/12 months), full or part time, male or female, race, number of PhD students produced, number of years in academe, senate committee memberships in the past ten years, administrative positions in the past ten years, and professional society activity. The resulting regression formulas were compared with analogous formulas from Betty's CCHE study (Bio/Phy for statistics, and Social Science for anthropology). The CCHE data were from 1970, and so the team adjusted the CCHE predictions for the 30% increase in UC-Berkeley salaries between 1970 and 1975. The Berkeley and CCHE formulas were then used to predict the female and minority salaries, and the team studied the residuals. (ii) Graphics methods involved plotting the rank of each faculty member in a department against the years since completion of PhD. If there was discrimination, the team argued, then the plotted lines for men will be mostly above the plotted lines for females and minorities.[37,38]

30.5   Federal Testimony on Affirmative Action

Back in May, the US House Subcommittee on Equal Employment Opportunity was scheduling hearings on the civil rights enforcement policies and procedures of relevant federal agencies: HEW, the US Department of Labor, and the US Civil Service Commission. The ASUC leadership thought of Betty. They asked her if she would consider providing some expert testimony in these federal public hearings: "Because of your expertise and historical involvement with Affirmative Action, I have taken the liberty of suggesting that you may be willing to testify before these committees."[39]

This was the kind of request to which Betty had become accustomed to saying "yes," and so she did, although she testified in writing rather than going to Washington, DC in person "since Berkeley is very far from Washington," as she told the labor department.[40] Betty's testimony was a four-page summary of her findings to date. She addressed, in relationship to universities, five points that were included in the "Notice of Hearing."

(1) Regarding "variations in the susceptibility of types of employment to uniform or quantifiable methods of evaluating and predicting performance." Betty's statement said there were good and accurate systems for evaluating faculty performance. It was possible to very accurately estimate salaries. Stud-

---

[37] 7/11/1975 letter from [presumably] Betty Scott to [UC-Berkeley Genetics Associate Professor Patricia St. Lawrence].

[38] "Salary Inequity Study" (draft report; UC-Berkeley), August 1975.

[39] 5/27/1975 letter from ASUC Academic Affairs Vice Presidents Edmundo Anchondo and Don Gillies to Betty Scott.

[40] 8/19/1975 letter from Betty Scott to US Department of Labor, Federal Contract Compliance Director Philip J. Davis.

ies showed that, when compared with men of equivalent training or position, women performed at least as well as men but tended to be paid less (about $1,800 a year less, on average, or over $7,000 in 2010 dollars). Affirmative action had not improved the salary inequities, nor the overall percentage of women on the faculty (which was still less than 3% in the highest ranked universities). Only a small percentage of the highest salaries went to women (only one of the highest 100 at Berkeley). Betty attached copies of her studies as documentation. [41]

(2) Regarding "variations in policies of recruitment and advancement and in other personnel practices." Betty's statement said: The percentage of women being hired as new faculty members was smaller than their percentage in the availability pool. The percentage of women among assistant professors was increasing, but their starting salaries were still lower than for men, and women got promoted less often.

(3) Regarding "methodologies usually used by institutions of higher education in the development of written affirmative action programs..." Betty's assessment was harsh. The campus appointed special or ad hoc committees on affirmative actions that had little or no effect. The plans were written by the administration and their counsels with the objective of having "the weakest possible plan with the weakest enforcement." The plans were not routed through the committee processes before submission to the government. UC-Berkeley was not the only university to have these "bad methods" of development. Betty wrote: "In summary, the method of development used [in affirmative action planning] was bad in attitude and in substance."

(4) Regarding "detail and adequacy of pertinent statistical data." Betty had a very negative opinion about the UC-Berkeley data: It was "inaccurate and incomplete," even as presented to the government. She continued: "A lot of money was spent (wasted) and there is little to show for it."

(5) Regarding "special circumstances, if any, in higher education..." and "other information relevant to achieving positive, results-oriented equal opportunities..." Here Betty was adamant and showed emotion:

> *There are no [sic] special circumstances in higher education that absolve it from affirmative action (I am very tired of hearing that fair employment is fine for every other institution except universities and colleges – this is false. I am sorry to say that affirmative action is very much needed in higher education. It is too bad for them still that they ignore women and minorities of talent – they will not employ them and do not pay them fairly. By ignoring some 40% of the talent, universities are not getting the best person to teach and to do research.) To achieve positive, results-oriented equal opportunity requires administrators who are positive and results oriented towards equal employment opportunities. Such persons*

---

[41]Note: The attached papers included "Women in Higher Education – The Facts of the Matter," "The Status of Academic Women on the Berkeley Campus," and "Application of Multivariate Regression to Studies of Salary Differences between Men and Women."

need an adequate results oriented staff, need to consult the committees and experts concerned before presenting programs. When they have drawn up their programs, the university should appoint a [sic] sympathetic, positive, results-oriented administrators to guide the program. At this university and at others there are striking differences within the university, and from time to time just according to who is in charge and carrying the ball. I think that the only changes in the approaches that are required are: the employment of positive, results-oriented administrators (with strong staffs) in the universities and in Washington throughout.[42]

Put simply, Betty concluded that affirmative action is a matter of leadership.

30.6 AAUP Pilot Studies Progress

Response to the AAUP's call for assistance to develop the Kit was encouraging. Administrations, AAUP chapters, or faculty committees at 16 institutions offered to cooperate and be considered as pilot sites. Eymonerie hoped the advisory committee would select the pilot study sites by mid October. She expressed the need to move faster on development of the Kit:[43] "Time continues to pass so rapidly and so little gets done... at least on my side."[44]

Faculty at several other universities conducted AAUP-funded pilot studies over the summer.[45,46] Betty's CCHE study was the most extensive salary analysis to date using linear regression methods, and she had already conducted some pilot studies at Berkeley. Now Betty was interested in trying out her methods on some other UC campuses (Riverside and San Francisco in particular), as well as other non-UC campuses: "I hope so even though more work would be needed."[47] Her try-outs on other campuses would provide indications of which methods are the most efficient, "in the sense that it gives good predications at the least cost."[48]

Betty sent a copy of her Berkeley report to St. Lawrence, saying: "The

---

[42]8/20/1975 statement of Betty Scott in public federal testimony re "implementation of the affirmative action requirement of the Executive Order as applied to employment at institutions of higher education."

[43]9/16/1975 memorandum from AAUP Staff Member Maryse Eymonerie to the advisory board (Exxon project) re the Kit consulting statistician's effort report.

[44]9/17/1975 letter from AAUP Staff Member Maryse Eymonerie to Betty Scott.

[45][Kit consulting statistician], "Effort Report to Exxon Education Foundation on the Development of a Higher Education Salary Kit," September 1975.

[46]9/19/1975 letter from Michigan State University Economics Assistant Professor Daniel H. Saks to AAUP Associate Secretary Maryse Eymonerie.

[47]9/20/1975 letter from Betty Scott to AAUP Staff Member Maryse Eymonerie and [Kit consulting statistician].

[48]9/20/1975 letter from Betty Scott to UC-Riverside Statistics Department Dr. Charles Huszar.

conclusion is not completely clear." Basically, Betty's team had a good, simple method that worked for statistics and anthropology department data. But they didn't have a good method for English department data, where "the standard predictors are not enough." Betty's English department colleague, Josephine Miles, suggested two additional factors be considered: (1) quality of the published book, and (2) difficulty younger people had publishing both books and peer-reviewed journal articles given the current climate. Betty suggested including an interaction term to account for the fact that younger people were expected to publish less than their older colleagues did at the same age.[49] St. Lawrence told the SEC of Betty's efforts, which she saw as "essential to the task of our committee."[50]

Advisory committee member Bergmann also led a pilot study toward development of a prototype Kit, this one at the University of Maryland – College Park. The study was published in the Autumn issue of the *AAUP Bulletin*. Bergmann and her collaborator reported that many academic salary equity studies had been conducted around the country, and in all cases women faculty were shown to be paid, on average, less then men; and there were unexplained differences in salaries after relevant factors had been taken into account. The Bergman model to predict salary had only three predictors: Year since degree, whether the person had a doctorate, and a department differential (dollars per year since degree to be added or subtracted depending on the department). Note that none of the predictors had to do with productivity or performance. Thus the model explained only 52% of the variance in men's salaries. Yet, the study illustrated to other institutions how to use administrative data and a regression equation to predict women's salaries based on men's salaries, with a suggestion of how to use the results of such a procedure to communicate salary equity issues to the institution's women faculty, administrators, and constituencies.[51]

By early October, the number of institutions offering assistance in terms of knowledge, expertise, or data sharing was "overwhelming." AAUP was on the fast track, hoping to complete the Kit in winter or early spring.[52]

---

[49]9/20/1975 letter from Betty Scott to UC-Berkeley Genetics Professor Patricia St. Lawrence. Note: Betty's assistant Amy Davis, who earned her PhD under UC-Berkeley Statistics Professor Eric Lehmann on "Robust Measures of Association," carried out the study and drafted the report.

[50]9/26/1975 memorandum from SEC Chair Patricia St. Lawrence to members re extension of the life of the committee and plans to have Betty speak to the committee.

[51]Barbara R. Bergmann and Myles Maxfield Jr., "How to analyze the fairness of faculty women's salaries on your own campus," *AAUP Bulletin*, Autumn 1975.

[52]Scott, Elizabeth L. "An Interview with Elizabeth Scott," an oral history conducted by Suzanne B. Riess, in "The Women's Faculty Club of the University of California, Berkeley, 1919-1982," Oral History Center, The Bancroft Library, University of California, Berkeley, 1983.

30.7   Equal Opportunity Data

By year's end, Betty had published a chapter in a book on *Women in Academia – Evolving Policies Toward Equal Opportunities*. The chapter discussed the problems of getting good data to set goals and then assess progress toward both long- and short-range goals. She discussed attaining equity in salary, and sex differences in intellectual abilities. Including many tables and graphs, Betty made the case, as recognized by the federal government, that equal opportunity research had to be statistical rather than based on personal histories, as the latter could be very biased.[53,54]

The UC-Berkeley academic senate budget committee published their annual report. It included some annual trend data. The percentages of women in ladder appointments in the six academic years between 1969–70 and 1974–75 were 4%, 14%, 13%, 24%, 24%, and 17%. In the past year, 63 men had been appointed to the professor series, and 13 women; 10 men had been appointed to the ranks of professor or associate professor, and 1 woman.[55,56] This was something to write home about. But in Betty's mind it was not nearly enough.

One of Betty's Policy CSAW collaborators, Lucy Sells, had done a study of incoming UC-Berkeley freshman showing only 8% of girls had four years of high school math in comparison with 57% of boys, and concluding girls had an aversion to math that held them back later in many academic fields and especially the sciences. An important result of Sells' research was a new program for girls at the Lawrence Hall of Science, located above the UC-Berkeley campus, called "Math for Girls." The afterschool program for girls 8 to 13 years old was designed to get them excited about learning math so that later at entrance to college they didn't have to take remedial math in order to try to catch up. As a result of the program, the percentage of girls having high school math in 1974 had tripled, rising from 8% to 24%.[57]

---

[53]Elizabeth L. Scott, "Developing criteria and measures of equal opportunities for women," in *Women in Academia – Evolving Policies Toward Equal Opportunities*, ed. E. Wasserman, A.Y. Lewin, and L.H. Bleiweis (New York: Praeger Publishers, 1975), 82.

[54]Martin A. Trow, "Appendix – Carnegie Commission on Higher Education National Survey of Faculty and Student Opinion," in *Teachers and Students*, ed. Martin A. Trow (New York: McGraw-Hill, 1975).

[55]11/24/1975 UC-Berkeley academic senate budget report presented on pp. 10-11 of the agenda.

[56]Note: For 1975–76, the percentage of women in ladder appointments went up to 21%. See 10/11/1976 Senate Report Appendix: 1975–76 Budget Committee Report, Table 1.

[57]Alan Cline, "U.C.'s hard look for a lady prof," *San Francisco Sunday Examiner* and *San Francisco Chronicle*, April 20, 1975.

31

Developing the Kit (1976)

This is not my favorite research topic – I just feel I must keep working on it.
-Elizabeth L. Scott

At the turn of the new year into 1976, Betty received a letter from Maryse Eymonerie. It was somewhat cryptic, having something to do with preparation of the AAUP Kit. Eymonerie wanted to know if Betty needed any funding for "assistants" and wrote, "Keep track of your own time and do not worry about the 20 days..."[1] When the advisory committee, now called the advisory board, met in Washington, DC in February, Betty wrote into her notes: "Emergency is to write good progress report... ELS will prepare a very rough draft in 30 days."[2]

Thus it was decided. Betty, "...who has done extensive work on the subject, and is currently involved in a comprehensive study of salary inequities at the University of California...," would prepare the preliminary report and Kit, aided by Stat Lab staff in California, in time for the AAUP annual meeting at the end of June. It was anticipated that, because the project was complex, additional work would likely be required, perhaps even additional pilot studies, before the Kit would be completed.[3]

Betty was supportive of the COPSS Visiting Lecturer Program in Statistics, and she had been in demand as a visiting lecturer. One university wanted her "very insistently and almost uncompromisingly."[4] The program produced a brochure with the names and bios of the participating visiting lecturers. Betty listed her lecture topics (along with level of difficulty) as: "(1) How to estimate the increase in skin cancer that will result from depletion of ozone (all levels); (2) Does cloud seeding decrease rainfall? (intermediate); (3) Is the universe expanding? (intermediate or advanced); (4) Status of women in higher

---

[1] 1/3/1976 letter from AAUP Staff Member Maryse Eymonerie to Betty Scott.

[2] 2/19/1976 handwritten notes by Betty Scott re meeting of the Kit advisory board to hear the statistical consultant's progress report.

[3] 2/24/1976 letter from AAUP Staff Member Maryse Eymonerie to Exxon Education Foundation Executive Director Frederick deWolfe Bolman.

[4] 5/19/1976 letter from Visiting Lecturer Program in Statistics Director H.T. David to Betty Scott.

485

education – the facts of the matter (all levels)." Betty didn't want her Status of Women topic to be emphasized: "This is not my favorite research topic – I just feel that I must keep working on it," she told the program director.[5]

31.1   More UC Affirmative Action

The UC administration continued its efforts, reporting that there were more, and more intense, actions toward affirmative action. For faculty, there were model recruitment programs for women and minorities that included more extensive outreach, communication, and advertising; there were stepped-up efforts to recruit from within; and departments were given larger budgets for recruitment. For staff, there were intensified efforts to recruit, train, and promote women and minorities. Betty would have taken special note that the data collection for ethnic identity was changed from visual identification to self-identification, and that a task force was appointed to oversee UC's procedures and practices in data management. The university was now devoting $1.25 M each year to affirmative action.[6]

The proportion of women and minorities hired in 1973–74 was 24%, and in 1974–75 had declined to 17% in spite of all of these efforts. This had not gone unnoticed by the UC-Berkeley academic senate. SWEM's research had shown an increase in availability pools, and goals had made little or no difference. The VC questioned whether there should be incentives or sanctions when there was lack of compliance with goals and timetables.[7]

Betty had the payroll data and used it to answer a question posed by UC-Berkeley Sociology Professor Arlie Hochschild about the percentage of women in temporary positions. Betty was thorough. She gave Hochschild numbers of both men and women, visiting and acting, head count and full time equivalent, for different ranks, for three years. She also gave Hochschild the weighted average monthly salary rates. Betty discovered that the trends for women in both temporary and regular positions were similar: lower proportions of women in higher ranks, less pay on average for women, and little change over time.[8]

Betty was still working on a draft report to aid UC-Berkeley law school admissions. She reasoned it could not be determined from the data whether the school was doing well at selecting from white males or white females, because "we are observing only the tip of the distribution."[9]

---

[5] 8/6/1975 letter from Betty Scott to Visiting Lecturer Program in Statistics Director H.T. David.

[6] "V. University Personnel Programs, 1," *Affirmative Action Personnel Programs* (University of California, 1976), 14.

[7] 1/27/1976 minutes of the UC-Berkeley academic senate, SWEM, pp. vi-vii.

[8] 2/27/1976 letter from Betty Scott to UC-Berkeley Sociology Professor Arlie Hochschild.

[9] 2/9/1976 letter from Betty Scott to UC-Berkeley School of Law Employee Fran Layton.

SEC Chair Patricia St. Lawrence set out to help Betty get the information needed to do her next pilot study on salary equity. What Betty needed was more detailed information on each faculty member, some of which was in the public domain (rank, step, salary, etc.) and some of which was only available in individual confidential personnel files (e.g., numbers of papers and books published). St. Lawrence asked the department chairs for the information, but the VC "intervened" and suggested to the academic senate budget committee that it would be "more efficacious" for the administration to provide the necessary information. He asked for their guidance about which information to supply and how, while preserving as much confidentiality as possible. He told the committee that the SEC needed access to personally identifiable information, at least St. Lawrence thought so.[10]

Perhaps not surprisingly, there was push back from some of the chairs who did not want to hand over the precise compensation of faculty in their departments. The chairs maintained this information was not public information; St. Lawrence thought the contrary. She asked the VC to solicit a legal opinion.[11]

If obtaining productivity information was important locally, it was even more important now that Betty had taken over the design and development of the national AAUP Kit. A week later, St. Lawrence reported back to the SEC on the status of attempts to obtain the productivity data. At a meeting, the VC had been "most supportive" and "anxious to find a way." Seven of the 19 departments solicited for information had responded. But: "Because of the freeze on Bio-Bibliographical forms and the concerns over our attempts to get information from departments, I have suspended operations. I shall try to find out this week whether there has been progress in resolving the matter," she told the committee. All things considered, she was pessimistic that the committee could complete its work that year, and thought the life of the committee should be extended.[12,13]

31.2   Faculty Club Merger Declined

By spring, there were about 2,000 members of The Faculty Club, and about 350 members of the WFC. Some of the women were calling for abandonment of a merger and exploration of alternatives. The proposal for a merger of the two clubs continued to be massaged, as people continued to question the specific

---

[10]3/4/1976 memorandum from VC Ira Michael Heyman to the UC-Berkeley academic senate budget committee re supplying the data needed for Betty's salary equity pilot studies.

[11]2/13/1976 letter from SEC Chair Patricia St. Lawrence to VC Ira Michael Heyman.

[12]3/4/1976 memorandum from VC Ira Michael Heyman to the budget committee re confidentiality issues in releasing faculty productivity information for a pilot salary study.

[13]3/15/1976 memorandum from SEC Chair Patricia St. Lawrence to UC-Berkeley Professors B. Heyns, S. Ervin-Tripp, J. Neyman, J. Noonan, and Betty Scott re present state of the salary equity study.

terms and asked for articulation of the benefits to the membership. In May, there was a call for votes by both the WFC and The Faculty Club.[14],[15]

Betty did resign from the WFC board of directors, a year before her two-year term expired. It is not clear why. Perhaps she just had enough.

However, Betty was still engaged. In July, after having left the board, Betty weighed in on the proposed agreement, directing her comments to the new board. She felt "very strongly" that the two clubs should merge but didn't like everything she saw in the proposed agreement, finding some of it to be "not workable" and having the potential to "destroy the merged Club" by causing a decline in membership. "We should take advantage of the merger to implement innovative advances that will extend and deepen the membership and the activities of the new Club, rather than submerging the Women's Faculty Club in the (Men's) Faculty Club and then changing its name to Berkeley Faculty Club," she wrote.

Betty saw two problems. One was that the membership qualifications were too restricted – this was at the core of the merger difficulties. The initiation fee was too high: "A high fee will be more harmful to women than to men because their salaries are lower. I regard the present proposal as discriminatory," she wrote. Also, there was a statement that said "new members should have to make a special contribution to pay for the facilities." This, Betty said, wasn't correct "and turns my stomach." She went on:

> *The Women's Faculty Club has no outstanding debt; there is nothing to pay any more except current maintenance and everyone should share equally the maintenance costs... The Men's Faculty Club does have a large debt outstanding but why should a new member have to pay anything special for earlier mismanagement? If past blunders must be paid for, then the costs should be shared equally. Statements that members of the Women's Faculty Club should be glad to assume one-third of the debt of the Men's Faculty Club in order to enter that facility, or that they should be taxed to enter the Men's Clubhouse, are not correct to put it mildly. While there have been members of the Women's Faculty Club (such as myself) who have been refused admittance or have been bodily and publicly ejected from the Men's Clubhouse because they are women, we can have no more nonsense about that building.*

In Betty's mind, the merger agreement should encourage new members. She even suggested some ways to encourage, e.g., have department members or spouse members. She suggested: Why not have a study to identify persons who signed someone else's name, asking why they don't join themselves?

The other problem in Betty's mind was that the operations were essentially the same as used under the joint operation and that had resulted in "errors" and "friction" which alienated members of the WFC: "The merger needs to be drawn so that no one will be asked to suffer such experiences a second time,"

---

[14]4/1976 WFC Newsletter, pp. 2-4.
[15]5/1976 The Faculty Club Newsletter 8:1, 4.

she wrote. She also thought the proposed agreement was too sketchy and more details needed to be spelled out, e.g., protections of the special funds for each club, establishment of two building funds, etc.[16]

Other women seconded some of Betty's opinions and added some opinions of their own, protesting the details of the merger, but not the merger itself. In the end, when members of the WFC would vote on December 1, the vote would be close and the proposed merger would fail (137 for, 142 against, with 9 abstentions).[17] The members of The Faculty Club had already met in October and were short 164 votes in order to approve the merger.

These votes marked the beginning of a time where the key word was "cooperation" between the two clubs.[18] At the time this book was written, there were still two faculty clubs, The Faculty Club and the Women's Faculty Club, although both had been opened to include both men and women members.

31.3   More Reports, Advocacy, Testimony

Reporting, advocating, and testifying were now at the core of Betty's academic existence. She was using every opportunity to advocate for women nationally.

In March, Betty complained that the ASA "Roster of Women in Statistics" only included their members and a few statistics students: This, she told the ASA executive director, was "shocking," "misleading and harmful" because the rosters were used in job searches and affirmative action availability counts. "You have harmed women statisticians by your actions," she wrote. Then she also chewed him out because the list of statistics departments in a previous year's edition of *The American Statistician* omitted the UC-Berkeley department, one of the largest in the country. She admonished the executive director: "I emphasize that this is not the first time that I have written to the ASA about this list...I repeat that the statistical associations have an obligation to send out accurate statistics about themselves and their members. One could argue that statisticians have more reason than other professions to provide accurate statistics. I urge you to take strong action to have the lists sent out from your office to be accurate and complete, once and for all."[19]

In April, Betty concluded some expert witness work for a San Francisco law firm working on gender equity in academe. The law firm used the CSAW report, and it declared her help to be "invaluable": "You were an outstanding

---

[16]7/10/1976 memorandum from Betty Scott to members of the board of directors of the WFC re "Proposed Merger Agreement and Proposed By-Laws," 4 pp.

[17]12/1976 WFC Newsletter, 1.

[18]1/1977 WFC Newsletter, 2.

[19]3/9/1976 letter from Betty Scott to ASA Executive Director Fred C. Leone.

expert witness for our side of the case, and it was a great pleasure to work with you. Our gratitude can hardly be adequately expressed."[20]

In June, Betty pushed the National Institute of Environmental Health Sciences (NIEHS) to advertise their branch chief positions: "Since HEW makes universities search openly, why not NIH?" she wrote. She also pushed them to nationally advertise summer jobs for students, and post doc positions. In a competitive moment, she wrote: "It really makes me cross to find three North Carolina statistics-type graduate students with jobs this summer of which California students were never even informed...I assert that Berkeley students are better than North Carolina students, any summer!" Betty also pushed for more women and minority scientists to be hired by NIEHS as regular staff, pointing out there were "essentially none" at the time.[21]

31.4   Berkeley Statistics Department Self-Evaluation

The UC-Berkeley administration decided to require each department to do an affirmative action self-evaluation. The statistics department's was drafted by then department chair Eric Lehmann. It said comments had been invited by department staff, students, faculty, affirmative action committee chair, personnel committee chair, Stat Lab director, and others. Lehmann wrote into the draft: "There seems to be unanimous agreement that this Department does not practice discrimination on the basis of sex... There is a strong imbalance in that a relatively small percentage of women go into statistics. We try to increase this number by giving preference to women applicants in the case of several applicants who are equally strong. But on the whole, this problem needs to be addressed much earlier, in schools and in the home."[22] Lehmann circulated the draft, saying suggestions were welcome.

Betty took issue with the draft, and sent an emotional response:

> *We do NOTHING to favor women in any way that I am aware. In particular, we do not give women applicants for student admission any favoritism at all. We do not recruit women students specially – we do not even write them special letters. So far as I can see, we do not make special effort to search out women staff members or women faculty. We do look at their files when they apply. I think the department can do many things to search out women and minorities. I am trying several things and have been doing so for some time. It would be much more effective if*

---

[20]4/22/1976 letter from Law Offices of Howard, Prim, Rice, Nemerovski, Canady & Pollak to Betty Scott.

[21]6/15/1976 letter from Betty Scott to University of Washington School of Medicine Pathologist Earl Benditt.

[22]Prior to 5/14/1976 draft letter from UC-Berkeley Statistics Professor and Chair Eric Lehmann to UC-Berkeley Provost Dean Park.

many persons did so, not just a woman faculty member. Anyway, I am getting very tired and it is 1 am just as it was last night and the night before.

Why didn't the draft say anything about salary inequity? Betty asked.

> *... and why isn't this study [of salaries] finished and redress made – because of flack from a couple of powerful old-line departments such as chem brought the whole thing to a standstill. This should be mentioned also since now the administration is afraid of unleashing the white backlash. I am angry about this; it is not that women and minorities are asking for something special and extra. It is that the Senate voted to have a study of whether salaries are fair: Does a white woman get the same salary that a white male of the same ability and attributes would get in her department? The answer is, generally speaking, NO. So, let's get these low women flagged and have the Budget committee look at their files. There are too many to "explain" away.*

Betty continued with commentary about the poor quality of data being used by the chancellor's office as a result of cost-cutting efforts. She then raised some issues that she considered even more important to include in the draft:

> *You should say something about the bad things we know are going on: This year the unemployment rate for new PhD's is TEN times as much for women as for men in the sciences (including math and statistics), although percentage seeking employment is same. Percentage of PhD [sic] to women is way up across the US. The neglect of Julia Robinson through the many years. The New York Times writes that she is a Lecturer in Math Dept here (and so did many other papers, I am told). But she is not. She does not even have a mail box there! The University has not behaved properly, year after year.*

Betty then mentioned the distressing and "disgraceful" treatment of women, one in the mathematics department and another at the Lawrence Berkeley Laboratory, specifically referring to lies being publicized about them. The statistics and math departments were both housed in Evans Hall: "You may think we do not have to worry about what happens upstairs [in math], but I think we must; they are not improving on their own," Betty wrote.[23]

---

[23]5/12/1976 letter from Betty Scott to UC-Berkeley Statistics Professor and Chair Eric Lehmann.

31.5   Mills College Conferences

Mills College, a small women's college less than four miles from the Berkeley campus, hosted a one-day conference, "Educating Women for Science: A Continuous Spectrum." The April conference was sponsored by Stanford University's center for teaching and learning. The objective was to create a climate "where it is as natural for women to take courses in math and sciences as it has been in the past in the humanities and fine arts; where women can view careers in science and technology as exciting and viable opportunities." Eight Bay Area women scientists were featured on a panel about their careers, including a physiologist, biologist, chemist, engineer, mathematician, physician, physicist, and scientific administrator. These talks were followed by 14 discussion groups, e.g., Lucy Sells led a discussion on "Admission and Attrition of Women in the Sciences."[24]

Betty was a panelist representing mathematics. She talked about her family's move to Berkeley, her decision to study astronomy, and the problems of women using the big telescopes at Mount Wilson and Palomar Observatories. She made sure to point out that women did have fair access to all telescopes at Lick Observatory. She talked about her gradual move into statistics that began with the outbreak of World War II, and her continuing interest in astronomy, especially doing astronomy research that had a "statistical flavor." She talked about women not being afraid to tackle controversial problems, because they were not biased by interests in personal gains (which historically were rarely available to them). She talked about her controversial work in cloud seeding and then affirmative action, especially her work on salary equity. She described her work for the CCHE, saying "We find that systematically women tend to enter at a lower salary and to go up less often and at a slower rate... The outcomes are startling! Women have a long way to go to attain equity, and many persons need to pay attention to the underlying problems."[25]

The participants, including Betty, found the conference to be excellent. Role models were advanced, opportunities revealed, and interests expanded. Mentoring happened. Self confidence was built. Betty observed:

> *I was alarmed by the number of older women who spoke to me about the problems they are having in getting their papers published (extra layers of approval) and, even more, in holding their jobs in these hard times. I really have no concrete ideas other than that I can sometimes help individual cases on their merits. But I do think that local organizations of*

---

[24] *Proceedings of the Conference on Educating Women for Science: A Continuous Spectrum*, Mills College, Oakland, CA, April 24, 1976.

[25] Elizabeth L. Scott, "Panel Talk," *Proceedings of the Conference on Educating Women for Science: A Continuous Spectrum*, Mills College, Oakland, CA, April 24, 1976, 32.

*professional societies for women should be encouraged. With cooperation
and professional airing of problems, we may be more helpful.*

Betty hoped there would be future conferences and she could collaborate with
the conference organizers in the future.[26] At the next year's (1977) conference
an attendee, who indicated that her present decision for a major was civil
engineering, said she got much out of the conference: "Role models, information, encouragement, empathy, choices, friendship, job connections, motivation, stimulation, the list is <u>infinite</u>!"[27]

In July 1976, women leaders at the UC Lawrence Hall of Science asked
Betty if she would be willing to include her biosketch in a career guide they
were developing. The objective was to encourage women and minorities to
enter careers in science and technology. The guide would be made available
to young women at the next women's conference at Mills College, but the
organizers hoped to find some funding for a wider distribution.[28,29] In the
questionnaire, Betty listed her bachelor's and doctoral degrees in astronomy
from UC-Berkeley, together with "post-doctoral training in Statistics over the
years, Sorbonne and elsewhere." She described her varied and multi-directional
research, and explained her definition of statistics. She described statistics as
a new science with expanding opportunities, especially in health sciences and
other applied areas. She advised that there were "innumerable directions in
which a person may work," either alone or as a member of teams. She gave
women the same advice she gave to all: Get a strong background in statistics,
probability, one or more areas of application, and data analysis using statistical
software. With respect to being a woman statistician, Betty wrote: "The main
difficulty of being a woman in Statistics is the shortage of women in the field.
If you think about this you tend to feel left out, [sic] I hope that, as more
women enter the field, this problem will disappear."[30]

---

[26]5/15/1976 letter from Betty Scott to Stanford University Center for Teaching and
Learning Associate Director Jean Fetter.

[27]3/13/1977 conference evaluation by a woman who indicated civil engineering as her
present decision for a major, women in science conference, Mills College.

[28]7/20/1976 letter from UC Lawrence Hall of Science, Research Center in Science Education, staff members Nancy Kreinberg and Cathy Shufro to Betty Scott.

[29]9/10/1976 letter from UC Lawrence Hall of Science, Research Center in Science Education, staff member Nancy Kreinberg to Betty Scott.

[30]Elizabeth L. Scott, "Career Guide Project Questionnaire," *UC Lawrence Hall of Science*, September 24, 1976. Note: These descriptions are consistent with others described in
the chapter on Championing Science, namely, (1) [?1977] "Brief Sketch of Activities" by
Betty Scott; and (2) 3/5/1977 Elizabeth L. Scott. In: New Directions for Women: Portraits
of 21 Women in Science. Participants in a Conference at Yale University.

31.6   Fighting to Hold Gains on Campus

In June, F.N. David wrote to Betty articulating the dangers of backsliding: "It is necessary always to take the 'best qualified person' for any opening, so if a woman leaves the chances are she will be replaced by a man and so we go backwards. I wanted to get it somehow (1) that if a woman leaves she is replaced by another woman, (2) that if a man leaves he will be replaced by either a woman or a man. How else ever to get anywhere?"[31]

Shortly thereafter, on August 2, the *LA Times* published a story on the struggles of women at UC Berkeley. It was important to feature Berkeley, they said, because it was the top public research university in the state, and they were being looked to as a model for affirmative action. The article reported that, according to leaders of the women's movement, women had "made few gains in their efforts to attain equality in faculty hiring and promotion at UC Berkeley, and even their slender achievements are threatened by a tightening job market and increased male resistance." The article quoted Hochschild, now chair of the SWEM, as saying, "We are just holding the line for some of the gains made in the 1960s and in some cases not even that." Susan Ervin-Tripp, who had recently moved from the rhetoric department to the psychology department, was quoted as saying the latest budget committee report indicated "...there were actually five fewer women in the faculty Academic Senate last fall than there were the year before," and that "...there was a higher percentage of women on the Berkeley faculty in the 1920s and 30s than there is today." The article pointed out that the Berkeley situation was not unique: The AAUP survey results showed there was an overall drop in percentage of women in universities across the country, and there was a gap in salaries between comparably positioned men and women that was over 5%. The women leaders reportedly thought the UC-Berkeley chancellor was doing his best, but the tightening job market was taking its toll. Betty was quoted as saying: "Clark Kerr said we were 10 years too late and I'm afraid he was right."

The article included a picture of Betty at her desk, with the caption, "Alone in Statistics: Elizabeth L. Scott, professor of statistics at UC Berkeley, is the only woman member of her department." It said Betty "...could recall a time when discrimination against women was more blatant, when she was not allowed to use certain pieces of astronomical equipment for her statistical studies because she would take valuable research time away from men."[32] This is the first time Betty was reported as having been turned away from equipment, a report that she would repeat in a 1978 questionnaire.[33] The

---

[31] 6/23/1976 handwritten note from UC-Riverside Statistics Professor and Chair FN David to Betty Scott re new hires and the dangers of backsliding.

[32] William Trombley, "Faculty struggle: Women fight to hold gains at UC Berkeley," *Los Angeles Times*, August 2, 1976.

[33] 6/10/1978 letter from Mina Edelston, attendee at the UC-Berkeley Conference for Women in Engineering and Computer Science on 5/13/1978, to Betty Scott.

article also quoted Betty as saying "There has been an attitude change. Many more people are encouraging women to enter academic life."

With these positive changes came backlash. The article reported the opposition was getting stronger, and there was a "mustering of male forces, gathering beneath the banner of merit," levying charges of reverse discrimination. Women countered "...that quality never has been the most important consideration, that most faculty members in the major research universities have been hired through an 'old boy' system of mentors, proteges and contacts. The merit argument, the women say, has provided a convenient cover for white male professors who don't much like the idea of competing with women and minorities for jobs." Ervin-Tripp was quoted as saying, "The most depressing thing is that we have a backlash without much progress...There is a cycle on these issues – you have to have a lot of felt pain before people will work hard." The article then discussed life-work integration issues, noting that the American professorate was set up under the assumption that it was the men who worked, and they had wives who cared for the homes and children, and who also sometimes helped the men with their research.

Finally, the article reported gains made. There was now a maternity leave policy, allowing for an option for up to one quarter of leave; permanent SWEM committee; SEC; and women's studies program. There were departmental affirmative action officers. There was prodding by the administration for departments to have unbiased, open faculty searches. The women were discouraged, and thought these gains were small. But even though the momentum for change had slowed, they were not about to give up.[34]

31.7   First Draft of Kit

It was confirmed. Betty would commit between 10 and 20 days to complete a preliminary report that would include a preliminary Kit. She would earn $150/day ($568/day in 2010 dollars). In addition, she could bill for clerical assistance, computer time, and a programmer, for a cost up to $2,000 (about $7,500 in 2010 dollars).[35]

Betty presented a first draft of the report and delivered it to the advisory board in June. It indicated the report would have three parts:

I. Recommended method for "flagging" women and minorities for whom there is apparent salary inequity,

II. Comparison of results and costs of several suggested methods for flagging women and minorities with apparent salary inequities,

---

[34]William Trombley, "Faculty struggle: Women fight to hold gains at UC Berkeley," *Los Angeles Times*, August 2, 1976.

[35]3/24/1976 letter from AAUP Staff Member Maryse Eymonerie to Betty Scott.

III. Technical background of suggested methods for studying salary inequity.

Actually, the seven-page draft only included part I. A revision of part II was still being typed, and part III hadn't yet been written.

The draft part I contained information on "how the faculty should be grouped for a study of salary inequities," and "what data should be collected for each faculty member." Betty recommended the faculty be grouped into the unit that made appointment or promotion decisions, which was usually the department. Since the objective was not to produce perfect estimates, but rather to flag, she recommended that only a few simple salary predictors be used for which data were generally easy to obtain: year of birth, highest degree, year of highest degree, sex, and whether the person was a minority.[36]

At this juncture, Betty said this about the study: "I think of it as the kit for the layman who has access to a packaged least squares program. If he or she does not have access to such a program, I suggest that the AAUP or someone designated by them could run the programs for a modest fee." She also suggested including a completed example in the appendix of the Kit, to ensure people would have their data lined up properly and so forth. Flexibility was built into the Kit, so users could choose their "favorite method."

Betty regarded the draft as "very preliminary." The main stumbling block was getting good data, and getting complete data: "I insist that there is too much danger of biased results with incomplete data, due to selection," she wrote. Another stumbling block was finding "a good method that will work everywhere: Academe is getting very uptight," she observed.[37]

Mary Gray's comprehensive report on activities of Committee W, which she chaired from 1974–76, included a paragraph on the Kit. The Kit was "in final form" and would "soon be ready for distribution." There had been a workshop on remedying salary inequities at the last AAUP annual meeting. Gray commended Eymonerie for "marshaling the project through the grant phase to the present final draft stage"; Barbara Bergmann for "her preliminary work"; and Betty for "her efforts in compiling the necessary data and translating it into a final, organized form." Gray wrote: "We are hopeful that the salary survey kit will provide an impetus for salary equity reviews which will particularly benefit those academic women and others who have suffered economic loss as a result of salary discrimination."[38]

---

[36] 6/1976 Betty Scott's first draft of Part I of the preliminary report about the Kit.

[37] 6/11/1976 memorandum from Betty Scott to advisory board of the Exxon project re first draft of Part I of the preliminary report about the Kit.

[38] 1975–76 Report of Committee W [AAUP], submitted by Chairperson Mary W. Gray, 9 pp.

31.8 Rank as a Salary Predictor

In the early fall of 1976, Betty had been so busy working on the Kit report, trying out different "schemes" with different data sets, that she hadn't had time to send an invoice to Eymonerie for her expenditures. At the beginning of October, she finally sent the list of expenditures, along with the wish that she could have found a "really tight way to estimate salary," but so far she had only found a way that worked "most of the time but not always."

Acknowledgment of Betty's expertise in the area of salary inequity was growing. Eymonerie asked Betty to review a salary inequity manuscript that sociologist/statistician Judith M. Tanur and sociologist Rose L. Coser, both at SUNY-Stony Brook, had written and submitted to the *AAUP Bulletin* for publication. Eymonerie considered the paper to be a step beyond methods used in the Kit.[39] Betty read the manuscript but didn't have time to give it a proper amount of thought or write a proper review. Instead, she added some thoughts to the cover letter of her invoice. When Betty was done, she realized she had written so much that it could serve as some sort of informal review.

The important part of this review is Betty's articulation of why it wasn't reasonable to use rank as a predictor in the regressions. Betty wrote:

> *Try looking at the number of years required to go up in rank and the number of years required to go up a given amount in salary roughly corresponding to the change in rank (all for a given field, given rate of publishing, etc.) and you will see that you are looking at two sides of the same coin – women take longer to be promoted (and have smaller probability of being promoted) and women take longer to find their salaries increased by, say, $4000. You will tend to underestimate the underpayment of salary if you use rank as a predictor. Of course, this is true of many other predictors that involve discrimination: Having Ph.D., getting book published, etc. but rank appears to be worse and more directly tied (after all, tends to be same persons determining both).*

Betty thought the title, which referred to "pockets of poverty," was perhaps too emotional, but the article was eventually published under that same title.[40]

---

[39] 9/19/1977 letter from AAUP Staff Member Maryse Eymonerie to Harvard University Economics Professor Robert Dorfman.

[40] 10/2/1976 letter from Betty Scott to AAUP Staff Member Maryse Eymonerie; Judith M. Tanur and Rose L. Coser, "Pockets of 'poverty' in the salaries of academic women," *AAUP Bulletin* 64, no. 1 (1978), 26.

31.9 Next Drafts

By Betty's admission, the October draft of Part I of the Kit report wasn't much different from the first draft. She now also had a draft of Part II but she really wasn't all that happy with it. She still hadn't found a very accurate way of predicting salaries, and so she decided to recommend one that was easy to use. She had even tried to use 1972/73 ACE survey results, but there were so many problems, including a response rate of only 49%. She asked the board: "Can we now say that we only want to flag women for higher consideration and assert that one of the enclosed is OK? I'd like to if only that I cannot get better results. What sould [sic] our attitude be?"[41]

The November advisory board meeting was productive. Betty's draft was now complete and suggestions were made. Betty took lots of notes. Give references to stat books. Don't say "minority," say "minority person" or "minority member." List other machine manufacturers, e.g., Hewlett-Packard, Texas-Instruments. State that you can do it yourself. Give up trying to get a precise estimate, stress that the objective is only to flag. Put in a paragraph about not including teaching or administration. Include a good bibliography of who has done studies. Omit section III. Include a sentence/paragraph on coding. Discuss "rule." Rationalize in terms of Type I and Type II errors. Say that method will miss many women, and describe who. Say something about white males. State that it is an unrealistic aspiration to have everyone paid above average... even to have everyone paid average except some above. Explain that the method provides a ballpark estimate of how many dollars are needed to redress salary inequity.

The above ideas and decisions were regarded as minor. The plan was to also obtain a limited review of the draft by selected representatives of Committees W, Z, and A (Academic Freedom and Tenure), and other experts. The plan was to complete the draft by December 30; create a nice printed product, perhaps with spiral binding; and distribute the report in early February to presidents/chancellors of all colleges and universities, each local AAUP chapter, representatives of professional women's groups, and other interested parties. St. Lawrence, a reviewer of the draft, was "impressed with the amount of work" that Betty had put into the report.

The minutes of the meeting included some additional suggestions, which Betty agreed to incorporate into the Kit. The Kit should include a way to do the salary evaluations without a computer or deep statistical knowledge so that it will be useful to institutions with limited resources. It should include more about strategy and approach to salary evaluation; and recommendations for any use of productivity variables, especially as relate to judging quality. Part III should be replaced by a list of studies conducted to date, including whether

---

[41] 10/28/1976 memorandum from Betty Scott to advisory board of the Exxon project re draft of Part I of the preliminary report about the Kit.

or not they were published. Appendix A should include examples. Values less than or equal to zero should be flagged so as to avoid manipulation. The Kit should include information on methods to estimate the amount of money needed to compensate for salary inequities.[42,43,44,45] Eymonerie followed up by sending a request to 100 organizations asking if their group had conducted any pertinent studies, or if they knew of any conducted by others, so she could prepare as complete a bibliography as possible. The response was "quite substantial."[46,47]

---

[42] 11/1976 handwritten notes by Betty Scott re meeting with Robert Dorfman, Mary Gray, and Maryse Eymonerie re review of Part I of the preliminary report about the Kit.

[43] 11/29/1976 memorandum from AAUP Staff Member Maryse Eymonerie to representatives of professional women's groups re request for bibliography information on relevant salary studies.

[44] 11/30/1976 memorandum from AAUP Staff Member Maryse Eymonerie to members of the advisory board re actions taken at the meeting of November 13, 1976.

[45] 12/16/1976 memorandum from SEC Chair Patricia St. Lawrence to Betty Scott re additional comments on Betty's draft Kit.

[46] 11/29/1976 memorandum from AAUP Staff Member Maryse Eymonerie to representatives of professional women's groups re request for bibliography information on relevant salary studies.

[47] 12/16/1976 letter from AAUP Staff Member Maryse Eymonerie to Betty Scott.

32

Completing the Kit (1977)

...there are so many things I admire in your character, that I really can't begin to list them all.
-Joyce Ann Grant

At the end of February, Betty was finishing up the AAUP Kit. She was still working on the bibliography of salary inequity studies. The bibliography wasn't yet in good shape. It was at a time before articles and books were available on the Internet: A person had to physically go to different libraries that held different journals to look up specific references; if a paper was not published but still in manuscript form, then it could be almost impossible to track down. Finishing the bibliography was a painful process, one that involved weeks and considerable detective work. Betty's nephew came into the office one day to help her, but there were still holes in the references, and she ended up having to throw some references out because she couldn't track down all of the details.[1]

At the same time, Betty was getting "a lot of" telephone calls to discuss using rank as a predictor of salary. It turned out that, in many studies where rank was used as a predictor, women appeared to be overpaid relative to men; and that when separate analyses were done for each rank, the results were similar. Betty called this "monkey business." She had found that using rank as a predictor tended to underestimate the underpayment of salary. "It seems to me that we have to discuss this problem very clearly and I tried in Part I [of the Kit] but I am not happy with the outcome," she told Maryse Eymonerie.[2]

Around the beginning of April, Betty sent the final draft of the Kit to Eymonerie, who in turn sent it to the advisory board for review and comment.[3] A month later, Betty sent copies of the computer output for two examples, one involving a large department and another involving small departments that needed to be combined, proposing that they be put into Appendix A. The output was on large continuous computer paper with perforated holes on

---

[1] 2/21/1977 letter from Betty Scott to AAUP Staff Member Maryse Eymonerie.

[2] 2/27/1977 letter from Betty Scott to AAUP Staff Member Maryse Eymonerie.

[3] 4/8/1977 memorandum from AAUP Staff Member Maryse Eymonerie to Kit advisory board members re asking for review and comment on the Kit.

each side, and Betty presented some options on how to include copies in the report. Also, Betty was still working to clean up the bibliography.[4]

SWEM gave its annual report to the UC-Berkeley academic senate in April. It recalled the Policy CSAW co-led by Elizabeth Colson and Betty that produced "a very exhaustive and scholarly report which has become a classic and model for subsequent reports of this nature." The annual report summarized the major affirmative actions taken by the campus in the years that followed, as well as the areas in which the committee had been ineffective. It concluded: "Briefly, in those areas where the Committee has been successful, this has been achieved to the extent that the Committee now serves as a commentator concerning the actions of the Administration with respect to affirmative action. However, information which would allow the Committee to report on the progress of the Berkeley Campus in achieving equality of opportunity is continually denied to the Committee." As a sign of its extreme frustration, the Committee posed this resolution: "The Committee respectfully requests that the Committee on Senate Policy review the status of the Committee on Status of Women and Ethnic Minorities to determine if this Committee has a viable function within the present Berkeley Campus structure."[5]

32.1   Praise and Potential for the Kit

The AAUP published the Kit in 1977 under the title "Higher Education Salary Evaluation Kit: A Recommended Method for Flagging Women and Minority Personnel for Whom There is Apparent Salary Inequity." It was 55 pages. A brief first section explained its purpose. Section II detailed the "Recommended Method for 'Flagging' Women and Minority Persons for Whom There is Apparent Salary Inequity" and included sections on how the faculty should be grouped, the data which should be collected for each faculty member, and the interpretation of apparent underpayment. Section III was a "Comparison of Results and Costs of Several Suggested Methods for Flagging Women and Minorities with Apparent Salary Inequities" and included sections on kinds of methods proposed, and applications of those methods in a pilot study. Five appendices comprised over half of the report and were named: example of computer calculation and data used; codes and coefficients for peer-group regressions; studies at universities and colleges; studies of professional societies; and general studies of salary inequity in academe.

The Kit was sent to college and university presidents and other academic administrators, local AAUP chapters, various conference officers, and members of AAUP committees Z and W. It was sent to various editors at the

---

[4]5/6/1977 letter from Betty Scott to AAUP Staff Member Maryse Eymonerie.
[5]4/4/1977 "Report of the SWEM," Berkeley Division of the UC Academic Senate, pp. 11-14.

Chicago Tribune, the *Christian Science Monitor*, the *Washington Post*, the *Wall Street Journal*, *Newsweek Magazine*, *Saturday Review*, the *Los Angeles Times*, *The London Times*, the *Associated Press*, *Time, Inc.*, *Forbes*, *Barrons*, *American Economics Review*, *American Economist*, and other newspapers and news magazines. It was sent to the news editor at *Scientific American*, the editor of the *Journal of Higher Education*, as well as editors of various professional association journals in a variety of fields including statistics, mathematics, anthropology, sociology, political science, library science, modern languages, information science, education, business/finance, psychology, and economics. The Kit was sent to various women's journals and magazines, including the *American Association of University Women Journal*, *Women Today*, *Spokeswomen*, *MS Magazine*, *Ladies Home Journal*, *Redbook*, and *Working Woman*; and it was sent to various African American magazines, including *Ebony*, *African-American*, and *Jet*. Finally, Eymonerie sent Betty 50 additional copies for her to distribute as she saw fit.[6] By the middle of July, the AAUP had distributed almost 5,000 copies of the Kit.[7]

The AAUP initially made copies of the Kit available upon request for $1.25 each.[8] Later the price went up to $5.00, with this marketing paragraph: "The 'Kit' provides instructions on how to identify and collect data, and compares the results and costs of several suggested methods for detecting disparities. Examples of statistical programs and studies from colleges and universities, professional associations, and general studies of salary inequity in academe are provided in appendixes. It has been used to resolve salary disputes at colleges and universities across the country."[9]

In July, Betty received a letter from the AAUP acting general secretary. Betty's term on Committee Z had ended, and he wanted to thank Betty for her service. He offered Betty "commendations for your fine work with the 'Higher Education Salary Evaluation Kit.' From all accounts thus far, it is being very well received."[10] Eymonerie called the Kit "the masterpiece."[11]

Bernice Sandler said: "The Higher Education Salary Evaluation Kit arrived, and oh, it looks very good. It should be helpful to many institutions throughout the country. We will mention it in our fall newsletter so that others may learn about it. It certainly is a job very well done."[12]

---

[6] 7/13/1977 letter from AAUP Staff Member Maryse Eymonerie to Betty Scott.

[7] 7/14/1977 letter from AAUP Staff Member Maryse Eymonerie to Betty Scott.

[8] 11/15/1977 memorandum from AAUP Association Secretary Lesley L. Zimic to AAUW (District of Columbia Branch) re some documents related to Title IX compliance.

[9] "Salary Evaluation Kit for Higher Education" marketing paragraph. Publication and date unknown.

[10] 7/13/1977 letter from AAUP General Secretary (Acting) Jordan E. Kurland to Betty Scott.

[11] 7/13/1977 letter from AAUP Staff Member Maryse Eymonerie to Betty Scott.

[12] 7/13/1977 letter from Association of American Colleges, Project on the Status and Education of Women, Director Bernice Sandler to Betty Scott.

The UC-Berkeley provost said: "I am slowly plugging my way through [the Kit] and hopefully will be better able to identify and correct these problems as they arise at Berkeley. Congratulations on your effort!"[13]

Committees W and Z organized several events at the annual AAUP meeting. One was a lunch meeting, sponsored by both committees, where Betty demonstrated how to use the Kit.[14] The AAUP regarded the Kit as augmenting their general policy on discrimination.[15]

Mary Gray submitted her last annual report as chair of AAUP Committee. It was for the year 1976–77. She began it like this:

> *While preparing this report during a recent trip to Italy, I chose one word to summarize my reflections: "Basta." This all-purpose word meaning "enough," "stop," "that will do," captures my sense of deja-vu having announced to you last year that I was delivering my last Committee W report. It also conveys my own sense of frustration because I am reporting again on unresolved issues that continue to affect the ability of women to function without sex discrimination in the academic world. Basta, Basta.*

An attorney herself, Gray continued:

> *Although the problems of academic women remain essentially the same as in past years, my thoughts about effective avenues for modifying discriminatory practices have sharpened and changed. It has become increasingly clear that the courts and federal agencies are reluctant to probe critically the inner workings of academic decision-making. I have previously expressed my feeling that while theoretically the courts offer a means for improving the status of individual academic women, the isolated successes are few, far between and subject to multiple appeal procedures. The Byzantine structure of academe does not lend itself to microscopic examination of the application and misapplication of standards.*

Then, explained Gray, there were the high costs of litigation that included psychic in addition to monetary damages, where

> *... every detail of one's academic career, every hastily written word, every disgruntled student, every disapproving colleague, come back through testimony or statements dredged up by a battery of defense lawyers to recount the mistakes of the past.*

Gray offered that one of the most useful avenues to use to attack discrimination in the future may be various analytic tools like the Kit:

> *Committee W has been instrumental in developing tools for self-help and in preparing policies for the guidance of faculty and administrators. One*

---

[13]8/23/1977 letter from UC-Berkeley Provost and Dean Roderic B. Park to Betty Scott.

[14]5/5/1977 memorandum from AAUP Associate Secretary and Associate Counsel Carolyn I. Polowy to Committee W and Conference Committees W re information update.

[15]11/15/1977 memorandum from AAUP Association Secretary Lesley L. Zimic to AAUW (District of Columbia Branch) re some documents related to Title IX compliance.

very important addition to the collection of AAUP anti-discrimination materials in [sic] now available and ready to be distributed. The Salary Evaluation Kit... should be a significant resource for evaluating faculty salaries which are artificially inflated by discrimination.[16]

Gray hoped "the enforcement of the Equal Pay Act would rely on such analyses [as the Kit] to broaden the scope of its protection to provide <u>substantially</u> equal pay for <u>substantially</u> equal work."[17]

32.2   Old Master at Purdue

Not surprisingly, Betty's profile as a role model for academic women expanded in 1977. This was a time of affirmative action outreach. It was a time of committee activity. There are a number of examples.

In March, Betty spoke at the Yale University Conference, "New Directions for Women: Portraits of 21 Women in Science." There Betty really gave the students "something to think about" and she "left everyone feeling exceptionally pleased to have met" her.[18]

In April, Betty was on National Public Radio within a five-part series, "Closer Look," in a segment on "Affirmative Action: Not a Black and White Issue." She was included along with the VC and several others from UC-Berkeley. Betty was introduced as a person who "keeps an eye on how many women she has for colleagues, and she doesn't like what she sees. Since 1970, the proportion of women there [at UC-Berkeley] has only moved from 4.1% to 6.9%." Betty responded: "You still have to go a long ways to see a woman teaching in this university. Most all of the students go to this university and never, ever have a woman professor, a woman associate professor, or even a woman assistant professor teaching them. There's a lack of role models, there's a lack of teaching, and it brings a lack of breadth into the teaching."[19]

In June, Betty was asked if she would be willing to serve in November as a visiting expert at an NSF-funded workshop at San Diego State University designed to attract young women to careers in science and engineering.[20] The plan was to bring together 32 professional women and 300 freshman

---

[16]Mary Gray, "Annual Report of Committee W, 1976-77," 9 pp.

[17]7/22/1977 letter from American University Mathematics Professor and Committee W Chair Mary Gray to Pat Cheatham (Washington, DC).

[18]3/8/1977 handwritten letter from Yale University Office on the Education of Women Representative Connie Gersich to Betty Scott.

[19]4/25/1977 transcript, "Affirmative Action: Not a Black and White Issue," within a 5-part series, "Closer Look." *National Public Radio*, Washington, DC.

[20]6/20/1977 letter from San Diego State University Women's Studies Chair Marilyn J. Boxer and San Diego State University Economics Assistant Professor Elyce J. Rotella to Betty Scott.

and sophomore women from 26 Southern California colleges and universities. Betty was happy to participate, and also suggested other women who might be willing.[21]

In July, Betty was asked if she would be willing to participate in a workshop series on women in science (math, physics, economics, and engineering) organized by Stanford University's center for research on women. Betty not only said "yes," but she suggested the organizers be in touch with Blum and Kreinberg who had organized similar programs, also in the Bay Area.[22] As a member of a panel, Betty discussed what she did, how she got where she was, and how she managed the multiple demands on her life.[23]

By September, Betty was getting overloaded. She declined an invitation to participate in a visiting women scientists program, organized by the Research Triangle Institute in North Carolina, where professional women would visit three to four high schools over the first quarter of the next year and encourage high school girls to pursue careers in science. This was even though she had participated in such programs previously and said she would be willing to do so again in the future. But not this time: She just had too many heavy administrative duties.[24] As she declined, Betty offered this advice:

> *I think the most important thing is to make clear that there are interesting job opportunities, for women, in the sciences. I think it is also very important to point out to women that 95% of them will have to support themselves, at least partially, and perhaps some children too. Also, they should realize that the opportunities for women in primary and secondary school teaching and in the humanities, will be miniscule during the next ten years. I think it is also important to give a flavor of what a scientist does, and the pleasure and troubles one has in being a scientist.*

Betty advised the program be extended to reach girls in the lower grades.[25]

Betty did, however, accept a smaller, more local invitation soon after. This was to speak in a seminar on careers in physical sciences, math, aeronautics, and engineering at West Valley Community College in Saratoga, California in October. Betty and other speakers were asked to tell the students how they chose what they were doing, what they liked and disliked, and what they

---

[21] 7/8/1977 letter from Betty Scott to San Diego State University Women's Studies Chair Marilyn J. Boxer and San Diego State University Economics Assistant Professor Elyce J. Rotella.

[22] 7/8/1977 letter from Betty Scott to Stanford University, Center for Research on Women, Student Coordinator for Women in Science Workshop Kismet Collins.

[23] 9/1977 letter from Stanford University, Center for Research on Women, Student Coordinator for Women in Science Workshop Kismet Collins and Stanford University Center for Research on Women Administrator Margaret Collins to students.

[24] 9/21/1977 letter from Betty Scott to Research Triangle Scientists Program, Visiting Women Scientists Program, Staff Member Iris R. Weiss.

[25] 9/22/1977 Research Triangle Institute Visiting Women Scientists Program Application Form completed by Betty Scott.

would suggest for further exploration.[26] The students rated her talk as "outstanding," and the conference leader called her talk "unique and valuable."[27] A speaker who represented United Airlines singled out Betty and one other of the 11 speakers as providing "brief, but fascinating glimpses of their work and careers that could not help but be inspiring to [the] students."[28]

Purdue University has an event called the "Old Master's Program." Each year a group of "successful and outstanding individuals" who have given of themselves and made significant contributions to their field are invited to the campus to share experiences and ideas with the students and inspire them.[29] In 1977, Betty was invited to be an Old Master. She spent three days on the Purdue campus in early November. Her time there was well received. Her personal student hostess said this:

> ... *We all certainly hated to see you leave......it was a pleasure and an eduction being your personal hostess. I learned so much about astronomy and cancer research and weather modification...I feel I could actually discuss these areas intelligently. All this gain in knowledge is due to your sharing of insights and information...If someone asked me what quality I liked best in my Old Master, I'd have to reply that intelligence and sincerity were the outstanding factors. However, there are so many things I admire in your character, that I really can't begin to list them all. In closing, I want to again express my appreciation and thanks for being an "Old Master." The program has been one of the highlights of my life.*[30]

Betty saved the certificate of participating in the Old Masters program.

32.3   More Salary Work and Kit Dissemination

In November, Betty was back at work on the UC-Berkeley personnel (PERS) data. As in previous years, she asked the chancellor's office for a listing of all of the newly appointed tenure ladder faculty members that included name, sex, ethnicity, rank and step, department, and highest degree with place where it

---

[26] 9/8/1977 letter from West Valley Community College Counselor Betty A. Michelozzi to Betty Scott.

[27] 10/28/1977 letter from West Valley Community College Counselor Betty A. Michelozzi to Betty Scott.

[28] 10/28/1977 letter from United Airlines Organizational Services Staff Representative Thomas A. Carwardine to West Valley Community College Counselor Betty A. Michelozzi.

[29] "Old Masters," *Purdue University*, accessed April 4, 2015, http://purdueoldmasters. org/aboutus/history.html. Note: The program was started in 1950, and by 2015 there were about 600 Old Masters, one of whom was Betty.

[30] 11/12/1977 handwritten letter from Purdue University student Joyce Ann Grant to Betty Scott.

was earned. She used these listings to compare women and minorities with white males from the same entering cohort on both entering salaries and progress through the ranks over time. In previous years, analyses on these listings had shown that "in spite of so-called Affirmative Action," women were entering at lower salaries and lower rank-steps than men.[31]

The administration gave Betty the data she requested, some of which was sensitive. It was therefore required that she keep it "in the strictest confidence." She was also advised of specific problems in the data: The reports could not be compared from one year to the next due to some changes in data collection, calculation, or programming; the FTE information was unreliable; headcount for a given unit was undercounted; and the weighted average salary calculation was incorrect.[32] Betty's assistant summed it up in a simple note: "Why are we trying to get PER's [sic] when they're inaccurate? Why can't we get Louise Taylor's data? Oh well!"[33]

In December, the AAUP sent Betty 60 more copies of the Kit in order to replenish her supply. Eymonerie reported she had received a "significant number of requests" for copies of studies listed in the Kit bibliography. The AAUP was ready to announce and advertise a new service to perform regression computations upon request; they would make these computations for entire institutions only, not for an individual unit. Eymonerie asked Betty about some of the technical details that she would need to pass on to those making the requests. [34]

32.4   Faculty Club Questionnaire

As the year 1977 came to a close, Betty filled out a questionnaire sent to her and others who were not members of The Faculty Club. Betty told The Faculty Club what she really thought. She usually ate lunch in the third floor Evans Hall conference room, or else in the WFC; but in the past academic year, she had eaten lunch at The Faculty Club more than ten times; also, she had gone there for a drink after seminars; and she had dinner there one to four times, which she described as "terrible!" In response to the question which asked her to say frankly why she was not a member, Betty wrote this:

> a) I am a member of the Women's Faculty Club – I think there should not be two clubs!!!

---

[31] 11/4/1977 letter from Betty Scott to UC-Berkeley Chancellor's Office Staff Assistant for Planning Louise Taylor.

[32] 11/7/1977 letter from UC-Berkeley Chancellor's Office Staff Assistant for Planning Louise Taylor to Betty Scott.

[33] 11/15/1977 handwritten note from UC-Berkeley Stat Lab Assistant Jeanne [?Lovasch] to Betty Scott.

[34] 12/5/1977 letter from AAUP Staff Member Maryse Eymonerie to Betty Scott.

b) I have been bodily ejected from MFC [The Faculty Club] because I am a woman faculty member – I will NEVER join after such MFC experiences.
c) Fee for joining.
d) Dues too high.

No, Betty did not want to be contacted about becoming a member of The Faculty Club.

33

Influencing Salaries (1978)

I am indeed distressed by the present use of statistical methods in adversary situations.
-Elizabeth L. Scott

At the turn of the year 1978, *Ms. Magazine* contracted with a journalist to investigate and report on equal pay for equal work. The author asked Betty for a copy of the AAUP Higher Education Salary Evaluation Kit, having seen it mentioned in the *Spokeswoman*. The author also asked Betty for a copy of her bio, promising to give appropriate credit to her in the article.[1,2]

Now Betty was working on logistics of the proposed Kit-related service to provide regression computations for those unable to do the computations themselves. She preferred being a cooperating partner, with the AAUP as the service provider. She agreed the service should be provided for all units of an institution, not just an individual unit. Betty drafted two sets of instructions for how to submit data, one on computer tape, and one on punch cards. She could also accept data in typed form, suggesting the UC-Berkeley Computer Center be engaged to digitize the typed data. She provided a rough estimate of costs.[3] Maryse Eymonerie thought Betty's instructions were "clear enough," agreed the service should be provided by AAUP in cooperation with UC, and wanted to start it "as soon as possible."[4]

In February, Eymonerie asked Betty to review a draft paper on part-time wages. Betty criticized the authors for not defining part-time faculty: Was it everyone who wasn't full time, or what? She also faulted the authors for lumping all part-time faculty together: Betty's studies had shown that part timers were a heterogeneous group when it came to wages, and she thought there should be separate analyses for the different subgroups, e.g., students in the same institution, professionals who teach as a primary occupation, professionals who teach just one or two courses, etc.[5]

---

[1] 12/29/1977 note from *Ms. Magazine* Mary Scott Welch to Betty Scott.

[2] 1/9/1978 letter from Betty Scott to *Ms. Magazine* Mary Scott Welch.

[3] 1/10/1978 letter from Betty Scott to AAUP Staff Member Maryse Eymonerie.

[4] 2/15/1978 letter from AAUP Staff Member Maryse Eymonerie to Betty Scott.

[5] 2/20/1978 letter from Betty Scott to AAUP Staff Member Maryse Eymonerie.

33.1   Persisting Salary Inequities at Berkeley

Betty continued to work to effect change at UC-Berkeley. Back in January, she had a meeting with the director of the office of planning and analysis. It must have been important, because she kept two copious hand-written sets of notes, one by herself and another by someone else.[6] Betty and the director discussed reasons given for each declined or denied appointment. Where were the promotions "being cut off"? Was it at the department, dean, or chancellor level? They discussed advancement patterns over time, and which data sources to use, with which strategies, for analyzing patterns in academic employment in order to determine possible inequities.[7]

Betty remained deeply concerned about the persistence of salary inequities. She wanted to know both how and why they kept occurring. One possibility, she thought, was that there were inequities in salaries at the initial appointments. Betty had conjectured that women and minorities would enter employment at higher salaries than white men given the pressures of affirmative action, but she found it to be the opposite case. In fact, she was seeing this pattern in all the data she looked at: It wasn't for just one campus or year. She challenged a statement made by UC to HEW that the number of appointments was too small "to establish statistically significant discrepancies": Even for the Berkeley campus alone, she pointed out, the differences between men and women for many of the past eight years were statistically significant, and when the eight years were looked at together, the p-value was "even smaller," i.e., 0.00161. This, she pointed out, meant: "we can say that the probability that there is a fair distribution of initial appointments between men and women, with regard to initial salary, is less than 2 in 1000. Such a small significance probability can hardly be due to chance and I think that efforts should be made to find the causes of inequity." She dismissed a number of possible explanations for why lower ranks were offered to women: the possibilities that men were going into harder fields (e.g., the sciences); women were being appointed closer to the time of the granting of their PhDs; and there were larger proportions of women at lower ranks.

Betty sent these arguments to the VC, along with three data tables, a set of graphs, and dense explanations of the peculiarities of the data that she used. The tables included new appointments to ladder ranks by sex and step (1975–1978), comparison of initial salaries of men and women in ladder rank appointments, and the same comparison but for ladder rank assistant professors. The graphs were of proportions of ladder rank appointments at each step or lower, by sex, and each of six academic years beginning with

---

[6]Note: Louise Taylor was UC-Berkeley Director of the Office of Planning and Analysis and Special Assistant to the VC. She served six chancellors and had a record of "extraordinary accomplishments." See "A Citation for Louise Taylor," *Berkeleyan*, accessed April 7, 2015, http://www.berkeley.edu/news/berkeleyan/1998/0304/awards.html.

[7]1/10/1978 handwritten notes by Betty Scott and handwritten notes by anonymous.

1974–75. She used Wilcoxon rank sum tests to compare populations of men and women.

The graphs and tests showed clearly that the salary distributions of women were lower than of men. Betty was especially concerned, because the salary at initial appointment "will tend to persist forever": Women brought in at lower salaries than men will "tend to remain behind them forever." She was also especially concerned, because these salary inequities were compounded, given that women got promoted more slowly than men. She confessed she didn't "have all the tables at hand to demonstrate these inequalities. This is because it is such a difficult and tedious process to obtain the necessary data and get it into relatively error-free shape. Nevertheless, the conclusions are clear," she told the VC. She hoped that his office would scrutinize each new case for appointment or promotion that came forward from the departments and take measures to both insure fair assignments of future salaries and rectify unfair assignments of past salaries. She copied everyone she could think of who might be influential, from the UC president down to the chairperson of the SEC.[8]

It had become a collaboration between Betty and the chancellor's office. Betty asked for data, and the administration gave it to her, asking her to keep it confidential. In turn, they asked Betty to inform them of any data errors she found so they could correct them; and they asked for her ideas to improve their summary data tables.[9] Betty saw her intense work on salary equity studies as "one way to examine how much effect affirmative action is having," and the results were "discouraging." There was evidence of "very little progress and even some declines" for women. Here is Betty's summary:

> *There certainly are more women being appointed at the Assistant Professor level but the proportion being appointed is not keeping up with the increasing proportion who are getting doctorates... In this University, not only at Berkeley but at every campus, we see that women come in with initial appointments which tend to be at lower salaries than for white males. This is true year after year and the results are highly significant. It is also true if you restrict attention to appointments at the assistant professor level or to appointments in particular fields such as biological sciences. We also notice that women are less likely to be promoted to associate professor (that is, more likely to be fired), less likely to obtain merit increases, and so forth.*

From Betty's perspective, she had good cooperation from the administration, but the data were in terrible shape "so it is very tedious and frustrating."[10]

The VC read Betty's letter and noticed that her conclusions were different from the ones included in the last year's affirmative action report. He pledged

---

[8]1/28/1978 letter from Betty Scott to VC Ira Michael Heyman.

[9]2/6/1978 letter from UC-Berkeley Planning and Analysis Director Louise Taylor to Betty Scott.

[10]3/11/1978 letter from Betty Scott to AAAS Scientific Manpower Commission Staff Member Betty Vetter.

to her that he would look into the discrepancies with an attempt to reconcile them, and he would keep her informed of the progress.[11] The director would get back to Betty a few months later with more thoughts and data.[12]

33.2   Conferences and Colloquia

Betty continued to participate on panels at women-in-science workshops. For example, she was invited to be a member of a panel on discrimination organized by the UCLA continuing education program in engineering and mathematics. Other panel members would include someone from the office of federal contract compliance, a UCLA academic affirmative action leader, and a department manager from TRW Defense and Space Systems Group. Betty was asked to be prepared to talk about age discrimination; how to detect and cope with discrimination; grievance procedures and class action suits; the difference between discrimination and sexism; how to size up an industry for discrimination; and HEW regulations that can affect legal action.[13]

In addition there were now conferences on math anxiety. In February Betty received a notice for one at Cal State Fresno. Her response to the announcement gives a glimpse of what she thought about mathematics training:

> ... I hope that you will provide information about what mathematics really is, what mathematicians, including applied mathematicians, typically do, what they think about and how this may be related to the many problems our society faces. I feel very strongly that most of our college age students, both women and men, had very poor training in mathematics in primary and secondary schools. It is not surprising that they do not understand mathematics – they do not even know what it is. The students whom we see in elementary statistics courses are often very anti-quantitative. They do not know much algebra and they hardly can do arithmetic. But these deficiencies are not that important! The important thing is for the student to make a fresh start doing something interesting whose usefulness he/she can visualize. Trying to internalize rote learning is not constructive in my opinion.[14]

Thus Betty was optimistic about possibilities for math learning in college.

Betty was also spending a fair amount of time reviewing national reports on affirmative action for evidence of "improper sampling procedure, incor-

---

[11]2/24/1978 letter from VC Ira Michael Heyman to Betty Scott.

[12]6/20/1978 letter from UC-Berkeley Planning and Analysis Director Louise Taylor to Betty Scott.

[13]10/26/1978 letter from UCLA University Extension Assistant Director Estelle Klingler to Betty Scott.

[14]2/22/1978 letter from Betty Scott to California State University-Fresno Office of the Dean of Natural Sciences.

rect analyses and misleading conclusions."[15] For example, Betty was asked to contribute to a report on the availability and utilization of women science doctorates being prepared by the Committee on the Education and Employment of Women in Science and Engineering (CEEWISE), a committee of the National Research Council (NRC). Could she write something on her work on salary differentials among men and women in science? She was told: Anything she could carve out time to prepare would be welcome; it could be long or short, rough draft or final form.[16]

Betty was still interested in the WFC even though she had resigned from the board. The club was losing members. Betty was sufficiently concerned that she asked for a list of people who discontinued their memberships. She thought: Some must have retired; or left Berkeley; or just resigned, because she knew they were still on campus.[17]

Ben's Forum was a group of women researchers at who used to participate in a research colloquium series that had been "vital" in the early life of the WFC. In April 1978, Betty and four others contacted the senior women professors asking if they would be interested in reviving the Ben's Forum: The sentiment was that there would be "so much to share..."[18,19,20]

33.3   Astronomy, Statistics, Engineering

Astronomy: The federal ERA was designed to guarantee equal rights for women. It was in front of the state legislatures for ratification. Time was getting short: The deadline for ratification was only a year away.

Now there was a resolution in front of the AAS council to support the ERA, specifically that the AAS would refuse invitations to hold meetings in states that had not ratified the ERA. Margaret Burbidge, as president of the AAS, distributed a survey asking members for their opinions on the resolution.

Back at Berkeley, the all-male astronomy faculty submitted a letter to the AAS condemning the action of the council. The department now had a record number (six) of women graduate students. In response to the faculty letter, three of the women sent a memorandum to "EVERYBODY" informing them

---

[15]3/14/1978 letter from Betty Scott to AAAS Scientific Manpower Commission Betty Vetter.

[16]4/25/1978 letter from NRC Staff Officer Leila Rosen Young to Betty Scott.

[17]2/22/1978 letter from Betty Scott to Physical Education Professor Doris White.

[18]1/1978 list of members of Ben's Forum (13 names, including Betty).

[19]4/15/1978 memorandum from Ben's Forum organizers (Marian Diamond, Sue Ervin-Tripp, Josephine Miles, Margaret Wilkerson, and Betty) to senior women professors asking if they would be interested in participating in a revived Ben's Forum.

[20]10/29/1980 memorandum from UC-Berkeley Physiology-Anatomy Professor Marian C. Diamond to friends.

of the situation and proposing to send a letter of their own to the AAS in support of the resolution. Everybody included Betty.

Betty was disturbed by the reaction of the women students to the "intense backlash" of some of the faculty. She went into action and sent a communication to the faculty, urging them to reconsider their letter. She wrote: "I urge the members of the faculty of the Astronomy Department to reconsider their letter of February 7th. This letter is too harsh and has been harmful to the Department and its students... I am shocked and distressed to see young women graduate students in Astronomy deeply hurt by the contents of this letter and frightened by its implications for their future." Then she reflected: "I cannot tell these women to forget their worries," citing the results of her gender equity studies. She put the following question in front of the department:

> *Why does this discrimination persist? Statistical studies cannot provide the full answer; it is the responsibility of the white male faculties, of the white male astronomers, who make the decisions, to write the letters. I urge you to look beyond yourselves, I urge you to understand the problems these women face and to consider their basic human rights (Amendments to the Constitution of the United States are concerned with human rights, not politics.) and to enrich astronomy by providing equal opportunities for women astronomers. Now is a time to change. Please do change your letter.*[21,22]

Betty hoped her colleagues would reconsider.

The faculty responded by saying their letter had been "misread." They weren't opposed to the ERA; instead, they wanted to reserve "for our private expression any position on the ERA." What they were opposed to was having the AAS take a position (which they viewed as a political action) without thoroughly consulting the membership. The department chair said the letter was written to reflect the views expressed in an open department meeting; the agenda for this meeting was posted in advance; although students were welcome, none attended; and so the letter was written on behalf of the faculty. "I am pleased to see so much interest around this issue, for it is an important one," he told Betty. He also told her there had been another meeting about the matter at which 30 people were in attendance, and there would be more discussion at the next department meeting.[23]

Betty wasn't placated. She wrote again to the chair:

> *I want to come back to my letter... As I tried to say in my letter, my concern is the troubled reaction of the women graduate students in Astronomy. They are worried, and, I think, rightly so by the feelings manifest*

---

[21]5/7/1978 letter from Betty Scott to University of Maryland Astronomy Professor Elske Smith.

[22]4/8/1978 letter from Betty Scott to colleagues [in the UC-Berkeley astronomy department].

[23]4/12/1978 letter from UC-Berkeley Astronomy Professor and Chair John E. Gaustad to Betty Scott.

in the faculty letter. My urgent plea to you is to ensure that a reasonable letter be written. The fright among your graduate students has not gone away. I hope that you will find it incumbent upon you to restore good feeling in the Astronomy Department. You may be surprised that I am concerned with graduate students in Astronomy, but I am concerned with all graduate students, most especially women in the sciences. I think that it is difficult to understand what happens to women in science. My interpretation is that in order to maintain your own sanity, you have to push out of your thoughts the discrimination against women that has occurred and still occurs. Then, when something happens such as the Astronomy faculty letter that you cannot ignore, it is a much harder blow.[24]

Betty had been a member of the AAS for 30 years and had authored 20 or so papers in astronomy. She viewed herself as being "on the edge of the discussion" about the AAS proposal but continued to feel deeply about women in astronomy. Betty thought the problems for these women were "very real," and urged Burbidge to form a special committee to monitor their employment opportunities. She told a colleague in Maryland: "The most frightening aspect from my point of view is the women graduate students... are really terrorized by the well-publicized backlash of the faculty members..."[25]

Statistics: Betty was now president of the IMS and facing a situation similar to the AAS. The IMS membership was divided on the issue of not holding meetings in states that had not ratified the ERA. Betty told Burbidge, "... I hope that the Council can maintain what I consider to be a very reasonable position not involving any sort of politics."[26]

Later in the year, in August, the ASA considered a resolution stating that it would only hold national or regional meetings in states that had ratified the ERA. The board passed the resolution for national meetings through 1984, but chose not to include regional meetings. The ASA Committee on Women wasn't happy that regional meetings were left out and pledged to bring the matter back for another vote at the February meeting. Some of the board members didn't support the boycott but said they supported the ERA.[27]

Betty was advised that the IMS should use the ASA resolution as the basis of a related resolution,[28] and she was looking for arguments for and against. Betty observed that the IMS membership was "badly split" on the idea, "with some members very stubborn... The Program Secretary volun-

---

[24]5/9/1978 letter from Betty Scott to UC-Berkeley Astronomy Professor and Chair John E. Gaustad.

[25]5/7/1978 letter from Betty Scott to University of Maryland Astronomy Professor Elske Smith.

[26]5/25/1978 letter from Betty Scott to UC-San Diego Physics Professor E. Margaret Burbidge.

[27]9/20/1978 letter from Bureau of the Census Research Center for Measurement Methods Chief Barbara A. Bailar to Betty Scott.

[28]8/14/1978 handwritten memorandum from Nancy Geller to Betty Scott re the ASA board of directors resolution regarding the ERA.

teered not to hold any national or regional meetings in a state which has not ratified ERA. However, several Council members urged that this be announced to the membership for discussion... and that the members present vote on the policy." Betty's term as IMS president had just ended, but she was still taking a leadership role on this matter, and was looking for the best way to deal with it.[29] She didn't know of any associations that had changed their bylaws; instead, they had passed resolutions. Noting a resolution IMS passed the previous year, Betty observed: "I read this to indicate that resolutions restricting the location of meetings are not so unusual, even in IMS."[30]

Betty also suggested the Caucus for Women in Statistics (CWIS) pass its own resolution. She thought it could be the same as the one ASA had, and IMS was considering passing; or it could be even stronger.[31]

By March 1979, there would be 15 states that had not yet ratified the ERA. Some 140 organizations had supported the ERA by passing resolutions not to hold meetings in these states.[32] Betty would urge the IMS not to meet in South Carolina, which was a state that did not ratify the ERA:

> *I think that it is time for the Deep South to grant equal rights to women... Some Eastern members of IMS would not be able to go to South Carolina – some because they decline to meet in non-ERA states, some because their employers will not pay travel funds to meetings in non-ERA states. I want to point out that the passage of ERA is not an empty thing in the Deep South. It is just in these states that women cannot obtain loans as easily as men, even though it is federal money being expended, as occurs in certain student loans; women cannot sign papers even pertaining to property which they have inherited; women do not appear as members of State and local committees which have authority to act...*

Betty preferred to meet at universities and suggested the IMS meet instead at Penn State.[33]

By September 1979, the IMS would follow the ASA's lead in deciding to meet only in states that had passed the ERA. Betty quickly informed the NOW National Action Center in Washington, DC.[34]

Engineering: Betty became concerned about the field of engineering when she gave the welcome at a conference at Berkeley on "Student Women Engineers and Computer Scientists." The UC-Berkeley engineering college was one

---

[29] 8/31/1978 letter from Betty Scott to Bureau of the Census Research Center for Measurement Methods Chief Barbara A. Bailar.

[30] 10/15/1978 letter from Betty Scott to Stanford University Statistics Professor Ingram Olkin.

[31] 11/11/1978 letter from Betty Scott to Bureau of the Census Research Center for Measurement Methods Chief Barbara A. Bailar.

[32] 3/1979 notice to Astronomical Society of the Pacific members, "A Poll About the Equal Rights Amendment and Future A.S.P. Meetings."

[33] 4/16/1979 memorandum from Betty Scott to IMS council members re mailings of March 30 on IMS meetings.

[34] 9/13/1979 letter from Betty Scott to NOW National Action Center in Washington, DC.

of the sponsors. Some 500 women students attended and were encouraged to become engineers and computer scientists. Betty advised the students about their participation in the conference: "You will widen your horizons, I trust, but I hope you will retain your own selves too!"

But there was a disconnect. It was alleged at the conference that admissions standards at Berkeley for engineering were so high that essentially none of the attendees were likely to be admitted, and that transfer was equally unlikely. Betty was troubled if the college was encouraging women to be engineers and at the same time had unrealistic requirements. She asked the college for admissions data and encouraged them to base their decisions on factors beyond GPA and/or test scores.[35,36]

Engineering responded by talking about their large number of applicants, and lauding the conference for bringing information to the women students that would help them "properly to prepare for entrance." The college rebutted the data presented at the conference which showed that admissions and transfer into their program were essentially improbable, and they pointed out that admissions committees also considered a number of other factors. The acting dean told Betty, "...it is my opinion that women who prepare for engineering studies are as well able to compete academically as their male counterparts, if not better able."[37] The next month engineering reported to Betty the results of an affirmative action self-study.[38]

Betty responded to the dean that the data she referenced in her letter to him were the data presented at the conference by the engineering college associate dean, and she saw no reason for him to present information that was incorrect. She pressed the dean for additional information about the "other criteria" they used for admissions. She suggested he set up special programs to advance women in engineering, and based on best practices of other universities. She challenged him to attract more women, which would in turn attract even more women. She suggested the college simplify their admissions procedures: This "would speed up the horrendous length of time that the college now wastes in making its decisions and would certainly facilitate attracting the most outstanding graduate students," she said bluntly.[39]

One of the students who attended the conference prepared a four-page questionnaire about the experience of women in math/science/engineering. Betty graciously completed it. One of the questions was "What procedures can you recommend to upgrade affirmative action programs for engineering

---

[35][?1978] handwritten welcome speech for the UC-Berkeley conference on working in engineering and computer science.

[36]5/22/1978 letter from Betty Scott to UC-Berkeley College of Engineering Dean Ernest S. Kuh.

[37]5/28/1978 letter from UC-Berkeley College of Engineering Acting Dean A.M. Hopkin to Betty Scott.

[38]6/21/1978 letter from UC-Berkeley College of Engineering Electronics Research Laboratory Staff Member Pamela Humphrey to Betty Scott.

[39]6/24/1978 letter from Betty Scott to UC-Berkeley College of Engineering Dean A.M. Hopkin.

fields in the schools? In the business sector?" Betty responded: "Only what I am doing already." Another question was "What hassles, struggles, stumbling blocks do you encounter in your current job?" Betty replied, "Same old problems." Yet another question was "In which ways is your current job enjoyable and challenging?" Betty said, "Most all." When asked if she had encountered tokenism, she said simply, "Yes."[40]

33.4   Idea of a Collaboration

Back in April of 1978, Betty received a letter and paper from University of Illinois Economics Professor Marianne Ferber (1923–2013) after seeing Ferber in Bethesda and hearing about a study Ferber was working on. Betty was curious about whether the unemployment rate of job seekers was higher for women than men. She told Ferber, "The indication would... be that the small differences in years in the labor market may be explained by unemployment differences, further reducing any need for an explanation that women withdraw because of family responsibilities." She told Ferber that each field should be considered separately. Then she approached Ferber and asked to collaborate: "In fact, I wonder whether you would consider doing, perhaps jointly, a study similar to those we have been making using CCHE and ACE Surveys. I would like to compare the results of the differentials between men and women for persons who remained in the several categories of universities and colleges with the differentials for Ph.D.s in other kinds of employment. It seems to me that one should be able to get this information from your surveys."[41]

Betty's idea of a collaboration using Ferber's data "basically" appealed to Ferber. Ferber followed up by sending Betty a copy of her questionnaire so Betty could see what data Ferber actually had. Ferber wanted to do more with Betty than just duplicate Betty's findings with the Ferber data set, and Ferber was concerned that her data were getting dated. These caveats aside, Ferber wondered to Betty how they should proceed.[42]

There were two things Betty thought could be done beyond duplicating the results she had obtained earlier with the CCHE and the ACE datasets, which were limited to women who had (non-student) university teaching positions. One was to study a broader group of women than just those who had succeeded in academe by having such positions. Betty wanted to do multivariate analyses to assess the causes of differences in salaries between men versus women, and

---

[40]6/10/1978 letter from UC-Berkeley Conference for Women in Engineering and Computer Science (5/13/1978) attendee Mina Edelston to Betty Scott. Note: Betty's completed questionnaire was attached.

[41]5/9/1978 letter from Betty Scott to University of Illinois Economics Professor Marianne A. Ferber.

[42]5/16/1978 letter from University of Illinois Economics Professor Marianne A. Ferber to Betty Scott.

she wanted to do this by type of employer; she also wanted to estimate in this broader sample what a women's salary would be if she were paid like an equivalent man. Betty thought the Ferber dataset was more up to date than any other dataset that had "anywhere near the same depth."[43]

Betty saw Ferber again in San Diego in August. They made plans for "what to do next," which would be to try to follow "the individual paths in the careers of women versus men. I hope that we can study more deeply where it is, and even why it is, that women's paths deviate from those of men," Betty told Ferber.[44]

33.5   Irritations

By May, Betty was complaining to the chancellor's office about the "heavy expenses" the statistics department was shouldering by helping UC-Berkeley with their affirmative action problems. She accordingly made a formal request for some special funds, especially to help with the salary inequity studies. She reported that AAUP funding had allowed her to do pilot studies on a few departments at Berkeley and to do the Kit. But 13 months had passed, and the administration hadn't given the SEC any additional data.

Betty had been conducting studies using the most recent CCHE faculty survey data; and university payroll information over a ten-year period beginning 1973–74. These were "difficult and exacting" tasks, with many problems arising due to poor data. They took much more time and energy than she thought. But she felt she had excellent cooperation from both the president's and chancellor's offices. Her studies focused on finding the causes of salary, promotion, tenure, appointment, and merit raise inequities; and identifying trends over time. For example, she found a statistically significant average difference in starting salaries between women and men of between $700 and $1,000 per year; in this matter, she urged "more careful surveillance" of departments, where the problems seemed to originate. She also found persisting unfair appointment rates, which were not even "Affirmative Action appointment rates which would eventually (in thirty or forty years) close the gap in the proportions we see in ladder-rank faculty of different ranks."

Betty presented more findings to the chancellor's office to justify her request for special funds, saying, "I find these conclusions very depressing and I feel obligated to try to pinpoint the causes." She wanted to study advancement patterns of regular faculty for each department over a 15-year period. She requested $4,815.40 (about $16,000 in 2010 dollars) to pay for 60% of

---

[43]6/17/1978 letter from Betty Scott to University of Illinois Economics Professor Marianne A. Ferber.

[44]8/22/1978 letter from Betty Scott to University of Illinois Economics Professor Marianne A. Ferber.

Senior Statistician Jeanne Lovasich's salary and $400 of computer time that they had "bootlegged." Lovasch's salary was considered to be high, but it was because she was an expert with special skills and much experience.[45]

At the same time, the SEC was running out of money. Chair Patricia St. Lawrence was embarrassed to keep asking for more money without producing products. She thought at least her committee should give an account of Betty's work and ask for better cooperation with the administration. She hoped the senate could lean on the administration to give the committee better access to personnel records. She was discouraged that there wasn't more support for moving the project over to SWEM so it could become permanent.[46]

By June, Betty was still working closely with the SEC. The committee's annual report for 1977–78 articulated the statistical methods the committee had chosen with Betty's help. It cited Betty as having "generously made available her valuable contributions" in the area of methods for flagging apparent salary inequities and thanked her for her "untiring efforts and enormous patience in carrying out the detailed analyses." Although it sometimes seemed to the committee that salary equity was at the bottom of the list of the university's affirmative action priorities, they were happy to report that the 13-month delay in getting data had recently ended: Data were flowing again from the administration to the committee.[47]

It was really hard to find grant funding for gender studies. NSF didn't have any provision in its budget to support unsolicited research proposals. Betty thought this was truly unfortunate: Funding agencies needed to allow for new areas of study that were unforeseen when their budgets were prepared, and for shifting priorities among areas of study. What if there is a proposal that is just too good to turn down, Betty wondered? Betty thought the NAS should take up this issue of support for unsolicited research proposals for future study. She thought the issue was especially important for women and minorities, as they had a higher chance of doing research in new areas.[48]

There were many disagreements about conclusions from gender equity studies. Betty didn't think crowded committee meetings were the place to try to settle such agreements; she seemed to prefer written correspondence that articulated the details of the disagreements.[49]

Betty was irritated and frustrated when researchers used statistics "in a sloppy way" and then came to the wrong conclusions.[50] She told William

---

[45] 5/11/1978 letter from Betty Scott to Chancellor's Office Professor W.A. Shack.

[46] 5/5/1978 handwritten note from SEC to Betty Scott.

[47] 6/5/1978 memorandum from SEC Chair Patricia St. Lawrence to members of the committee (Susan Ervin-Tripp, Laura Nader, John Noonan, Paul Takagi, Curtis Hardyck, and Betty Scott) re draft and outline of annual report to the senate (with attached report).

[48] 10/21/1978 letter from Betty Scott to NAS Human Resources Studies Director Dorothy Gilford.

[49] 3/27/1978 letter from Betty Scott to Higher Education Resource Services Executive Director D. Lilli Hornig.

[50] 5/29/1978 letter from Betty Scott to University of Chicago Statistics Professor and Social Sciences Dean William Kruskal.

Kruskal: "I am indeed distressed by the present use of statistical methods in adversary situations. Partly, my difficulty is that here is an arena in which there is no effort to establish the truth or to get as close to the truth as we can. Rather, the attitude seems to befuddle the judge and to employ any misuse of statistics (such as using rank as a predictor) which might help your own side." She thought there was "entirely too much chance in academic promotion decisions." Betty was concerned with the "very strong collinearity" (i.e., near perfect correlation) of some of the predictor variables in both her skin cancer studies and her gender equity studies: "It [collinearity] haunts us also in salary studies which is one reason why I do not like to put sex in as one of the predictors. However, I must admit that this 'model' is most convenient for convincing non-believers," she told Kruskal.[51]

Betty and Kruskal discussed a United Airlines pay inequity case, where she suggested use of methods similar to ones outlined in the Kit. Focus should be on a specific class of employees, paying close attention to "the chain of decisions which determine salary when selecting individuals to be grouped," just as her approach focused on departments or groupings of similar departments. Betty also warned against using a predictor like rank, "which would have essentially the same bias as salary."

It is interesting to see how Betty combined statistical reasoning with common sense in coming to conclusions. She wrote: "I prefer to demark negative residuals [differences between observed and predicted values] as cases of possible underpayment which should be investigated carefully. At the same time, when I find a department where ten out of twelve of the women and minority members appear to be underpaid no matter what reasonable prediction equation is employed, I cannot help but suspect that discrimination exists. From the point of view of the university or the company which will be asked to adjust such salaries and from the point of view of the woman/minority person who appears to be underpaid by thousands, or even tens of thousands of dollars year after year, I think that here is where the problem lies."

Now Betty was looking at the stochastic process of salary over time, from initial appointment through merit increases and rank promotions. Her primary research question was "When we examine this process and see the pattern for white males compared to the pattern for women and for minority persons, is the process a result of discrimination against a few individuals or is it the result of a general shift downward in the patterns of all women and minority persons?" She conjectured that the shifts were general, more pronounced for some departments than others, and basically unchanged by affirmative action. "The situation is no better in the seven years since 1970 than in the seven years before 1970, which were pretty bad years," she told Kruskal.[52]

---

[51]3/30/1978 letter from Betty Scott to University of Chicago Statistics Professor and Social Sciences Dean William Kruskal.

[52]5/29/1978 letter from Betty Scott to University of Chicago Statistics Professor and Social Sciences Dean William Kruskal.

Betty, the VC, and his director of planning and analysis continued their dialog about institutional data and gender inequity analysis strategies. The dialog was particularly intense beginning in the summer and into the fall, with over a dozen people being copied on the correspondence. Betty and the director clearly were not on the same page about data (definitions, details, accuracy, etc.) and methodological details. In one letter, the director told Betty: "Since the issue of inequities in appointments and merit and promotion actions is such an important one, an analysis of each process approached in the detail outlined here [in the letter] seems necessary for clear identification of problem areas."[53] Betty responded she was "shocked" by the director's assertions: "Frankly, your letter just does not make any sense at all!...I just do not understand why you wrote the letter that you wrote. I would welcome the opportunity to see any revisions in the fundamental data but I stand on our computations and interpretations."[54] The two researchers met at the end of June to get their "shared data base in order," and they continued their deep conversation and correspondence over data and methodology.[55,56]

In August, Betty sent to SWEM some tables produced from the October 31, 1977 payroll data. These were similar to tables prepared for the committee in past years. They included head counts and FTEs in teaching positions, by type of position, sex, and year. Once again, the finding was that women were paid less, even in the same type of position. Betty's summary indicated some increase in proportions of TAs, assistant professors, associate professors, and professors who were women; but it also indicated some decrease in proportions of instructors, associates, and lecturers who were women. She offered: "I do not see why this should be happening."[57,58]

The VC questioned Betty's conclusions. He didn't think the data were accurate or reliable enough to warrant them: "Under the circumstances, I think it is advantageous that you continue to work closely with [the director of planning and analysis] in verifying the accuracy of the information before drawing any conclusions about the data."[59]

Betty found the VC's letter "distressing" and waited three weeks before answering. "The main content of the letter is not sensible," she told him, and she went into detail as to why. Basically, she was using historical definitions, because she wanted to make valid comparisons over time. "It is well known in Statistics that if one is allowed to select the measure on which the inter-

---

[53] 6/20/1978 letter from UC-Berkeley Planning and Analysis Director Louise Taylor to Betty Scott.

[54] 6/26/1978 letter from Betty Scott to UC-Berkeley Planning and Analysis Director Louise Taylor.

[55] 8/25/1978 letter from UC-Berkeley Planning and Analysis Director Louise Taylor to Betty Scott.

[56] 9/11/1978 letter from Betty Scott to UC-Berkeley Planning and Analysis Director Louise Taylor.

[57] "What Jeanne Produced from the PER Tape" (white paper), n.d.

[58] 8/3/1978 letter from Betty Scott to UC-Berkeley Physical Education Professor and SWEM Committee Chair Helen Eckert.

[59] 8/20/1978 letter from VC Ira Michael Heyman to Betty Scott.

pretation will be based, if one is sufficiently clever, (s)he can find a measure appropriate to any conclusion. Thus, proper statistical philosophy requires that the measure be specified in advance and not altered," she told the VC. She also said she was aware of the problems with the data provided her by the administration: "We have been struggling through the years to improve the quality of data; it is better than it used to be. I think that it is fair to say that no one is more aware of the problems in the PER data than we are; if we had access to better data, we would use it," Betty insisted.

Betty tenaciously wrote this very strong statement to the VC:

> *We obviously cannot claim that the numbers we provide to the Senate Committees are accurate. However, I'll wager three martinis that our "weighted monthly averages" are more accurate than those of Systemwide, whoever Systemwide may be. I trust that you realize that neither Mrs. Lovasich nor I enjoy performing computations such as weighted average salary. When and if the University Administration can construct reasonably accurate and complete information which no longer requires exhaustive verification, we would be glad to stop verifying. Since we have not yet arrived at such a happy state, I have sent a request for obtaining the October 31, 1978 PER computer tapes and any new tapes which may be more suitable for our purposes.*

In the next sentence, she asked for funding to reimburse the statistics department for the time Lovasich spent working on the computations.[60]

Betty may have sounded miffed, but she was not deterred from her mission of analyzing the PER data for salary equity. In November, she sent a list of data queries to the director of planning and analysis, asking her to examine each one and sign off on each correction. Once the corrections were agreed upon and made, then Betty could redo the analyses.[61,62] One of the VC's assistants sent Betty some rosters to support her salary equity research, asking her to please keep them in strict confidence.[63]

The VC responded to Betty's tenacious letter in a conciliatory fashion. "I am glad you are aware of the problems with the PER data... and that you are attempting to rectify these inaccuracies with your own computational corrections... With all our recent communications over how we gather data, I would hope that we now can get back to the business of working together to iron out the kinks we have discovered and shall discover in the data." He

---

[60]10/10/1978 letter from Betty Scott to VC Ira Michael Heyman.

[61]"What Jeanne Did With New Appointments" (white paper), n.d.

[62]11/2/1978 letter from Betty Scott to UC-Berkeley Planning and Analysis Director Louise Taylor.

[63]11/18/1976 letter from UC-Berkeley Office of the Chancellor Senior Administrative Analyst Lynn Bailiff to Betty Scott. Note: It is likely that this letter was from 1978 and misdated, rather than 1976, as it refers to "rosters of individuals appointed to Ladder-Rank Faculty and equivalent titles at UC Berkeley during the 1974–75, 1975–76, and 1976–77 Academic Years."

authorized Betty be given the 1978 PER data and $2,000, expressing the wish that he could have come up with more.[64]

33.6   Kit Promotion and Experience

The AAUP was mindful in promoting the Kit. For example, when the association secretary sent one of his periodic letters to Committee W section contacts, he specifically mentioned the Kit and offered additional copies if the campus hadn't yet done a salary equity evaluation.[65] When the AAUP General Secretary (executive director) invited the administrative law judge from the department of labor to speak at the AAUP annual meeting, he included a copy of the Kit.[66] When the AAUP Associate Secretary wrote a follow-up letter to the White House Assistant Director, Domestic Policy Staff for Education and Women's Issues, he included a copy of the Kit.[67]

Betty also personally promoted the Kit. In fact, She was overwhelmed with invitations to speak or write. For example, she accepted an invitation from the Rutgers Council of AAUP to speak to their local Committee W about her salary equity studies, including the Kit. She suggested her talk be titled "How to Estimate Whether Affirmative Action is Even Fair." She told Rutgers: "... the evidence which I have indicates that women in academe are not getting their fair share of jobs, tend to get a lower salary, have a lower probability of being retained, and receive smaller merit increases when they are retained – all of this in comparison with what happens to white males in the same field."[68] Rutgers told Betty after her talk: "The Rutgers University Committee W of AAUP feels that through you we have made a real contribution this year by alerting more people in this area to the inequities faced by women and minorities. The procedures you have devised to ferret out the glossed over and blurred discriminatory practices should give impetus to overcoming these methods. We are grateful to you for sharing your precious time with us, and we hope the fresh awareness of the problem that you provided will cause members of your audience here to work harder to improve women's salaries. We are delighted that we had a chance to meet you and thank you more than we

---

[64]11/13/1978 letter from VC Ira Michael Heyman to Betty Scott. Note: Heyman specifically mentioned California Proposition 13, which decreased property taxes, as constraining the amount of funds he could give Betty for her work on the PER data.

[65]5/8/1978 memorandum from AAUP Association Secretary Lesley L. Zimic to Committee W contacts re annual letter.

[66]5/12/1978 letter from AAUP General Secretary Morton S. Baratz to Department of Labor Administrative Law Judge Edith Barnett.

[67]6/16/1978 letter from AAUP Association Secretary Lesley L. Zimic to The White House, Domestic Policy Staff for Education and Women's Issues, Assistant Director Elizabeth A. Abramowitz.

[68]3/4/1978 letter from Betty Scott to Rutgers University Committee W Chairperson Professor Anne Brugh.

can express, for a resoundingly successful program."[69] For another example, a faculty member at Humboldt State University asked Betty to contribute a chapter to a book on women in science and math targeted at high school and college levels. But Betty declined, feeling she was "badly over-extended with teaching and administration and other reports, some of which are already over-due," and did not "have time to do this properly."[70,71] In her talks, Betty was known to interject humor, which would "give the participants some much-needed perspective."[72]

By September, the Kit had been used at various institutions across the continent. Then Betty received a letter from a researcher at the Society of Wildlife and Parks in Montreal that caused her much concern. The letter identified an error in Example 1 of the Kit. Betty looked at the example closely, and sure enough, there was an error. She wrote to the researcher: "You are quite right that the sum of residuals in dollars for the 57 white male members should sum to 0. It took us a long time to find out what happened and I still have no explanation of why it happened. I very much regret the error involved, especially since this is example #1."

Betty admitted it took her and her staff a long time to find out that it was a programming error. The programmer should have written that entries "greater than 57" should be assigned a weight of 0 in the regression computations; but instead she wrote "greater than or equal to 57... When the error was found, the programmer changed that line [of code] but did not recompute the entire example as certainly should have been done!" She went on: "From my point of view it was a disaster, essentially ruining the example – and I am very sorry it occurred."[73]

Betty sent a letter of reprimand to the programmer: "As you can see, we are distressed by what happened. I presume that you made the change so I want to call this to your attention and ask you to be careful to double check that whenever an error is discovered, all its consequences are corrected."[74]

Mary Gray had been chair of AAUP Committee W for six years when, in 1978, she decided to step down. AAUP regretted Gray's decision: "the extraordinary energy and effort she has expended in behalf of women academicians everywhere deserves our highest praise." But Mary would stay involved over

[69] 5/5/1978 letter from Rutgers University Committee W Chairperson Professor Anne Brugh to Betty Scott.

[70] 3/14/1978 letter from Betty Scott to AAAS Scientific Manpower Commission Betty Vetter.

[71] 3/10/1978 letter from Betty Scott to Humboldt State University Mathematics Department employee Phyllis Z. Chinn.

[72] 5/18/1978 letter from UC-Berkeley Center for Continuing Education for Women Associate Director Sheila M. Humphreys.

[73] 9/4/1978 letter from Betty Scott to FAPUQ [Society of Wildlife and Parks] Thierry Wils.

[74] 9/4/1978 letter from Betty Scott to National Institute of Environmental Health Sciences Statistician.

the next eight years as a consultant and then resume the chairmanship of Committee W in 1986 for six more years.[75,76]

Back in 1978, Betty's term on AAUP Committee Z had ended. She had provided such "dedicated service" that, now in October, AAUP invited her to serve a three-year term on Committee W. It was a natural appointment, as Committees Z and W had collaborated on development of the Kit.[77]

Betty's first reaction was to say "no": She was overextended, so it wasn't reasonable to add another assignment. She added: "Also I have been experiencing rather stubborn treatment in my present efforts to delve deeper into the difficulties faced by women in higher education. I wish the problem would go away and let me turn my full attention to problems in cosmology which, even though they are controversial and even emotional, do not descend to personalities." But then she had two experiences that week that convinced her more work was needed on the status of women in academe. So her final answer was "yes," she wanted to help.[78]

Also in October, Committee W board member Barbara Bergmann reported to Betty the experience of the University of Maryland-College Park in using the Kit. Bergmann talked about diagnosis and remedies. She felt diagnosis should include estimates of both statistical differences between salaries of men and women, and monetary requirements that would be needed "to bring the women's scatter around the line of regression into parity with the men's scatter." She felt the Kit should provide more details on remedies; these would be useful both to institutions and the courts. Bergmann told Betty: "Flagging women under the line of regression is a non-solution because there are lots of men there too, and because it assumes that women above the line of regression have no problem... What is needed is some guideline for the distribution among all women of the amount of the deficiency which was found in the diagnosis part of the study. Across-the-board percentage or absolute increases is one such. Adjusting departmental scatters is another. Still others would fold in estimates of relative deprivation based on measures of merit." Bergmann then suggested perhaps Mary Gray could include something on remedies in the law review article she was writing; and that remedies could be added to a second edition of the Kit.[79] The correspondence between Betty and Bergmann regarding a second edition of the Kit made its way to the AAUP association secretary, who in turn presented the idea to Committee W members for discussion at their next meeting.[80]

---

[75]7/12/1978 memorandum from AAUP Association Secretary Lesley F. Zimic to Conference and Chapter Committees W re Mary Gray's resignation as chair of Committee W (attached report of Committee W, 1977–1978).

[76]Mary Gray, "Curriculum Vitae," December 2011.

[77]10/3/1978 letter from AAUP Associate Secretary Lesley Francis Zimic to Betty Scott.

[78]10/21/1978 letter from Betty Scott to AAUP Associate Secretary Lesley Francis Zimic.

[79]10/17/1978 letter from University of Maryland Economics Professor Barbara R. Bergmann to Betty Scott.

[80]11/30/1978 memorandum from AAUP Association Secretary Lesley L. Zimic to Committee W re possible second edition of the Kit.

Many campuses around the country were using the Kit. A few uncovered problems and thoughtfully brought them to Betty's attention. Some suggested a second edition that provided additional details. At Berkeley, Betty was using methods that required additional productivity data that were, unfortunately, difficult to collect as they came from inaccurate or incomplete vitae. She wrote: "The result is that we actually provide estimates of underpayment by a set of several methods which will allow the appropriate committees and administration to look at the situation this way and that way." But Betty wasn't sure this was the best way to proceed, as the reviews took up considerable amounts of time, which she thought was unfair both to the women and minorities being reviewed and the committee members doing the reviewing.[81]

One of Bergmann's primary comments about the Kit was that the flagging was imprecise. Betty pointed out that the simplicity of the flagging method did serve the purpose of providing a rough estimate of underpayment, which could then be investigated further. Betty explained, there was the expectation that more individuals would be flagged by using Kit methods than by using statistical significance tests. She was discovering that in some universities there were underpayments to women and minorities across the university, whereas in others underpayments were clustered in certain departments or colleges; that all women and minorities in the trouble spots should be investigated, not just the ones flagged; and that causes of the patterns should be investigated and remedies for underpayments found. Betty told Bergmann the Berkeley team had been focusing on progression of women through the ranks, and comparing the progression with that of white males in the same field. They were finding that women were often not recommended by their departments for merit awards, which made their cases invisible to their administrations.

Bergmann commented that there was a large amount of spread around the regression curve. Betty thought the only solution was to use more predictor variables. UC-Berkeley had been collecting data on more predictor variables for four years and the work was ongoing. In Betty's experience, it took "unbelievable" time and effort to collect the additional data. Because the records were incomplete and it had to be done by hand, it was low priority.

Bergmann also mentioned that she planned to analyze the differences between actual and predicted salaries "to look for anomalous bunching." Betty thought this was a great idea and asked Bergmann to keep her informed.

Betty gave two warnings. One was against using rank as a predictor. "This should not be allowed," she told Bergmann, because "salary and rank are too much the same thing. Decisions to increase the rank-step and decisions to increase the salary are the same thing, done by the same person. Any discrimination in setting the salary is equivalent to discrimination in setting the rank-step... [This] is just the monkey-business that one must guard against."

---

[81]11/5/1978 letter from Betty Scott to Montreal's FAPUQ [Society of Wildlife and Parks] Thierry Wils.

Another was against splitting the sample up into such small parts that statistical power would "disappear": "This trap should be avoided."

Betty thought administrations were using the inadequacies of their own data as a "whipping boy" to invalidate their own institutional salary studies. But her experience indicated that the flagging was only slightly affected by a few inaccuracies in the data for white males; It also showed that the estimates of underpayment were only slightly affected by a few missing data points for women or minorities.

Betty thought there should be a follow-up to the Kit, maybe in the form of an article, or maybe in the form of another Kit, she wasn't sure which.[82] A few days later, the AAUP contacted Betty and asked her if she would contribute her special expertise in collaboration with Mary Gray and Committee W to work on the formulation of salary inequity remedies. In fact, they hoped Betty would be willing to lead the project which would involve a revision of the Kit.[83]

Betty agreed this would be an important project, but:

> *I am depressed, however, that this is a very lopsided situation. There are so many ways for the administration to argue, complain, delay, bemoan the shortage of funds and otherwise be frustrating. Also, I am very badly overextended with a mountain of deadlines, quite a few of which I am already behind on. Thus, I think it is not realistic for me to agree to head any project as of now.*

Betty was willing to work on the project, but not lead it.[84]

33.7   Affirmative Action Report Review

As 1978 was ending, UC-Berkeley distributed their latest affirmative action report for review and comment. Betty read the report carefully and provided a five-page commentary, which she said was all she could do given the deadline. She started out by saying that "The general flavor of the Report seems too congratulatory"; there had been some good progress, "but not everything is good." She also said the report didn't include all of the information needed to verify many of the statements made that there had been progress.

Betty gave nine detailed examples. She wanted more complete reporting on grants awarded to women and minorities in comparison to males. She asked why, among many nominations, there was only one appointment ("literally a drop in the ocean of possibilities") made to the Berkeley program of postdoctoral fellowships. She wondered why only positive and no negative staff devel-

---

[82]11/5/1978 letter from Betty Scott to University of Maryland Economics Professor Barbara Bergmann.

[83]11/9/1978 letter from AAUP Staff Member Lesley Francis Zimic to Betty Scott.

[84]11/16/1978 letter from Betty Scott to AAUP Staff Member Lesley Francis Zimic.

opments had been reported; she had seen changes made that sounded "good on paper" but "what actually is happening is very different...Something is wrong!" She wanted to know why a report on academic probationary service hadn't been circulated to committees (Betty hadn't seen it and couldn't find a single woman faculty member who had seen it) to ask if the new rules were better than the old ones. She fussed about some of the data presented in the tables, asking how one could tell if there was improvement for women in promotions given that the tables didn't show a correct comparison ("If anything, the discrimination is worse than it was 20 years ago"). She wanted to know which definition of headcount was used and, saying that "either definition inflates the proportion of women and minorities because they are more likely to be appointed part time"; she asked to see the FTE data. She asked that more recent data be used for availability; this was important because the pools had been increasing rapidly; and "they do comment on some deficiencies but they forget that they are not comparing the correct figures." Finally, closest to Betty's heart, she asked why the report didn't consider salary or appointment inequities, saying: "The University is dragging its feet in these domains. I can remember that the University was required to report on these topics some 8 years ago but the report has not been completed yet."[85]

Basically, Betty thought the report should include all of the evidence (including definitions in addition to data) to back up the statements; some actions that hadn't gone (in her opinion) far enough should be explained; and negative developments should be reported in addition to positive developments. She advocated for valid and up-to-date comparisons; more transparency; and better statistical thinking. Statistician Betty knew what questions to ask and asked them. When she saw something she thought was wrong, she challenged the administration to acknowledge it and do something about it.

---

[85]Elizabeth L. Scott, "Comments on the Report on the Status of Affirmative Action Personnel Programs, 1978" (white paper), 1978.

Part IX

Conclusion

34

Final Decade of Leadership (1979–1988)

No matter when, no matter what, ask the incredible Betty Scott.
-Jerzy Neyman

Betty spent the next, and final, ten years of her life doing all she could to promote women in science or academe. She talked to colleagues about issues on women, was a role model by advancing in her career, created and expanded national advocacy networks, and exchanged gender equity information with others. She encouraged women students; wrote articles or papers about the status of women; gave speeches, talks, or workshops on women; and supported new programs for the career development of women.[1]

Betty continued to look for advocacy data[2] and be concerned about its quality.[3] She supported "reliable alternative paths for admission to college"[4] and studies to explain why women were winning so few NSF awards.[5] She recruited WFC members[6] and promoted WFC business practices that would benefit young women scholars.[7] She advocated for human rights in general[8] and federal legislation to promote women in the sciences in particular.[9,10,11,12]

---

[1] [?6/1978] Math and Science Network Questionnaire completed by Betty Scott.

[2] 2/10/1979 letter from Betty Scott to Education Testing Service Staff Member Patricia Lund Casserly.

[3] 6/15/1980 letter from Betty Scott to AAAS Manpower Commission Staff Member Betty Vetter.

[4] n.d. white paper, "Comments by Elizabeth L. Scott."

[5] 1/20/1984 letter from Betty Scott to NRC/NAS Office of Scientific and Engineering Personnel Acting Director of Fellowships J. Chester McKee and NSF Graduate Fellowship Program Director Joan R. Sagransky.

[6] 3/13/1981 typed note from Betty Scott to Pat Harrington.

[7] 2/9/1980 letter from Betty Scott to WFC Board of Directors Members.

[8] 1/9/1981 letter from UC-Berkeley Statistics Professor Peter J. Bickel to Stanford University Statistics Professor and IMS Committee on Human Rights Chair P. Switzer, with attached 1/8/1981 letter from UC-Berkeley Statistics Professor Lucien Le Cam to UC-Berkeley Statistics Professor Peter J. Bickel.

[9] 4/22/1980 telegram from Betty Scott to US Senate Labor and Human Resources Committee Honorable Alan Cranston.

[10] 9/26/1980 letter from Betty Scott to US House Subcommittee on Science, Research and Technology Representative George E. Brown, Jr.

[11] 10/17/1980 letter from US Senate Labor and Human Resources Committee Honorable Alan Cranston to Betty Scott.

[12] 11/4/1980 letter from US House Subcommittee on Science, Research and Technology Representative George E. Brown, Jr. to Betty Scott.

Betty's national efforts stood out even among some of her AAUP associates, making her an ideal person to serve on other national advisory committees to advance women in science.[13,14] All of this took a huge amount of effort, but Betty still took the time to make personal connections[15] and write letters of support or congratulations for able individuals.[16,17,18]

By persistently applying pressure, Betty was instrumental in promoting quality improvements in data and data reporting on the Berkeley campus and at the system level. Staff responsible for institutional data benefited from the eye of a professional statistician. The executive assistant to the UC president was particularly grateful for Betty's work: "...the support I did not feel in this office has always been offset by the admiration I felt for your work on salaries...Each point you raised years ago is beginning to show in campus policies. 'With the help of our friends...' in the Beatles' language."[19]

34.1   More Women's Studies Publications

Betty authored or co-authored five more women's studies publications between 1979 and 1981.[20]

In 1979, Betty and Jeanne Lovasich published a paper, "Historical Trends in the Appointment and Promotion of Women Faculty," in a UC-Berkeley women's center volume. Betty used the opportunity to talk about how she got involved with women's issues back in 1968:

> *Of course, I was very well aware of these problems, but I think that, like many other women, I already had enough other problems some of which were very clearly not my responsibility but responsibilities of the system.*

---

[13] 11/17/1980 letter from UC Vice President of the University William B. Fretter to NSF Director John Slaughter.

[14] 5/20/1980 memorandum from AAUP Associate Secretary Lesley Lee Francis to Members of Committee W re Women in Science and Technology (S568).

[15] 1/4/1983 letter from Betty Scott to University of Illinois Mathematics Department Professor Stephen Fortnoy's wife.

[16] 11/14/1980 letter from UC-Berkeley Center for Continuing Education for Women Staff Member Sheila Humphreys to Betty Scott.

[17] 11/7/1980 letter from Betty Scott to UC-Berkeley Chancellor I.M. Heyman.

[18] 11/4/1983 letter from Betty Scott to UC Executive Assistant to the President Gloria Copeland. Note: Betty wrote, "There is a strong tendency for women who have been active in any sort of affirmative action to be blackballed, and to be transferred to a back burner or even to be put out. But sometimes this is not what happens, and you are a shining example. You are really very special in many ways, and it is so nice that you continue up, up!"

[19] n.d. handwritten note from UC Executive Assistant to the President Gloria Copeland to Betty Scott.

[20] Note: In 1978 Betty published two proceedings articles that are currently not available in library or online holdings: one to compare the status of men and women in higher education exemplifying the biological sciences; the other on methods to study career discrimination in academe exemplifying Asian/Pacific Americans.

> *The way I had learned to manage was to try not to pay too much attention to the problems, and to try to go around the barn, so to speak, in order to do what one wants to do. When I saw that so many other women had similar difficulties, some perhaps even stronger than I had – I am in my present position more by good luch [sic] than by good management – I thought maybe it is time to do something to see what's really going on.*

Betty then presented an evidence-based comparison of the situation before and after 1970 at UC-Berkeley. She was proud that her work included salaries: "...I can go back to 1972 because since 1972, with excellent cooperation from the Chancellor's office, we have been examining data once each year from the academic payroll for the Berkeley campus." She thought more attention should be paid to ensuring fair treatment for women all along the pathway from searches to merit increases once hired; and more study was needed to determine causes of, and solutions to, women's problems.[21]

Also in 1979, Betty published a paper, "Linear Models and the Law: Uses and Misuses in Affirmative Action," in the *ASA Proceedings of the Social Statistics Section*. Among the most common misuses was including in the models groups of departments or institutions that were heterogeneous; "large numbers of predictor variables so that the reliability of the conclusion is destroyed"; sex as an additive constant; and sex, rank, and salary variables together.[22] Betty received many requests for reprints, and she declared she "never had such a popular paper – perhaps because the title is so provocative."[23] Her topic was of interest to editors of *The American Statistician*[24] and *Journal of the American Statistical Association* (JASA);[25] the latter expressing interest in it as an invited paper with commentary, which appealed to Betty.[26]

In 1980, Betty published a comment in *JASA* in response to an article by Harvard University Statistics Professor Donald Rubin, "Using Empirical Bayes Techniques in the Law School Validity Studies." Rubin proposed a method to predict a student's performance in law using a standardized test score and the undergraduate grade point average. The editor solicited Betty's comment, and she was glad to provide it because she thought Rubin's "claims

---

[21] J.L. Lovasich and E.L. Scott, "Historical trends in the appointment and promotion of women faculty," in *The Impact of the Bakke Decision and Proposition 13 on Equality for Women in California Higher Education*. ed. CCEW Women's Center (Berkeley: University of California, 1979), 92.

[22] E.L. Scott, "Linear models and the law: Uses and misuses in affirmative action," in *Proceedings of the Social Statistics Section, American Statistical Association*, American Statistical Association (Washington, DC, 1979), 20.

[23] 6/23/1981 letter from Betty Scott to Carnegie Mellon University Statistics Professor and *JASA* Invited Papers Editor George T. Duncan.

[24] 7/31/1979 letter from University of Georgia Quantitative Business Analysis Professor and *The American Statistician* Editor John Neter to Betty Scott.

[25] 5/21/1980 letter from Carnegie Mellon University Statistics Professor and *JASA* Invited Papers Editor George T. Duncan to Betty Scott.

[26] 6/7/1980 Letter from Betty Scott to Carnegie Mellon University Statistics Professor and *JASA* Invited Papers Editor George T. Duncan. Note: The invited paper did not appear.

were exaggerated."[27] The *JASA* editor thought her discussion was "lively and thought-provoking and hence a valuable contribution to the paper."[28]

Also in 1980, Betty teamed up with Mary Gray to publish "A 'Statistical' Remedy for Statistically Identified Discrimination."[29] This study was commissioned by AAUP Committee W.[30] The April 1979 Committee W meeting had included an agenda item discussing an amendment of the Kit to include remedies in the particular context of legislation.[31] Noting "a pattern of misuse and error in the use" of the Kit, a plan was made for Gray to draft an article for publication that would describe experience with the Kit and offer advice for future users, including an appeal process related to perceived salary inequities.[32] Gray teamed up with Betty to draft the follow-up article,[33,34] which was published in the AAUP publication, *Academe*. The article explained the use of regression to identify systemic salary discrimination, and then what the remedy should be of such discrimination once identified.

The Kit had recommended an individualized remedy, where individuals flagged would be further investigated on a case-by-case basis for discrimination versus lack of merit. There were methodological and human problems associated with this recommendation. The methodological problem was that, given the way regression works, if the salaries for all women who fell below the regression line were adjusted up to the regression line, then discrimination for women wouldn't necessarily be completely remedied; moreover, men whose salaries fell below the regression line could also have a discrimination claim. The human problem was that the review of flagged cases would rely on subjective qualitative factors, could be costly and painful for the individuals being reviewed, and would overlook the fact that outstanding women above the regression line could also be victims of salary discrimination.

In recognition of these problems, Gray and Betty recommended in their article that a statistical remedy would be "more appropriate" than an individual remedy. A statistical remedy would give the same amount to all women across the board rather than different amounts to individuals based on a case-by-case review: "What regression analysis does is to identify a class discrimination, and thus the remedy should also be a class one." A statistical remedy would match the methodology: When a systematic discrimination against women

---

[27] 1/18/1979 letter from Betty Scott to Bell Laboratories Statistician and *JASA* Editor C.L. Mallows.

[28] 2/22/1980 letter from Carnegie Mellon University Statistics Professor and *JASA* Associate Editor George T. Duncan to Betty Scott.

[29] M.W. Gray and E.L. Scott, "A 'statistical' remedy for statistically identified discrimination," *Academe* 66, no. 4 (1980), 174.

[30] 1/18/1980 memorandum from AAUP Associate Secretary Lesley Lee Francis to Members of Committee W re the statistical remedy paper.

[31] 4/6-7/1979 Committee W Agenda, Item XI.

[32] 4/18/1979 minutes of the Committee W Meeting, April 6-7, 1979.

[33] 11/23/1979 letter from Betty Scott to University of North Carolina Pharmacology Professor Betsy J. Stover.

[34] 11/7/1979 memorandum from AAUP Associate Secretary Lesley Francis to Members of the Council re Committee W Update.

was revealed, then the whole class of women should be remedied, perhaps with a pool of funds set equal to the algebraic sum of the salary residuals among the women. This approach would be clear and fair. Gray and Betty encouraged women to pursue it on their campuses.[35]

AAUP made plans to distribute a copy of the article with future mailings of the Kit.[36] Six months later, Committee W remained hopeful the Kit and article would catalyze "real progress" to eliminate salary inequities for faculty women.[37] After its publication, the *JASA* editor urged Betty and Gray to rework the paper for *JASA*.[38]

Twenty-four years later, Gray observed: "What is appropriate for a statistically identified problem is a statistically based remedy. Thus Betty and I wrote [this] article explaining that if the average difference between men's and women's salaries as shown by a regression model is $2000, then the salary of each woman should be increased by that amount. Sorry to say, this is not an idea that has been widely accepted. Women faculty are still paid less on the whole, there are still occasional regression-based studies, there are spot remedies, and often the very best women faculty continue to be underpaid."[39]

In 1981, Betty and NRC Staff Officer Nancy Ahern published a study, *Career Outcomes in a Matched Sample of Men and Women PhDs: An Analytical Report*.[40] This large study was conducted as an activity of the NRC's Committee on the Education and Employment of Women in Science and Engineering (CEEWISE) with the Commission on Human Resources, and it was funded by the NSF. The aim was to "determine whether the sex differences in career outcomes of PhDs observed in earlier studies diminish or disappear when males and females in a sample are closely matched by education, experience, and type of employment" using data from the 1979 Survey of Doctorate Recipients.[41] Betty had been appointed to CEEWISE in November 1979 as a paid senior statistical consultant to be a co-author of this study.[42] Betty and Ahern grappled with matching criteria,[43] inclusion criteria and comparisons

---

[35]M.W. Gray and E.L. Scott, "A 'statistical' remedy for statistically identified discrimination," *Academe* 66, no. 4 (1980), 174.

[36]"Report of Committee W, The Status of Women in the Academic Profession, 1979-1980."

[37]11/10/1980 memorandum from AAUP Associate Secretary Lesley Lee Francis to members of the AAUP council re Committee W on the Status of Women.

[38]1/19/1981 letter from The University of Chicago Statistics Professor and *JASA* Editor Paul Meier to Betty Scott.

[39]M.W. Gray, "Promoting equity," in *Past, Present, and Future of Statistical Science*, ed. X. Lin, C. Genest, D.L. Banks, G. Molenberghs, D.W. Scott, and J. Wang (Boca Raton: Chapman & Hall/CRC, 2014), 130.

[40]N.C. Ahern and E.L. Scott, "Career Outcomes in a Matched Sample of Men and Women PhDs: An Analytical Report," *National Academy Press* (Washington, DC, 1981), 99.

[41]7/24/1981 letter from NAS Executive Director William C. Kelly to NSF Associate Study Director Morris Cobern.

[42]12/17/1979 letter from NAS [unknown] to Betty Scott.

[43]2/13/1980 letter from NRC CEEWISE Staff Officer Nancy Ahern to Betty Scott.

to be made,[44] variable selection and effects on residuals,[45] and other matters. After a report had been drafted, Betty suggested various expansions or refinements of the study[46] and her appointment to CEEWISE was extended for a second year,[47] although NSF declined to provide support for additional regression analyses.[48] Results indicated that "objective factors alone do not explain why women PhDs earn less than men."[49] One reviewer called the study "admirable...thorough and well-planned...It lays to rest several assumptions about women scholars that I have long felt unwarranted. It is also depressing to find the women so clearly behind for what don't seem explicable reasons. No great surprises here, but we need this study badly."[50] A number of print media covered the report, including *Science*.[51]

In 1983, Betty contributed to the oral history of the WFC. She was asked to talk about her involvement with the club; and the club's mission, membership, and role. She shared memories about her family and education, the Berkeley mathematics and astronomy departments, lunch groups at the WFC, attempts to merge the two faculty clubs, location of the women's center in the WFC, and the organization of the WFC.[52]

34.2 More Honors

Betty gave a speech at the AAUP Georgina M. Smith Awards dinner. It was presumably 1979, the year the award was established in memory of the Rutgers University mathematics professor. The award is given to "...a person who has provided exceptional leadership in a given year in improving the status of academic women or in academic collective bargaining and through that work has improved the profession in general."

Mary Gray won the award for a long list of accomplishments: "Those of us who know her, recognize her extraordinary ability, but stand somewhat in

---

[44]3/8/1980 memorandum from Betty Scott to NRC CEEWISE Staff Officer Nancy Ahern re survey of women scientists in industry.

[45]10/9/1980 letter from NRC CEEWISE Staff Officer Nancy Ahern to Betty Scott.

[46]1/16/1981 letter from NRC CEEWISE Staff Officer Nancy Ahern to Betty Scott.

[47]1/28/1981 letter from NAS Associate Executive Officer Robert W. Johnston to Betty Scott.

[48]2/27/1981 letter from NRC CEEWISE Staff Officer Nancy Ahern to Betty Scott.

[49]10/16/1981 media advisory: Research Council Committee Finds Objective Factors Alone Do Not Explain Why Women PhDs Earn Less than Men.

[50]4/14/1981 letter from Tufts University Liberal Arts Dean Nancy S. Milburn to NAS Executive Director William C. Kelly.

[51]C. Norman, "Sex discrimination persists in academe," *Science*, (1981).

[52]Scott, Elizabeth L. "An Interview with Elizabeth Scott," an oral history conducted by Suzanne B. Riess, in "The Women's Faculty Club of the University of California, Berkeley, 1919-1982," Oral History Center, The Bancroft Library, University of California, Berkeley, 1983.

awe of the enormous range of her contributions and her tireless dedication to fostering both the growth and welfare of faculty women in particular, and the highest ideals of the professoriate in general. To paraphrase a statement which Ted Walden of Rutgers made about Georgina Smith, Mary Gray has accomplished more in selfless service through these last ten years of her career, than most persons could accomplish in three life-times."[53,54] At the time of this book, Gray was still working hard to elevate the status of academic women.

At the awards dinner, Betty contextualized her big telescopes story:

> *Thank you – It is a great pleasure to be here and I want to say how much I appreciate the opportunity to participate in the Forum this afternoon and in this Awards Dinner (Georgina Smith). It is inspiring to participate in a dinner honoring Georgina Smith and I hope that we can carry on the "good fight." There has indeed been progress but there is a way to go yet. Let me give you a personal example. Actually, it is not often that one can point to something as surely discrimination – usually we cannot distinguish between the consequences of our own lack of ability and those of an unfair authority.*
>
> *My degrees are in astronomy. As a graduate student I was very much interested in the distribution of galaxies and the expansion of the universe. Professor Shane, the chairperson of astronomy at Berkeley, arranged that I could spend the summer at the Mt. Wilson Observatory to do some work under Hubble, the great authority in this area. And I did this – it was a wonderful experience – I measured some galaxies and some stars on photographic plates and I counted galaxies on plates taken by Hubble and other astronomers. But I did not assist him at any telescopes, I did not go near any telescopes although some of my student colleagues did. I knew that women were not allowed to use the big telescopes at Mt. Wilson Observatory – the 60", the 100", and later the 200"– there were no women as astronomers there. Oh, there were women as computers, women as plate measurers, doing essentially what I was doing and they, in kindness, warned me that I was foolish trying to get a PhD, that I would be overqualified, that there was no hope for me to be an astronomer. I accepted all this in a sort of "you can't fight city hall" attitude.*
>
> *The situation at Mt. Wilson and Mt. Palomar (now the Hale Observatories) continued like this – Then, in the late sixties, people started to notice that in the observing book for the 60" and for the 100" there was the name of Geoffrey Burbidge, but he was not an observer, it was*

[53] "Georgina M. Smith Award," AAUP, accessed November 8, 2015, http://www.aaup.org/about/awards/smith-award.

[54] 6/1979 Georgina M. Smith Award speech given by Committee W Chair Dorothea Hubin at the AAUP Annual Meeting. Note: The presumption that Betty gave the speech in 1979 is based on the location of the handwritten speech in Betty's files, as well as her close collaborations with Gray. Note: Georgina M. Smith was known for improving the status of academic women through new collective bargaining strategies.

Margaret Burbidge who was observing. Several astronomers approached the Director to request that Margaret Burbidge be allowed to observe in her own name. And so the Council of the Observatory decided to write a woman astronomer to observe in her own name any telescope except the 200" and the 40" Schmidt. But they did not invite Margaret Burbidge, they invited Vera Rubin – at first Vera was annoyed – "It's too bad I am not black – they could kill 3 birds with 1 stone." But she used the 100" and, a couple of years later, the Schmidt. But no woman has yet used the 200" in her own name – there are still no women astronomers on the staff of the Hale Observatories. Women are making progress, but there is a long way to go yet to obtain equality.

As we were saying this afternoon, we can point to some marked improvements in the status of women in the past few years, especially the increase in women and minorities at the assistant professor level, but care must be used in assessing the gains. I am reminded of a story I heard last week about the Texas mule driver...

Betty concluded by declaring, "Life is complicated but not uninteresting."[55]

In 1980, the Chicago Chapter of the ASA selected Betty for their Outstanding Statistician Award.[56] She was asked to submit a photograph but said she couldn't send a new one because she had just broken her nose.[57] She was the 14th recipient of this award, and the first woman.[58] William Kruskal called her talk "splendid" and "stimulating": "It refreshed my concerns for how society should approach questions like the supersonic plane – aerosol – ozone – cancer one where the data are conflicting and often poor, the theory largely unformed or nebulous, and the range of possible conclusions so wide. Of course that's a description of most serious public policy questions."[59]

In 1981, Betty received an honor from The Center for the Study, Education and Advancement of Women (formerly the UC-Berkeley Women's Center). She was given a plaque for her "extensive contributions to women in higher education."[60] A congratulatory note from a woman in economics said: "I personally am grateful for your hard work in striving toward equal rights for women as an everyday reality. I hope my generation can live up to the promise and example set by you."[61]

1984 was a heady year for Betty. In addition to being the fifth, and first

---

[55][1979?] handwritten speech given by Betty at the Georgina M. Smith Award dinner.

[56]"Dr. Elizabeth L. Scott will receive Outstanding Statistician Award at May luncheon," *Parameter, ASA Chicago Chapter* 22, no. 9 (1980).

[57]2/20/1980 letter from Betty Scott to Standard Oil Company Statistician Fred S. Wood.

[58]List, Outstanding Statistician of the Year Award recipients.

[59]5/21/1980 letter from University of Chicago Statistics Professor William Kruskal to Betty Scott.

[60]April 1981 WFC Newsletter.

[61]1981 handwritten note to Betty Scott from a woman in the UC-Berkeley economics department.

woman, president of the Bernoulli Society,[62] on November 3, 1984 a day-long conference was held at Berkeley in honor of Betty. The conference was organized by David Blackwell, David Brillinger, Lucien Le Cam, and others. Princeton University Physics Professor Philip James Edwin Peebles came to talk on "Statistics in the realm of the nebulae." Close friend and colleague Professor F.N. David, who was then retired from her position at UC-Riverside and working as a consultant at the Berkeley Forestry Service station, talked on "The development of health statistics." Professor Margaret L. Kripke, who is an immunologist, skin cancer expert, and advocate for women in science, came from the University of Texas Cancer Center to talk on "Skin cancer carcinogenesis." Biostatistics Professor Virginia F. Flack came from UCLA to talk on "Modeling integrated circuit product yields." Mathematics Professor Howard G. Tucker came from UC-Irvine to talk on "Statistical instruction, consulting and research in the provinces." Professor John L. Kelley, mathematician from UC-Berkeley, talked on "Reflections on academia (aside from teaching and research)." There was coffee, lunch, and a cocktail party. There was a banquet hosted by Blackwell at the WFC where Eudey, then president of California Municipal Statistics, gave a talk on "The way it used to be."[63]

34.3 Neyman's Stroke

Neyman had passed away on August 5, 1981; former student Mark Eudey was the executor of Neyman's estate.[64] But then Miller Research Professor of Statistics and Mathematics Jack Kiefer (1924–1981) died five days later. Neyman's office was in the corner, Kiefer's a few doors down the hall, and Betty's a few doors down from that. Former PhD student Dennis Pearl recalls her saying: "Death is moving down the hall... I'd better move my office..." Betty in fact lived seven years longer than Neyman and Kiefer. But she did reportedly suffer a stroke a few years before her death that, according to Dennis Pearl, "took her out for awhile," causing distortion and discoloration on one side of her face; the nerves didn't fully come back. But Betty otherwise seemed as healthy as before and continued to work as hard as ever.[65]

The death notice sent to the department and others was simple: "Professor Neyman was one of the founders of our science, and his passing marks the end of an era. We are all the poorer."[66] Ingram Olkin summarized: "Jerzy Neyman

---

[62]7/30/1984 letter from UC-Berkeley Statistics Professor and Chair David Freedman to UC-Berkeley Letters & Science Dean Peter J. Bickel.

[63]11/3/1984 agenda, "Conference in Honor of Professor Elizabeth L. Scott," UC-Berkeley.

[64]8/5/1981 certificate of death, Jerzy Neyman.

[65]2/17/2014 personal communication from Ohio State University Statistics Professor Dennis Pearl.

[66]8/5/1981 memorandum from [?JB] to the UC-Berkeley statistics department re announcing Neyman's death.

has been a dynamic force in statistics for over 50 years. It is hard to imagine an area in statistics that he did not influence. But more than that, Neyman fought for the profession in fostering and nurturing students, in the teaching of statistics, in helping others to form departments of statistics, in trying to influence governments to use good statistical procedures."[67]

Neyman's son Mike, a cameraman and film editor for small companies related to the Hollywood film industry, expected that Betty would travel soon after the death to see Neyman's relatives in Poland.[68] Betty knew the details of Neyman's heritage relating to his official name "Splawa-Neyman" in the Polish book of nobility. She and Neyman had once gone together to look for his ancestral home in Eastern Europe. Neyman had a habit of sending his relatives in Poland money for food every Christmas and Easter; after his death, Betty wanted to send money in his stead. However, the year after his death there had been a "crackdown" and it was then illegal for Betty to send anything other than a letter; so she asked David Kendall in England if he might be able to send money or food to Neyman's family on her behalf.[69,70]

Condolences poured in to Betty from mutual friends and colleagues across the globe. Some friends in France wrote: "I think it must be terrible and so difficult for you! You did so much, and you was [sic] so much for him! You are in the de Neyman-family, and we shall always be happy to receive you, when you will come in Europe."[71] C.D. Shane reminded Betty how he was instrumental in getting Neyman to Berkeley and how then Betty's "...own very successful career got under way. Our joint work with the galaxy counts became one of the most happy and fruitful experiences of my directorship. I look back on this time with much satisfaction."[72]

Betty reminisced: "I could never keep up with [Neyman]...Perhaps the best thing is that he was so active and alert...and still kept up with other activities. We were arguing away as usual..."[73]

When he died, Neyman was director of the Stat Lab and the principal investigator (PI) on several grants. Betty became PI on the ONR contract, and Lucien Le Cam on the Department of Health, Education, and Welfare (HEW) grant, with Betty as co-PI.[74,75] This meant the burden of seeking

---

[67] 8/6/1982 letter from Stanford University Statistics and Education Professor Ingram Olkin to Writer Constance Reid.

[68] 8/6/1981 letter from Mike Neyman to Neyman's cousin Roman Lutoslawski.

[69] 1/9/1982 letter from Betty Scott to University of Cambridge Mathematical Statistics Professor David G. Kendall.

[70] 1/29/1982 letter from University of Cambridge Mathematical Statistics Professor David G. Kendall to Betty Scott.

[71] 9/29/1981 handwritten letter from [Dominique] to Betty Scott.

[72] 8/8/1971 handwritten letter from Former Lick Observatory Director C.D. Shane to Betty Scott.

[73] 9/18/1981 letter from Betty Scott to Columbia University Statistics Professor Herbert Robbins.

[74] 8/12/1981 letter from Betty Scott to UC-Berkeley ONR Research Resident Representative Elmer G. Keith.

[75] 8/13/1981 letter from UC-Berkeley Statistics Professor and Acting Director of the Stat Lab Lucien Le Cam to DHEW Regular Research Programs, Program Manager Edward Gardner.

renewal or new funding then fell squarely on Betty's and Le Cam's shoulders. Betty now worried more about bending to the interests of the funders, specific terminology for data analysis, and incorporating intuition into the proposals to indicate how they expected the research to proceed.[76]

Betty was asked to write an obituary for Neyman for the journal *Biometrics*. She declined, saying: "It would be very difficult for me to do this because many of the papers were joint with me so that I would be somehow describing my own work." She suggested they contact some other faculty member at Berkeley or Stanford, perhaps Lincoln Moses, or a former student, or Professor Douglas Chapman at the University of Washington in Seattle.[77] She did agree to send to Cambridge University Mathematical Statistics Professor David G. Kendall various biographical materials and a bibliography. She had to do quite a bit of searching for the information Kendall wanted. She had to update the bibliography which was seven years out of date for books and four years out of date for journal articles, and also to include the Polish articles. She also looked for photographs for Kendall. Interestingly, even though Betty and Neyman went together to Lick Observatory many, many times, "almost every week for two years," she wasn't able to come up with any photographs that had a background related to astronomy.[78] Four years later, Betty agreed to write an entry about Neyman in the *Encyclopedia of Statistical Sciences*.[79]

Betty received many condolences upon the death of her partner Neyman. A few excerpts given below indicate the importance of their partnership.

NIH Public Health Service Statistician Rodney Wong wrote from Bethesda, Maryland:

> *I am deeply grieved to hear about Jerzy Neyman's death. I wanted to share with you this feeling of loss and my deep appreciation for what I have learned from you two. I don't mean the statistics courses especially, but more the examples of integrity of beliefs which I see in both you and Jerzy Neyman. I remember also many late afternoons working in my office on the fourth floor and hearing you call down the hall to him, that it was time to go home – and realizing that it was later than I thought, and it was time for me to go home, too. I feel that you had a very special,*

---

[76] 9/1982 handwritten letter from Betty Scott to UC-Berkeley Stat Lab Staff Member Shyrl and UC-Berkeley Statistics Professor Lucien Le Cam.

[77] 9/18/1981 letter from Betty Scott to Oregon State University Statistics Professor Paula H. Kanarek.

[78] 11/5/1981 letter from Betty Scott to Cambridge University Mathematical Statistics Professor David G. Kendall. Note: Margaret Stein supplied the Polish bibliography for Jerzy Neyman. The reference to Kendall's article is: D.G. Kendall, M.S. Bartlett, T.L. Page, "Jerzy Neyman, 16 April 1894 - 5 August 1981," *Biographical Memoirs of Fellows of the Royal Society* 28, (1982), 379.

[79] Elizabeth L. Scott, "Jerzy Neyman," *Encyclopedia of Statistical Sciences* 6, (1985), 215.

caring relationship to Neyman, and that makes my memories of Berke-
ley something special and wonderful too. Please don't be offended by this
personal note. It is just that since leaving Berkeley I have realized that
the true beauty and dignity of life lies in the compassion and caring that
we have for each other, and I see that quality in the way you cared for
Neyman as well as your students, of whom I am one – and a grateful
one.[80]

Columbia University Associate Professor of Biostatistics Agnes Berger, a collaborator of Neyman's, wrote from New York:

I grieve with you for him, a loved and honored friend. The pain of his loss
is shared by so many around the world, by all who knew him, by all whose
lives were touched by his ideas, his spirit, his generosity and kindness, his
energy and courage. A man of vision, he was, above all, a noble spirit, who
knew no compromise, who labored relentlessly for the good he believed in,
who willingly bent down to straighten, to strengthen, even the lowest blade
of grass. You must try to find comfort in the knowledge that you were so
much a part of what he has accomplished, that he depended and relied on
your support which was always there, never failing. I hope remembering
this can give you strength and help you to carry on, as he would want you
to.[81]

Former collaborator and University of Chicago Biology Professor Thomas Park wrote from Chicago:

The news of Jerry's death is very sad. The international community of
scholars has lost one of its most distinguished members. I have lost a
friend; a person who helped me mature in more ways than one and a
person for whom I had great personal affection. I shall always remember
our sessions together: work, conversation, food and drink, and humor.
You were an integral and significant part of these meetings... I must also
remark that you phrased, ever so gently, a poignant tribute to Jerry when
you used the analogy of granite and snow.[82]

Ruth Gold, a friend of Agnes Berger, also wrote from New York:

You have my deepest sympathy for the loss of your remarkable and won-
derful pal. It was only today that we heard the tragic news about Neyman.
I think that he was the only truly great man that I have ever known. Be-
cause of my fortunate friendship with Agnes, I was privileged to see and
enjoy Neyman, and you, under informal conditions from time to time. I
will always treasure those precious glimpses of this most unusual man, as

---

[80]Undated handwritten letter from NIH statistician Rodney Wong to Betty Scott.

[81]8/13/1981 handwritten letter from Columbia University Biostatistics Associate Professor Agnes Berger.

[82]8/22/1981 letter from University of Chicago Biology Professor Thomas Park to Betty Scott.

well as of the genius. One time, during a lunch with Neyman and others at a meeting, he was talking with great pride and delight about your accomplishments and with his characteristic smile and twinkle quoted someone who was impressed with your own ability to do anything, and do it superbly well:

"No matter when, no matter what,
Ask the incredible Betty Scott."

I hope that the incredible Betty Scott will not find it presumptuous of me to say that I grieve for and with you.[83]

Dinah Boerma, [presumably] the wife of Food and Agriculture Organization of the United Nations Director-General Addeke Hendrik Boerma, wrote from Vienna with the suggestion that Betty write a book about Neyman:

Perhaps you yourself! There was no doubt that he looked upon you as his main collaborator. He never tired of praising your personal charms and professional brilliance. Moreover, I feel certain that you were the one person in the world who best understood his loveable [sic], illuminated, compassionate, stubborn, argumentative, and unforgettable personality. From this distance I don't really need to believe that my dear Steady has gone. But for you, and all those who saw him frequently, the shock of his sudden end, and the void he has left, must be very hard to bear. That he was able to participate in the work he loved so dearly, right up to the end, may bring a thread of comfort. It is the way he would have chosen, knowing that its continuation was in such good hands.[84]

Joel Brodsky, who earned a PhD at Berkeley under Bickel, and his partner wrote from Japan: "We hope you are coping well with the tremendous changes you must be enduring. But Betty, we know you as a very strong person, and believe you will surmount all with style. We enclose lots of positive energy."[85] Barbara H. Mikulski wrote on behalf of herself and husband, University of Maryland Statistics Professor Witold Mikulski, who earned his PhD at Berkeley under Lehmann: "...I am truly sorry for your loss of such a friend, and I hope you can adapt yourself to the change it must make in your own life and work..."[86]

---

[83] 8/14/1981 handwritten letter from Ruth Gold to Betty Scott.

[84] 9/9/1981 handwritten letter from Dinah Boerma to Betty Scott. Note: It is not clear how Betty knew the Boermas.

[85] 9/27/1981 handwritten letter from Joel Brodsky and Joanie Swetman to Betty Scott.

[86] 12/12/1981 handwritten letter to Betty Scott from Barbara H. Mikulski, wife of University of Maryland Statistics Professor Piotr Witold Mikulski.

34.4   Censored Interview?

In 1982, Betty provided a brief interview on the status of women faculty to
UC Clip Sheet and Science Editor Sylvia Paull who was producing programs
on women and science. These programs were broadcast on "Science Editor,"
a UC radio program which was distributed nationwide by CBS and included
on the Voice of America and Armed Forces Overseas Radio.[87] Paull's draft,
unedited script covered Betty's work on salary equity, without specifically
mentioning UC:

Standing Still

*S.E: Just because a woman is hired by a university to teach science doesn't
necessarily mean she'll be earning the same salary as her newly hired
male colleague. Recently the National Academy of Science published a
report showing that in the U.S. most colleges and universities pay their
women faculty less than men. A member of the study group, which was
actually conducted by the NAS Committee on the Education and Employment of Women in Science and Engineering, was Elizabeth Scott,
professor of statistics at the University of California, Berkeley. She
summarizes the findings:*

*SCOTT: I'm saying that women are still starting off at a lower salary after
the Ph.D. than men who are comparable, very comparable. Because men
are the same age, same year they got the degree, same field, and usually
the same subfield, and they, at the same university or a very similar
university. . . sometimes you can't at the same university.*

*S.E: As for these starting salaries, women earn anywhere from 700 to
$1,000 less per year than their male counterparts, depending on their
field.*

*Anncr: Well, what's the reason? Were these women trained differently, or
did their qualifications vary?*

*S.E: In some cases, Scott says, women transferred from a research position
to a faculty one because they decided later on in their careers that they
would rather teach than do just research.*

*Anncr: So then they would get a lower starting salary than a man who had
transferred directly from a faculty position at another university.*

*S.E: Right. But even comparing these women with men who also transfer
from research positions to academic ones, the women still earn less to
start with.*

*Anncr: Well, one way to remedy the problem is by organizing some effort to
ensure that women earn similar pay rates. Of course, in these difficult
economic times, money for funding such efforts might be hard to get.*

---

[87]Mike Castro, "Controversial comments destined for CBS radio: UC professor seeks probe
of 'censored' interview," *The Sacramento Bee* (Sacramento, CA), September 9, 1982.

S.E: Another approach, Scott suggests, is to make people aware that a problem even exists.

SCOTT: I think we really need a change in attitude. I think that what happens to, for example, I was really very surprised when at a meeting with the policy committee a few weeks ago, that most members of that committee have never, aren't aware of the problem, don't really kind of believe that it exists, don't know of any other universities that have made any salary changes, you know. It's a very curious situation.

S.E: The UC professor visits high schools to encourage young women and minorities to take science courses. But she says her work is voluntary and it's up to the individual high school teacher to take the initiative to even invite her to speak.

Anncr: And what's been the response to her outreach efforts?

S.E: Unfortunately, Scott finds that attitudes among young women haven't really changed.

SCOTT: Well, I still find among high school girls, you know, so here they are, fourteen years old at least, I don't know how old they are. They still think that they're going to get married and somebody's going to take care of them all their lives. The world just doesn't work like that any more, especially in California.

S.E: The UC statistician adds that the only way to change these attitudes is through well-organized outreach efforts, but given lack of federal support for such programs, she doesn't think the prospect of attracting women and minorities into science looks very good.

Anncr: What about her own situation in the field?

S.E: Interestingly enough, Scott says that her own salary still does not compare with others in the field, although she publishes many papers and has been teaching for years. She argues, however, that it's hard to show discrimination from one individual case, although in the past, there was blanket discrimination against women in certain of the scientific fields.

SCOTT: My Ph.D. is in astronomy. I wanted to be an astronomer, but it's even harder to get a position in astronomy for women, and actually, that particular time, it was very hard. Like, for example, the biggest telescope, the 100-inch telescope, then later the 200-inch telescope, which was just being built at that time, women were not allowed to use it. So it's not very often that you can quote and really prove that discrimination really exists against women. But here's an example. Women were expressly forbidden to use this telescope.

S.E: Which is why, says the UC professor, she developed an expertise in the statistical aspects of astronomy, because she wasn't permitted to handle the larger telescopes for making observations. Elizabeth Scott, professor of statistics at UC-Berkeley.

Betty expected to be able to edit her remarks, but Paull surprised her with the news that the special assistant to the president (Paull's supervisor) had already rejected the entire script on the grounds that the university should not air its dirty linen in public.[88] Betty's interview was, in her own words, "...deleted for reasons that had nothing to do with its scientific merit; the real reasons are a matter of controversy and keep changing."[89]

Betty didn't think much of the matter at first: "I don't run City Hall, after all."[90] But when Betty learned that the special assistant to the president was also "censoring" other programs and articles that might imply "deficiencies in the University" or use words that were "too graphic,"[91] she approached the chair of the senate policy committee:

> *This troubles me. The person doing the censoring is not a scientist and is not trained in research. I am told, third hand, the [sic] the object is to protect the University image. There will be nothing on studies of women's issues, nothing on studies of sex problems nor on crime nor on unpleasant symptoms of disease, and so forth. Also, reports on the research being done by certain specified faculty members are taboo. I feel that the Senate should look at the extent of the censoring being done and the kind of censoring being done...I am dubious that any censoring is necessary or desirable.*

Betty suggested that the senate may want to appoint a special committee to investigate.[92] Patricia St. Lawrence suggested the senate academic freedom committee should be the investigating committee.[93]

There were casualties and consequences. Paull lost her Science Editor and University Explorer assignments and filed a grievance;[94,95] the Peabody Award-winning producer of Science Editor likewise was demoted.[96,97] The local California State Employees' Association issued a press release, "University of California Public Relations Chief Censors Faculty Research," where it reported that, after Paull indicated the censorship to Betty and others and challenged it, Paull was "forbidden to speak with UC faculty; removed of all radio

---

[88]5/25/1982 letter from UC Clip Sheet and Science Editor Sylvia Paull to Betty Scott. Draft of "Standing Still" attached. Note: The secondhand account of Betty's motivation for focusing on statistical aspects of astronomy does not conform exactly to Betty's previous personal accounts; the script would have been given to Betty to review had the interview not been censored.

[89]4/6/[1982] memorandum of call from Vivian Ouslander to Betty Scott re the deletion of Betty's interview.

[90]Matt Soyster, Untitled white paper, 18.

[91]Dan Brekke, "UC public affairs chief censors science program," *Daily Californian* (Berkeley, CA), September 10, 1982.

[92]7/28/1982 letter from Betty Scott to UC-Berkeley Mechanical Engineering Professor and Committee on Senate Policy Chair Stanley Berger.

[93]9/9/1982 memorandum from UC-Berkeley Genetics Professor [Patricia St. Lawrence] to Betty Scott re which committee should investigate the censorship.

[94]8/12/1982 letter from UC Communications Operations Coordinator Valena M. Williams to UC Editor Sylvia Paull.

[95]8/20/1982 Formal Grievance by UC Editor Sylvia Paull.

[96]Richard Saltus, "Accused of censorship, UC denies topic objectionable," *San Francisco Examiner* (San Francisco, CA), September 17, 1982.

[97]"Employees claim UC censors shows," *News-Press* [unknown], September 1982.

assignments and given clerical duties; and placed on a 15-day 'investigatory leave,' an action usually taken in cases involving criminal charges."[98,99,100]

The special assistant who rejected Betty's interview went to see her and attempted to explain the censorship as editing and judgment. Betty wrote to her file: "I do not know what to make of this. Does not make any sense; that is, the picture as a whole does not make sense."[101,102] The special assistant told the newspaper the script was rejected because it contained inappropriate material and expressed Betty's personal views rather than her research. Betty disagreed.[103] The special assistant was promoted to assistant vice president and stayed in touch with Betty, providing copies of newspaper articles about the censorship issue, as well as a highly detailed critique of sections of Paull's script: "When I left your [Betty's] office... I was convinced that you believed that there was another plausible explanation for the disposition of the Science Editor script... I don't ask that you accept it as the only explanation but that you consider it a logical alternative to what you have been told by Ms. Paull."[104]

Betty's response indicated she hadn't been swayed. She still wasn't able to sort out the different versions of what happened; she would have corrected her hazy statements had the matter been handled in a usual way.[105] Betty was caught in the middle, but felt...

> *...more than before that the Senate should look into what is happening... the senior employees in the Office of Public Information are now quite frankly insisting that they will select and judge all persons and topics to be interviewed... I call this censorship... The actions of the Office of Public Information during the last six weeks do not make sense to me. Why did the two top employees react so strongly when they saw my letter to you? We all know that the Senate moves slowly and might not move at all. Why didn't they keep quiet and thus avoid grievance suits and newspaper and radio publicity? Why did they talk about my draft interview using statements that are incorrect and far from appropriate?*[106]

[98] 9/9/1982 press release, California State Employee's Association, Chapter 41: "University of California's Public Relations Chief Censors Faculty Research."

[99] Dan Brekke, "UC public affairs chief censors science program," *Daily Californian* (Berkeley, CA), September 10, 1982.

[100] Mike Castro, "Controversial comments destined for CBS radio: UC professor seeks probe of 'censored' interview," *The Sacramento Bee* (Sacramento, CA), September 9, 1982.

[101] 9/13/1982 To the File by Betty Scott re a meeting with the special assistant to the president.

[102] "Employees claim UC censors shows," *News-Press* [unknown], September 1982.

[103] Dan Brekke, "Woodard explains censorship: Broadcast quotes 'inappropriate,'" *The Daily Californian* (Berkeley, CA), September 15, 1982.

[104] 9/22/1982 letter from [UC Communications and Public Affairs Assistant Vice President] to Betty Scott.

[105] 9/23/1982 letter from Betty Scott to [UC Communications and Public Affairs Assistant Vice President].

[106] 9/24/1982 letter from Betty Scott to UC-Berkeley Mechanical Engineering Professor and Committee on Senate Policy Chair Stanley Berger.

Betty felt so strongly she wrote to the UC-Berkeley Chancellor in addition to the senate policy committee chair, telling him she thought the UC offices of public information and public relations should be separated, and there should be guidelines to "allow and encourage the free access to reports of the scientific research of University faculty and researchers."[107] Betty continued: "As a faculty member, I think that we are entitled to have the science interviews broadcast without any changes imposed by a public relations officer to improve their notions about the University image."[108]

Paull was eventually dismissed.[109],[110] In the process, the California State Employees Association filed a federal lawsuit on behalf of Paull and the demoted producer. The suit asked that any censorship be stopped on the basis of constitutional protections to free speech, and that Paull and the producer be reinstated to their previous duties.[111] Betty became a headliner: "Teacher says UC Berkeley censored her."[112] A group of faculty called for "an immediate and urgent investigation"; the UC president began one; and meanwhile, Betty turned down an offer from the assistant vice president to be re-interviewed.[113] Betty reported that nothing was said about poor writing on the part of Paull in their first communications; this allegation came "only at the end."[114] The assistant vice president maintained that "...Paull was perfectly free to redo the interview, but that, as presented, it made Scott appear 'bumbling and anecdotal.' In intervening, she says, 'I thought I was saving Elizabeth Scott's skin.'"[115] The grievance committee sided with the university, as did the federal judge who said: "...programs such as 'Science Editor'...were not subject to the same constitutional guarantees as other forms of media because it performed primarily a 'public relations' as opposed to a 'public information' function.[116]

According to Betty, both the Science Editor series, which had run for 34 years, and the University Explorer series, which had run for 49 years, were discontinued; and the Clip Sheet distribution was reduced. The UC committee on academic freedom's investigation of the office of communications and public affairs concluded the academic freedom of Betty and others was not

---

[107]9/26/1982 letter from Betty Scott to UC-Berkeley Chancellor I.M. Heyman.

[108]12/4/1982 letter from Betty Scott to UCSF Psychiatry Professor and UC Committee on Academic Freedom Professor Carroll M. Brodsky.

[109]9/17/1982 memorandum from [UC Communications Operations Coordinator] to Sylvia Paull re intent to dismiss.

[110]Dan Brekke, "UC employee fired in 'censorship' case," *The Daily Californian* (Berkeley, CA), September 20, 1982.

[111]"Suit accuses UC of censorship," *Tribune/Today* [unknown], September 28, 1982.

[112]"Teacher says UC Berkeley censored her," [unknown, n.d.]

[113]Dan Brekke, "'Urgent' investigation demanded by faculty," *The Daily Californian* (Berkeley, CA), September 29, 1982.

[114]12/4/1982 letter from Betty Scott to UCSF Psychiatry Professor and UC Committee on Academic Freedom Professor Carroll M. Brodsky.

[115]Matt Soyster, Untitled white paper, 18.

[116]Gary Lee Seto, "Judge rejects censorship case: Opinion says UC radio program was only edited," *UCLA Daily Bruin* (Los Angeles, CA), January 12, 1983.

violated.[117] "I am sorry that they [the committee] are condoning actions that are more complex than I could have realized...I think that these actions are most unfortunate for the University and for the people of California," Betty wrote.[118,119] Basically, she felt "clobbered by the Senate."[120]

Thirty years later, Paull recalled the help she received from Betty:

> *Elizabeth Scott spoke in my defense in both an internal administrative hearing and in a lawsuit filed against the University of California by myself. Someone on the committee notified [the assistant vice president], who eventually fired me for insubordination when I contacted faculty and media about the situation. [The assistant vice president] was fired a few months later when a Daily Cal reporter discovered she had lied about having a college degree (from UC Riverside)...I later met with Scott, who was a courageous and rebellious person for the time.*

Paull reported that the lawsuit went on for five years, and that leaving UC turned out to be wonderful for her career.[121] She filled in some details:

> *Scott's interview was the first to be censored... [It was followed by two others.] I thought long and hard before writing her a letter (pre-Internet) about the script being censored, but by that time, the second script on the menstrual cycle had also been censored, and it seemed like a serious issue to me...I just thought...that she would like to know what had happened. And I thought, too, that she might help because from her interview, she seemed like a fighter. I also was well aware that I would probably get fired. After I sent the letter, I got a call from her a few days later and she asked me to have lunch with her at the Women's Faculty Club. At lunch, I told her about all the censorship and she said she would look into it...*

Betty fought for Paull, even though this was the only time the two women ever talked face-to-face.[122]

---

[117][n.d.] "Faculty Committee Finds No Evidence of Censorship in Academic Freedom Case."

[118]4/6/[1983] memorandum of call from Vivian Ouslander to Betty Scott re the deletion of Betty's interview.

[119]Richard Saltus, "UC officials defend radio interview cuts," *San Francisco Examiner* (San Francisco, CA), September 17, 1982.

[120]4/20/1983 letter from Betty Scott to Sylvia Paull.

[121]Sylvia Paull, e-mail to author, May 13, 2012.

[122]Sylvia Paull, e-mail to author, May 14, 2012.

34.5 Revelations

Archival records for the years 1979–1988 in the Elizabeth L. Scott Collection in the Bancroft Library at UC-Berkeley contain some interesting revelations from the years 1979–1988. Here are a few such revelations.

In 1979, Betty thought more effort should be put into getting a "better understanding of why women meet roadblocks in sciences – why and where and what to do to improve." She also thought it was "important to make more young women appreciate possibilities of science."[123]

Betty wrote many papers with Neyman. He had been submitting papers under the name J. Neyman because he thought that writing Jerzy Neyman would have sounded pretentious, as it wasn't an English name. Betty recalled: "Sometime when we were publishing joint papers, a comment came out that women were required to write out their first names even though men were not [sic] but this was a convenience for me when I wanted to count papers according to sex, or to count appointments, etc. Then he started writing Jerzy Neyman and Elizabeth L. Scott!"[124]

People began associating facts about big telescopes with Betty, some of which weren't entirely accurate. In the December 1979 WFC Newsletter, the following statement was credited to Betty: "There are many women who do not have enough training to do what they want. For years, there has been a rule at Mt. Hamilton forbidding women to use the larger telescopes. In the past, one woman professor used the telescope by presenting her husband's name."[125] In this case, the rule was at Mt. Wilson, not Mt. Hamilton. Mary Shane and Betty were quick to point this out, both declaring that "...the use of the large telescopes on Mt. Hamilton by women has never been discouraged." Betty thought the comment was confused with one she had made earlier in a film "...deploring the difficulties women had in the past observing at Mt. Wilson..." Betty then pointed out: "My own Aunt Phoebe used [the telescopes on Mt. Hamilton] before 1912."[126,127] This is the only time there is a record of Betty talking publicly about her Aunt Phoebe, which she was apparently prompted to do out of embarrassment and because she valued her relationship with the Shanes.

In 1980, Betty completed a higher education career questionnaire that revealed some additional details about her background and attitudes. Her mother had done some graduate work. During Betty's childhood, her mother had worked full time for six years and one day a week for the remaining years. Betty's family considered her education to be highly important, both so that

[123] 9/10/1979 "Math/Science Network Member Survey" completed by Betty Scott.
[124] 1/9/1982 letter from Betty Scott to Cambridge University Mathematical Statistics Professor David G. Kendall.
[125] 12/1979 WFC Newsletter.
[126] 1/1980 WFC Newsletter.
[127] 2/1980 WFC Newsletter.

she could provide for herself, and as an end in itself: "Education is important as an aid in being a complete person," Betty inserted. Her career goal was to "do interesting research." She might consider a move if presented with a more attractive job offer, whereas she was neutral on moving for more money and better benefits; but she liked her community and therefore didn't want to move. She worked for other than economic reasons "because I like to do things," and she planned to work indefinitely. She saw her work as including frequent social responsibilities (e.g., entertaining); she organized and planned, invited guests, bought items, and cooked; and she employed a special person to help her serve and clean up.

In the same questionnaire, Betty acknowledged three pillars to her success: hard work and effort; a good deal of luck, timing, and networking; and native intelligence. She saw herself at work as being mostly intelligent, open, friendly, concerned, sincere, and just; she tried to be sensitive, pleasant, and compassionate; but she didn't see herself as being dominant. She was productive, having over the past five years published over 15 articles and given over 100 professional presentations. She did extensive traveling and actively engaged in professional development. She had a broad view of family that included her parents, relatives, and friends; and she considered them to be highly important in her life. It was her responsibility to take time from work when needed to take care of her mother, with whom she lived; she was satisfied with the division of labor in her household; and she felt her career had benefited her, and not cost her. As a young adult, she didn't have expectations to marry, have children, or care for elderly parents. She did not fall victim to impostor syndrome or feel she had a life crisis in the last year. She felt there wasn't enough time to take care of her job, family, and self. She acknowledged having mentors or sponsors, all helpful male research collaborators at least 30 years her senior. She also acknowledged having several role models, "depending on the situation." On a scale for satisfaction with life in general of 1 (not at all satisfied) to 7 (highly satisfied), she gave herself a rank of "5."[128]

34.6   Betty's Fatal Stroke

On July 1, 1988, Betty transitioned to Professor Emeritus. She remained fiercely intellectually curious. She wrote a Faculty Research Grant Application, asking for funds to resume study of the spacial distribution of galaxies that she began in 1952 and stopped in the mid 1960s due to lack of observations. Now there was a flood of new data available, in new parts of the sky,

---

[128]1980 "Higher Education Career Questionnaire," Graduate School of Education and Human Development, University of Rochester (New York).

of new wavelengths, using new techniques. Now there was also a flood of new theory as observed currently and also at times near the "origin." She wrote:

> *I want to produce proper estimates and I want to compare the conse-*
> *quences and even to test one model against another. Such tests are dif-*
> *ficult theoretically and involve heavy computations. During the last few*
> *years I have been working on theoretical arisint [sic] with very large bod-*
> *ies of data, many dimensions, and different types of multivariate data*
> *sets. This last year I have been collating the various observational sets*
> *and the various theories. Several Berkeley astronomers are leaders in this*
> *field. I am, I think, now ready to start the appropriate empirical studies.*

Betty requested $600 for data sets and $300 for assistance in special programming.[129]

Statistics Professor Nancy Flournoy, then at American University, had in 1986 become the first woman to direct the NSF program in statistics. She noticed there were few women submitting and getting NSF grants or attending annual meetings of the IMS: At the time, the IMS and ASA meetings were separate, and only five of the 200 IMS attendees were women. In 1988, Flournoy wanted to changes this: "...I thought I couldn't stand being there [at the IMS meeting] again with no women. I thought women needed to be included in all the discussions that went on." In response, she teamed up with University of Alabama Statistics Professor Lynne Billard to create an annual Pathways to the Future workshop that brought together junior and senior women in statistics to learn and share. The workshops were "aimed at helping the recent woman graduate as she sets out on her career." A number of "senior" women in statistics were invited to attend as a way of enhancing the experience for the young academic women in statistics.

Pathways was administered outside of the IMS meetings with support of NSF grants. Initially, Flournoy approached NSF Division of Mathematical Sciences Director Judith Sunley to request funding to get women to the IMS meeting. The grant application deadlines had already passed, so Sunley pointed to a wall of manuals, said to look for a loophole, and if Flournoy found one then Sunley would sign. Flournoy gave it her best to find a loophole. Billard was encouraged to write the proposal and gave the initiative its name. A loophole wasn't found, but Sunley signed anyway. NSF funded the program for 17 years (1988–2004); the Office of Naval Research developed an interest in the program and co-funded it from 1989–1995. Pathways was copied by other organizations including the IMS, which developed a new researchers workshop, and the AMS, which was given copies by NSF of Billard's grant proposals to use. A female NSF program director began to require what was deemed by Pathways leaders to be an unreasonable amount of documentation of effectiveness, until eventually the program was discontinued; however, over

---

[129]4/28/1988 UC-Berkeley Faculty Research Grant Application for 1988–89 by Betty Scott regarding the project "Spacial Distribution of Galaxies: Clusters versus Voids versus Sheets versus Bubbles versus Wormholes."

the years, more than 100 women participated in the workshops, many of whom became or were leaders in statistics professional associations, universities, and other organizations.

Billard had spent time in the Bay Area as a Stanford University visiting assistant and associate professor in 1974, and later as a UC-Berkeley research fellow in 1979. She knew Betty had "data" because the two had talked about it over the years. In 1988, Billard invited Betty to give the opening talk at the first Pathways workshop to be held on August 13, 1988 at the IMS annual meeting in Fort Collins, Colorado. According to Billard, Betty was delighted to be asked and changed her plans so she could participate and present her research on women in statistics. In addition to Betty's keynote talk, Sunley and her staff "addressed issues involving the gaining of federal support for research projects." The workshop always addressed tenure issues, how to react to negative referee reports, teaching responsibilities, and so forth. Flournoy recalled that the men were "terribly threatened" by the meeting: Texas A&M University Distinguished Professor Emanuel "Manny" Parzen and others were listening at the door; when the women inside the room realized this, they grabbed Manny and brought him inside so he could report back to the others.

The first Pathways workshop was held shortly after Betty retired from active faculty duty and a few months before her death. Flournoy recalled that Betty "...gave a great talk...that set the agenda for subsequent conferences – the last talk she gave before she died...Betty's talk was profound. She was wonderful, and it was a wonderful experience. She set the tone for the future for women in statistics." Billard recalled that Betty's presentation "...was a real eye-opener." Flournoy added that Betty's talk "...was as real eye opener even to me...I think Elizabeth hit us all with her data like a ton of bricks. We were really stunned. After providing undeniable evidence of gender discrimination, she insisted we face it bolding [sic] and presented a series of recommended action items to deal positively with issues and situations." When asked if Betty looked ill, Flournoy said "no": "Betty was so glad she was talking to us. It was great. It was like, now I can finally let it go, because someone else is working on this. It's almost like Betty knew she was dying. The talk had a feeling of handing over. It had a feeling of a big sigh of relief." Billard added: "Yes, well, Elizabeth was palpably escatic/overjoyed that someone was taking this up for her...can't remember her exact words, but her 'relief' was very clear...it came across that phone call loudly, but also at the first Pathways too."[130,131,132,133,134,135,136,137,138,139] Billard continued to direct the Pathways workshops after Betty's death in December 1988, including giving the opening talks in the spirit of Betty.[140]

Juliet Shaffer heard that Betty was taking medication for some kind of health problems. She recalls the day she learned Betty had a stroke. Betty had religiously brought baked goods to the lab as treats every Saturday at 3:00 pm. According to Shaffer, "I heard early one Saturday that she [Betty] had a stroke. I told everyone at the lab: 'Not [any baked goods] today.' She was driving with her gardener back to her house. Fortunately he was able to get control of the car. Eric [Lehmann] and I visited her in the hospital a few times. Her admin assistant... was getting everything ready for ELS to come home. She had a heart attack when she was still in the hospital."[141]

Betty passed away at 2:45 pm on December 20, 1988 at Alta Bates Hospital in Berkeley. She had been a UC-Berkeley professor for 40 years. The immediate cause of death was pulmonary embolus, with intracerebral hemorrhage also contributing. She had a craniotomy a month earlier, on November 23. Betty was 71 years old when she died. Her remains were cremated. She was outlived by her mother, whom Betty lived with and cared for. Betty never married.[142] According to Shaffer, "At the memorial at the chapel, we were all waiting and we heard a voice that sounded just like hers [Betty's]. We turned around and it was her mother, she was in her 90s."[143]

F.N. David said this in Betty's obituary: "Her weakness was in trying to do too much. She did not know how to say no and in consequence was always behind in her commitments which necessitated long hours of work when she should have been resting and relaxing. Many of her friends told her this but she continued on her own indomitable way until the end. She received many honours [sic] and was known throughout the statistical world."[144]

Upon Betty's death, the UC-Berkeley Office of Public Information issued a press release that called her "... an internationally renowned contributor to the theory of statistics..." and "A feisty academician who argued strongly

---

[130]Nancy Flournoy, e-mail to author, January 24, 2014.

[131]Nancy Flournoy, e-mail to author, January 25, 2014.

[132]7/8/1988 letter from University of Georgia Professor and Head of Statistics Lynne Billard to Ohio State University Statistics Professor Sue E. Leurgans.

[133]Nancy Flournoy, e-mail to author, March 16, 2016.

[134]Lynne Billard, e-mail to author, March 16, 2016.

[135]Lynne Billard, e-mail to author, March 18, 2016.

[136]Lynne Billard, e-mail to author, March 22, 2016.

[137]Nancy Flournoy, e-mail to author, March 23, 2016.

[138]Lynne Billard, e-mail to author, March 25, 2016.

[139]Calvin C. Moore, "Women mathematicians at Berkeley – The early years," *Association for Women in Mathematics Newsletter* 36, no. 4 (2006), 23.

[140]Lynne Billard, e-mail to author, March 23, 2016.

[141]Juliet Shaffer, e-mail to author, August 5, 2014.

[142]12/21/1988 State of California Certificate of Death for Betty.

[143]Juliet Shaffer, e-mail to author, August 5, 2014.

[144]Florence N. David, "Obituary: Elizabeth Scott, 1917–88," *Journal of the Royal Statistical Society A (Statistics in Society)* 153, no. 1 (1990), 100.

for her ideas..." It pointed out that her pioneering study of gender salary inequity in academe "...had a major impact on universities worldwide." It remembered her as "...a superb teacher and a warm person with a very powerful intellect and a strong will."[145]

Lucien Le Cam, Susan Ervin-Tripp, David Blackwell, Elizabeth Colson, Erich Lehmann, and Laura Nader constituted a committee to write a Memorial Resolution for Betty. LeCam wrote this into the obituary:

> *She was a role model and a mentor for many aspiring young female scientists whose careers she followed with active interest but she was also a mentor for many aspiring male statisticians. She will be remembered for her readiness to help and for her extreme generosity by the students and staff of the Department. She will be remembered by deans as a feisty chair of the Department and champion of its studies. She cared greatly about Berkeley, the Statistics Department and the Laboratory and strove to enhance their reputation. She gave a great deal of time to various campus committees and played a prominent role in the design and maintenance of Evans Hall where Statistics is presently housed. Members of the Department used to joke that Betty owned Evans Hall. She was faithful [sic] member of the Berkeley Senate and will be remembered by faculty colleagues across campus for the integrity and courage with which she spoke out at Senate meetings to challenge what she regarded as inequities or stupidity. She was prepared to fight in behalf of others for what she believed was right and to provide time, labor and evidence to effect needed changes through the usual procedures. She believed that in Academia, of all places, action ought to be guided by thought and evidence and she brought to that the standards of a scientist.*[146,147]

In her advocacy, Betty collaborated toward positive action; made evidence-based recommendations; used statistical methods; was precise and accurate; used channels; worked hard and persevered; kept monitoring; and involved men.[148] Because she brought the standards of a scientist to her advocacy work, Betty was listened to and not dismissed.

The LA Times called Betty an "accomplished astronomer" who "made major contributions to the understanding of the spatial distribution of galaxies, the likelihood of success in attempts to modify the weather, and ozone layer depletion and its link to skin cancer." But they also called her an "ar-

---

[145] 12/21/1988 press release announcing Betty Scott's death.

[146] Lucien Le Cam, Draft Obituary, Elizabeth Leonard Scott.

[147] David Blackwell, Elizabeth Colson, Susan Ervin-Tripp, Lucien Le Cam, Erich Lehmann, and Laura Nader, "Elizabeth Leonard Scott, Statistics: Berkeley, 1917–1988, Professor Emerita," *California Digital Library*, accessed July 17, 2016, http://texts.cdlib.org/view?docId=hb4t1nb2bd;NAAN=13030&doc.view=content&chunk.id=div00061&toc.depth=1&brand=calisphere&anchor.id=0.

[148] Amanda L. Golbeck, "Four leadership principles for statisticians: A note on Elizabeth L. Scott," in *Leadership and Women in Statistics*, ed. Amanda L. Golbeck, Ingram Olkin, and Yulia R. Gel (Boca Raton: Chapman & Hall/CRC, 2016), 31.

dent feminist," stating that her studies conclusively showed "women faculty members at major universities were paid substantially less than men"; she helped "reveal the 'gender gap' in salaries for women" and "start the trend toward equal pay for equal work." Le Cam was quoted as saying, "She was not afraid to stand for what she believed in."[149]

Betty worked for gender equity across a span of 20 years. She was discouraged time and again by not seeing any major gains. Nevertheless, she was as optimistic as she could be and couldn't bring herself to give up. She had to keep trying. Equivalence was her vision.

Kjell Doksum recalled: "Betty started out studying astronomy, and talked about the sexism... So she met Jerzy Neyman who was not a sexist, and ended up in statistics. However, Neyman and Scott had a routine. He would say: Betty, please make the coffee, it tastes so much better when made by a woman. Then she would say: You are such a CP (chauvinist pig). They would both smile."[150]

[149] "Obituary: Elizabeth Scott, UC Professor and Feminist," *LA Times* (Los Angeles, CA), December 22, 1988.

[150] Kjell Doksum, e-mail to author, December 4, 2014.

Publications of Elizabeth L. Scott

Astronomy Publications

Bartlett TJ, Scott EL and Panofsky HAA. Comet, Cosik-Peltier, elements and ephemeris. Harvard College Observatory Announcements, No. 471, 1939.

Bartlett TJ, Panofsky HAA and Scott EL. Comet, Vaisala, elements and ephemeris. *Harvard College Observatory Announcements*, No. 482, 1939.

Kaster KP, Bartlett TJ, Scott EL and White R. Comet Hassel, elements and ephemeris. *Harvard College Observatory Announcements*, No. 480, and *Union Astronomique Internationale Circulaire*, No. 762, 1939.

Bartlett TJ, Scott EL and Panofsky HAA. Note on periodic comet Vaisala. *Publications of the Astronomical Society of the Pacific* **51**(301):173-174, 1939.

Panofsky HAA and Scott EL. Comet Rigollet, elements and ephemeris. *Harvard College Observatory Announcements*, No. 500 and *Union Astronomique Internationale Circulaire*, No. 791, 1939.

Fell PE, Scott EL, White R, Irvin JHB and Panofsky HAA. Comet Friend, elements and ephermeris. *Harvard College Observatory Announcements*, No. 512, and *Union Astronomique Internationale Circulaire*, No. 799, 1939.

Scott EL. Comet Okabayasi, elements and ephemeris. *Harvard College Observatory Announcements*, No. 540, and *Union Astronomique Internationale Circulaire*, No. 826, 1940.

Scott EL and Panofsky HAA. Comet Okabayasi-Honda, improved elements and ephemeris. *Harvard College Observatory Announcements*, No. 571, and *Union Astronomique Internationale Circulaire*, No. 847, 1941.

Scott EL. Note on comet Okabayasi (1940 e). *Publications of the Astronomical Society of the Pacific* **53**:34, 1941.

Scott EL and Stahr ME. Comet du Toit-Neujmin, elements and ephemeris. *Harvard College Observatory Announcements*, No. 597, and No. 603, 1941.

Cunningham LE and Scott EL. Comet Pajdusakova-Rotbart, elements and ephemeris. *Harvard College Observatory Announcements*, No. 749 and *Union Astronomique Internationale Circulaire*, No. 1042, 1946.

Cunningham LE and Scott EL. Comet Pajdusakova-Rotbart, improved elements and ephemeris. *Harvard College Observatory Announcements*, No. 751 and 752, *Union Astronomique Internationale Circulaire*, No. 1049, 1946.

Scott EL. Distribution of the longitude of periastron of spectroscopic binaries. *Astrophysical Journal* **109:** 194-207, 1949.

Scott EL. Further note on the distribution of the longitude of periastron. *Astrophysical Journal* **109:** 446-451, 1949.

Scott EL. Statistical studies relating to the distribution of the elements of spectroscopic binaries. In *Proceedings of the Second Berkeley Symposium on Mathematical Statistics and Probability.* pp 417-435. University of California Press, Berkeley, 1951.

Neyman J and Scott EL. A theory of the spatial distribution of galaxies. *Astrophysical Journal* **116:**144-163, 1952.

Neyman J, Scott EL and Shane CD. On the spatial distribution of galaxies: A specific model. *Astrophysical Journal* **117:**92-133, 1953.

Neyman J and Scott EL. Frequency of separation and of interlocking of clusters and galaxies. *Proceedings of the National Academy of Sciences* **39**(8):737-743, 1953.

Neyman J and Scott EL. On the problem of expansion of clusters of galaxies. In *Studies in Mathematics and Mechanics* (Anniversary Volume for Professor von Mises) pp 336-345. Academic, New York, 1954.

Scott EL, Shane CD and Swanson MD. Comparison of the synthetic and actual distribution of galaxies on a photographic plate. *Astrophysical Journal* **119:**91-112, 1954.

Neyman J, Scott EL and Shane CD. The index of clumpiness of the distribution of images of galaxies (abstract). *Astrophysical Journal* Supplement 8, **1:**269-293, 1954.

Neyman J and Scott EL. Spatial distribution of galaxies – Analysis of the theory of fluctuations. *Proceedings of the National Academy of Sciences* **40**(10):873-881, 1954.

Scott EL. Distribution of certain characteristics of clusters of galaxies, with particular reference to the hypothesis of an expanding universe (abstract). In *Proceedings of the International Congress of Mathematicians 1954* **2:**303-304, 1954.

Neyman J and Scott EL. On the inapplicability of the theory of fluctuations to galaxies. *The Astronomical Journal* **60**(2):33-38, 1955.

Neyman J and Scott EL. The distribution of galaxies. *Scientific American* **195**:187-200, 1956.

Neyman J, Scott EL and Shane CD. Statistics of images of galaxies with particular reference to clustering. In *Proceedings of the Third Berkeley Symposium on Mathematical Statistics and Probability* **3**:75-111. University of California Press, Berkeley, 1956.

Scott EL. The brightest galaxy in a cluster as a distance indicator. *The Astronomical Journal* **62**(152):248-265, 1957.

Neyman J and Scott EL. Statistical approach to problems of cosmology. *Journal of the Royal Statistical Society. Series B (Methodological)* **20**(1):1-43, 1958.

Neyman J and Scott EL. Large scale organization of the distribution of galaxies. *Handbuch der Physik* **53**:416-444, 1959.

Mayall NU, Scott EL and Shane CD. Statistical problems in the study of galaxies. *Bulletin de l'Institut International de Statistique* **37**:35-53, 1960.

Lovasich JL, Mayall NU, Neyman J and Scott EL. The expansion of clusters of galaxies. In *Proceedings of the Fourth Berkeley Symposium on Mathematical Statistics and Probability* **3**:187-227. University of California Press, Berkeley, 1961.

Neyman J and Scott EL. Estimation of the dispersion of the redshift of field galaxies. *The Astronomical Journal* **66**(3):148-155, 1961.

Neyman J and Scott EL. Field galaxies: Luminosity, redshift and abundances of types. Part I. Theory. In *Proceedings of the Fourth Berkeley Symposium on Mathematical Statistics and Probability* **3**:261-276. University of California Press, Berkeley, 1961.

Neyman J and Scott EL. Magnitude-redshift relation of galaxies in clusters in the presence of instability and absorption. *The Astronomical Journal* **66**(10):581-589, 1961.

Neyman J, Page T and Scott EL. Foreword: Conference on the Instability of Systems of Galaxies. *The Astronomical Journal* **66**(10):533-535, 1961.

Neyman J, Page T and Scott EL, Eds. Proceedings of the Conference on the Instability of Systems of Galaxies (monograph). *The Astronomical Journal* **66**(10):533-636, 1961.

Neyman J and Scott EL. Diameters, magnitudes and distances of galaxies. Research Report, Berkeley Statistical Laboratory, 1961.

Scott EL. Distribution of galaxies on the sphere: Observed structures (groups, clusters, clouds). In *Problems of Extra-Galactic Research*, Proceedings from IAU Symposium 15 (GC McVittie, editor) pp 269-293. MacMillan, New York, 1962.

Neyman J and Scott EL. Contribution to the study of the abundance of multiple galaxies. In *Studies in Mathematical Analysis and Related Topics: Essays in Honor of George Pólya*. (G Szego, C Loewner, S Bergman, MM Schiffer, J Neyman, D Gilbarg and H Soloman, editors) pp 262-269. Stanford University Press, 1962.

Neyman J and Scott EL. Luminosity of galaxies in clusters and in the field (abstract). *The Astronomical Journal* **67**:582-583, 1962.

Neyman J, Scott EL and Zonn W. Abundances of morphological types among galaxies in clusters and in the field (abstract). *The Astronomical Journal* **67**:583, 1962.

Abell GO, Neyman J and Scott EL. Subclustering of galaxies (abstract). *The Astronomical Journal* **69**(8):529, 1964.

Neyman J and Scott EL. Problem of selection bias in the statistics of galaxies. *Bulletin de l'Institut International de Statistique* **40**(Book 2):1026-1050, 1964.

Neyman J and Scott EL. Field galaxies and cluster galaxies: Abundances of morphological types and corresponding luminosity functions. In *Confrontation of Cosmological Theories with Observational Data* (MS Longair, editor) pp 129-140, 1974.

Scott EL. Statistics in astronomy in the United States. In *On the History of Statistics and Probability* (DB Brown, editor) pp 319-331. Dekker, New York, 1976.

General Statistical Methods Publications

Neyman J and Scott EL. Consistent estimates based on partially consistent observations. *Econometrics* **16** (1):1-32, 1948.

Scott EL. Note on consistent estimates of the linear structural relation between two variables. *The Annals of Mathematical Statistics* **21**(2):284-288, 1950.

Neyman J and Scott EL. On certain methods of estimating the linear structural relation. *The Annals of Mathematical Statistics* **22**(3):352-361, 1951.

Scott EL. Testing Hypotheses. In *Statistical Astronomy* (by RJ Trumpler and HF Weaver) Chapter 1.8, pp 200-225, 229-230. University of California Press, Berkeley, 1953.

Neyman J and Scott EL. Correction for bias introduced by a transformation of variables. *The Annals of Mathematical Statistics* **31**(3):643-655, 1960.

Neyman J and Scott EL. Asymptotically optimal tests of composite hypotheses for randomized experiments with noncontrolled predictor variables. *Journal of the American Statistical Association* **60**(311):699-721, 1965. (Also in *Studies in Mathematical Statistics: Theory and Applications* (1968) pp 89-99. Akademiai Kiado, Budapest, Hungary.)

Scott EL. Subclustering. In *International Symposium on Classical and Contagious Discrete Distributions* (GP Patil, editor) pp 33-44. Statistical Publishing Society, Calcutta, 1965.

Neyman J and Scott EL. On the use of $C(\alpha)$ optimal tests of composite hypotheses. *Bulletin de l'Institut International de Statistique* **41**:477-497, 1967

Neyman J and Scott EL. Outlier proneness of phenomena and of related distributions. In *Optimizing Methods in Statistics; Proceedings* (JS Rustagi, editor) pp 413-430. Academic, New York and London, 1971.

Neyman J and Scott EL. Processes of clustering and applications. In *Stochastic Point Processes: Statistical Analysis, Theory and Applications* (PAW Lewis, editor) pp 646-681. Wiley, New York, 1972.

Le Cam L, Neyman J and Scott EL, Eds. *Proceedings of the Sixth Berkeley Symposium on Mathematical Statistics and Probability.* University of California Press, Berkeley. (*1 Theory of Statistics*, 760 pp; *2 Probability Theory*, 605 pp; *3 Probability Theory*, 711 pp; *4 Biology and Health*, 353 pp; *5 Darwinian, New-Darwinian and Non-Darwinian Evolution*, 369 pp; *6 Effects of Pollution on Health, Monograph*, 599 pp), *1972*.

Scott EL. Correlation and suggestions of causality: Spurious correlation. In *Statistical Ecology: Multivariate Methods in Ecological Work* (L Orloci, CR Rao and WM Stiteler, editors) **7**:237-251. International Cooperative, Fairland, Maryland, 1979.

Scott EL and Neyman J. In *Encyclopedia of Statistical Sciences* **6**:215-223. Wiley, New York, 1985.

Dawkins SM, Schuepp M and Scott EL. Statistical analysis of irregular effects. In *IVth International Vilnius Conference on Probability Theory and Mathematical Statistics, 1986*.

Weather Modification Publications

Jeeves TA, Le Cam LM, Neyman J and Scott EL. On the methodology of evaluating cloud seeing operations. Berkeley Statistical Laboratory Report to Division of Water Resources, Department of Public Works of the State of California, 1953.

Scott EL. Weather modification operations in California (co-editor). State Water Resources Board, Bulletin 16, 1955.

Neyman J, Scott EL and Vasilevskis M. Chapter V. Santa Barbara randomized cloud seeding project. Evaluation of seeding operations, Santa Barbara and Ventura Counties in 1957, 1958, and 1959. In *Interim Report on the Santa Barbara Project*. Department of Water Resources, Sacramento, California, V1-V56, 1960.

Neyman J, Scott EL and Vasilevskis M. Randomized cloud seeding in Santa Barbara. *Science* **131**(3407):1073-1078, 1960.

Neyman J, Scott EL and Vasilevskis M. Statistical evaluation of the Santa Barbara randomized cloud seeding experiment. *Bulletin of the American Meteorological Society* **41**(10):531-547; (Spanish translation of this paper appeared in 1960 in *Boletín de Técnicas y Applicaciones del Muestro*, Departamento de Muestro, Mexico **7**:11-36), 1960.

Neyman J and Scott EL. Unbiased estimation based on transformed variable, with particular reference to cloud seeding experiments. In *Proceedings of the Fifth Annual Conference on Design of Experiments ARD* pp 353-372, 1960.

Neyman J and Scott EL. Design of cloud seeding experiments. *Bulletin de l'Institut International de Statistique* **38**:31-41, 1961.

Neyman J and Scott EL. Further comments on the final report of the Advisory Committee on weather control. *Journal of the American Statistical Association* **56**:580-600, 1961.

Neyman J and Scott EL. Some outstanding problems relating to rain modification. In *Proceedings of the Fifth Berkeley Symposium on Mathematical Statistics and Probability* **5**:293-326. University of California Press, Berkeley, 1967.

Neyman J and Scott EL. Appendix: Planning an experiment with cloud seeding. In *Proceedings of the Fifth Berkeley Symposium on Mathematical Statistics and Probability* **5**:327-350. University of California Press, Berkeley, 1967.

Neyman J and Scott EL. Note on the weather bureau ACN project. In *Proceedings of the Fifth Berkeley Symposium on Mathematical Statistics and Probability* **5**:351-356. University of California Press, Berkeley, 1967.

Neyman J and Scott EL. Note on techniques of evaluation of single rain stimulation experiments. In *Proceedings of the Fifth Berkeley Symposium on Mathematical Statistics and Probability* **5**:371-384. University of California Press, Berkeley, 1967.

Neyman J and Scott EL. Rationale of statistical design of a rain stimulation experiment. In *Project Skywater Proceedings: Skywater Conference II*. pp 193-250. United States Department of the Interior, Bureau of Reclamation, Denver, Colorado, 1967.

Neyman J, Scott EL and Wells MA. Influence of atmospheric stability layers on the effect of ground-based cloud seeding, I. Empirical results. *Proceedings of the National Academy of Sciences* **60**(2):416-423, 1968.

Neyman J, Scott EL and Wells MA. Statistics in meteorology. *Review of the International Statistical Institute* **37**(2):119-148, 1969.

Neyman J, Scott EL and Smith JA. Areal spread of the effect of cloud seeding at the Whitetop experiment (reports). *Science* **163**(3874):1445-1449, 1969.

Lovasich JL, Neyman J, Scott EL and Smith JA. Timing of apparent effects of cloud seeding (reports). *Science* **165**(3896):892-893, 1969.

Neyman J, Scott EL and Smith JA. Whitetop experiment (reports). *Science* **165**(3893):618, 1969.

Lovasich JL, Neyman J, Scott EL and Smith JA. Wind directions aloft and effects of seeding on precipitation in the Whitetop experiment. *Proceedings of the National Academy of Sciences* **64**(3):810-817, 1969.

Lovasich JL, Neyman J, Scott EL and Smith JA. Statistical aspects of rain stimulation: Problems and prospects. *Review of the International Statistical Institute* **38**(1):155-170, 1970.

Lovasich JL, Neyman J, Scott EL and Wells MA. Further studies of the Whitetop cloud-seeding experiment. *Proceedings of the National Academy of Sciences* **68**(1):147-151, 1971.

Lovasich JL, Neyman J, Scott EL and Wells MA. Hypothetical explanations of the negative apparent effects of cloud seeding in the Whitetop experiment. *Proceedings of the National Academy of Sciences* **68**(11):2643-2646, 1971.

Neyman J, Osborn HB, Scott EL and Wells MA. Re-evaluation of the Arizona cloud seeding experiment. *Proceedings of the National Academy of Sciences* **69**(6):1348-1352, 1972.

Neyman J and Scott EL. Some current problems of rain stimulation research. In *Proceedings of the International Symposium on Uncertainties in Hydrologic and Water Resource Systems 6* (CC Kisiel and L Duckstein, editors) pp 1167-1244. University of Arizona, Tucson, 1973.

Neyman J, Scott EL and Wells MA. Downwind and upwind effects in the Arizona cloud-seeding experiment. *Proceedings of the National Academy of Sciences* **70**(2):357-360, 1973.

Scott EL. Problems in the design and analysis of weather modification studies. In *Third Conference on Probability and Statistics in Atmospheric Science.* pp 65-72. American Meteorological Society, Boston, Massachusetts, 1973.

Neyman J and Scott EL. Rain stimulation experiments: Design and evaluation. In *Proceedings of the WMO/IAMAP Scientific Conference on Weather Modification*, No. 399, pp 449-457. World Meteorological Organization, Geneva, Switzerland, 1974.

Dawkins SM, Neyman J, Scott EL and Wells MA. Chapter 13 – Preliminary statistical analysis of precipitation amounts; Chapter 14 – Data base and methodology of analysis; Numerical result of preliminary analysis. In *The Final Report on the Pyramid Lake Pilot Project 1970 to 1975* (P Squires, editor) **1**. Desert Research Institutes, Reno, Nevada, 1977.

Dawkins SM, Neyman J and Scott EL. The Grossversuch III Project. In *Transactions of Workshop of Total-Area Effects of Weather Modification, August 8-12, 1977, Fort Collins, Colorado* (KJ Brown, RD Elliot and MW Edelstein, editors) pp 510-525. North American Weather Consultants, Goleta, California, 1978.

Dawkins SM and Scott EL. Comments on the paper by R Braham, "Field experimentation in weather modification." *Journal of the American Statistical Association* **74**:70-77, 1979.

Scott EL. Modelling weather modification experiments. In *Statistics: Applications and New Directions, Proceedings of the Indian Statistical Institute Golden Jubilee International Conference* (J Roy and JK Ghosh, editors) pp 521-530. Indian Statistical Institute, Calcutta, 1984.

Dawkins SM, Neyman J and Scott EL. Preliminary analysis of Grossversuch IV hail damage on experimental days. In *Proceedings of the Berkeley Conference in Honor of Jerzy Neyman and Jack Kiefer* (L Le Cam and RH Olshen, editors) pp 109-121. Wadsworth, Monterey, California, 1985.

Bioscience and Health Publications

Neyman J, Park T and Scott EL. Struggle for existence. The Tribolium model: Biological and statistical aspects. In *Proceedings of the Third Berkeley Symposium on Mathematical Statistics and Probability* 4:41-79. University of California Press, Berkeley, 1956.

Neyman J and Scott EL. On a mathematical theory of populations conceived as conglomerations of clusters. In *Proceedings of the Cold Spring Harbor Symposia on Quantitative Biology* **22:** 109-120, 1957.

Neyman J and Scott EL. Stochastic models of population dynamics. *Science* **130**(3371):303-308, 1959.

Neyman J and Scott EL. A stochastic model of epidemics. In *Stochastic Models in Medicine and Biology: Proceedings of a Symposium* (J Gurland, editor) pp 45-83. University of Wisconsin Press, Madison, 1964.

Alberts WW, Feinstein B, Levin G, Wright EW Jr., Darland MG and Scott EL. Stereotaxic surgery for Parkinsonism: Clinical results and stimulation thresholds. *Journal of Neurosurgery* **23:**174-183, 1965.

Neyman J and Scott EL. Statistical aspect of the problem of carcinogenesis. In *Proceedings of the Fifth Berkeley Symposium on Mathematical Statistics and Probability* **4:**745-776. University of California Press, Berkeley, 1967.

Scott EL. Summary of panel discussion: Planning a comprehensive study of effects of pollution on health. In *Proceedings of the Sixth Berkeley Symposium on Mathematical Statistics and Probability* **6:**571-574. University of California Press, Berkeley, 1972.

Bloomfield P, Cole P, Craig RA, Haynes HA, Scott EL, Setlow RB, Straf ML and Woolsey TD. Estimates of increases in skin cancer due to increases in ultraviolet radiation caused by reducing stratospheric ozone. In *Environmental Impact of Stratospheric Flight* pp 177-221. National Academy Press, Washington, D.C., 1975.

Brillinger DR and Scott EL, Eds. *Conference on Forecasting Pollution Held in Berkeley, California, May 17 and 18, 1974.* State of California Air Resources Board, Sacramento, California, Monograph, 118 pp, 1975.

Scott EL and Straf ML. Ultraviolet radiation as a cause of cancer. In *Origins of Human Cancer, Book A* (HH Hiatt, JD Watson and JA Winston, editors) **4:**529-546. Cold Spring Harbor Conferences on Cell Proliferation, 1977.

Scott EL and Wells MA. Estimating the increase in skin cancer caused by increases in ultraviolet radiation. In *Research in Photobiology: Proceedings of the Seventh International Congress on Photobiology* (A Castellani, editor) pp 621-635. Plenum, London, 1977.

Scott EL. Comparison of two treatments when there may be an initial effect. In *Essays in Statistical Science* (J Gani and EJ Hannan, editors) *Journal of Applied Probability: Essays in Statistical Science* **19:** 253-264, 1982.

Scott EL. Response of mice to varying times of ultraviolet radiation with implications toward the response of human beings. In *Probability Models and Cancer, Proceedings of an Interdisciplinary Cancer Study Conference* (L Le Cam and J Neyman, editors) pp 221-243. University of California, Berkeley, *1982*.

Scott EL. Combining estimates of the increase in skin cancer. *A Fetschrift for Erich L. Lehmann in Honor of His Sixty-Fifth Birthday* (PJ Bickel, K Doksum and JL Hodges, Jr., editors) pp 395-408. Wadsworth, Belmont, California, 1982.

Scott EL. *Assigned Share for Radiation as a Cause of Cancer.* National Academy Press, Washington, D.C. (With Oversight Committee on Radioepidemiologic Tables). Monograph, 210pp, 1984.

Scott EL. Epidemiology of skin cancer under increasing ultraviolet radiation. In *Indo-US Workshop on Global Ozone Problem* (AP Mitra, editor) pp 235-243. Insdoc, New Delhi, 1984.

Pearl DK and Scott EL. The anatomical distribution of skin cancers. *International Journal of Epidemiology* **15**(4):502-506, 1986.

Pearl DK and Scott EL. Relative increases in skin cancer rates associated with increasing exposure to ultraviolet radiation. Research Report, Berkeley Statistical Laboratory, 1986.

Women's Studies Publications

Blumer H, Newman F, Ervin-Tripp S, Colson E and Scott EL. Report of the Subcommittee on the Status of Academic Women on the Berkeley Campus. 78 pp, Academic Senate, University of California, Berkeley, 1970.

Darland MG, Dawkins SM, Lovasich JL, Scott EL, Sherman ME and Whipple JA. [erroneously published under Scott EL only]. Women in higher education: The facts of the matter. In *Proceedings of the Twelfth Annual Meeting of the Council of Graduate Schools* (J Ryan, editor) pp 55-64. Council of Graduate Schools, Washington, D.C., 1972.

Darland MG, Dawkins SM, Lovasich JL, Scott EL, Sherman ME and Whipple JA. Application of multivariate regression to studies of salary differences between men and women faculty. In *Proceedings of the Social Statistics Section, American Statistical Association*, pp 120-132. American Statistical Association, Washington, D.C., 1973.

Scott EL. Developing criteria and measures of equal opportunities for women. In *Women in Academia: Evolving Policies Toward Equal Opportunities* (A Lewin, E Wasserman and L Bleiweiss, editors) pp 82-114. Praeger, New York, 1975.

Scott EL. Career Profiles: Mathematician. In *Educating Women for Science: A Continuous Spectrum* (J Fetter, editor) Center for Teaching and Learning, Stanford University, Stanford, California, 1976.

Scott EL. *Higher Education Salary Evaluation Kit.* American Association of University Professors, Washington, D.C., 55 pp, *1977*.

Lovasich JL and Scott EL. Comparison of the status of men and women in higher education, with an application to the biological sciences. *Proceedings of the Workshop on Women in Biomedical Science*, NIH Bethesda, Maryland, May 1978.

Scott EL. Methodology for studying career discrimination, with an application to women in higher education and a preliminary study of Asian/Pacific Americans with doctorates. *Proceedings of Workshop on the Status of Employment, Unemployment and Underemployment of Asian/Pacific Americans*, Stanford University, August 1978.

Lovasich JL and Scott EL. Historical trends in the appointment and promotion of women faculty. In *The Impact of the Bakke Decision and Proposition 13 on Equality for Women in California Higher Education.* pp 92-107. CCEW Women's Center, University of California, Berkeley, 1979.

Scott EL. Linear models and the law: Uses and misuses in affirmative action. In *Proceedings of the Social Statistics Section, American Statistical Association*. pp 20-26. American Statistical Association, Washington, D.C., 1979.

Scott EL. Rubin's empirical Bayes computations are not useful for law school admissions (comment). *Journal of the American Statistical Association* **75**(372):821-823, 1980.

Gray MW and Scott EL. A 'statistical' remedy for statistically identified discrimination. *Academe* **66**(4):174-181, 1980.

Ahern NC and Scott EL. *Career Outcomes in a Matched Sample of Men and Women Ph.D.'s: An Analytical Report.* National Academy Press, Washington, D.C., 99 pp, *1981*.

Scott EL. Oral History. In *The Women's Faculty Club of the University of California, Berkeley, 1919-1982* (SB Reiss, editor) pp 148-169. Bancroft Library, University of California, Berkeley, 1983.

Index

A

AAAS, *see* American Association for the Advancement of Science

AAS, *see* American Astronomical Society

AAU, *see* The Association of American Universities

AAUP, *see* American Association of University Professors

AAUW, *see* American Association of University Women

AAVSO, *see* American Association of Variable Star Observers

ACE, *see* American Council on Education

Adams, Walter Sydney, 35, 36

Admissions, financial support, 393

Affidavit, 371–372

Affirmative Action, 323–380
  AAUP kit development, 486–487
  campus attitudes toward, 457–459
  coordinator, 327–328, 330, 343, 350, 358, 369, 370, 393–394, 408, 417–418, 420, 432, 441
  delays, 367–369
  Eisenberg Plan, 343, 349, 351, 353–354, 357–358, 362–362, 397
  federal testimony on, 480–482
  a lot of power (August–December 1971), 335–351
  nominated for, 468–469
  not easily erased overnight (June–July 1971), 323–333
  report review, 530–531

salary data and, 406–407
statistical scrutiny of, 455–457
time for action (March–June 1972), 367–380
weak, grudging, incomplete (January–February 1972), 353–365

Ahern, Nancy, 539

Air Force, US, 27, 64, 67, 78, 428

Almost alone in statistics (1955–1988), 127–141
  administrator and professor, 134–136
  colleague Juliet Popper Shaffer, 137–139
  flexibility and resilience, 140–141
  new statistics department (1955), 128–131
  teaching, 131–134

α geminorium, 60, 61

Alphabetical order of authors, 97–98

American Association for the Advancement of Science (AAAS), 371
  activity, salary data and, 401–403
  bylaws, 403
  committee, 402
  Committee on Women, 373, 374, 391–394
  council resolution, 341
  ideas for, 339–341
  leadership position with, 340
  membership in, 123
  Newcomb Cleveland Prize, 105
  Office for Women's Equality, 341

575